Molecular Nanowires and Other Quantum Objects

NATO Science Series

A Series presenting the results of scientific meetings supported under the NATO Science Programme.

The Series is published by IOS Press, Amsterdam, and Kluwer Academic Publishers in conjunction with the NATO Scientific Affairs Division

Sub-Series

I. **Life and Behavioural Sciences**	IOS Press
II. **Mathematics, Physics and Chemistry**	Kluwer Academic Publishers
III. **Computer and Systems Science**	IOS Press
IV. **Earth and Environmental Sciences**	Kluwer Academic Publishers
V. **Science and Technology Policy**	IOS Press

The NATO Science Series continues the series of books published formerly as the NATO ASI Series.

The NATO Science Programme offers support for collaboration in civil science between scientists of countries of the Euro-Atlantic Partnership Council. The types of scientific meeting generally supported are "Advanced Study Institutes" and "Advanced Research Workshops", although other types of meeting are supported from time to time. The NATO Science Series collects together the results of these meetings. The meetings are co-organized bij scientists from NATO countries and scientists from NATO's Partner countries – countries of the CIS and Central and Eastern Europe.

Advanced Study Institutes are high-level tutorial courses offering in-depth study of latest advances in a field.
Advanced Research Workshops are expert meetings aimed at critical assessment of a field, and identification of directions for future action.

As a consequence of the restructuring of the NATO Science Programme in 1999, the NATO Science Series has been re-organised and there are currently Five Sub-series as noted above. Please consult the following web sites for information on previous volumes published in the Series, as well as details of earlier Sub-series.

http://www.nato.int/science
http://www.wkap.nl
http://www.iospress.nl
http://www.wtv-books.de/nato-pco.htm

Series II: Mathematics, Physics and Chemistry – Vol. 148

Molecular Nanowires and Other Quantum Objects

edited by

Alexandre S. Alexandrov
Loughborough University,
Physics Department,
Loughborough, United Kingdom

Jure Demsar
"Jozef Stefan" Institute,
Ljubljana, Slovenia

and

Igor K. Yanson
B. Verkin Institute for Low Temperature Physics and Engineering,
National Academy of Sciences of Ukraine,
Kharkiev, Ukraine

SPRINGER SCIENCE+BUSINESS MEDIA, B.V.

Proceedings of the NATO Advanced Research Workshop on
Molecular Nanowires and Other Quantum Objects
Bled, Slovenia
7–9 September 2003

A C.I.P. Catalogue record for this book is available from the Library of Congress.

DOI 10.1007/978-1-4020-2093-3

Printed on acid-free paper

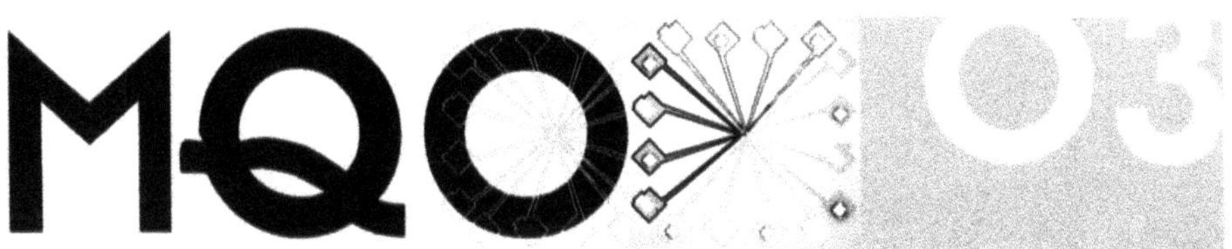
MQO 03
molecular nanowires and other quantum objects
bled, slovenia
september 7 - 10, 2003

TABLE OF CONTENTS

CARBON NANOTUBES

SUPERCONDUCTING NANOSTRUCTURES

POLARONS

COMPLEX QUANTUM DOTS

FUNDAMENTALS OF NANOSCALE

LOW DIMENSIONAL QUANTUM OBJECTS

PREFACE

There is a growing understanding that the progress of the conventional silicon technology will reach its physical, engineering and economic limits in near future. This fact, however, does not mean that progress in computing will slow down. What will take us beyond the silicon era are new nano-technologies that are being pursued in university and corporate laboratories around the world. In particular, molecular switching devices and systems that will self-assemble through molecular recognition are being designed and studied. Many laboratories are now testing new types of these and other reversible switches, as well as fabricating nanowires needed to connect circuit elements together. But there are still significant opportunities and demand for invention and discovery before nanoelectronics will become a reality. The actual mechanisms of transport through molecular quantum dots and nanowires are of the highest current experimental and theoretical interest. In particular, there is growing evidence that both electron-vibron interactions and electron-electron correlations are important. Further progress requires worldwide efforts of trans-disciplinary teams of physicists, quantum chemists, material and computer scientists, and engineers.

The NATO Advanced Research Workshop "Molecular Nanowires and Other Quantum Objects" brought together 40 experts in molecular and other nanowires, carbon nanotubes, mesoscopic superconductors and semiconductors, and theorists in the field of strongly correlated electrons and phonons from 14 NATO, NATO-Partner and Mediterranean-Dialogue countries working in 32 university and corporate laboratories. Topics for discussion included molecular nanojunctions and electronics, mesoscale semiconductors and superconductors, carbon nanotubes, low dimensional conductors, polarons and strongly-correlated electrons in nanoobjects, quantum theory of nanoscale, including first-principle simulations, new techniques for making mesoscopic sensors and detectors. The framework of the meeting allowed participants to become acquainted with a diverse cross-spectrum of available skills, and to engage in exchanging their often-complementary experiences and views. Many participants acknowledged the program as very stimulating and well-organised. These resulting proceedings of the NATO ARW on 'Molecular Nanowires and Other Quantum Objects' represent a tiny but important part of the activities in the field.

It is our pleasant duty to acknowledge the wide assistance we received. We greatly appreciate the Science Programme of NATO in Brussels for funding the Workshop. In addition, the members of the organizing committee Alexander F. Andreev, Viktor V. Kabanov and R. Stanley Williams deserve special thanks for their efforts in organizing the final program. Furthermore, we would like to acknowledge Dragan Mihailovic and Nika Simcic (Quantum Materials Group at Jozef Stefan Institute) for their support and help before and during the workshop. Finally, we would like to thank R. Stanley Williams (Hewlett Packard Labs) for letting us use the photo of the HP 64-bit molecular memory as the conference logo.

Sasha Alexandrov
Jure Demsar
Igor Yanson

December, 2003

CHARACTERIZATION OF NANOSCALE MOLECULAR JUNCTIONS

Artur Erbe, Zhenan Bao, David Abusch-Magder, Donald M. Tennant, and Nikolai Zhitenev
Lucent Technologies, Bell Labs, 600 Mountain Avenue, Murray Hill, NJ, 07974, USA
aerbe@lucent.com

Abstract Two methods for formation of metal-molecule-metal junctions are demonstrated. The use of different techniques allows us to contact short molecules approaching the single molecule limit as well as longer polymers in single layers. The sample geometries also enable us to test gate dependencies on some of the tested structures.

Keywords: Nanoscale Molecular Junctions, metal-molecule-metal junctions

1. Introduction

Exploring the electronic possibilities of nanoscale organic materials has become an important challenge as modern lithographical techniques approach ultimate limits. In this regime, the properties of single or few molecules can dominate the behavior of whole devices. Recent experiments on nanoscale molecular junctions show a large variety of results [1, 2, 3, 4, 5]. Values of the conductivity as well as energy scales of the states participating in current transport vary over a broad range. Differences in the properties of the molecules themselves can only explain some of these variations. This fact indicates that details of the contact formation to the molecules play an important role in the behavior of the whole junction and control of the contact properties is a highly important prerequisite for successful testing of molecular behavior on this scale. We present electrical measurements of various types of molecules using two recently developed contacting techniques. Both methods allow us to contact single or a few molecules electrically and characterize them under varying conditions. The first technique is based on shadow stencil mask evaporation similar to the well-known metallic Single Electron Transistor (SET) fabrication [6]. The mask is defined by electron beam lithography (EBL) in direct contact with the substrate. The second technique presented here uses two closely spaced electrodes defined by EBL that are covered by the molecules.

A.S. Alexandrov et al. (eds.), Molecular Nanowires and Other Quantum Objects, 1–12.

Electrical characterization is done on molecules bridging the distance between the electrodes. This method is especially useful for characterization of longer polymers. We performed measurements on a variety of molecules using both techniques. Distinct features are found in the I-V-characteristics at low temperatures indicating that single or a few molecules are contacted. Some of those features can be affected by changes in applied gate voltage.

2. Fabrication

The measurements reported in this paper were performed on "short" (three rings of phenyl or thiophene in series) molecules and "long" (Poly-thiophene and Poly-fluorene) polymers. Two different techniques have to be developed in order to reliably connect to both kinds of molecules. In case of the short molecules we can rely on the formation of a Self Assembled Monolayer (SAM) on a gold surface in order to separate the two contacting gold electrodes. Therefore a shadow evaporation technique can be used to contact these molecules. The long polymers do not form a well ordered monolayer. Thus they are contacted by two EBL defined electrodes separated by a very small distance.

The lengthscales of short single molecules (1 to 2 nm) are not accessible by traditional fabrication techniques. Furthermore the SAM is destroyed if the structures are processed with standard lithography after deposition of the molecules. Using a shadow stencil mask for the definition of the contacts we avoid exposure of the SAM to any harmful chemistry.

Shadow masks based on resist bilayers (PMMA/MMA, for example) have been used for the fabrication of nanoscale structures like metallic SETs for a long time [6]. The double layer resist is attacked by the solvents of the molecules (usually THF (TetraHydroFuran) in case of our molecules) and can be destroyed during self assembly of the monolayer or contaminate the SAM. We therefore choose a combination of SiO_2 and Si_3N_4 as mask material. This double layer is structured using EBL. The pattern of the mask is transferred to the double layer using two consecutive dry-etch steps. A detailed description of the fabrication process is shown in figure 1.

After the definition of the mask gold contacts are evaporated perpendicular to the sample surface. This creates two gold fingers on the left- and right-hand side of the SiO_2 -bridge separated by 50-100 nm. In order to form a SAM on the gold, the sample is soaked in a molecular solution for several hours. A second evaporation step is performed on the tilted sample in order to connect both contacts through the monolayer under the bridge (a schematic drawing is shown in figure 2).

The tilting angle of the sample determines the overlap between the upper and lower gold and thus the area of the junction. Monitoring the conductivity between source and drain during this second evaporation step and changing

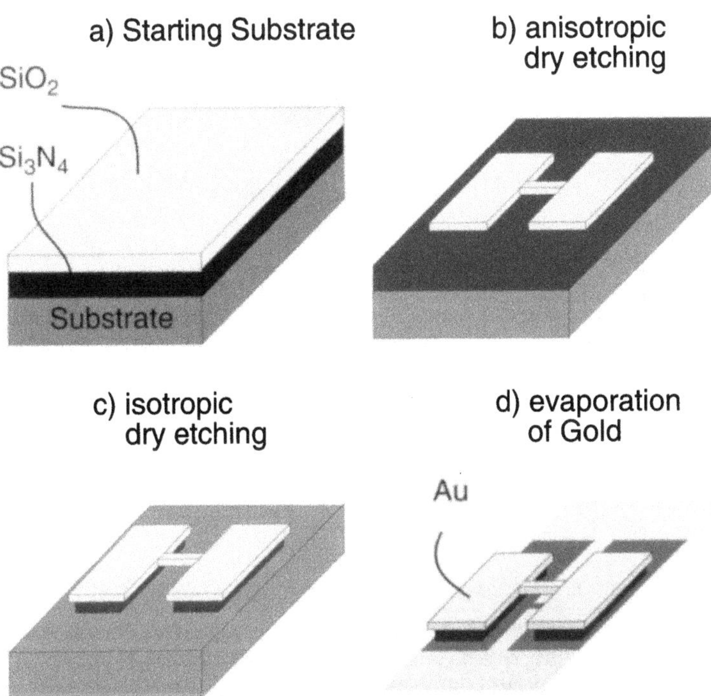

Figure 1. Fabrication of a shadow mask with smallest dimensions around 100 nm. **a)** The substrate is covered with PECVD deposited Si_3N_4 (200 nm) and SiO_2 (100 nm). **b)** Definition of the shadow mask using EBL. The resist serving as a mask for the etch steps is PMMA. The transfer into the SiO_2 -layer is done by anisotropic reactive ion etching (etch gas CHF_3). **c)** Isotropic reactive ion etching (etch gas CF_4) removes the Si_3N_4 under and around the places where the SiO_2 layer was opened. A small bridge (50-100 nm wide) is left in the center of the structure. **d)** Definition of the metallic contacts by vertical evaporation of gold.

the tilting angle can ensure that the evaporation is stopped at the first onset of conductance and thus single or only a few molecules are contacted. A tantalum gate electrode (width 5μm) is defined in the center of the structure and covered by a gate oxide (50 nm of SiO_2) before the fabrication of the mask. A micrograph of the resulting structure can be seen in figure 3.

This method was used for molecules P3 and T3 (the molecules are shown in figure 4).

Thiol endgroups provide attachment to gold for self assembly and ensure good mechanical and electrical bond to the top gold contact. The central ring in the P3 molecule is rotated by about 45° around the backbone of the molecule. It is expected that this leads to differences in conductance behavior compared to T3, which is mostly flat [7]. Both molecules are completely conjugated, so a rather high conductance along the molecule is expected.

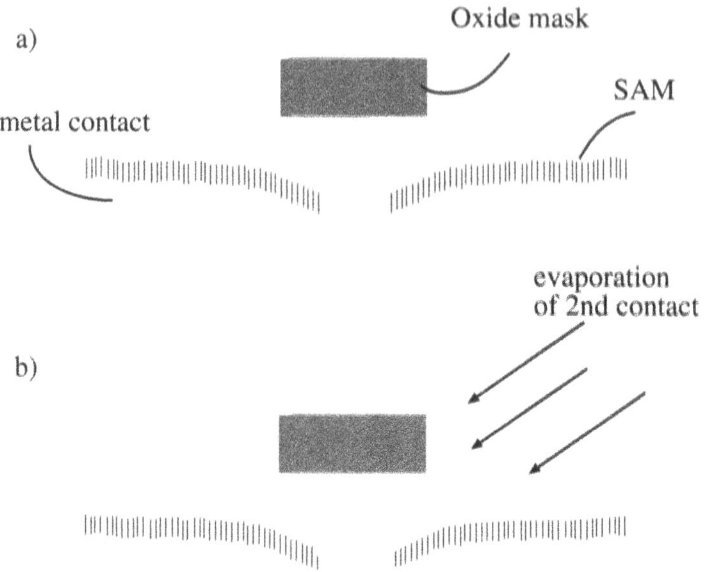

Figure 2. Evaporation on the shadow mask under different angles in order to produce nanoscale junctions **a)** Contacts are evaporated vertically through the shadow mask. A SAM is formed on both contacts. **b)** Evaporation on the tilted sample results in a connection between the two contacts through the monolayer. During this step the conductance of the junction can be monitored in order to ensure that contact is made to single or few molecules. The tilting angle and thus the overlap between top and bottom electrode can be changed during the second evaporation step.

In order to contact the polymers (length distribution centered around 50 nm) metal electrodes (15-20 nm of $Au_{0.6}Pd_{0.4}$) are defined by EBL with distances ranging from 5 to 20 nm. Typical structures are shown in figure 5.

The gate electrode in this setup is given by the substrate (highly doped Si) which is insulated by 90 nm of thermally grown SiO_2 .

3. Results

The measurements were performed in a low temperature probe station allowing for easy simultaneous access to many junctions. Thus we were able to measure a large amount of working samples although the yield of the production technique was still rather low. The main failure mechanism in the shadow mask technique results from shorts between the contacts and the gold layer on top of the mask. Three typical examples for the behavior of the remaining junctions at 4.2 K are given in the following discussion.

One group of curves strongly resembles Coulomb blockade behavior (shown in figure 6). These curves show a conduction gap around V_{sd}=0. The width of the gap ($\pm$100 mV) corresponds to Coulomb blockade on a metallic cluster

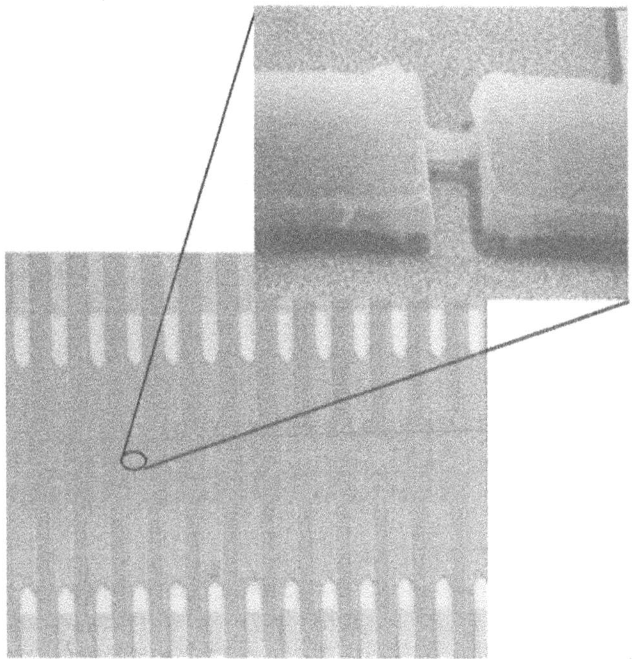

Figure 3. Micrograph of several shadow mask junctions produced in parallel. On the top and bottom the connecting pads leading to the small structures in the center can be seen. The mask material is covering these leads over most of the sample area in order to minimize shorts between the leads and the top gold-layer. Only the areas where EBL and optical lithography overlap make the connection between the big pads and the nanoscale electrodes. The gate electrode lies underneath the center part of the junctions (dark region in the center of the picture). Upper right: SEM micrograph of the central part of the mask after vertical evaporation of gold.

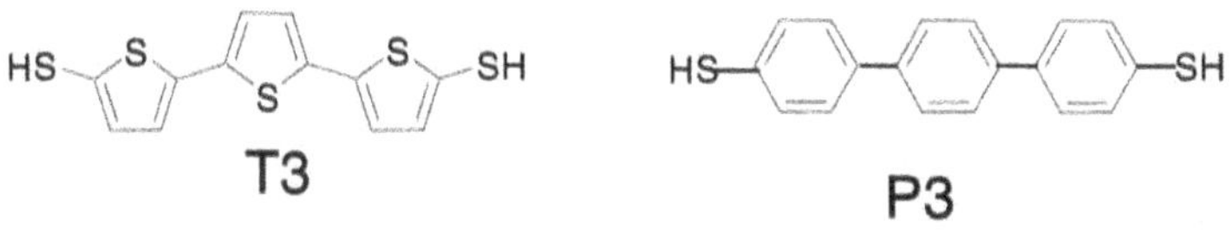

HS S S S SH
T3

HS SH
P3

Figure 4. Characterized molecules: T3 consists out of 3 thiophene rings, P3 out of 3 phenyl rings. Both molecules are designed with thiol endgroups on both ends

of 4–6 nm diameter. Similar curves are measured when small metal clusters are deposited on purpose on top of the SAM before the second evaporation step (This measurement was done in the geometry discussed in [8]). It is

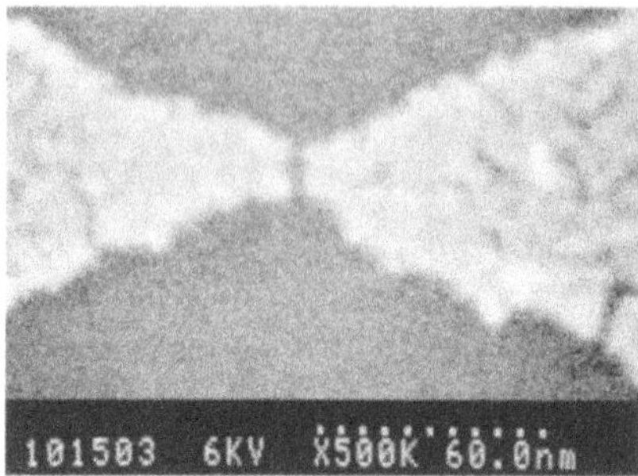

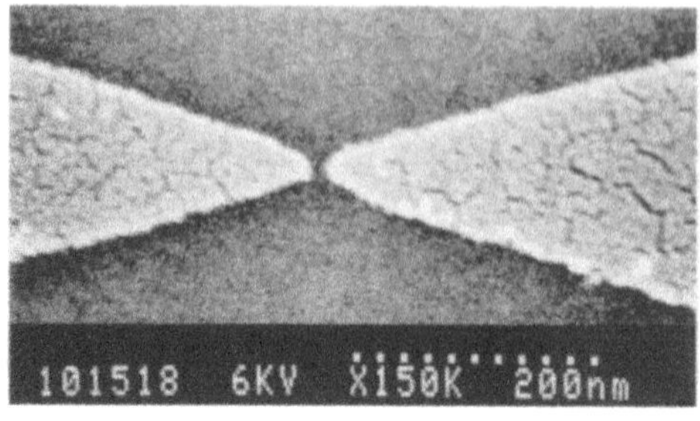

Figure 5. SEM-micrograph of two EBL defined contacts with gaps of few nanometers.

therefore likely that in this situation metal clusters accidentally form during the second evaporation step and are measured in series with the molecules.

The measurements shown in figure 7 show a lower conductivity. This can indicate that the area of the measured junction is reduced compared to junctions characterized in figure 6.

The I-V-curves exhibit a series of clear periodic steps. The energy scale for these steps is smaller than the energy scale calculated from the gap in figure 6 by a factor of 2–3. Some of these measurements show also a non zero conductivity at zero bias. Energy scales extracted from temperature dependence reveal an even lower energy scale, in the range of a few meV, similar to the energy scale obtained from larger junctions (see below). Hence the behavior differs from the Coulomb blockade behavior on small clusters and the properties have to be attributed to a different mechanism. The distance between consecutive steps depends on the chosen molecule. In figure 7 b) a measurement on T3 is shown for comparison. The period of the steps is about 100 mV in the case of T3, while it is about 50 mV in the case of P3. The overall conductance is comparable and very low. Qualitatively samples measured in a quartz fiber geometry on the same molecules show a very similar behavior [8]. On the other hand there are some obvious differences in the quantitative behavior. The most remarkable difference is the overall resistance. In the shadow mask geometry overall resistance of a single junction (estimated from the linear background) is around 100 GΩ, while typical resistance in the tip geometry is about 200 MΩ. The second quantitative difference is the spacing of the steps which is 22–50 mV for the T3 molecule in the quartz tip geometry. Both differences possibly result from the different coupling of the molecules to the electrodes in the two geometries. It can be seen in figure 3 that the bottom gold underneath

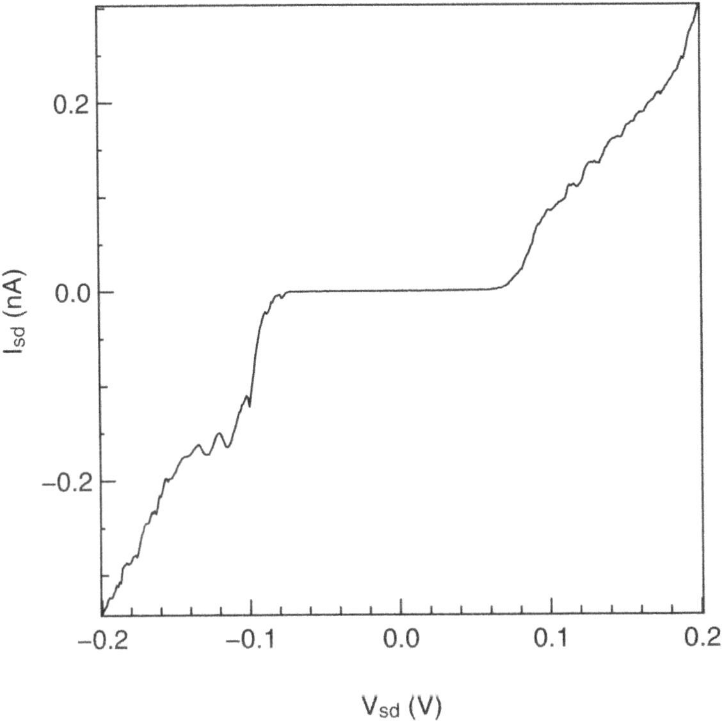

Figure 6. I-V-curve obtained from the shadow mask technique on P3 at 4.2 K. A conductivity gap extends from $V_{sd} = -0.1$ V to $V_{sd} = 0.1$ V. No further conductivity steps can be seen.

the shadow mask is rough due to the processing steps performed on the SiO_2 before evaporation of the gold. The coupling to the top electrode is probably weakened due to this roughness. This results in a reduced conductivity and can also explain the different energy scale of the resonances. One possible explanation for the resonances is tunneling through some low-energy states in the monolayer due to interaction between molecules and the metal. The features may also be explained by coupling of electronic states to molecular vibrations [8]. The energy scale for both mechanisms changes if the coupling to the metal changes. In case of molecular vibrations the mechanical boundary conditions change as well as the coupling of the different modes to the contacts. The tunneling through the low energy states depends on the height of the barrier between metal and molecules and thus also on the coupling.

In figure 7 a) we also present the gate dependence of one of the planar junctions. Some of the resonances react in a linear way to the applied gate voltage (as indicated by the arrow), some others react in less regular way. This result also resembles the behavior found in [8] qualitatively. Only a few

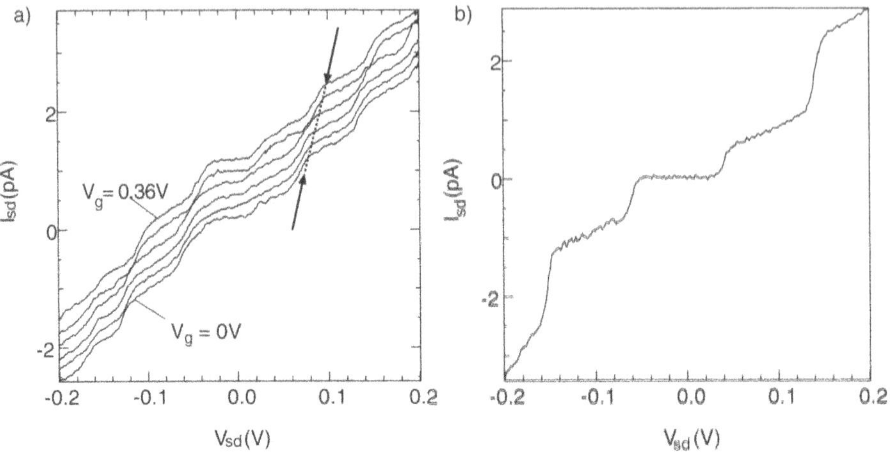

Figure 7. **a)** Measurement of P3 at 4.2 K. Strong resonances in the I-V-characteristics show that single or a few molecules are contacted. Measurements with applied gate voltage are offset vertically for clarity. **b)** Measurement of T3 at 4.2 K. Different periodicity of the steps is found compared to the measurements on P3.

samples show this clear gate dependence, while in most samples no gate effect is present. This behavior can also be explained by the strong surface roughness of the gold in the planar junction. In this situation contact formation to a molecule deposited on the top of the bottom electrode is much more likely than connecting a molecule on the sidewalls. Molecules on the top are electrostatically screened from the gate underneath the bottom electrode and do not show gate dependence.

In figure 8 a very smooth nonlinear curve is shown. Only a small suppression of conductivity around $V_{sd} = 0$ can be seen, all other nonlinearities seen are averaged out. The overall conductance is larger than in figures 6 and 7. It is therefore likely that a larger area is probed in this measurement.

If the tilting angle during second evaporation is too large or the top metal is too thick too many molecules may bind to the top layer and contribute to the conduction. This can explain a smooth nonlinearity as seen in the measurement. The energy scale attributed to the suppression of conduction close to $V_{sd} = 0$ is of the order of few meV. A similar energy scale is extracted from the temperature dependence.

Measurements on EBL-fabricated junctions were done on Poly-thiophene and Poly-fluorene (the length of both polymers is about 50 nm). The polymers are designed with thiol endgroups on both ends. Therefore some of the polymers will bind to both electrodes with one end on each electrode if the inter-electrode distance is below 50 nm. For preparation of the polymer film a

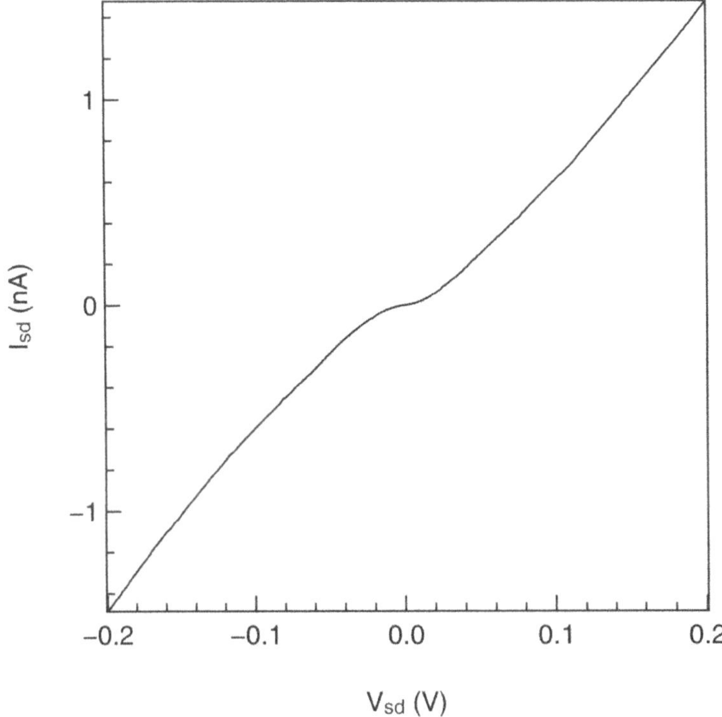

Figure 8. *I-V*-curve obtained from the shadow mask technique on P3 at 4.2 K. A smooth nonlinear curve is obtained when the junction size is too large to contact only few molecules.

small droplet of polymer solution was cast on the electrodes and washed away after a short soaking time with THF, Acetone and IPA. The cleaning steps remove all polymers that are not bonded to the electrodes with at least one end. After this step the current can only flow through polymers that are connected to both electrodes. A typical measurement is shown in figure 9.

The measurements show a conductivity gap at zero bias which increases its width to ± 2 V at 4.2 K. This behavior cannot be changed by application of a gate voltage varying from -10 V to +10 V. It is therefore likely that we see coulomb blockade on metal structures that are much smaller than the size of the molecules. The charging energy estimated from this behavior corresponds to a size of the metallic structure of about 0.5 nm. The resolution of the pictures shown in figure 5 is not good enough to resolve such fine structures but it is plausible that they result from irregularities during the lift-off process.

A slight overexposure of the pattern shown in figure 5 leads to a metallic short (resistance below 100 Ω). Applying a large source-drain bias to these

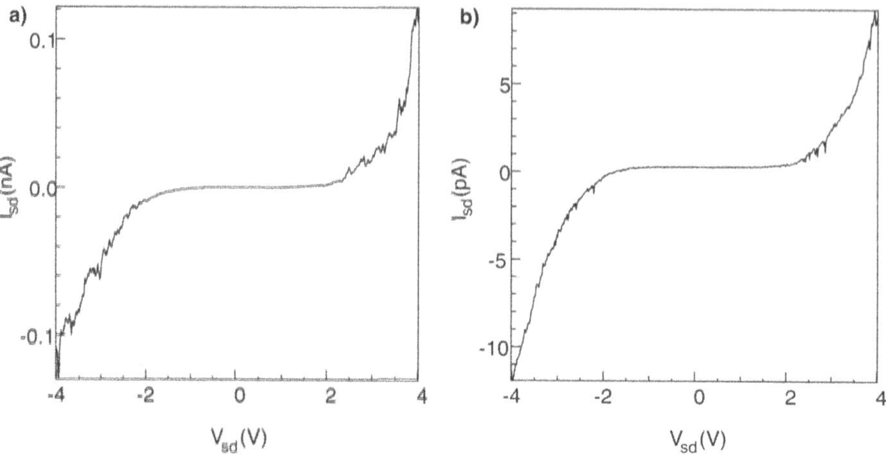

Figure 9. Current transport on nanojunctions fabricated by EBL and covered with single layers of Poly-thiophene. **a)** Source-drain sweep at room temperature. The curve exhibits a conductivity gap around zero V_{sd}, the gap extends from -2 to +2 V. **b)** At 4.2 K overall conductivity is reduced by a factor of about 10.

junctions this short can be broken by electromigration [9]. This break was performed at room temperature after the polymer layer was deposited on top of the electrodes. A source-drain curve measured during the break is shown in the inset in figure 10. Typical values of the resistance after breaking the junction are higher than 100 GΩ. I-V-characteristics of the majority of the junctions are similar to the behavior shown in figure 9. A few junctions, however, show some gate-induced conductivity and no conductivity at zero V_G. An example is shown in figure 10.

No conduction was measured at zero and positive bias, while a very low conduction could be turned on by application of strong negative bias. Even at the highest value of the gate voltage the conductance is much lower than in the measurements shown in figure 9. This can indicate that transport is given over a much larger distance, possibly along the backbone of the polymer. The behavior is consistent with the hole doped transistors that have been observed on macroscopic Poly-thiophene FETs [10]. Samples with Poly-fluorene (with thiol endgroups) or Poly-thiophene without the thiol endgroups did not show this behavior. Unfortunately the effect is still very irreproducible and the fabrication of break junctions needs more research in order to be a reliable tool for measurements of these polymers.

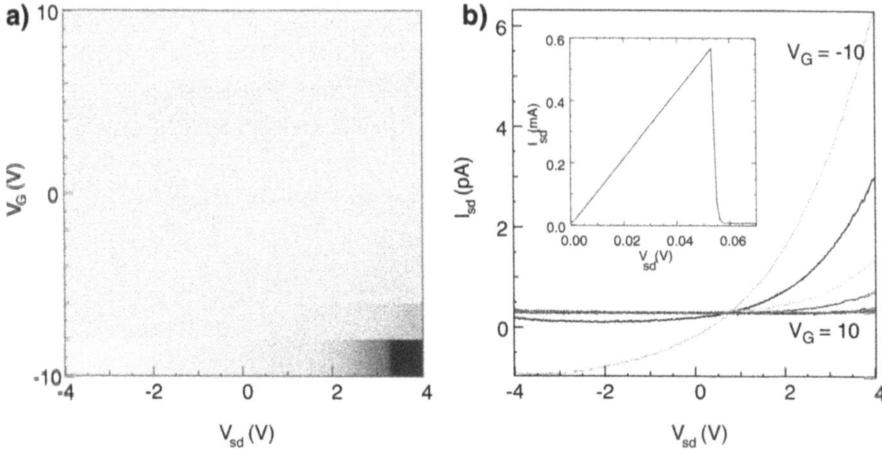

Figure 10. Current transport through Poly-thiophene at room temperature. **a)** Greyscale plot of the current as a function of source-drain bias V_{sd} and gate voltage V_G. It can be clearly seen that current transport occurs only at gate voltages $V_G > 4$ V. **b)** Source-drain curves taken at V_G varying from -10 V to $+10$ V in steps of 2 V. At the highest gate voltage ($V_G = 10$ V) transport is also observed at negative V_{sd}. Inset: Break by electromigration at room temperature. V_{sd} is ramped up slowly, the current drops abruptly to zero when a gap is opened.

4. Summary

We have shown measurements of single or a few molecules in a recently developed shadow mask technique. The current transport through these junctions shows at low temperatures strong periodic nonlinearities that can be affected by an applied gate voltage. Comparison to measurements on a quartz tip geometry shows that the current level as well as the reliability of the gate electrode suffer from the roughness of the gold deposited on the already processed SiO_2 substrate. In recent experiments we are trying to minimize this effect by using different substrates. In junctions fabricated by electromigration gate induced hole doping of possibly single layers of Poly-thiophene was shown. Although this is a very preliminary result it shows that it should be possible to study the conduction mechanisms along single strands of polymers using the break-junction technique.

References

[1] M.A. Reed et al., Science **278**, 252 (1997).

[2] C. Kergueris et al., Phys. Rev. B **59**, 12505 (1999).

[3] W. Liang et al., Nature **417**, 725 (2002).

[4] J. Reichert et al., Phys. Rev. Lett. **88**, 176804 (2002).
[5] J. Park et al., Nature **417**, 763 (2002).
[6] T.A. Fulton and G.J. Dolan, Phys. Rev. Lett. **59**, 109 (1987).
[7] J.M. Seminario, A.G. Zacarias and J.M. Tour, J. Am. Chem. Soc. **120**, 3970 (1998).
[8] N.B. Zhitenev, H. Meng and Z. Bao, Phys. Rev. Lett. **88**, 226801 (2002).
[9] H. Park et al., Nature **407**, 57 (2001)
[10] A. Tsumura, H. Koezuka and T. Ando, Appl. Phys. Lett. **49**, 1210 (1986).

CONTROLLED ELECTRON TRANSPORT IN SINGLE MOLECULES

I. M. Grace, S. W. Bailey and C. J. Lambert
Department of Physics, Lancaster University,
Lancaster, United Kingdom
i.grace@lancaster.ac.uk

J. Jefferson
Qinetiq, Sensors and Electronic Division,
Great Malvern, Worcestershire, United Kingdom

Abstract Using a combination of the first principles DFT code SIESTA, along with a Greens function scattering technique, the conductance properties of candidate molecular switches are calculated. These molecules are based on flourinone structures with added benzene rings. Results are shown for the effect on electron transport when various changes are made to the molecule such as removing side groups and twisting the molecule.

Keywords: molecular electronics, transport

1. Introduction

Molecular electronics in the sense of electronic devices, whose active region consist of a few molecules, or even a single molecule, has been proposed at the conceptual level for many years[1]. Recent results which have demonstrated that this concept is rapidly becoming a viable technology include the discovery of negative differential resistance[2] and a molecular memory effect [3]. Theoretical models have focused on the benzene-1,4-diathiolate molecule [4] and the phenyl-ethelene oligomers known as Tour wires[5]. The realization of single molecule electronics requires an understanding of the interplay between molecular electronic structure and metallic contacts. At the experimental level, it requires improving the yield achieved for devices which utilize self assembled monolayers as the active region of the device. The aim of this paper is to survey the transport properties of a range of candidate molecules of varying sizes (from approximately 20 to 80Å). This includes studying the

A.S. Alexandrov et al. (eds.), Molecular Nanowires and Other Quantum Objects, 13–20.

effect of adding extra benzene rings to each end of the molecule which is one method that can be used to lengthen it. One feature of these molecules is the presence of side groups which are added to allow solubility. In this paper we investigate the effect of these groups on transport. Molecules by their nature are not rigid structures and therefore it is also of interest to investigate the effect of conformation changes. These changes could occur through a variety of means such as the transfer of charge or by the application of an external field. In this paper we examine these effects by altering the initial orientation of the molecule, reducing the structure to a local energy minimum and then compute its transport properties using a recursive Greens function technique.

The molecules studied are basic flourinone structures with extra benzene rings added (for an example, see figure. 1) synthesized by a group at the University of Durham. The aim of designing these molecules is to address the difficulties in matching the gap between metal contacts that can currently be achieved (typically 100Å). Usually device contacts are put in place both before and after deposition of the molecule monolayer, which risks the partial destruction of the monolayer when the top contact is deposited. To overcome this problem a method is being developed were both contacts are deposited first which has a small insulating layer between them. Molecules are then introduced at the end of the process scheme with no further invasive metal deposition. These structures have a high yield with typically 80-100Å edge gaps. Therefore the focus of this work is on molecules which have similar sizes to this gap.

2. Model and Method

Here a hierarchical theoretical approach is used to calculate the transport through a molecule contacted between two semi-infinite leads. This consists of three stages the first of which is calculating the relaxed geometry of the molecule. This, in principle could be achieved using any first principles quantum chemistry code. Here we use the density functional theory code SIESTA[6], combined with a supercell approach. We use the local density approximation parameterized by Perdue and Zunger[7], nonlocal norm-conserving pseudopotentials[8] and valence electrons described by a single-ζ basis set. This stage of the calculation also gives useful information about the energetics of the isolated molecule.

One of the main problems in this calculation is finding the correct contact region between the molecule and the two leads. All of the molecules studied here are terminated by thiol end groups (-SH) and it is known[9] that when the thiol group is brought into contact with a gold substrate the hydrogen atom desorbs and the sulfur atom bonds with three gold atoms arranged in an equilateral triangle. To determine the bonding between the surface of the gold lead and the end of the molecule self consistently, the molecule is extended to include several layers of the gold leads at each of its ends. The optimal distance between the sulfur atom and the three gold atom contact area is found by minimizing this distance with respect to the energy.

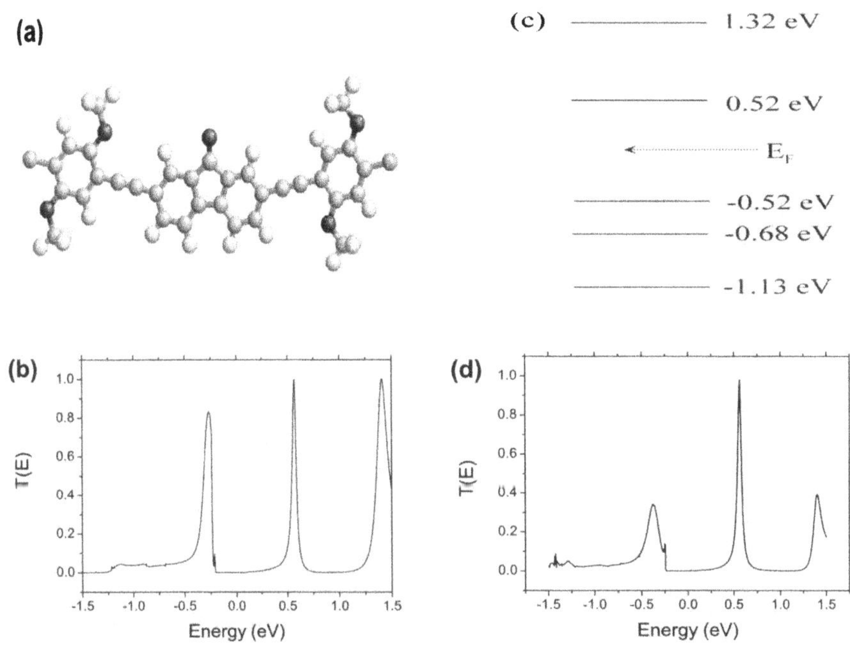

Figure 1. (a) geometry of the CSW-362 Molecule. (b) Zero-bias transmission of the CSW-362 molecule. (c) Energy Levels of the isolated CSW-362 molecule. (d) The transmission of the CSW-322 molecule which is the CSW-362 molecule (figure 1) with the side groups removed

Once the relaxed structure has been calculated, the tight binding hamiltonian for this extended molecule and contacts can be extracted using an extension to the SIESTA code. The final stage in the hierarchy is to utilize this material specific hamiltonian in a transport calculations, using single electron, scattering codes[10] developed at Lancaster. This technique determines the quantum mechanical scattering matrix of a phase coherent region connected to ideal external reservoirs. At zero temperature the energy dependant conductance is given by the Landauer-Büttiker[11] formula $G(E)=(2e^2/h)T(E)$, where $T(E)$ is the total transmission coefficient evaluated at the energy E, and is the probability of an electron being transferred from one lead to the other. In the linear response limit of a small applied bias, E is the Fermi energy.

In what follows, the contact region of the extended molecule consists of layers of a triangle of three gold atoms with triangular leads whose cross-section also contains a triangle of three gold atoms. This allows a simple continuation between the semi infinite leads and the ends of the molecule. By increasing the number of gold atoms that are self-consistently included as part of the molecule contacts, the conductance across the molecule is found to converge, due to the fact that the charge transfer effects at the gold-molecule interface decay over a small number of slices into the bulk gold lead.

3. Results

The first molecule we study is named CSW-362 (see Figure 1(a)) and has the chemical formula $C_{33}H_{24}S_2O_5$. This molecule has two benzene rings at either end, which contain the side groups of an oxygen atom with a methyl attached. After relaxing the geometry of this structure its length was found to be approximately 23.84Å. The resulting zero bias conductance computed for this molecule is shown in Figure 1(b). This has resonant peaks on each side of the Fermi energy (0 eV) and shows a conducting gap of approximately 0.8eV. Generally, in a mesoscopic system that has discrete energy levels (e.g. a molecule) which is contacted between two continuous reservoirs, the transport is expected to show resonant peaks. The energy levels of the isolated molecule predicted using SIESTA can be seen in figure 1(c). The conductance shows a resonance peak for the HOMO (Highest Occupied Molecular Orbital) level at approximately 0.4 eV below the Fermi level and the LUMO (Lowest Unoccupied Molecular Orbital) peak is 0.5eV above the Fermi level. These are similar to the HOMO and LUMO levels of the isolated molecule. When the molecules interact with the gold leads, the molecular levels are broadened and the main mechanism behind the transport is found to be resonant tunnelling between the leads.

All of the molecules that we study have side groups incorporated, mainly to aid the solubility of these long structures. Removing the side groups from the CSW-362 molecule gives the simpler CSW-322 molecule whose chemical formula is $C_{28}H_{13}S_2$. In this case the four side groups are simply replaced by single Hydrogen atoms. The transmission plot for this molecule can be seen in figure 1(d) and shows similar positions for the resonant peaks that occur in the molecule with side groups. However, the size of the HOMO peak has been reduced by approximately a factor of two. This suggests that adding side groups may not greatly alter the chemical properties of the molecule, but could play an important role in determining the transmission across a molecule.

One of the main aims of this study is to investigate long molecules that could possibly be incorporated in a single molecule electronic device. The CSW-362 molecule is much shorter than the expected contact gap and one method

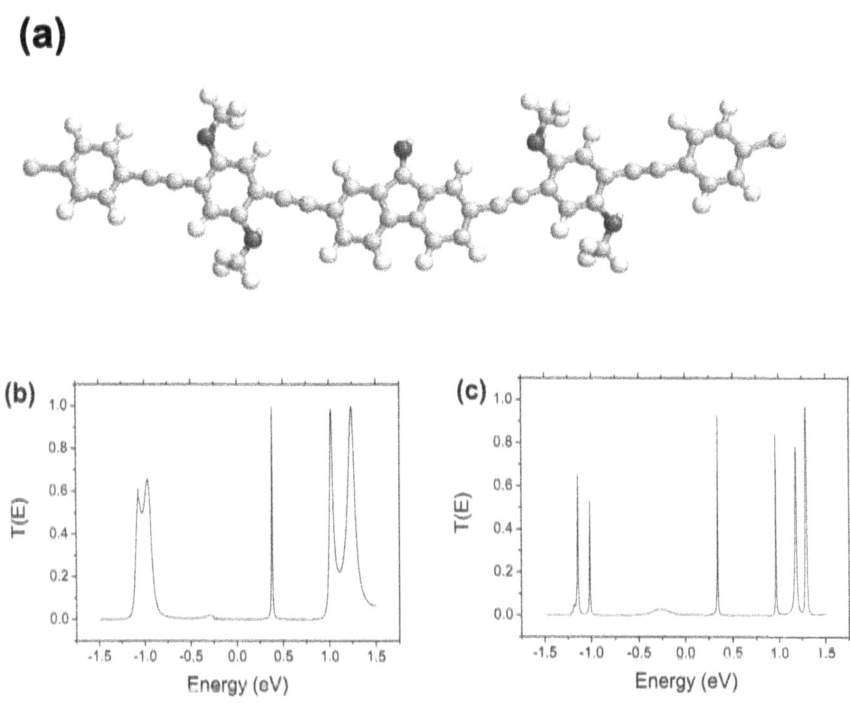

Figure 2. (a) The CSW-363 Molecule. (b) Transmission across the CSW-363 molecule. (c) Transmission across the molecule with a larger contact area.

employed in designing a longer molecule is to add benzene rings to each of the ends. The second molecule we look at is named CSW-363 and can be seen in Figure 2(a). It is simply the CSW-362 molecule with two extra benzene rings added, the calculated length has now been increased to approximately 37.93Å and its chemical formula is $C_{49}H_{32}S_2O_5$. The effect on the transport of increasing the length of the molecule by this method can be seen in figure 2(b). A similar conductance profile to the smaller molecule can be seen, however introducing these benzene rings has shifted the position of both the HOMO and LUMO resonance peaks and increased the conducting gap between them to approximately 1.25eV.

So far the conductance has been calculated using a small contact area consisting of triangular leads formed from slices of three gold atoms. Increasing the number of atoms in a slice from three to twelve and maintaining the gold (111) surface should give a more detailed description of the contact region. A comparison between the transport for a large contact region can be seen in figure

2(c). While the position and size of the resonant peaks appears to be unaltered by changing the contact area, the width of the resonances is much smaller when a larger contact region is added to the extended molecule. This shows that using a larger contact region doesn't have a large effect on the transport properties, and therefore a small contact region can give the major features of transmission across the molecule.

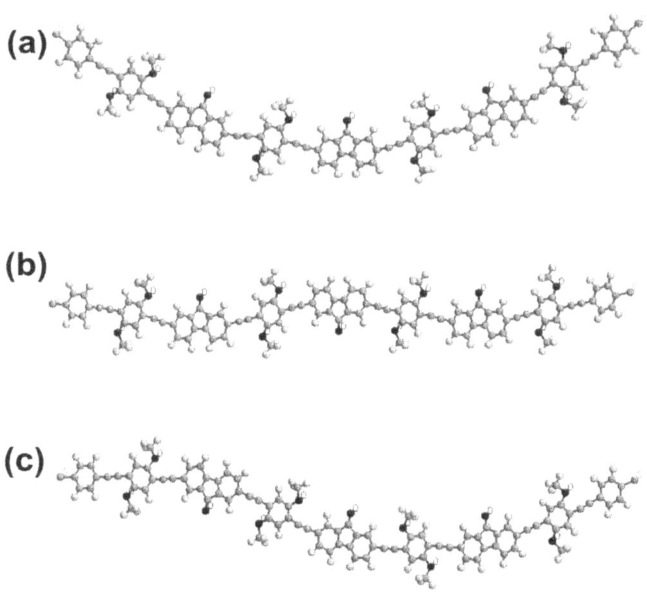

Figure 3. Three different configurations of the CSW-364 Molecule ((a)–(c)) achieved by rotating parts of the molecule by 180 degrees.

The final molecule we study named CSW-364 (see figure 3) is a much longer molecule than those studied to date and is partly formed by doubling the CSW-363 molecule. It has a length approaching the target size between metal contacts. Depending on the initial orientation of various parts of the molecule its length can vary between approximately 68Å to 74Å. Figure 3 shows three possible orientations of this molecule which have been relaxed to a local minimum, although these are only a selection of a number of possible configurations formed by rotating parts of the molecule. Calculating the ground state energy of the three different geometries gives respectively -21409.44eV, -21409.4eV and -21410.7eV for molecules (a), (b) and (c) in Figure 3. Thus the configuration of molecule (c) is the most energetically stable.

The calculated transmission across the CSW-364 molecule for each of the three selected geometries can be seen in Figure 4. The transmission profiles

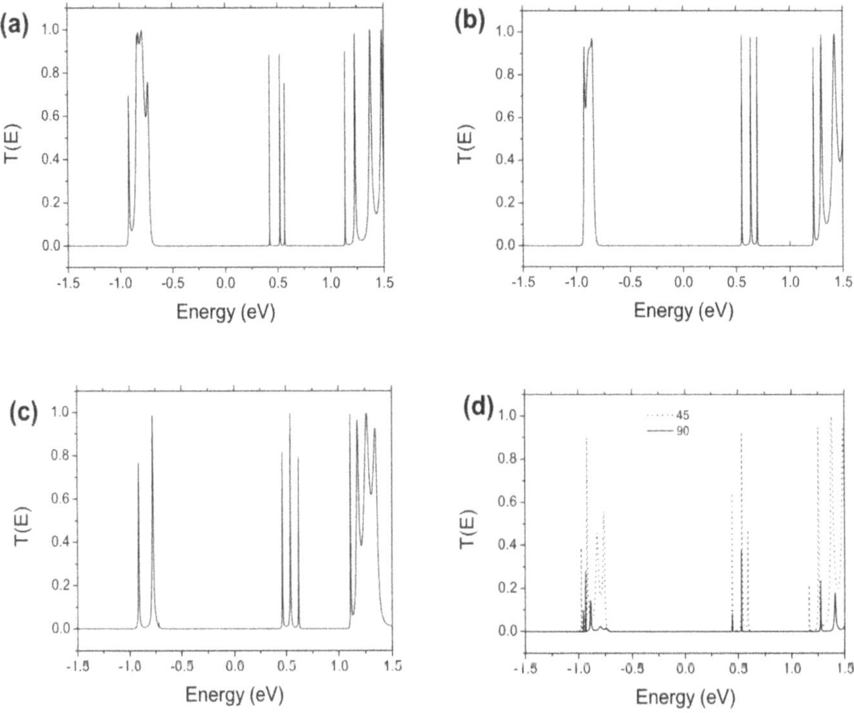

Figure 4. Shows the transmission across the CSW-364 molecule for each of the three configurations (a), (b) and (c) in figure 3. (d) The effect on transmission of the rotation of the end benzene ring in the CSW-364 (b) molecule by 45 degrees and 90 degrees

are quite similar for each configuration, in each case the HOMO peak is positioned at approximately -0.9eV and there are three peaks close together at the LUMO level, at approximately 0.5eV. Therefore, for this much longer molecule, formed by effectively doubling the CSW-363 molecule, the conducting gap has increased to approximately 1.4eV. However, there are noticeable differences between the different configurations, i.e. the size of the conducting gap does change. In the case of molecule (a) it is approximately 1.22eV, for (b) it is 1.39eV and (c) is 1.24 eV. This suggests that single-molecule transport can be significantly altered by changing its conformation.

For the case of rotating the benzene rings about the central axis of the molecule we only consider one of the end benzene rings. The energy barrier for the rotation of a single ring is quite small, approximately 122meV. The transport properties for rotations of 45 and 90 degrees can be seen in figure 4 (d), this shows that rotation can cause the magnitude of the transmissions to decrease in both the HOMO and LUMO resonances.

4. Conclusions

In conclusion, a hierarchical, theoretical approach has been employed to calculate the transport properties of large molecules. It has been shown that increasing the length of a molecule or adding side groups can significantly alter the transport properties, as can conformation changes. In the present paper we have confined our calculation to linear transport. In the near future we aim to generalize these calculations to look at non-equilibrium transport and to search for current induced conformation changes.

References

[1] A. Aviram and M.A. Ratner, Chem. Phys. Lett. 29, 277 (1974)

[2] J. Chen, M. A. Reed, A. M. Rawlett, and J. M. Tour, Science 286, 1550 (1999)

[3] M. A. Reed, J. Chen, A. M. Rawlett, D.W. Price and J.M. Tour, Appl. Phys. Lett. 78, 3735 (2001)

[4] P. S. Damle, A. W. Ghosh, and S. Datta, Chem. Phys. 281, 171 (2002)

[5] J. Taylor, M. Brandbyge, and K. Stokbro, Phys. Rev. B 68, 121101(R) (2003)

[6] J. M. Soler, E. Artacho, J. D. Gale, A. García, J. Junquera, P. Ordejón and D. Sánchez-Portal, J. Phys.: Condens. Matter 14, 2745-2779 (2002)

[7] J.P. Perdew and A. Zunger, Phys. Rev. B 23, 5048 (1981)

[8] N. Troullier and J.L. Martin, Phys. Rev. B 43, 1993 (1991)

[9] N.B. Larsen, H. Biebuyck, E. Delamarche, and, B. Michel, J. Am. Chem. Soc. 119, 3107 (1997)

[10] S. Sanvito, C. J. Lambert, J. Jefferson, and A. Bratkovsky, Phys. Rev. B, 59, 11936 (1999).

[11] M.Buttiker, Y. Imry, R. Landauer, and S. Pinhas, Phys. Rev. B 31, 6207 (1985).

SINGLE-MOLECULE CONFORMATIONAL SWITCHES

Pavel Kornilovitch
Hewlett-Packard Company, MS 321A, Corvallis, OR 97330, USA
pavel.kornilovich@hp.com

Abstract A class of bistable stator-rotor molecules is described. The stationary part connects the two electrodes and facilitates electron transport between them. The rotary part, which has a large dipole moment, is attached to an atom of the stator via a single sigma bond. Hydrogen-like bonds formed between the rotor's oxygen and stator's hydrogen atoms make the symmetric orientation of the dipole unstable. The rotor has two potential minima with equal energy for rotation about the sigma bond. The dipole can be flipped between the two states by an external electric field. The two rotor-orientation states have different current-voltage characteristics, so they are distinguishable electrically. Theoretical results on conformation, energy barriers, retention times, switching voltages, and current-voltage characteristics are presented for a particular family of stator-rotor molecules.

Keywords: molecular electronics molecular rectifiers molecular switches bistability conformational switching

1. Molecular Electronics

The exponential growth of the computer speed and storage capacity poses the question of the ultimate limit of miniaturization of electronic elements. The workhorse of the contemporary integrated circuit technology is the metal-oxide-semiconductor field effect transistor. Although its operating principle is scaleable down to below 10 nanometers (nm) the key performance parameters become sensitive to the presence of just one additional atomic layer [1]. The necessary stability of the transistor performance will require fabrication of the entire macroscopic device with atomic precision. The enormous fabrication costs will likely make such a top-down approach impractical. It is imperative to begin looking for alternative ways to fabricate complex electronic structures, which will complement the silicon technology in the future.

Molecular electronics is one attractive possibility that takes advantage of the great uniformity between the shapes of individual molecules with the same chemical structure. Also, molecules are synthesized in bulk quantities, which

A.S. Alexandrov et al. (eds.), Molecular Nanowires and Other Quantum Objects, 21–28.

makes this part of the fabrication process really inexpensive. The challenge therefore shifts to the ways to organize the molecules in regular structures and integrate them with other parts of the circuit. Another clear advantage of molecules is their very small size.

An important consideration is the speed of development of the conventional semiconductor, optical, magnetic, and emerging polymer storage technologies. With currently achievable storage densities of $\sim 10^9$ bit/cm^2 and continuing exponential growth, the best of existing technologies are likely to achieve 10^{10} bit/cm^2 by 2010, and perhaps even 10^{11} bit/cm^2 with multi-layer architectures. If by that time molecular electronics is to provide qualitative improvements over those density levels, it has to aim at 10^{12}-10^{13} bit/cm^2 as a potential goal. The latter densities correspond to a linear bit size of 3 to 10 nanometers. Given the typical size of the molecules as 1 to several nanometers, such a small memory bit must be represented by just a few or even a single molecule. The above argument emphasizes the importance of single-molecule means of operating small currents and charges.

The small sizes of molecules are already being matched by the small sizes of wires. Regular arrays of 8 nm-wide platinum wires and 20 nm-wide silicon wires were recently fabricated by Melosh et al [2] using the superlattice technique [3]. Another possibility is to use carbon nanotubes as electrical wires. Unfortunately, no reliable method of arranging nanotubes in regular structures is known at present. Also, no technique to arrange molecules in three-dimensional structures has been proposed.

The key assumption of molecular electronics is that molecules can perform non-trivial and useful electronic functions. This idea was pioneered by Aviram and Ratner in 1974 [4], who proposed a design for the single-molecule rectifier of current (diode). Since molecules can be very complex, there exist no fundamental reasons of why just one molecule cannot be a transistor, a switch, a memory bit, or even a more complex circuit. Since the Aviram-Ratner paper, several designs for single-molecular devices have been proposed theoretically, including rectifiers [5, 6], switches [7, 8, 9, 10, 11, 12], and transistors [13]. Experimentally, electrical properties of molecules have been studied by mechanically-controlled break junctions [14, 15, 16], electro-migration break junctions [17, 18], scanning tunneling microscopy (STM) [19, 20, 21], and by other methods [22, 23, 24]. Out of the standard electronic operations, the use of molecules to store charge (single-molecule capacitors) and to conduct current (single-molecule wires) is perhaps best understood. The more complicated functions of rectifying, switching, and amplifying current are less understood. In the rest of this paper, we present some recent ideas about molecular switching.

2. Single-Molecule Switches

The molecular switch is a highly non-trivial element of molecular electronics. Switching implies bistability (or even multiple stability) of the molecule. The latter can be realized via bistability of electronic structure, for example as two different charge states. Alternatively, switching can be achieved via bistability in the molecular shape. The shape bistability is harder to realize but it has several potential advantages over the charge bistability including non-destructive readout and greater stability against temperature and quantum fluctuations.

Several conformational switching mechanisms have been described in the literature. The groups of Heath and Stoddard proposed that the movement of a large molecular ring sliding up and down along the dumbbell component of a pseudo-rotaxane is responsible for the resistance change observed in their devices [25, 26]. The ring movement between two stable "stations" is triggered by an oxidation reaction of the TTF group. For the rotaxane-based memory devices, a low switching time of $\sim 10^{-2}$ seconds was reported. Donhauser et al [27] observed conformational switching of Tour wires, apparently caused by temperature fluctuations and changes in the molecule-electrode geometry. Moresco et al [28] used STM with the goal to change conjugation in four-legged porphyrin derivatives, which affected the molecules' resistance. Schumaker proposed to use a combination of electric field and light impulses to change the shape and chirality of chiropticene molecules [8, 29]. Emberly and Kirczenow [11] proposed a molecular switch based on the strong orientation dependence of the molecule-electrode overlap matrix element, following the switching mechanism discussed in Reference [30].

Here we describe a novel design for the single-molecule electrical switch, which utilizes conformational changes that can be induced in the molecule by external electric field. The principal scheme of the switch is illustrated in Figure 1a. The molecule consists of two main parts, the stator and the rotor. The stator connects two electrodes and facilitates transfer of electrons between them. The rotor is a side group with a large dipole moment, which is attached to the stator via a single or triple covalent bond. A key design feature is the instability of the symmetric (perpendicular) orientation of the rotor relative to the stator. The instability occurs because the charged ions of the rotor (the same ions that are responsible for the dipole moment) interact electrostatically with the electro-positive hydrogen atoms of the stator. In particular, if an oxygen ion forms the negative pole of the dipole, it is attracted to one of the hydrogen atoms of the stator. (This mechanism is similar to the formation of a hydrogen bond, but is weaker.) The same applies to the nitrogen atom of the amino group $-NH_2$ if the latter is used as the positive pole of the dipole. As a result, in the symmetric configuration the dipole has a higher energy than when tilted, due to the greater rotor-hydrogen separation. That causes the dipole to tilt away from the perpendicular orientation. Since the two directions are equivalent, the dipole can tilt either way. Thus the molecule becomes bistable. Total molecule energy as a function of the stator-rotor angle assumes a double-well shape. The two equilibrium configurations are mirror images of each other.

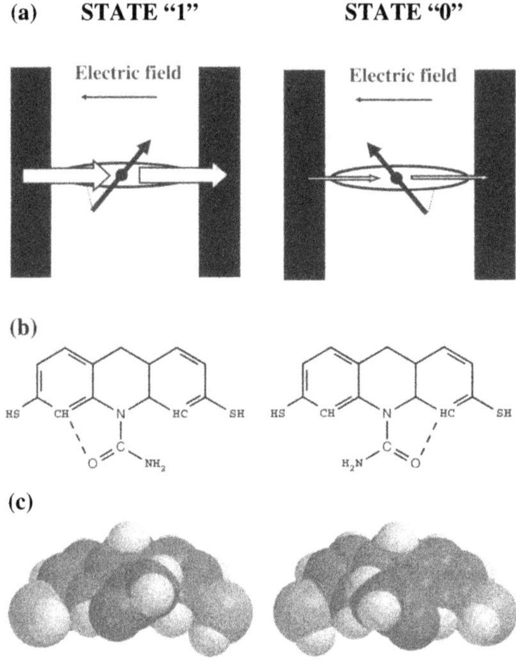

Figure 1. (a) The operating principle of the stator-rotor molecular switch. The stator is the thin oval in the center. The dipole group, represented by the solid arrow, is unstable against slipping in one of the two potential minima caused by the formation of hydrogen bonds between the stator and the rotor. Open arrows indicate the direction of the electron flow. For electrons going from left to right the two states are clearly nonequivalent, which is indicated by different widths of the open arrows. Thus the two states are distinguishable electrically. (b) A bistable stator-rotor molecule 9-hydro-10-acridinecarboxamide-2,7-dithiol shown in states "1" and "0". Electrostatic bonds formed between the oxygen of the amide group (-$CONH_2$) and the hydrogens of the stator are indicated by dashed lines. (c) Space-filling models of the two states.

Apart from providing bistability the rotor performs two other critical functions. First, it makes the two states distinguishable. As in any system with broken symmetry neither of the two equilibrium states is mirror-symmetric. Therefore both states distinguish the opposite directions of current flow. The physical reason for that is the electric field generated by the tilted dipole, which lifts the mirror symmetry of the electron density of the stator. The second function performed by the rotor is interaction with an external electric field. The dipole tends to align with the field. Thus a strong enough field can change the energy balance and flip the rotor between the two states in a controlled manner.

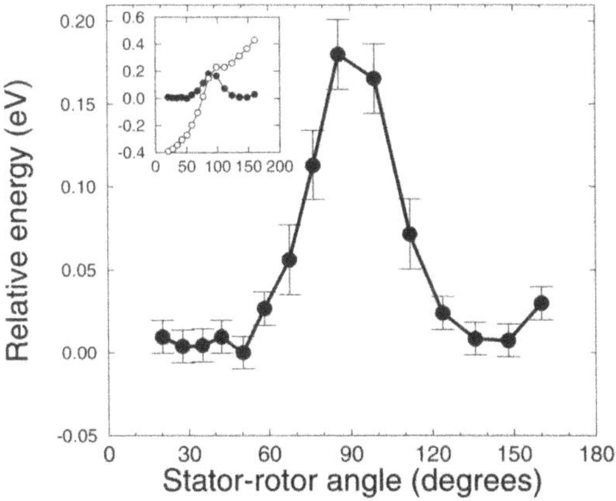

Figure 2. The relative energy of the molecule shown in Figure 1, as a function of the stator-rotor angle θ. The height of the energy barrier is (0.18 ± 0.02) eV. The error bars have been estimated from the energy variance during geometry minimization. Inset: the same data (solid circles) compared with the energy of the same molecule in an external field of $0.5V/\mathring{A}$.

A molecular switch built according to the principles described above is shown in Figure 1b and 1c. The rotor is represented by the amide group $-CONH_2$ that has a large dipole moment $d = 4.0$ Debye $\approx 0.8\, e{\cdot}\mathring{A}$. The rotor is attached to the stator via a C-N sigma bond. The stator consists of three fused rings, the two side ones being conjugated benzenes. Thiol clips are included on both ends of the stator to facilitate attachment to gold electrodes. The molecule bistability has been analyzed by performing constrained density-functional calculations of the total energy as a function of the stator-rotor angle θ [31]. The results are shown in Figure 2. The two energy minima occur at $\theta \approx 40°$ and $140°$ (tilt angle of $50°$). The barrier height that separates the minima is found to be $\Delta E = 0.18$ eV. Interaction with an external electric field F modifies the θ-dependence of the energy as follows:

$$E(F, \theta) = E(\theta) - Fd\cos\theta, \tag{1}$$

where the first term is the function plotted in Figure 2. At zero temperature, the rotor will flip between the states when the field reduces the energy barrier to zero. Equating the θ-derivative of the last equation to zero, one obtains the switching field $F_{\text{switch}} = 0.5$ V/Å. At non-zero temperatures, temperature fluctuations might result in dipole flipping at lower fields.

Let us now discuss distinguishability of the two states. We have performed quantum-mechanical transport calculations for the two stable states of the switch. The results are shown in Figure 3. The two I-V curves differ substantially in the 3 to 4 volts window. It should be noted that the current difference is a non-linear effect because the linear term is the same for any two states that are mirror images of each other. That is why the two I-Vs overlap between 0 and 2 volts. The actual non-linear process that leads to an I-V difference is reorganization of the stator electron density caused by the source-drain field. This reorganization depends on the internal field generated by the dipole. The internal field is oriented differently in the two states. A more detailed discussion is presented in Reference [9].

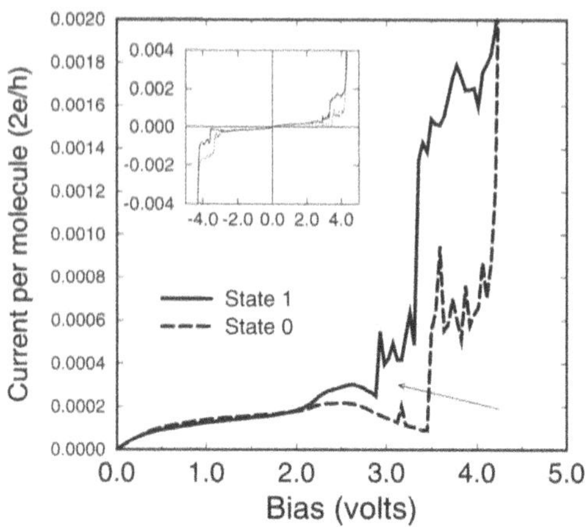

Figure 3. I-V characteristics of the two states of the stator-rotor switch. The major "window of distinguishability" lies in the interval of 3 to 4 volts. An arrow indicates the region where the currents in the two states differ because of the asymmetric localization of the highest occupied molecular orbital. The inset shows the same I-V characteristics in a symmetric voltage interval.

Finally, we address temperature stability of the stator-rotor molecules. It is determined by an energy barrier ΔE and the frequency Ω of the vibration mode that attempts to cross the barrier. In our case, the relevant mode is "rocking" of the dipole around its equilibrium. The corresponding frequency was calculated [31] to be $\Omega = 97$ cm^{-1} $= 1.8 \cdot 10^{13}$ rad/s. The retention time can be found from the Kramers formula:

$$\tau = \frac{1}{\Omega}\, e^{\Delta E / k_B T}, \qquad (2)$$

where T is the temperature. Using $\Delta E = 0.18$ eV one obtains $\tau = 58$ ps at room temperature (T = 300 K), $\tau = 33$ ms at T = 77 K, $\tau = 1$ hour at T = 54 K, and $\tau = 30$ years at T = 40 K. Temperature stability is going to be an issue for this type of switches if they are to be used at room temperature. The principal way to improve this characteristic is to increase ΔE. This could be achieved by enhancing the hydrogen-rotor interaction. We have found, for instance, that replacing –H with –OH on the stator brings the hydrogen atoms closer to the rotor and increases their ionicity. As a result the barrier grows to 0.6 eV. However, those switches are more difficult to synthesize. Also, the larger the barrier the harder it is to switch with the available electric filed. That, in turn, could be overcome by increasing the dipole moment of the rotor (by increasing the rotor's length, for example).

3. Discussion

If molecular electronics is to take over the existing memory and computing technologies, it has to overcome many challenges. One of those is the development of truly single-molecular methods of operating electrical signals. The switching mechanism discussed in this paper should be regarded as a design concept. The concrete molecular realization presented here is just an illustration. Effective optimization process is impossible without systematic feedback from experiment. Still, even a simple concept-level analysis has resulted in a number of useful observations.

Low speed and competition between stability and switching seem to be the major problems. Room temperature stability requires energy barriers as large as 1.2 eV. This, in turn, requires switching fields as high as several volts per angstrom (essentially, atomic-scale). The molecules are unlikely to survive such strong fields for long. Another possibility is to rely on temperature assisted switching: the available external field reduces the barrier and then a temperature fluctuation takes the molecule over the barrier. Such a process, however, is inherently slow. There exist at least two principal strategies to overcome some of the difficulties. The first is to engineer more complex molecules with larger dipoles, higher polarizabilities, and so on. This leads to increasingly complex chemical synthesis. The other is to reduce the operating temperature of the device. However impractical this approach might seem, this is definitely a way towards understanding the fundamental properties of the stator-rotor switches as well as molecular electronic elements in general [32]. These types of issues are quite similar to those of the electronic field-operated switch introduced by Aviram [33].

Another direct consequence of conformational switching is the presence in molecules metal ions, dipoles, and other charged groups. This necessarily leads to a long-range electrical interaction between the molecules, which may alter their electrical response, and perhaps significantly. The inter-molecule interaction is an important issue that is yet to be addressed with all seriousness. Single-molecule measurements seem to be the key for separating intra-molecular and collective effects. Proliferation of novel experimental techniques that occurred in the past six years [14, 15, 16, 17, 18, 21, 22, 23, 27, 34] ensures that significant progress in this area will be achieved in the near future.

References

[1] T.J. Walls, V.A. Sverdlov, and K.K. Likharev. Physica E **19**, 23 (2003).

[2] N.A. Melosh et al., Science **300**, 112 (2003).

[3] Y. Chen and R.S. Williams. Nanoscale patterning for the formation of extensive wires, UP Patent 6,407,443 (2002).

[4] A. Aviram and M.A. Ratner. Chem. Phys. Lett. **29**, 277 (1974).

[5] J.C. Ellenbogen and J.C. Love. Architectures for molecular electronic computers: 1. Logic structures and an adder designed from molecular electronic diodes. Proceedings of the IEEE **88**, 386 (2000).

[6] P.E. Kornilovitch, A.M. Bratkovsky, and R.S. Williams. Phys. Rev. B **66**, 165436 (2002).

[7] I.D.L. Albert, T.J. Marks, and M.A. Ratner. J. Am. Chem. Soc. **120**, 11174 (1998).

[8] L.B. Aubin et al. Org. Lett. **3**, 3413 (2001).

[9] P.E. Kornilovitch, A.M. Bratkovsky, and R.S. Williams. Phys. Rev. B **66**, 245413 (2002).

[10] A.S. Alexandrov, A.M. Bratkovsky, and R.S. Williams. Phys. Rev. B **67**, 075301 (2003).

[11] E.G. Emberly and G. Kirczenow. The Smallest Molecular Switch, cond-mat/0301296 (2003).

[12] P. Orellana and F. Claro. Phys. Rev. Lett. **90**, 178302 (2003).

[13] A.M. Bratkovsky and P.E. Kornilovitch. Phys. Rev. B **67**, 115307 (2003).

[14] M.A. Reed et al., Science **278**, 252 (1997).

[15] J. Reichert et al., Phys. Rev. Lett. **88**, 176804 (2002).

[16] R.H.M. Smit et al., Nature **419**, 906 (2002).

[17] H. Park et al., Nature **407**, 57 (2000).

[18] J. Park et al., Nature **417**, 722 (2002).

[19] L.A. Bumm et al., Science **271**, 1705 (1996).

[20] S. Datta et al., Phys. Rev. Lett. **79**, 2530 (1997).

[21] X.D. Cui et al., Science **294**, 571 (2001).

[22] J.G. Kushmerick et al., Phys. Rev. Lett. **89**, 086802 (2002).

[23] N.B. Zhitenev et al., Nanotechnology **14**, 254 (2003).

[24] C.P. Collier et al., Science **285**, 391 (1999).

[25] C.P. Collier et al., J. Am. Chem. Soc. **123**, 12632 (2001).

[26] Yi Luo et al., Chem. Phys. Chem. **3**, 519 (2002).

[27] Z.J. Donhauser et al., Science **292**, 2303 (2001).

[28] F. Moresco et al., Phys. Rev. Lett. **86**, 672 (2001).

[29] http://www.calmec.com, and references therein.

[30] P.E. Kornilovitch and A.M. Bratkovsky, Phys. Rev. B **64**, 195413 (2001).

[31] SPARTAN version 5.0, Wavefunction, Inc. 18401 Von Karman Avenue, Suite 370, Irvine, CA 92612 U.S.A.

[32] Wenyong Wang, Takhee Lee, and M.A. Reed, Phys. Rev. B **68**, 035416 (2003).

[33] A. Aviram, J. Am. Chem. Soc. **110**, 5687 (1988).

[34] Wenjie Liang et al., Nature **417**, 725 (2002).

DIPOLE INTERACTIONS IN NANOSYSTEMS

Philip B. Allen
Department of Applied Physics and Applied Mathematics
Columbia University, New York, NY 10032
and
Department of Physics and Astronomy
State University of New York, Stony Brook, NY 11794-3800 *
philip.allen@stonybrook.edu

Abstract The dipole-dipole interaction influences nanoscopic matter by fixing the patterns of permanent, displacive, and induced dipole moments, subject to constraints of molecular size and other short range interactions. Prediction of these arrangements is a challenging problem. The eigenvector of maximum eigenvalue of the dipole-dipole interaction matrix can provide insights and sometimes a complete solution. As an example, the octahedral tilt instabilities of perovskite-type crystals is shown to optimize dipolar interactions. Therefore this instability can be designated as antiferroelectric, because dipole-dipole interactions are a dominant driving force.

Keywords: dipole interaction, Clausius-Mossotti, ferroelectric, antiferroelectric, octahedral tilt, nanosystem

1. Introduction

In condensed matter one finds three kinds of electrical dipoles: (1) permanent (as in a molecular cluster of polar molecules, such as CH_3Cl), (2) diplacive (as in $BaTiO_3$), and (3) induced (a cluster of benzene molecules in an external field). The unifying feature is that the dipole moments feel each other. Their magnitudes and orientations depend on where the other dipoles are, in a self-consistent fashion. This paper argues that one should study the linear algebra, and particularly the eigenvector of maximum eigenvalue, of the dipole-dipole matrix Γ which relates the induced electric field $\vec{F}_{i,\mathrm{ind}}$ at the i'th site $\vec{R}_i$ to the

*permanent address

A.S. Alexandrov et al. (eds.), Molecular Nanowires and Other Quantum Objects, 29–38.

dipoles $\vec{\mu}_j$ at sites $\vec{R}_j$:

$$F_{i\alpha,\text{ind}} = \sum_{j\beta} \Gamma_{i\alpha,j\beta} \mu_{j\beta}, \tag{1}$$

$$\Gamma_{i\alpha,j\beta} = \frac{3R_{ij\alpha}R_{ij\beta} - \delta_{\alpha\beta}R_{ij}^2}{R_{ij}^5}, \tag{2}$$

where $\vec{R}_{ij} = \vec{R}_i - \vec{R}_j$. The insights obtained from the eigenvectors and eigenvalues are illustrated below in some simple examples, and in an application to real materials with spontaneous electrical polarity, the octahedral tilt instability of ReO_3-type perovskite crystals.

Of course, it must be acknowledged that not all properties of polar matter have much to do with the dipole moments of their molecules. For example, the water molecule has a dipole in vapor phase, and electrical polarity is a dominant effect in the interactions of water molecules with each other and with dissolved species. However, for near-neighbor pairs, it is not a good approximation to replace the polar charge distribution of water by a single dipole. A better approximation would use two dipoles, located along the two bonds, each pointing away from the oxygen toward one of the two hydrogens. The sum of these two dipoles is the total measured dipole moment of the water molecule. Also, it would be not a very good approximation to assume that the two dipole moments of the water molecule are unchanged in their various chemical environments. Induced polarity should be treated on top of permanent or displacive polarity, and this is done in the case of the octahedral tilt. A complete theory would not make any type of dipolar approximation, but would deal with the actual varying charge distribution of the molecules in detail. Modern density functional theory does this well, but slowly, which motivates a search for a less complete but still sensible theory.

2. One Polarizable Molecule

The energy of a single molecule is $\mu^2/2\alpha - \vec{\mu} \cdot \vec{F}_{\text{ext}}$ where α is the polarizability and $\vec{F}_{\text{ext}}$ is an externally applied field. The actual dipole moment $\vec{\mu}$ is fixed by the condition that the energy is minimum, which gives $\vec{\mu} = \alpha \vec{F}_{\text{ext}}$.

3. More than One Polarizable Molecule

Now consider N molecules, approximated as points, at chosen fixed positions $\vec{R}_i$. It costs energy $\mu_i^2/2\alpha$ to create a dipole $\vec{\mu}_i$ on the ith site, but one gets back energy $-\vec{\mu}_i \cdot \vec{F}_i$ from the total field $\vec{F}_i = \vec{F}_{i,\text{ind}} + \vec{F}_{i,\text{ext}}$ at the site, where $\vec{F}_{i,\text{ext}}$ may be spatially inhomogeneous, varying from site to site. The total energy, in a standard vector space notation, is

$$U(\vec{\mu}_1, \ldots, \vec{\mu}_N) = \frac{1}{2}\left\langle \mu \middle| \left(\frac{1}{\alpha} - \Gamma\right) \middle| \mu \right\rangle - < \mu | F_{\text{ext}} > . \tag{3}$$

When there are N sites, each with a dipole, then $|\mu >$ is a $3N$ column vector containing all the components $\mu_{i\alpha}$ for $\alpha = x, y, z$ and $i = 1, \ldots, N$. Notice that the interaction energy $-\vec{\mu}_i \cdot \vec{F}_{i,\text{ind}}$ is reduced by a factor of 2 in the total energy, to avoid double counting. With no external field, the system is stable if $1/\alpha - \Gamma$ is a positive operator. Equivalently, all eigenvalues of Γ should be less than $1/\alpha$.

Just as for the single molecule, the induced dipoles $\vec{\mu}_i$ should minimize the total energy, $\partial U / \partial \mu_{i\alpha} = 0$. The solution is

$$|\mu >= \left(\frac{1}{\alpha} - \Gamma\right)^{-1} |F_{\text{ext}} > \tag{4}$$

This is a generalized version of the Clausius-Mossotti law [1, 2, 3]. An alternate notation, $|\mu >= \sum (1/\alpha - \gamma)^{-1} < \gamma | F_{\text{ext}} > |\gamma >$ uses an expansion in eigenvectors, where $|\gamma >$ is the eigenvector of Γ with eigenvalue γ. If the overlap $< \gamma_{\max} | F_{\text{ext}} >$ is not zero, the eigenvector of largest eigenvalue $\gamma_{\max}$ may dominate the response.

If $\gamma_{\max}$ increases to become equal to $1/\alpha$, then the system becomes unstable. It will develop a spontaneous polarization, with a pattern $|\mu >$ proportional to the corresponding eigenvector $|\gamma_{\max} >$. Higher order terms not considered here are needed to determine the magnitude of the spontaneous moment.

4. Two Dipoles

For pedagogical purposes, let us examine the 6×6 matrix Γ which describes the interaction between two dipoles separated by distance $|\vec{R}| = a$; to be specific, let $\vec{R}$ point along the $\hat{z}$ axis. Even without writing the matrix, we can find all six eigenvectors by simple physical reasoning. The eigenstates of Γ are those patterns of dipoles such that the induced field at each dipole is parallel and proportional to the dipole itself. Consider the patterns in Fig. 1.**a**-**d**. The field $\vec{F}_1$ lies along the z-axis if $\vec{\mu}_2$ points along z, and lies in the xy plane antiparallel to $\vec{\mu}_2$ if $\vec{\mu}_2$ lies in the xy plane. Therefore vectors lying along $\hat{z}$ separate from those lying in the xy plane, and by cylindrical symmetry, eigenvectors in the xy plane are doubly degenerate. The degenerate pair can be chosen so that one lies along $\hat{x}$ and one along $\hat{y}$. To save space, we project the problem onto the 2d xz plane. The resulting 4×4 matrix is

$$\Gamma = \frac{1}{a^3} \begin{pmatrix} 0 & 0 & 2 & 0 \\ 0 & 0 & 0 & -1 \\ 2 & 0 & 0 & 0 \\ 0 & -1 & 0 & 0 \end{pmatrix} \tag{5}$$

The eigenvalues are (2,1,-1,-2) in units of $(1/a^3)$, and the corresponding eigenvectors, shown in Fig.1.**a-d**, are

$$|a>= \begin{pmatrix} 1 \\ 0 \\ 1 \\ 0 \end{pmatrix}; \; |b>= \begin{pmatrix} 0 \\ -1 \\ 0 \\ 1 \end{pmatrix}; \; |c>= \begin{pmatrix} 0 \\ 1 \\ 0 \\ 1 \end{pmatrix}; \; |d>= \begin{pmatrix} -1 \\ 0 \\ 1 \\ 0 \end{pmatrix}. \tag{6}$$

These eigenvectors all have different symmetries. Under inversion and mirror reflection in the central xy plane, the vectors have (λ_i, λ_m) = $(--)$, $(+-)$, $(-+)$, and $(++)$ symmetry, respectively.

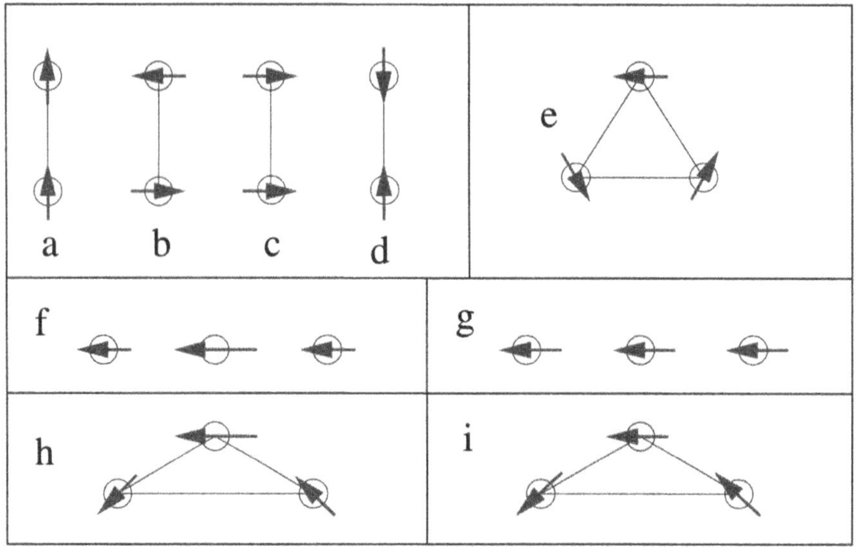

Figure 1. Dipole patterns of two and three interacting dipoles. In **a-d** the eigenvectors of Γ are shown for two dipoles. In **e,f,h**, the eigenvector of largest eigenvalue is shown for three dipoles. In **e**, the dipoles all have equal length. In **g** and **i** the optimal patterns (not eigenvectors!) for fixed length dipoles are shown, nearly agreeing with the eigenvectors of **f** and **h**. The vectors in **h** and **i** differ in angle by about 3°.

The isotropic Heisenberg interaction $U = J\vec{\mu}_1 \cdot \vec{\mu}_2$ for two spins has exactly the same eigenvectors but very different eigenvalues. For $J < 0$ (ferromagnetic exchange), states **a** and **c** are low energy states, equal in energy, and states **b** and **d** are high energy states, also equal in energy. For $J < 0$ (antiferromagnetic exchange), the reverse is true. For magnetic systems (the O_2 molecule is a nice example), the exchange energy is much larger than the dipole-dipole interaction. The ground state of O_2 has two outer electrons, polarized ferromagnetically $(S = 1)$, and only a tiny spin-orbit-induced preference for the moment to point perpendicular to the axis. By contrast, electrical dipoles have a large dipole-dipole interaction which is intrinsically anisotropic and strongly prefers state **a**.

5. Three Dipoles

In the two-dipole problem of the previous section, the eigenstates all had the property that each molecule's moment $\vec{\mu}_i$ (where $\mu_{i\alpha} =< i\alpha|\gamma >$) had the same magnitude $|\vec{\mu}_i|$. This brings a nice benefit: these patterns tell us also about the energetics of the problem of fixed size permanent dipoles.

Fig. 1.**e-i** shows geometries and corresponding dipole patterns for three molecules. The physical situations of interest are (1) polarizable molecules with no permanent moment, (2) permanent dipoles which can orient in space but not change their magnitude, and (3) permanent dipoles which are also polarizable and change their magnitude. We omit category (3) for now. Category (1) is mathematically simplest because linear algebra determines what happens in an external field, and whether spontaneous polarity appears. Fig. 1.**e,f,h** show the eigenvectors of largest eigenvalue of Γ for (**e**) the equilateral triangle, (**f**) the linear arrangement, and (**h**) an obtuse isosceles triangle. For the equilateral case, the eigenvector is determined by symmetry, but for the other two cases, there are two independent vectors odd under the perpendicular mirror, so a 2×2 matrix is needed, and the central atom (or molecule) has a larger $|\vec{\mu}_{\text{ind}}|$ than the end atoms.

Because $|\vec{\mu}_i|$ is no longer the same for each molecule in the maximal eigenvector, it is not the solution of the problem of a fixed moment. Instead, for fixed moments we want to know the pattern $|\mu >$ which minimizes

$$U(\vec{\mu}_1, \ldots, \vec{\mu}_N) = -\frac{1}{2} < \mu|\Gamma|\mu > - < \mu|F_{\text{ext}} >, \tag{7}$$

subject to N nonlinear constraints $|\vec{\mu}_i| = \mu_0$. This is a much harder mathematical problem. Fortunately, the solution is guaranteed [4] to have a smooth relation to the maximal eigenvector of Γ. The solutions to the fixed moment problem in zero field are shown in Fig. 1.**g,i**, and are close to the unconstrained solutions in Fig. 1.**f,h**.

6. Infinite Stack of Dipoles

Consider the arrangement $|\mu >$ of Fig. 2, with equally spaced dipoles $\vec{\mu}$ of equal magnitude, all pointing in the $\hat{z}$ direction which is also the direction of the stack of molecules. It is clear that this pattern is the maximal eigenvector of the corresponding Γ. The eigenvalue is $4\zeta(3)/a^3$ where a is the spacing of dipoles and $\zeta(3) = 1.202057\ldots$ is the Riemann zeta function. Since the moments are equal, this pattern is also the solution to the problem of fixed dipoles on a 1-d

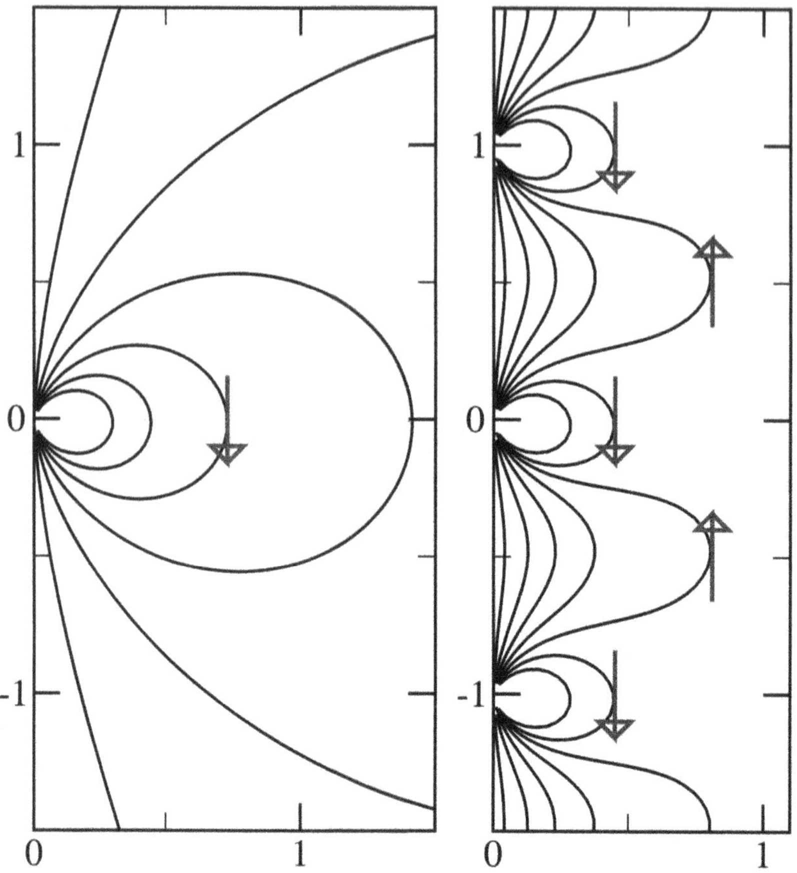

Figure 2. Faraday flux lines indicating the field strength and direction (indicated by arrows) for a single dipole pointing up (left panel) and for a periodic stack of dipoles pointing up (right panel). In the left panel the field falls as $1/r^3$ in all directions. In the right, the field decays exponentially in the directions transverse to the stack.

line. It is interesting to compute the electric field of this stack, which is the negative gradient of the electrostatic potential $\Phi(\rho, z)$ (where $\rho = \sqrt{x^2 + y^2}$ is the radial distance). By methods described elsewhere [4] the potential has the asymptotic form

$$\Phi(\rho, z) = -\frac{4\pi\mu}{a^2} \frac{\sin(2\pi z/a) e^{-2\pi\rho/a}}{\sqrt{\rho/a}} \tag{8}$$

This result explains why fixed dipoles particularly like to form one-dimensional stacks. The dipole-dipole attraction within the stack is very big, and the interaction with other dipoles falls very rapidly with distance. The asymptotic form Eq.8 is already accurate to 2% at $\rho = a$.

7. Cubic lattice of Dipoles

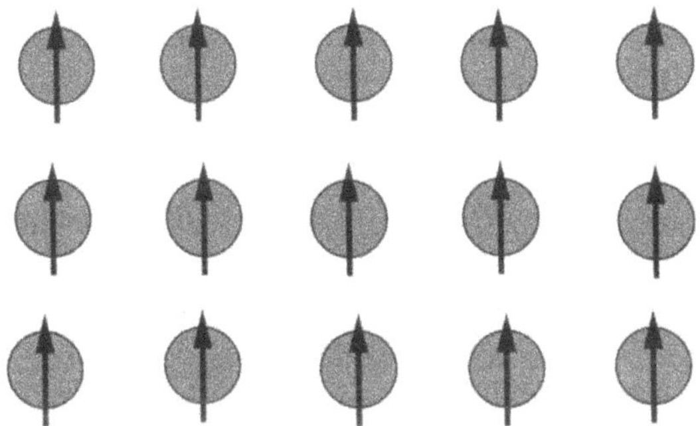

Figure 3. Infinite periodic lattice of cubic symmetry, showing the ferroelectric eigenstate studied by Clausius and Mossotti. Also the 2-d analog is an eigenstate. These are not necessarily the maximal eigenstates.

An eigenvector of this problem was found independently by Clausius [1] and Mossotti [2], and is illustrated in Fig. 3. If all dipoles point along a cube axis, with equal magnitude, it is clear that the induced field does the same thing, so this is an eigenvector. The Clausius-Mossotti arguments show that the eigenvalue is $\gamma_{CM} = 4\pi n/3$ where n, the density of molecules, is $1/a^3$ for the simple cubic lattice. A good pedagogical discussion is given by Aspnes [3]. One should also ask whether this is the maximal eigenvalue. From Fig. 2, it is clear that for the simple cubic lattice illustrated in Fig. 3, the answer is "no." One can further lower the energy by reversing the direction of alternate stacks of dipoles, producing a slightly larger eigenvalue in an antiferroelectric pattern. Luttinger and Tisza [5] proved that this pattern had the extremal eigenvalue. For diamond structure, a more complicated extremal antiferroelectric eigenstate was found by White *et al.* [6]. These authors also showed that for *fcc* and *bcc* structures, the Clausius-Mossotti ferroelectric eigenstate is maximal.

8. Octahedral Tilt

In this last section a brief discussion is given of recent work to be explained more fully elsewhere [7, 8, 9]. We have found that dipole-dipole interactions play an unexpectedly big role in the "octahedral tilt" instabilities common in perovskite materials [10, 11, 12, 13]. The motif is evident in Fig. 1.**e**, showing the tendency of a triangular plaquette to polarize circumferentially. In the

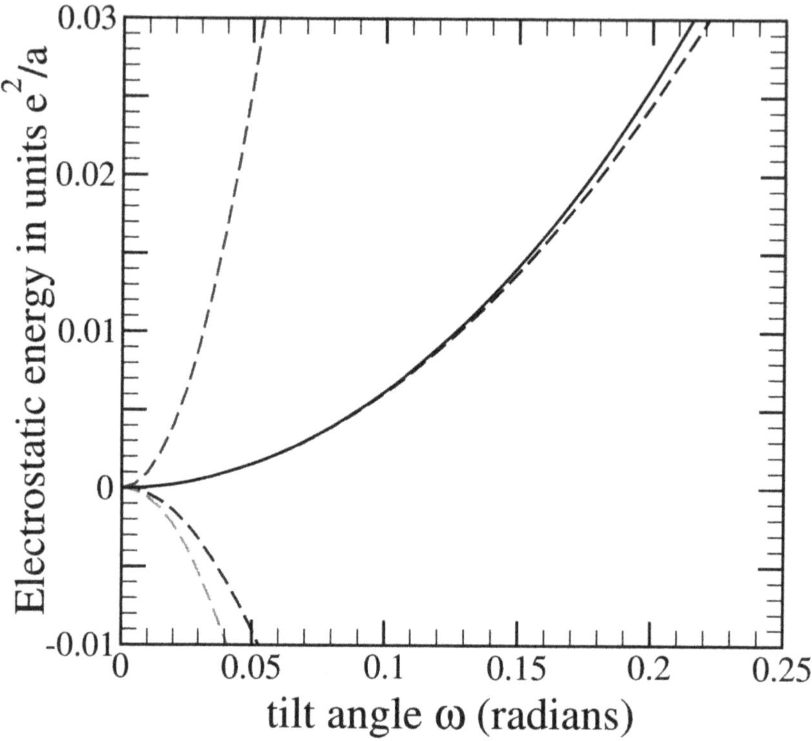

Figure 4. Energy versus tilt angle for (π, π, π) periodic tilts around the (111) axis of BX_3 (ReO_3 structure type) perovskites. The solid line is the total electrostatic energy (units Z^2e^2/a, where Z is the anion charge) computed for rigid ions comprising rigid octahedra, with a corresponding hard-core contraction perpendicular to (111) planes. The dashed curves , from top to bottom, are electrostatic energy for one displaced anion, dipole-dipole attraction energy of interacting displacements, and volume contraction energy. The sum of these is the dashed curve, coinciding with the exact energy up to fourth order terms in the Taylor series.

perovskite tilt, when the axis is (111), the displaced anions create the same triangular pattern of displacive dipoles. The moment points from the new anion position to the undistorted cubic anion position. These moments interact because they create an induced field at each undisplaced site. After some thought one can conclude that this is an eigenstate of Γ, and after some calculation one can prove that it is the maximal eigenstate. There is no better way to place dipoles on the X sites of ABX_3 perovskite than in the pattern corresponding to this tilt. The eigenvalue $\gamma_{\max} = 14.461/a^3$, is 15% bigger than γ_{CM}. The consequence is that in a rigid ion picture, the BX_3 simplification of the ABX_3 lattice is only marginally stable under electrostatic and hard sphere forces. When the anion polarizability is added to the theory [7], the structure is almost immediately unstable even for very small polarizability. The maximum eigenvalue is accidentally 5-fold degenerate. Three of the partners permit (π, π, π) alternating tilts in any linear combination around any cube axis. These generate the distortions labeled $a^- b^- c^-$ in Glazer notation [10]. The Glazer distortions with + signs require tilts with $(\pi, \pi, 0)$ periodicity, which happen also to be eigenvectors of Γ, with slightly smaller than maximal eigenvalue. It seems appropriate, given the role that dipolar interactions play, to regard all these distorted ground states as multiple-sublattice antiferroelectrics. The other two partners are related to the cooperative Jahn-Teller ground state of $LaMnO_3$, which thus also has an unexpected contribution from polar interactions in its distorted ground state.

9. Summary

Maximal eigenstates of the dipole-dipole interaction tensor Γ are useful and relevant to several problems: response of polarizable clusters to external fields, causes and patterns of spontaneous distortion of polar assemblies, and the self-organization of polar molecules in various media. It is surprising how robust and useful this concept seems to be. Given that short range repulsive forces are large, and that near-neighbor electrostatics may involve higher multipoles, there is no obvious reason why a pattern like the perovskite octahedral tilt should have to conform to the maximal eigenvector. The fact that it does should encourage us to use this tool in other problems.

Acknowledgments

I thank Dr. Y.-R. Chen, C. P. Grey, and S. Chaudhuri for help and stimulation. The work was supported in part by NSF grant no. DMR-0089492, and in part by a J. S. Guggenheim Foundation fellowship. Work at Columbia was supported in part by the MRSEC Program of the NSF under award no. DMR-0213574.

References

[1] R. Clausius, *Die Mechanische Warmtheorie* II, 62 Braunschweig (1897).

[2] O.F. Mossotti, Memorie Mat. Fis. Modena 24, 49 (1850).

[3] D. E. Aspnes, Am. J. Physics **50**, 704 (1982).

[4] P. B. Allen, J. Chem. Phys. (in press); cond-mat/0307209.

[5] J. M. Luttinger and L. Tisza, Phys. Rev. **70**, 954 (1946).

[6] S. J. White, M. R. Roser, J. Xu, J. T. van der Noordaa, and L. R. Corruccini, Phys. Rev. Lett. **71**, 3553 (1993).

[7] P. B. Allen, Y.-R. Chen, S. Chaudhuri, and C. P. Grey, manuscript in preparation.

[8] S. Chaudhuri, P. Chupas, M. Wilson, P. Madden, and C. P. Grey, manuscript in preparation.

[9] Y.-R. Chen, V. Perebeinos, and P. B. Allen, Phys. Rev. B (in press); cond-mat/0302272.

[10] A. M. Glazer, Acta Cryst. B**28**, 3384 (1972).
[11] P. M. Woodward, Acta Cryst. B**53**, 32 (1997).
[12] C. J. Howard and H. T. Stokes, Acta Cryst. B**54**, 782 (1998).
[13] N. W. Thomas, Acta Cryst. B**54**, 585 (1998).

CHARGE AND SPIN TRANSPORT IN ORGANIC NANOSYSTEMS: RECTIFICATION, SWITCHING, AND SPIN INJECTION

A.M. Bratkovsky
Hewlett-Packard Laboratories, 1501 Page Mill Road,
Palo Alto, California 94304
alexmb@exch.hpl.hp.com

Abstract Devices for nano- and molecular size electronics should allow for an efficient current rectification and switching. A few molecular scale devices are analyzed here on the basis of first-principles and model approaches. Current rectification by molecular quantum dots can produce the rectification ratio $\lesssim 100$. Current switching due to conformational changes in the molecules is slow, on the order of a few kHz. Fast switching ($\sim$THz) may be achieved, at least in principle, in a molecular quantum dot with attractive carrier-carrier interactions due to strong coupling with vibrational excitations. The observed switching in many cases is extrinsic, caused by changes in molecule-electrode geometry or metallic filament formation through the film. New possibilities are opening up by using a spin injection from ferromagnetic electrodes into semiconductors. It is especially interesting in organic materials, where the spin coherence lifetime is extremely long.

Keywords: molecular electronics, molecular rectifiers, molecular quantum dot, molecule-electrode geometry effect, extrinsic molecular switching, molecular doping, molecular switches, Coulomb blockade, strong electron-vibron interaction, correlation mechanism of current hysteresis, spin injection, spin accumulation, spin coherence, ultrafast spin-injection devices

1. Introduction

Current strong interest in molecular electronics is largely driven by hopes that molecules can be used as nanoelectronics components [1]. The chemical synthesis and characterization have developed tremendously since the first speculations about making molecular electronic devices [2]. To beat alternative technologies for e.g. dense (and cheap) memories, one should aim at 10^{12}-10^{13} bit/cm^2, which correspond to linear bit sizes of $3 - 10$ nanometers. The allowances for a corresponding assembly, especially of hybrid structures

A.S. Alexandrov et al. (eds.), Molecular Nanowires and Other Quantum Objects, 39–56.

(molecules on silicon CMOS), are actually on the order of a fraction of an Angstrom, so actually a *picotechnology* is required [3]. The hope is that perhaps a directed self-assembly will do the trick, but self-assembly on a large scale is impossible without defects, since the entropic factors work against it. Above some small defect concentration threshold the *mapping* of an algorithm on such a self-assembled network becomes *impossible*. There are also fundamental problems with the operation of molecular components, which are poorly understood, especially when it relates to interfacing the molecules. Organic molecules are, as a rule, very resistive ($\sim$ 100MΩ $-$ 1GΩ range, or more). Power dissipation and a speed of operation is a big concern. Therefore, it is difficult to expect that organic molecules will displace silicon any time soon. The first moletronic applications would most likely come in the area of chemical and biological sensors. This paper describes some specific molecular systems, which can be used as the rectifiers and switches, their main characteristics, and related physical problems. We finish with a brief discussion of new possibilities opening up with regards to spin injection into semiconducting and organic materials.

It is worth mentioning that carbon nanotubes (CNT) are studied extensively due to their relatively simple structure and many unique properties, including very high conductance. Although CNT are big carbon molecules, research into using CNT as electronic components circuits should be viewed as a special topic [4].

We shall describe below some new molecular rectifiers and switches that all have the same general design, which we shall call a molecular quantum dot (molQD), Fig. 1. The molecular quantum dot consists of a central *conjugated* unit (containing half-occupied, and, therefore, extended $\pi-$orbitals). Frequently those are formed from the p-states on carbon atoms, which are not *saturated* (i.e. they do not share electrons with other atoms forming strong $\sigma-$bonds, with typical bonding-antibonding energy difference about 1Ry). Since the $\pi-$orbitals are half-occupied, they form the HOMO-LUMO states (HOMO is the Highest Occupied and LUMO the Lowest Unoccupied Molecular Orbitals). The size of the HOMO-LUMO gap is then directly related to the size of the conjugated region d, Fig. 1, by a standard estimate $E_{\text{HOMO-LUMO}} \sim \hbar^2/md^2 \sim 2-5$ eV. In a molecular quantum dot the central conjugated part is separated from electrodes by insulating groups with saturated $\sigma-$bonds, like e.g. the alkane chains [5], Fig. 1(c). There are two main possibilities. If the length of at least one of the insulating groups $L_{1(2)}$ is not very large (a conductance $G_{1 \text{ or } 2}$ is not much smaller than the conductance quantum $G_0 = 2e^2/h$), then the transport through the molQD will proceed by resonant tunneling processes. If, on the other hand, both groups are such that the tunnel conductance $G_{1(2)} \ll G_0$, the charge on the dot will be quantized. Then we will have another two possibilities: (i) the interaction of the extra carriers on the dot is *repulsive* $U > 0$, and we have a Coulomb blockade [6], or (ii) the effective interaction is *attractive* $U < 0$, then we would obtain the current *hysteresis*. Coulomb blockade in molecular quantum dots has been demonstrated in Refs. [7]. In these works, and in Ref. [8], the three-terminal active molecular devices have been fabricated and successfully tested.

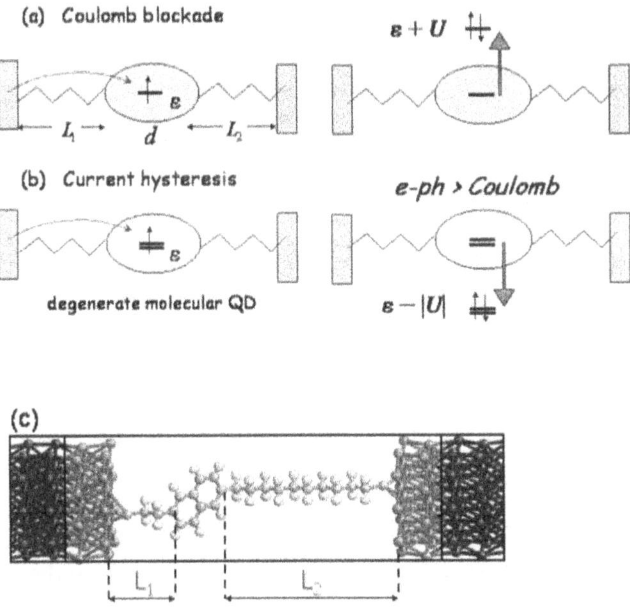

Figure 1. Schematic of the molecular quantum dot with central conjugated unit separated from the electrodes by wide-band insulating molecular groups. First electron tunnels into the dot and occupies an empty (degenerate) state there. If the interaction between the first and second incoming electron is repulsive, $U > 0$, then the dot will be in a Coulomb blockade regime (a). If the electrons on the dot effectively attract each other, $U < 0$, the system will show current hysteresis (b). (c) Schematic diagram of typical Au-molecular rectifier-Au device structure with Naphthalene conjugated central group.

2. Molecular quantum dot rectifiers

Aviram and Ratner speculated about a rectifying molecule containing donor (D) and acceptor (A) groups separated by a saturated σ-bridge (insulator) group, where the (inelastic) electron transfer will be more favorable from A to D [2]. The molecular rectifiers actually synthesized, $C_{16}H_{33} - \gamma Q3CNQ$, were of somewhat different $D - \pi - A$ type, i.e. the "bridge" group was conjugated [9].

Although the molecule did show rectification (with considerable hysteresis), it performed rather like an anisotropic insulator with tiny currents on the order of 10^{-17}A/molecule, because of the large alkane "tail" needed for LB assembly. It was recently realized that in this molecule the resonance does not come from the alignment of the HOMO and LUMO, since they cannot be decoupled through the conjugated $\pi-$bridge, but rather due to an asymmetric voltage drop across the molecule [10]. Rectifying behavior in other classes of molecules is likely due to asymmetric contact with the electrodes [11, 12], or an asymmetry of the molecule itself [13].

To make rectifiers, one should avoid using molecules with long insulating groups, and we have suggested using relatively short molecules with "anchor" end groups for their self-assembly on a metallic electrodes, with a phenyl ring as a central conjugated part [14]. This idea has been tested in Ref. [15] with a phenyl and thiophene rings attached to a $(CH_2)_{15}$ tail by a CO group. The observed rectification ratio was $\lesssim$10, with some samples showing the ratio of about 37.

We have recently studied a more promising rectifier like $-$S-$(CH_2)_2$-Naph$-$$(CH_2)_{10}-S-$ with a theoretical rectification $\lesssim$ 100[20]. This system has been synthesized and studied experimentally [16]. To obtain an accurate description of transport in this case, we employ an ab-initio non-equilibrium Green's function method [17]. The present calculation takes into account only elastic tunneling processes. Inelastic processes may substantially modify the results in the case of strong interaction of the electrons with molecular vibrations, see Ref. [18] and below. There are indications in the literature that the carrier might be trapped in a polaron state in saturated molecules somewhat longer than those we consider in the present paper [19]. One of the barriers in the present rectifiers is short and relatively transparent, so there will be no appreciable Coulomb blockade effects.

The structure of the present molecular rectifier is shown in Fig. 1(c). The molecule consists of a central conjugated part (naphthalene) isolated from the electrodes by two insulating aliphatic chains $(CH_2)_n$ with lengths $L_1(L_2)$ for the left (right) chain. The principle of molecular rectification by a molecular quantum dot is illustrated in Fig. 2, where the electrically "active" molecular orbital, localized on the middle conjugated part, is the LUMO, which lies at an energy Δ above the electrode Fermi level at zero bias. The position of the LUMO is determined by the work function of the metal $q\phi$ and the affinity of the molecule $q\chi$, $\Delta = \Delta_{\rm LUMO} = q(\phi - \chi)$. The position of the HOMO is given by $\Delta_{\rm HOMO} = \Delta_{\rm LUMO} - E_g$, where E_g is the HOMO-LUMO gap. If this orbital is considerably closer to the electrode Fermi level E_F, then it will be brought into resonance with E_F prior to other orbitals. It is easy to estimate the forward and reverse bias voltages, assuming that the voltage mainly drops on the insulating parts of the molecule,

$$V_F = \frac{\Delta}{q}(1+\xi), \qquad V_R = \frac{\Delta}{q}\left(1+\frac{1}{\xi}\right), \tag{1}$$

$$V_F/V_R = \xi \equiv L_1/L_2, \tag{2}$$

where q is the elementary charge. A significant difference between forward and reverse currents should be observed in the voltage range $V_F < |V| < V_R$.

The current is obtained from the Landauer formula

$$I = \frac{2q}{h}\int dE\,[f(E) - f(E+qV)]\,T(E). \tag{3}$$

We can make qualitative estimates in the resonant tunneling model, with the transmission $T(E)$ given by the Breit-Wigner formula

$$T(E) = \frac{\Gamma_L \Gamma_R}{(E - E_{MO})^2 + (\Gamma_L + \Gamma_R)^2/4}, \tag{4}$$

E_{MO} is the energy of the molecular orbital. The width $\Gamma_{L(R)} \sim t^2/D = \Gamma_0 e^{-2\kappa L_{1(2)}}$, where t is the overlap integral between the MO and the electrode, D is the electron band width in the electrodes, κ the inverse decay length of the resonant MO into the barrier. The current above the resonant threshold is

$$I \approx \frac{2q}{\hbar}\Gamma_0 e^{-2\kappa L_2}. \tag{5}$$

We see that increasing the spatial asymmetry of the molecule (L_2/L_1) changes the operating voltage range linearly, but it also brings about an *exponential* decrease in current [14]. This severely limits the ability to optimize the rectification ratio while simultaneously keeping the resistance at a reasonable value.

To calculate the I-V curves, we use an ab-initio approach that combines the Keldysh non-equilibrium Green's function (NEGF) with pseudopotential-based real space density functional theory (DFT) [17]. The main advantages of our approach are (i) a proper treatment of the open boundary condition; (ii) a fully atomistic treatment of the electrodes and (iii) a self-consistent calculation of the non-equilibrium charge density using NEGF. The transport Green's function is found from the Dyson equation

$$\left(G^R\right)^{-1} = \left(G_0^R\right)^{-1} - V, \tag{6}$$

where the unperturbed retarded Green's function is defined in operator form as $\left(G_0^R\right)^{-1} = (E + i0)\hat{S} - \hat{H}$, H is the Hamiltonian matrix for the scatterer (molecule plus screening part of the electrodes). S is the *overlap* matrix, $S_{i,j} = \langle \chi_i | \chi_j \rangle$ for non-orthogonal basis set orbitals χ_i, and the coupling of the scatterer

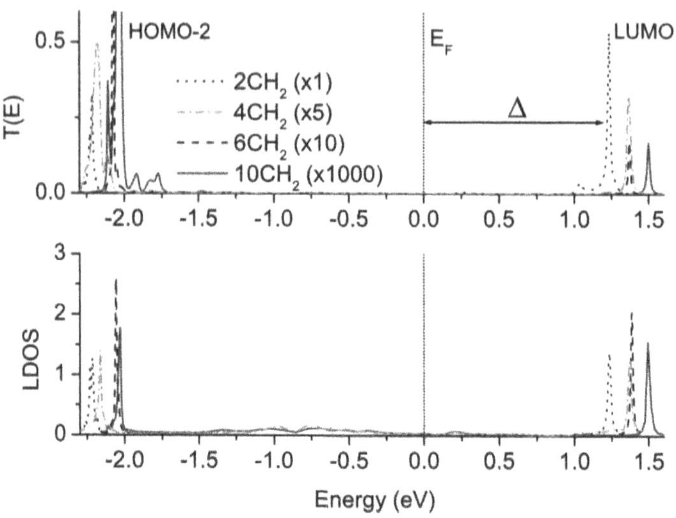

Figure 2. Transmission coefficient versus energy E for rectifiers $-S-(CH_2)_2-C_{10}H_6-(CH_2)_n-S-$, $n = 2, 4, 6, 10$. Δ indicates the distance of the closest MO to the electrode Fermi energy ($E_F = 0$).

to the leads is given by the Hamiltonian matrix $V = \text{diag}[\Sigma_{l,l},\ 0,\ \Sigma_{r,r}]$, where l (r) stands for left (right) electrode. The self-energy part $\Sigma^<$, which is used to construct the non-equilibrium electron density in the scattering region, is found from $\Sigma^< = -2i\text{Im}\left[f(E)\Sigma_{l,l} + f(E+qV)\Sigma_{r,r}\right]$, where $\Sigma_{l,l(r,r)}$ is the self-energy of the left (right) electrode, calculated for the semi-infinite leads using an iterative technique [17]. $\Sigma^<$ accounts for the steady charge "flowing in" from the electrodes. The transmission probability is given by

$$T(E) = 4\text{Tr}\left[\text{Im}\left(\Sigma_{l,l}\right) G^R_{l,r} \text{Im}\left(\Sigma_{r,r}\right) G^A_{r,l}\right], \tag{7}$$

and the current is obtained from Eq. (3).

The calculated transmission coefficient $T(E)$ is shown for a series of rectifiers $-S-(CH_2)_m-C_{10}H_6-(CH_2)_n-S-$ for $m = 2$ and $n = 2, 4, 6, 10$ at zero bias voltage in Fig. 2. We see that the LUMO is the molecular orbital transparent to electron transport, lies above E_F by an amount $\Delta = 1.2 - 1.5$eV. The transmission through the HOMO and HOMO-1 states, localized on the terminating sulfur atoms, is negligible, but the HOMO-2 state conducts very well. The HOMO-2 defines the threshold reverse voltage V_R, thus limiting the operating voltage range. Our assumption, that the voltage drop is proportional to the lengths of the alkane groups on both sides, is quantified by the calculated potential ramp. It is close to a linear slope along the $(CH_2)_n$ chains

[20]. The forward voltage corresponds to the crossing of the LUMO(V) and $\mu_R(V)$, which happens at about 2V. Although the LUMO defines the forward threshold voltages in all molecules studied here, the reverse voltage is defined by the HOMO-2 for "right" barriers $(CH_2)_n$ with $n = 6, 10$. The I-V curves are plotted in Fig. 3.

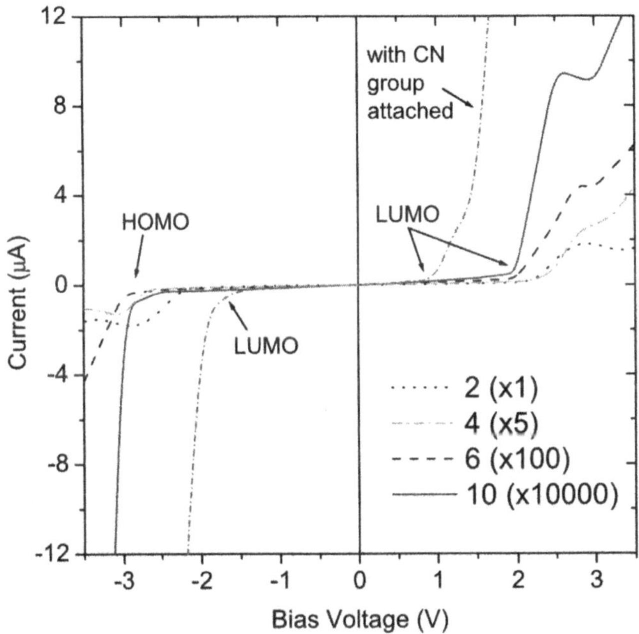

Figure 3. I-V curves for naphthalene rectifiers $-S\text{-}(CH_2)_2\text{-}C_{10}H_6-(CH_2)_n-S-$, $n = 2, 4, 6, 10$. The short-dash-dot curve corresponds to a cyano-doped (added group -C≡N) $n = 10$ rectifier.

We see that the rectification ratio for current in the operation window I_+/I_- reaches a maximum value of 35 for the "2-10" molecule ($m = 2$, $n = 10$). Series of molecules with a central *single phenyl* ring [14] do not show any significant rectification. One can manipulate the system in order to increase the energy asymmetry of the conducting orbitals (reduce Δ). To shift the LUMO towards E_F, one can attach an electron withdrawing group, like –C≡N [20].

The molecular rectification ratio is not great by any means, but one should bear in mind that this is a device necessarily operating in a ballistic quantum-mechanical regime because of the small size. This is very different from present Si devices with carriers diffusing through the system. As silicon devices become smaller, however, the same effects will eventually take over, and tend to diminish the rectification ratio, in addition to effects of finite temperature and disorder in the system.

3. Molecule-electrode contact and an extrinsic molecular switching

We have predicted some while ago that there should be a strong dependence of the current through conjugated molecules (like the Tour wires [1]) on the geometry of molecule-electrode contact [24, 25]. The apparent "telegraph" switching observed in STM single-molecule probes of the three-ring Tour molecules, inserted into a SAM of non-conducting shorter alkanes, has been attributed to this effect [22]. The theory predicts very strong variation of the current through the molecule on the tilting angle between a backbone of a molecule and a normal to the electrode surface. Other explanations, like rotation of the middle ring, charging of the molecule, or effects of the moieties on the middle ring, do not hold. In particular, switching of the molecules *without* any NO_2 or NH_2 moieties have been practically the same, as with them.

The simple argument in favor of the "tilting" mechanism of the conductance lies in a large anisotropy of the molecule-electrode coupling through π-conjugated molecular orbitals (MOs). In general, we expect the overlap and the full conductance to be maximal when the lobes of the p-orbital of the end atom at the molecule are oriented perpendicular to the surface, and smaller otherwise, as dictated by the symmetry. The overlap integrals of a p-orbital with orbitals of other types differ by a factor about 3 to 4 for the two orientations. Since the conductance is proportional to the square of the matrix element, which contains a product of two metal-molecule hopping integrals, the total conductance variation with overall geometry may therefore reach two orders of magnitude, and in special cases be even larger. In order to illustrate the geometric effect on current we have considered a simple two-site model model with $p-$orbitals on both sites, coupled to electrodes with $s-$orbitals [25]. For non-zero bias the transmission probability has the resonant form (4) with line widths for hopping to the left (right) lead Γ_L (Γ_R). The current has the approximate form (with $\Gamma = \Gamma_L + \Gamma_R$)

$$I \approx \begin{cases} \frac{q^2}{h}\frac{\Gamma^2}{t_\pi^2} V \propto \sin^4\theta, & qV \ll E_{\mathrm{LUMO}} - E_{\mathrm{HOMO}}, \\ \frac{8\pi q}{h}\frac{\Gamma_L\Gamma_R}{\Gamma_L+\Gamma_R} \propto \sin^2\theta, & qV > E_{\mathrm{LUMO}} - E_{\mathrm{HOMO}}, \end{cases} \tag{8}$$

where θ is the tilting angle, Fig. 4.

The tilting angle has a large effect on the I-V curves of benzene-dithiolate (BDT) molecules, especially when the molecule is anchored to the Au electrode in the top position, Fig. 5. By changing θ from 5° to just 15°, one drives the I-V characteristic from the one with a gap of about 2V to the ohmic one with a large relative change of conductance. Even changing θ from 10° to 15°

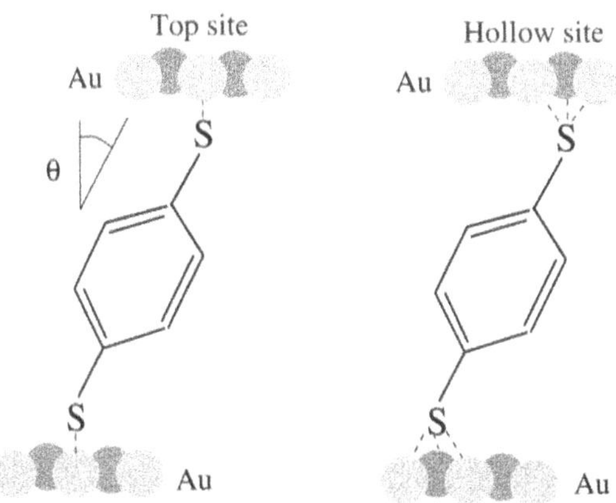

Figure 4. Schematic representation of the benzene-dithiolate molecule on top and hollow sites. End sulfur atoms are bonded to one and three surface gold atoms, respectively, θ is the tilting angle.

changes the conductance by about an order of magnitude. The I-V curve for the hollow site remains ohmic for tilting angles up to 75° with moderate changes of conductance. Therefore, if the molecule in measurements snaps from the top to the hollow position and back, it will lead to an apparent switching [22]. It is realized that the geometry of contact strongly affects coherent spin transfer between molecularly bridged quantum dots [26].

It is worth noting that another frequently observed *extrinsic* mechanism of "switching" in organic layers is due to electrode material diffusing into the layer and forming metallic "filaments". Formation and dissolution of those *conducting filaments* was known for decades to result in apparent "switching" in organic and inorganic thin films [23].

4. Molecular switches

For many applications one needs an *intrinsic* molecular "switch", i.e. a bistable voltage-addressable molecular system with very different resistances in the two states. There is a trade-off between the stability of a molecular state and the ability to switch the molecule between two states with an external perturbation (we discuss an electric field, switching involving absorbed photons is impractical at a nanoscale). Indeed, the applied electric field, on the order of a typical breakdown field $E_b \lesssim 10^7$V/cm, is much smaller than a typical atomic field $\sim 10^9$V/cm, characteristic of the energy barriers. Small barrier would be a subject for sporadic thermal switching, whereas a larger barrier $\sim 1 - 2$eV would be impossible to overcome with the applied field. One may only change the relative energy of the minima by external field and, therefore, redistribute the molecules statistically between the two states. An intrinsic disadvantage

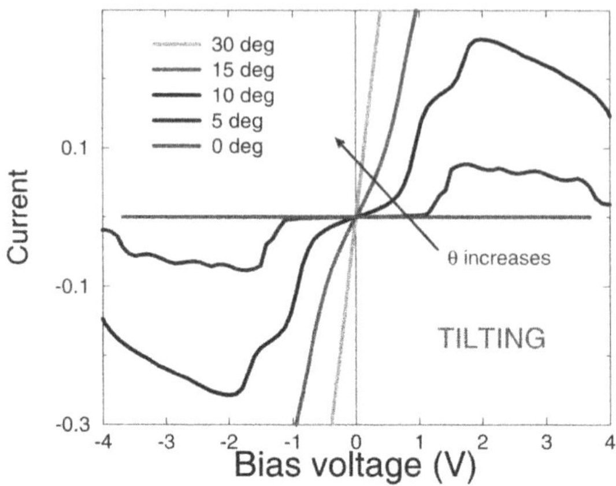

Figure 5. Effect of tilting on I-V curve of the BDT molecule, Fig. 4. Current is in units of $I_0 = 77.5\mu$A, θ is the tilting angle.

of the conformational mechanism, involving motion of ionic group, exceeding the electron mass by many orders of magnitude, is a slow switching speed ($\sim$kHz). In case of supramolecular complexes like rotaxanes and catenanes [21] there are two entangled parts which can change mutual positions as a result of redox reactions (in solution). For the rotaxane-based memory devices, a low switching speed of $\sim 10^{-2}$ seconds was reported. We have considered a bistable molecule with -$CONH_2$ dipole group [27]. The barrier height is $E_b = 0.18$ eV. Interaction with an external electric field changes the energy of the minima, but estimated switching field is enormous, ~ 0.5V/A. At non-zero temperatures, temperature fluctuations might result in statistical dipole flipping at lower fields. The I-V curve shows hysteresis in the 3 to 4 Volts window for two possible conformations. One can estimate the thermal stability of the state as 58 ps at room temperature, and 33 ms at 77 K.

Fast molecular switching

Much faster switching compared to the conformational one, described in the previous section, may be caused by coupling to the vibrational degrees of freedom, if the vibron-mediated attraction between two carriers on the molecule is stronger than their direct Coulomb repulsion [18], Fig. 1(b). The attractive energy is the difference of two large interactions, the Coulomb repulsion and the phonon mediated attraction, on the order of $1\mathrm{eV}$ each, hence $|U| \sim 0.1\mathrm{eV}$.

If we assume that the coupling to the leads is weak, so the level width $\Gamma \ll |U|$, we can can find the current from [29]

$$I(V) = I_0 \int_{-\infty}^{\infty} d\omega \, [f_1(\omega) - f_2(\omega)] \, \rho(\omega), \qquad \rho(\omega) = -\frac{1}{\pi} \sum_{\mu} \mathrm{Im} \hat{G}_{\mu}^{R}(\omega), \tag{9}$$

where $|\mu\rangle$ is a complete set of one-particle molecular states. Here $I_0 = e\Gamma$, $\rho(\omega)$ is the molecular DOS, $\hat{G}_{\mu}^{R}(\omega)$ is the Fourier transform of the Green's function $\hat{G}_{\mu}^{R}(t) = -i\theta(t) \left\langle \left\{ c_{\mu}(t), c_{\mu}^{\dagger} \right\} \right\rangle$, $\{\cdots, \cdots\}$ is the anticommutator, $c_{\mu}(t) = e^{iHt} c_{\mu} e^{-iHt}$, $\theta(t) = 1$ for $t > 0$ and zero otherwise. We calculate $\rho(\omega)$ *exactly* for the Hamiltonian, which includes both the Coulomb U^C and e-ph interactions,

$$\begin{aligned} H = & \sum_{\mu} \varepsilon_{\mu} \hat{n}_{\mu} + \frac{1}{2} \sum_{\mu \neq \mu'} U_{\mu\mu'}^{C} \hat{n}_{\mu} \hat{n}_{\mu'} \\ & + \sum_{\mu,q} \hat{n}_{\mu} \omega_q (\gamma_{\mu q} d_q + H.c.) + \sum_q \omega_q (d_q^{\dagger} d_q + 1/2). \end{aligned} \tag{10}$$

Here ε_{μ} are one-particle molecular energy levels, $\hat{n}_{\mu} = c_{\mu}^{\dagger} c_{\mu}$ the occupation number operators, c_{μ} (d_q) annihilates electrons (phonons), ω_q are the phonon (vibron) frequencies, and $\gamma_{\mu q}$ are e-ph coupling constants (q enumerates the vibron modes). We apply the standard Lang-Firsov polaron unitary transformation, integrating phonons out. The electron and phonon operators are transformed as $\tilde{c}_{\mu} = c_{\mu} X_{\mu}$, and $\tilde{d}_q = d_q - \sum_{\mu} \hat{n}_{\mu} \gamma_{\mu q}^{*}$, respectively, $X_{\mu} = \exp\left(\sum_q \gamma_{\mu q} d_q - H.c. \right)$, and the transformed Hamiltonian is $\tilde{H} = \sum_i \tilde{\varepsilon}_{\mu} \hat{n}_{\mu} + \sum_q \omega_q (d_q^{\dagger} d_q + 1/2) + \frac{1}{2} \sum_{\mu \neq \mu'} U_{\mu\mu'} \hat{n}_{\mu} \hat{n}_{\mu'}$, where $U_{\mu\mu'} \equiv U_{\mu\mu'}^{C} - 2 \sum_q \gamma_{\mu q}^{*} \gamma_{\mu' q} \omega_q$ is the interaction of polarons, which we simplify as $U_{\mu\mu'} = U$. The molecular energy levels are subject to a polaron level shift , $\tilde{\varepsilon}_{\mu} = \varepsilon_{\mu} - \sum_q |\gamma_{\mu q}|^2 \omega_q$. The retarded GF becomes

$$G_{\mu}^{R}(t) = -i\theta(t) [\left\langle c_{\mu}(t) c_{\mu}^{\dagger} \right\rangle \left\langle X_{\mu}(t) X_{\mu}^{\dagger} \right\rangle + \left\langle c_{\mu}^{\dagger} c_{\mu}(t) \right\rangle \left\langle X_{\mu}^{\dagger} X_{\mu}(t) \right\rangle]. \tag{11}$$

The phonon correlator is simply

$$\left\langle X_{\mu}(t) X_{\mu}^{\dagger} \right\rangle = \exp \sum_q \frac{|\gamma_{\mu q}|^2}{\sinh \frac{\beta \omega_q}{2}} \left[\cos\left(\omega t + i \frac{\beta \omega_q}{2} \right) - \cosh \frac{\beta \omega_q}{2} \right], \tag{12}$$

where the inverse temperature $\beta = 1/T$, and $\left\langle X_{\mu}^{\dagger} X_{\mu}(t) \right\rangle = \left\langle X_{\mu}(t) X_{\mu}^{\dagger} \right\rangle^{*}$. The remaining GFs $\left\langle c_{\mu}(t) c_{\mu}^{\dagger} \right\rangle$, are found from the equations of motion *exactly*. Finally, for the simplest case of a coupling to a single mode with the characteristic frequency ω_0 and $\gamma_q \equiv \gamma$ we find [18]

$$G_\mu^R(\omega) = \mathcal{Z} \sum_{r=0}^{d-1} Z_r(n) \sum_{l=0}^{\infty} I_l(\xi) \left[e^{\frac{\beta\omega_0 l}{2}} \left(\frac{1-n}{\omega - rU - l\omega_0 + i\delta} + \frac{n}{\omega - rU + l\omega_0 + i\delta} \right) + (1-\delta_{l0}) e^{-\frac{\beta\omega_0 l}{2}} \left(\frac{1-n}{\omega - rU + l\omega_0 + i\delta} + \frac{n}{\omega - rU - l\omega_0 + i\delta} \right) \right], \tag{13}$$

where $\mathcal{Z} = \exp\left[-\sum_{\mathbf{q}} |\gamma_q|^2 \coth \frac{\beta\omega_q}{2} \right]$ is the familiar *polaron narrowing* factor, $Z_r(n) = \frac{(d-1)!}{r!(d-1-r)!} n^r (1-n)^{d-1-r}$, $\xi = |\gamma|^2 / \sinh \frac{\beta\omega_0}{2}$, $I_l(\xi)$ is the modified Bessel function, and δ_{lk} is the Kroneker symbol. The important feature of the DOS, Eq. (9), is its nonlinear dependence on the occupation number n. It contains full information about all possible correlation and inelastic effects in transport, in particular, all the phonon sidebands.

Nonlinear rate equation and switching

In the present case of MQD weakly coupled with leads one can apply the Fermi-Dirac golden rule to obtain an equation for n. Equating incoming and outgoing numbers of electrons in MQD per unit time we obtain the self-consistent equation for the level occupation n as $(1-n) \int_{-\infty}^{\infty} d\omega \{\Gamma_1 f_1(\omega) + \Gamma_2 f_2(\omega)\} \rho(\omega)$ $= n \int_{-\infty}^{\infty} d\omega \{\Gamma_1[1-f_1(\omega)] + \Gamma_2[1-f_2(\omega)]\} \rho(\omega)$, where $\Gamma_{1(2)}$ are the transition rates from left (right) leads to MQD, and $\rho(\omega)$ is found from Eqs. (13) and (9). For $d = 1, 2$ the kinetic equation for n is linear, and the switching is *absent*. Switching appears for $d \geq 3$, when the kinetic equation becomes non-linear. For instance, in the limit $|\gamma| \ll 1$, $T = 0$, where we have $b_r = a_r$, $\mathcal{Z} = 1$, the remaining interaction is $U = U^C < 0$, we recover the negative-U model [28], and the kinetic equation for $d = 4$ is

$$2n = 1 - (1-n)^3 \tag{14}$$

in the voltage range $\Delta - |U| < eV/2 < \Delta$. The current is simplified as $I/I_0 = 2n$. The current-voltage characteristics show a *hysteretic* behavior for $d = 4$, Fig. 6. When the voltage increases from zero, 4-fold degenerate MQD remains in a low-current state until the threshold $eV_2/2 = \Delta$ is reached. Remarkably, when the voltage *decreases* from the value above the threshold V_2, the molecule remains in the high-current state down to the voltage $eV_1/2 = \Delta - |U|$ well below the threshold V_2. The bistability region shrinks down with temperature, and the hysteresis loop practically closes at $T/\Delta = 0.01$, Fig. 6. One should see if with any of these molecular switches the extrinsic switching can be prevented due to metallic filaments formation through the film [23].

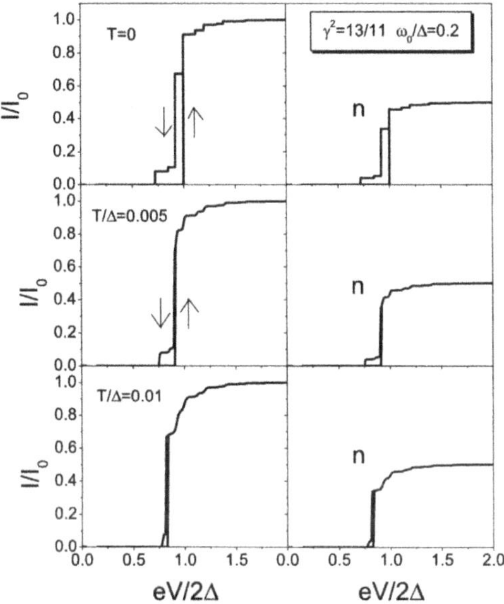

Figure 6. The I-V curves for tunneling through the molecular quantum dot with the electron-vibron coupling constant $\gamma^2 = 13/11$.

5. Efficient spin injection

Manipulation of a spin of an electron instead of its charge is expected to lead to breakthroughs in developing solid state ultrafast scalable devices due to relatively large spin-coherence lifetime of electrons in semiconductors [30]. The efficient spin injection into nonmagnetic semiconductors has been demonstrated from metallic ferromagnets [31, 32] and magnetic semiconductors [30]. Importantly, spin injection has been achieved into *organic* materials where the spin transport remains coherent over very large distances on the order of 100-1000 nm [33].

Normally, a wide Schottky barrier with a height $\Delta = 0.5 - 0.8$eV at a metal-semiconductor interface prevents spin injection from ferromagnet (the injection corresponds to a reverse biasing of a Schottky contact). The current through the FM-S-FM structure is negligibly small when $w \gtrsim 30$ nm, Fig. 7. To increase the spin injection current, a thin heavily n^+–semiconductor interfacial layer should be used, which practically produces an ohmic contact [31, 34]. It was shown recently that an efficient spin injection in FM-S structure is possible when the contact is ohmic[32]. We consider a heterostructure, which contains the left and right δ–doped layers satisfying the following conditions [34]: the

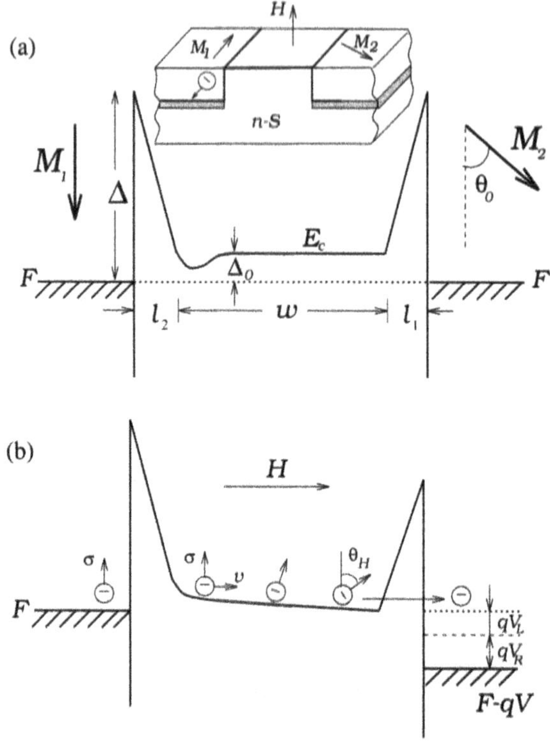

Figure 7. Energy diagram of the FM-S-FM spin injection valve, where V_L (V_R) is the fraction of the total bias dropping across the left (right) δ-layer, F the Fermi level, E_c the bottom of a conduction band in S. The spins, injected from the left, drift in the semiconductor layer and rotate by the angle θ_H in the external magnetic field H. Inset: schematic of the device with an oxide layer separating the FM films from the semiconducting substrate.

thicknesses $l_{1(2)} \lesssim 2$ nm and the donor concentration $N_d^+ \gtrsim 10^{20}\text{cm}^{-3}$ (see the energy diagram in Fig. 7).

The current is given by a standard expression [35]

$$J_\sigma = \frac{q}{h} \int dE[f(E - F_{\sigma 0}) - f(E - F)] \int \frac{d^2 k_\parallel}{(2\pi)^2} T_\sigma, \qquad (15)$$

where $T_{k\sigma}$ is the transmission probability, $f(E - F)$ the Fermi function, $v_{\sigma x}$ the x−component of velocity $v_\sigma = \hbar^{-1}|\nabla_k E_{k\sigma}|$ of electrons with spin σ in the ferromagnet in a direction of current, the integration includes a summation with respect to a band index. We allow for an accumulation of spin polarized electrons in a semiconductor. The electron spin density in the semiconductor is given by $n_\sigma = (1/2)N_c \exp[(F_{\sigma 0} - E_c)/T]$, where $F_{\sigma 0}$ is the quasi Fermi levels for $\sigma =\uparrow (\downarrow)$ in the semiconductor near the interface. The injected spin current per unit area (15) is found as [34]

$$J_{\sigma 0} = j_0 d_{\sigma 0}\left(2n_{\sigma 0}/n - e^{qV/T}\right),\ j_0 = \alpha_0 q n v_T e^{-4\kappa_0 l/3}, \tag{16}$$

where $\alpha_0 = 1.6(\kappa_0 l)^{1/3}$, $\kappa_0 \equiv 1/l_0 = (2m_*/\hbar^2)^{1/2}(\Delta - \Delta_0 + qV)^{1/2}$, $v_{t0} = \sqrt{2(\Delta - \Delta_0 + qV)/m_*}$, $v_{\sigma 0}$ the velocity in the ferromagnet, $v_T = \sqrt{3T/m_*}$ the thermal velocity, and the spin factor $d_{\sigma 0} = v_T v_{\sigma 0}(v_{t0}^2 + v_{\sigma 0}^2)^{-1}$. At zero bias voltage $F_{\sigma 0} = F$, and the current is zero.

If we neglect for a moment a spin accumulation in the semiconductor, i.e. assume $n_{\sigma 0} = n/2$, then, according Eq. (16), spin polarization of injected current at FM-S interface $P = (J_{\uparrow 0} - J_{\downarrow 0})/(J_{\uparrow 0} + J_{\downarrow 0})$ would be formally the same as the polarization in the ferromagnet renormalized by the tunnel barrier [35]

$$P_F = \frac{d_{\uparrow 0} - d_{\downarrow 0}}{d_{\uparrow 0} + d_{\downarrow 0}} = \frac{(v_{\uparrow 0} - v_{\downarrow 0})(v_{t0}^2 - v_{\uparrow 0}v_{\downarrow 0})}{(v_{\uparrow 0} + v_{\downarrow 0})(v_{t0}^2 + v_{\uparrow 0}v_{\downarrow 0})}. \tag{17}$$

Importantly, this P_F refers to the electron states in FM *above* the Fermi level at $E = E_c > F$, i.e. the high-energy equilibrium electrons, which may be highly polarized (see below). Due to spin accumulation in a semiconductor near the FM-S boundary $n_{\sigma 0} = n/2 + \delta n_{\sigma 0}$, where $\delta n_{\sigma 0}$ is a nonlinear function of the current J, $\delta n_{\sigma 0} \propto J$ at small current densities[36]. This renormalizes the current polarization P and makes it also a nonlinear function of J. The current in spin channel σ is given by the usual expressions [36, 37, 38].

$$J_\sigma = q\mu n_\sigma E + qD\nabla n_\sigma, \quad dJ_\sigma/dx = q\delta n_\sigma/\tau_s, \tag{18}$$

where D and μ are diffusion constant and mobility of the electrons, respectively, E the electric field, $J(x) = \sum_\sigma J_\sigma = \text{const}$, $n(x) = \sum_\sigma n_\sigma = \text{const}$, and $E(x) = J/q\mu n = \text{const}$, $\delta n_\sigma = n_\sigma - n/2$ and τ_s is spin-coherence lifetime of electrons in a semiconductor. We find from Eqs.(18) that, neglecting the right barrier for a moment,

$$\delta n_\uparrow(x) = C\frac{n}{2}e^{-x/L},\ L = \frac{1}{2}\left(L_E + \sqrt{L_E^2 + 4L_s^2}\right), \tag{19}$$

where $L_s = \sqrt{D\tau_s}$ and $L_E = \mu|E|\tau_s$ are spin- diffusion and drift lengths[36]. Here C defines the *accumulation* of spin polarized nonequilibrium electrons in the semiconductor near the interface, $P_n(0) = [n_\uparrow(0) - n_\downarrow(0)]/n = C$. By substituting $\delta n_\uparrow(x)$ into Eqs. (18) and (16), we find that

$$J_{\uparrow 0} = \frac{J}{2}\left(1 + C\frac{L}{L_E}\right) = \frac{J}{2}\frac{(1+P_F)(\gamma - C)}{(\gamma - CP_F)} \tag{20}$$

where $\gamma = \exp(qV/T) - 1$. From Eqs. (20),(16) one obtains a quadratic equation for C with a physical solution

$$P_n(0) = C = P_F\frac{\gamma L_E}{\gamma L + L_E}. \tag{21}$$

As expected, at small bias $qV \ll T$ and total current density $J \lesssim J_s \equiv qnL_s/\tau_s$ the *spin accumulation* $P_n \propto J$ [36] and $P_n(0) = (L_s/v_T\tau_s)(J/J_s)|P_F| \ll |P_F|$. At small currents the spin penetration depth $L = L_s$, whereas at large currents, $J \gtrsim J_s$, $L = L_E = \mu E\tau_s = J\tau_s/qn_0 \gg L_s$, i.e. $L \propto J$. At large current $J \gg J_s$ and the voltage, $qV \gtrsim T$, $P_n(0)$ saturates at the value given by the spin polarization in the ferromagnet for $E \simeq E_c > F$, $P_n(0) = P_F$, Eq.(17). Similar result was obtained phenomenologically in [36]. Thus indeed the spin accumulation P_n is a nonlinear function of the current J. We note that the spin polarization of the current at FM-S interface

$$P = \frac{J_{\uparrow 0} - J_{\downarrow 0}}{J_{\uparrow 0} + J_{\downarrow 0}} = C\frac{L}{L_E} = P_F\frac{\gamma L}{\gamma L + L_E}, \tag{22}$$

also depends on J. At large currents and voltage, $qV \gtrsim T$, the polarization P saturates at the maximal value $P = P_F$. Unlike the spin accumulation (21), P does not vanish at *small* currents, but tends to a finite value $P = P_F/(1+R) \ll P_F$, where $R = v_T\tau_s/L_s \gg 1$. Note that the polarization of injected current in the tunneling MIM structures equals P_F at small currents and *decreases* with increasing current, which is quite the opposite to the present behavior of spin injection in semiconductors [35]. By substituting $\delta n_\uparrow(x)$ into Eqs. (18), one can see that both polarizations P and P_n fall off in the semiconductor over the length L.

One may construct e.g. a spin-valve magnetoresistive sensor with two ferromagnetic electrodes separated by a semiconductor spacer layer, with δ-doped layers at both interfaces. With a slight modification of the previous formalism, we find the current through the structure

$$J = J_0\left(1 - P_R^2\cos^2\theta\right)\left(1 - P_L P_R\cos\theta\right)^{-1}, \tag{23}$$

where J_0 is the constant, $P_{L(R)}$ the electron (current) polarization in the L (R) electrodes, θ the angle between the *spin* of the electron incident on the right contact and $\vec{M}_2$ (in the counterelectrode FM_2). The maximal (minimal) current is reached at $\theta = 0$ (π) for parallel (antiparallel) moments on the electrodes, and their ratio is $J_{\max}/J_{\min} = (1 + P_LP_R)/(1 - P_LP_R)$, as in tunneling MIM structures [35]. Instead of varying θ by changing a mutual orientation of magnetic moments on electrodes (traditional read-head design), one may rotate the traversing electron spin in external magnetic field H. In general, $\theta = \theta_0 + \theta_H$, where θ_0 is the angle between the magnetizations M_1 and M_2, and θ_H the spin rotation angle. The spin processes with a frequency $\Omega = \gamma H$, where H is the magnetic field normal to the spin direction, and $\gamma = q/(m_* c)$ is the gyromagnetic ratio. For the semiconductor spacer $\sim$ 50nm of GaAs and $t_{tr} \sim 5 \times 10^{-11}$s the angle $\theta_H = \pi$ at $H \simeq 250$ Oe. Therefore, the device is capable of detecting the magnetic field varying with high frequency $\omega \sim 1/t_{tr} = 0.1 - 1$ THz at rather small bias voltage.

6. Conclusions

We have described various possible molecular-scale devices, like a rectifier and a switch, which may operate as single-molecule devices, at least in principle. The examples illustrate advantages (size, density) and disadvantages (limited voltage range, relatively small rectification ratio, large resistance) of these resonant tunnel devices [20]. The conformational switching is slow ($\sim$kHz), since it involves a motion of relatively heavy ionic groups. The fast molecular switch ($\sim$THz) is possible, at least in principle, which involves the electron tunneling through a multiple degenerate state on the molecule, which provides for an attractive correlations between two electrons dynamically loaded on the molecule. One needs to systematically study the candidate systems, which might exhibit such electron correlations. The corresponding current hysteresis should be observed in the I-V curve, which incidentally has a shape very close to that of a Coulomb blockaded system [18]. New interesting possibility is provided by an electron *spin injection* into semiconductiong and organic materials. The spin polarization and spin accumulation are strong functions of the current[34]. Importantly, the spin-flip time in organics is very long [33], since the nuclear charge (of e.g. carbon) is small and the spin-orbit coupling in minute. With the described band engineering, an efficient spin injection at room temperature should be possible with the use of elemental ferromagnets.

References

[1] J.M. Tour, Acc. Chem. Res. **33**, 791 (2000).

[2] A. Aviram and M.A. Ratner, Chem. Phys. Lett. **29**, 277 (1974).

[3] C. Joachim, Nanotechnology **13**, R1 (2002).

[4] A. Bachtold *et al.*, Science **294**: 1317 (2001); P.G. Collins *et al.* Science 292, 706 (2001); T. Rueckes *et al.*, Science **289**, 94 (2000).

[5] E.E. Polymeropoulos and J. Sagiv, J. Chem. Phys. **69**, 1836 (1978).

[6] D.V. Averin and K.K. Likharev, in: *Mesocopic Phenomena in Solids*, edited by B.L. Altshuler *et al.* (North-Holland, Amsterdam, 1991).

[7] H. Park *et al.*, Nature **407**, 57 (2000); J. Park *et al., ibid.* **417,** 722 (2002); W. Liang *et al., ibid.* **417**, 725 (2002).

[8] N.B. Zhitenev, H. Meng, and Z. Bao, Phys. Rev. Lett. **88**, 226801 (2002).

[9] A.S. Martin, J.R. Sambles, and G.J. Ashwell, Phys. Rev. Lett. **70**, 218 (1993); R.M. Metzger et al., J. Am. Chem. Soc. **119**, 10455 (1997).

[10] C. Krzeminski *et al.*, Phys. Rev. B **64**, 085405 (2001).

[11] C. Zhou, M.R. Deshpande, M.A. Reed, L. Jones II, and J.M. Tour, Appl. Phys. Lett. **71**, 611 (1997).

[12] Y. Xue, S. Datta, S. Hong, R. Reifenberger, J. I. Henderson, and C. P. Kubiak, Phys. Rev. B **59**, 7852 (1999).

[13] J. Reichert *et al.*, Phys. Rev. Lett. **88**, 176804 (2002).

[14] P.E. Kornilovitch, A.M. Bratkovsky, and R.S. Williams, Phys. Rev. B **66**, 165436 (2002).

[15] S. Lenfant *et al.*, Nanoletters **3**, 741 (2003).

[16] Shunchi Chang *et al.*, Appl. Phys. Lett. **83**, 3198 (2003).

[17] J. Taylor, H. Guo and J. Wang, Phys. Rev. B **63**, R121104 (2001); *ibid.* **63**, 245407 (2001); A.P. Jauho, N.S. Wingreen and Y. Meir, Phys. Rev. B **50**, 5528 (1994).

[18] A.S. Alexandrov and A.M. Bratkovsky, Phys. Rev. B **67**, 235312 (2003).

[19] C. Boulas, J.V. Davidovits, F. Rondelez, and D. Vuillaume, Phys. Rev. Lett. **76**, 4797 (1996).

[20] B. Larade and A.M. Bratkovsky, Phys. Rev. B **68**, 15 Dec (2003); cond-mat/0304379.

[21] C. P. Collier *et al.*, Science **285**, 391 (1999); *ibid.* **289**, 1172 (2000).

[22] Z. J. Donhauser *et al.*, Science **292,** 2303 (2001); Z. J. Donhauser *et al.*, Jpn. J. Appl. Phys. **41**, 4871 (2002).

[23] R.H. Tredgold and C.S. Winter, J. Phys. D **14**, L185 (1981).

[24] A.M. Bratkovsky and P.E. Kornilovitch, Phys. Rev. B **67**, 115307 (2003).

[25] P.E. Kornilovitch and A.M. Bratkovsky, Phys. Rev. B **64**, 195413 (2001).

[26] M. Ouyang and D.D. Awschalom, Science 301, 1074 (2003).

[27] P.E. Kornilovitch, A.M. Bratkovsky, and R.S. Williams, Phys. Rev. B **66**, 245413 (2002).

[28] A.S. Alexandrov, A.M. Bratkovsky, and R.S. Williams, Phys. Rev. B **67**, 075301 (2003).

[29] Y. Meir and N.S. Wingreen, Phys. Rev. Lett. **68**, 2512 (1992).

[30] S. A. Wolf *et al,* Science **294**, 1488 (2001);

Semiconductor Spintronics and Quantum Computation, edited by D. D. Awschalom *et al.* (Springer, Berlin, 2002).

[31] A. T. Hanbicki *et al.*, Appl. Phys. Lett. **82**, 4092 (2003); P. R. Hammar *et al.*, Phys. Rev. Lett. **83**, 203 (1999).

[32] H. Ohno *et al.*, Jpn. J. Appl. Phys. **42**, L1 (2003).

[33] V. Dediu *et al.*, Sol. State Commun. **122**, 181 (2002); E.Arisi *et al.*, J. Appl. Phys. **93**, 7682 (2003).

[34] A.M. Bratkovsky and V.V. Osipov, cond-mat/0307030; cond-mat/0307656; cond-mat/0309473; cond-mat/0310258.

[35] A. M. Bratkovsky, Phys. Rev. B**56**, 2344 (1997).

[36] A.G. Aronov and G.E. Pikus, Sov. Phys. Semicond. **10**, 698 (1976).

[37] Z. G. Yu and M. E. Flatte, Phys. Rev. B **66**, R201202, 235302 (2002).

[38] S. M. Sze, *Physics of Semiconductor Devices* (Wiley, New York, 1981).

[39] E. I. Rashba, Phys. Rev. B **62**, R16267 (2000).

FABRICATION OF CARBON NANOTUBE FIELD EFFECT TRANSISTORS BY SELF-ASSEMBLY

Emmanuel Valentin[1], Stephane Auvray[2], Arianna Filoramo[1,2], Aline Ribayrol[1], Marcelo Goffman[2], Julie Goethals[1], Laurence Capes[1], Jean-Philippe Bourgoin[2] and Jean-Noel Patillon[1]

Laboratoire d'Electronique Moléculaire

[1]*Motorola Labs, Espace Technologique Saint Aubin, 91193 Gif Sur Yvette Cedex, France*

[2]*CEA Saclay, DSM/DRECAM/SCM, 91191 Gif-sur-Yvette, France*

jbourgoin@cea.fr

Abstract We describe here the realization by self assembly of high quality single wall carbon nanotube field effect transistors (CNTFET). We solve the random deposition issue thanks to a high yield selective placement of single wall carbon nanotubes (SWNTs) on predefined region. This has been realised by the use self-assembled monolayers (SAMs) which modify the surface properties of a prepatterned substrate. The process has been optimized in order to avoid the formation of bundles and to obtain the suitable high densities necessary for the realisation of integrated devices. Then we show that such positioned SWNTs can be electrically contacted to realize high performance transistors, which very well compare with state-of-the-art CNTFETs.

Keywords: Single wall carbon nanotube, self assembly, transistors

1. Introduction

The CMOS technology is the base of the present day information technology. Its evolution is presently ruled by Moore's Law which have been proven true for the last 30 years. Moore's Laws however predict that the CMOS technology will reach fundamental limits in terms of miniaturization, concurrently to a dramatic increase of the production units. In this framework Molecular Electronics is increasingly studied as a candidate alternative technology for two main reasons. First it inherently deals with the size of molecular objects, which is foreseen as a possible answer to the miniaturization problem. Second, it is a natural field for the use of self-assembling techniques, supposedly the best way to reduce the fabrication costs.

A.S. Alexandrov et al. (eds.), Molecular Nanowires and Other Quantum Objects, 57–66.

Among the objects molecular electronics builds upon, carbon nanotubes (CNTs or NTs) occupy a special place. Indeed, since their discovery in 1991 by Ijima [1], the CNT have generated great interest and research studies. Their physical and chemical properties have been investigated for different purpose applications; starting from their interesting mechanical strength characteristics or for hydrogen storage in fuel cells as well as for their electronic and emissive features. In particular they exist as semiconducting or metallic wires and they have been used to demonstrate molecular devices like transistors, diodes or single electron transistors (SETs). A sudden acceleration of the field occurred with the demonstration of room temperature SET [2], and of NT transistors showing gain above unity [3]. This demonstration was immediately applied to the realization of logical gates mimicking the CMOS ones but with a lateral channel extension reduced to 1 nm [3, 4]. However these demonstrations are still based on a random deposition of nanotubes on a substrate. To fully take advantage of the unique electrical properties of CNTs in device/circuit applications, it is very desirable to be able to selectively place the CNTs at specific locations on a substrate. The locations of interest would be where electrodes are or will be successively positioned, so that electrical inputs and outputs to and from the CNTs can be realized.

The approach presented in this work is based on the use of self-assembled monolayers (SAMs) to modify the surface properties of certain regions of a substrate. This in turn affects the interactions between the sidewalls of a CNT and the surface so that the CNTs are preferentially attracted there. This kind of technique has been already proposed in literature by Liu et al.[5] or by Choi et al.[6], relying respectively on a local chemical functionalisation of the surface or on a electrostatic anchoring of surfactant covered NTs on amino-silane functionalised surfaces. However, the yield obtained by the processes used in these reports was not effective for the realization of large scale ICs. Our method conversely allows it for the first time achieving both the control of the density of deposition and its selectivity for isolated single wall carbon nanotubes (SWNTs). Indeed, using SWNTs dispersion in organic solvant and deposition on optimised SAM, we achieved highly selective deposition of high density isolated SWNTs in predefined areas of the substrate. In the following, we first describe the principal steps of our deposition technique, we discuss the influence of the used solvent for the nanotube solution and present results for self-assembled deposition of nanotubes on pre-patterned substrates. Finally, we report about the performance the CNTFETs fabricated with this selectively localized SWNTs.

2. Methodology of selective deposition

The process can be summarized as follows: electron-beam lithography is performed on PMMA deposited on the SiO_2 surface of a Si substrate. After a cleaning step, a monolayer of aminopropyltriethoxysilane (APTS) is deposited to form a "sticky patch" in regions opened in the resist. Exposure to Ethylene-diamine (EDA) is used to increase the surface concentration and the orientation of APTS [7] and consequently to improve interactions with SWNTs. The gas deposition was favoured instead of silanisation from an aqueous solution since it yields a much better control of the layer thickness [6]. Once the sticky patch has been formed it is expected that the adhesion of CNTs would be enhanced in the functionalised regions with the APTS "sticky patch". Two ways can be chosen to selectively place SWNT.

The first one consists in working with an aqueous solution where the Sodium-dodecyl-sulfate (SDS) surfactant is used to disperse SWNT. In this case, the sample is exposed to the SWNT suspension and then the resist is lifted off to enhance selectivity by removing any nanotube that would have been adsorbed on the PMMA (see figure 1a). However, in this case we faced the following bottleneck: the density of adsorbed tubes on the surface was too low for the realisation of integrated circuits. It was observed that increasing the concentration of NT in the solution leads mainly to the deposition of bundles on the surface. The problems to be solved are (i) to improve the dispersion of NT in the solution and (ii) avoid the competition, in the electrostatic anchoring on the amino-silane surface, between the SDS micelles present in the solution and the SDScovered tubes.

We developed the second approach to avoid aqueous solvents and these drawbacks. The main point is that nanotubes are dispersed in N-Methyl Pyrrolidone (NMP). This solvent has been observed in ref. [8] to allow dispersing NTs without any kind of surfactant. Therefore, it can solve point (i) and eliminates point (ii). However, due to its interaction with the PMMA resist, it has been necessary to modify the process as shown in figure 1b.

The main difference with respect to the previous sequence (figure 1a) is the reversed order of the last two steps. So, in this case the resist is removed before the exposition to the nanotube solution. A priori, one could think that in principle the selectivity of the placement will be partially lost. This is however not the case as shown below.

The nanotubes used are laser-ablation grown samples from TU Dresden [9]. The Dresden samples consist of single wall CNTs, and have undergone a purification stage as described in [10]. Then they were dispersed in NMP solvent. The solution was sonicated and a centrifugation was performed. The CNT concentration in NMP was varied from 0.1mg/ml to 0.005mg/ml and the variation of the density of deposited CNTs was recorded. The substrates for that study consisted of 200 nm thick thermally grown SiO_2 on Si, covered with a monolayer of APTS. The CNTs in NMP solution were deposited on the surface for one minute. AFM experiments allowed us to quantify the density of CNT deposited.

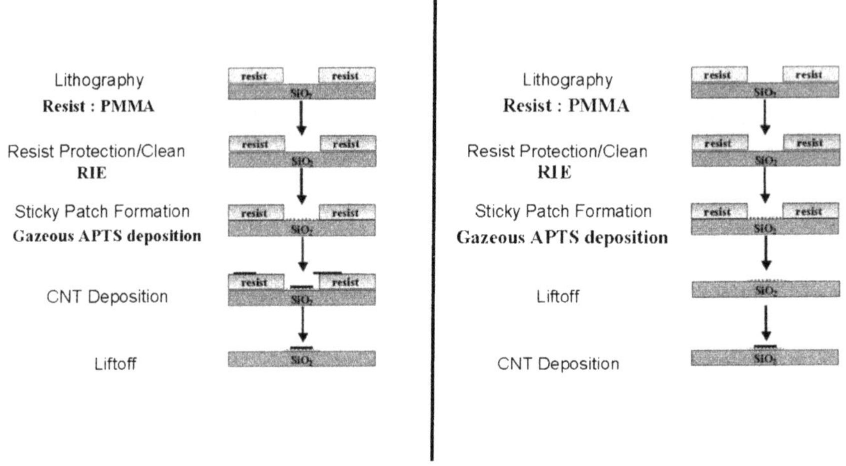

Figure 1. The principal steps of the process for selective placement of NTs: in figure 1a (left panel) the lift-off is performed after the carbon nanotube deposition while in figure 1b it is performed before (right panel).

3. Results and discussion about selective placement of SWNTs

For the sake of comparison we report in figure 2 the results obtained, without any kind of optimisation, by using an aqueous SDS solution (figure 2a) and by using a NMP solution (figure 2b) for the same CNT concentration and exposition time. Contrary to aqueous solvents, the adsorption process on APTS seems to be independent of any charge effect (thought they cannot be completely excluded [11]). Indeed, in the NMP case, the NH2 conversion of the silane group to NH3+ by exposition to HCl vapor does not seem to be relevant for the deposition yield. The adsorption is likely due to an interaction between the amine group of the APTS and the nanotube, as shown by Dai [12].

At a concentration of 0.1mg/ml, the density is 150 CNT on a $4\mu m^2$ area. This result is comparable to that of Liu et al [5], with 240 CNT on a $6.25\mu m^2$ area. 60 to 70 % of the nanotubes are less than 1.6 nm high, indicating individual singlewall nanotubes or small bundles. Therefore, the bundling problems observed with surfactants in aqueous solvents (point (i) above) seem to be much less significant in NMP.

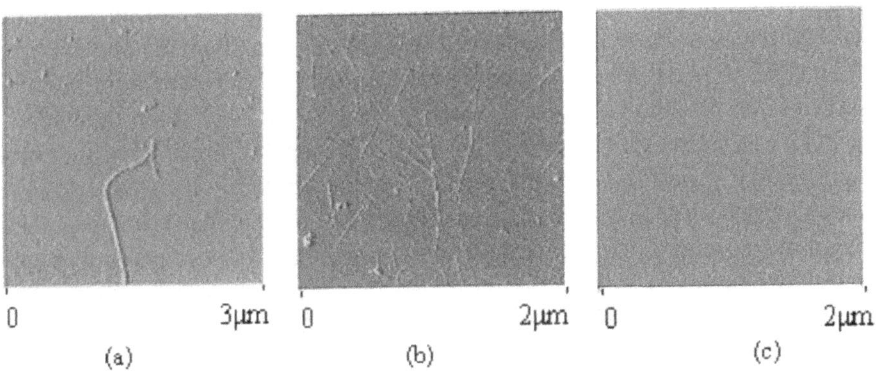

Figure 2. (2a and 2b) AFM images of the APTS treated substrate after the exposition to the two kind of nanotube solution. For the same experimental conditions (nanotube concentration and exposition time) in figure 2a (left panel) it is shown the substrate exposed to an aqueous solution with SDS surfactant while in figure 1b it is reported the case for a NMP solution. In figure 2c it is shown the AFM image of a bare SiO_2 substrate after the exposition to the same NMP solution as in figure 2b, no NTs deposition is observed.

Several parameters have been varied, such as the deposition time, centrifugation speed, sonication time and CNT concentration. The distribution of the nanotube diameter appears similar for all concentrations, indicating that there is no significant reduction of the bundles in diluted solutions. Concerning the centrifugation, the CNTs solutions were centrifugated for 10 minutes at different speeds up to 28000rpm. Contrary to the case of CNTs in aqueous solvents, the rotation speed seems to have no significant effect on the dispersion of nanotubes. Concerning the sonication, we found that after 24 hours the tubes were up to 1 to 2 μm long, but they were severely shortened to less than 400nm long if sonicated for 36 hours. Finally, as expected, by increasing the deposition time we observed accordingly an increase of the density of CNTs deposited.

Our aim was to achieve selective placement, therefore these experiments have been repeated on patterned substrates. As already explained, the removing of the resist has to be performed before the CNTs deposition. In this case, the selectivity is given only by the different behaviour of the APTS treated regions with respect to the non-treated ones. This point has been checked by a preliminary experiment where we exposed a non-silanized and a silanized substrate were exposed to the same nanotube solution. The results are reported in figure 2c (non-treated) and 2b (treated). No deposition is observed for non-silanized sample.

An important parameter for the yield of deposition is the geometry of the pattern. We have initially considered the simple stripes geometry and performed a study by varying the width of the stripes. Continuous stripes 500 nm-, 200 nm- and 100 nm-wide have been patterned in PMMA, then silanized and exposed to the CNT in NMP solutions (see figure 3 for the 100 nm trenches). Suitable densities can be achieved for any width by varying the deposition time and/or the NTs concentration in the solution within reasonable limits. For the same experimental conditions, the NT density increases roughly by a factor of two when the line-width is increased from 100nm to 200nm. Finally, it appears that longer tubes (length $>> 1\ \mu m$) are better aligned than shorter ones and that the quality of alignment is improved for narrower stripes, as already observed for aqueous solutions [6]. Moreover, we found that 100 nm-wide stripes represent the best way to limit the number of aligned SWNT to one, which is crucial to reliably study electrical transport in individual SWNT [13].

Furthemore, to reliably control the fabrication of a large number of SWNT based transistors on the same wafer, it was necessary to evaluate : i) the statistics of deposition on a given pattern and ii) a 'proximity' effect i.e., a possible combined interaction of patterned areas when they are in close vicinity to each other on the deposition yield.

We therefore checked the SWNTs depositions on groups of 100nm wide and $2\mu m$ long stripes spaced by 1, 3 and $20\mu m$ realized on the same substrate. AFM observation showed a placement yield superior to 85% in all groups. The constant density of deposition obtained for all group clearly indicated that no 'proximity' effects were involved in the placement process. This result obviously opens the road to the controlled fabrication of a large number of SWNT devices by self-assembly.

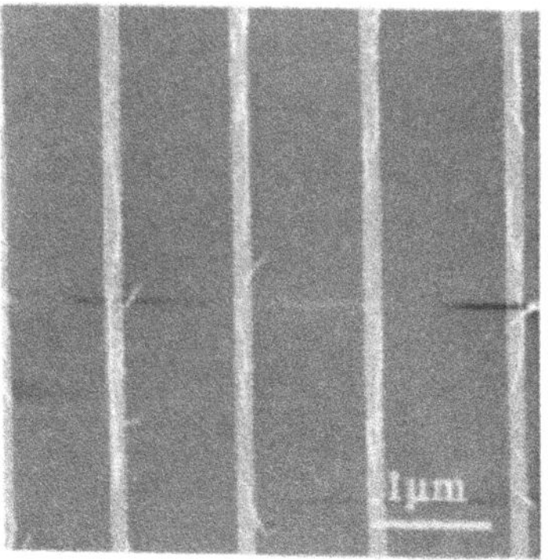

Figure 3. 100 nm wide trenches, interspace 1 μm.

An additional test was performed with the realization of crossed SWNT configuration. Finite size crosses 100nm wide and typically 750nm long arms were submitted to SWNT deposition. Typical deposition result and connection of crossed SWNTs are reported in figure 4.

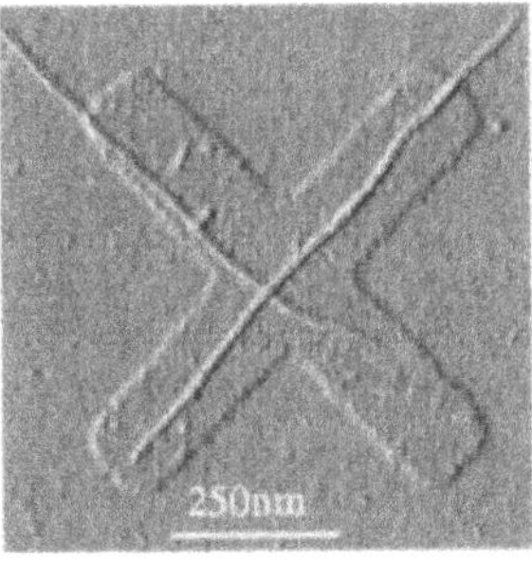

Figure 4. Left: AFM image of a cross pattern with two nanotubes aligned in each direction of the cross. No combing technique was applied during deposition. Right: 3D view of a connected crossed device.

4. Devices fabrication and performances

Since we define the deposition areas by e-beam lithography, the patterning of electrodes on top of precisely localized SWNT is therefore simple. No specific and tedious AFM imaging to locate SWNTs is required as is the case of randomly deposited SWNTs. The sample was made by first depositing the tube on a silane pattern prepared on a substrate fitted with position markers. After the nanotube deposition, a new resist bilayer (MMA/MAA copolymer first followed by PMMA) was spun on the sample and baked (typically 165°C 15 minutes for each layer). The contacts were subsequently patterned using e-beam lithography.

We compared device performances with two types of electrodes: (i) a 'classical' configuration with 0.2 nm Ti and 40 nm Au; (ii) an optimized configuration hereafter called 'TiC' with deposition of 20 nm Ti / 20 nm Pt and application of a Rapid Thermal Annealing (RTA) at 650°C in inert ambient gas to convert the contacts to titanium carbide [14]. As the electrodes are fabricated on top of SWNTs deposited on 200 nm of SiO2 film grown on a silicon wafer, the wafer itself was used as the gate electrode ('back gate' configuration). Electrical measurements were performed in vacuum.

'Classical' carbon nanotube field effect transistors (CNTFETs) behave as p-type FETs; i.e., the dominant carriers are holes. For this kind of device the transconductance (dI_d/dV_g) was in the 10^{-9} A/V range and the current modulation occurred through 4 orders of magnitude (see figure 5 left side). I-V curves obtained in 2-probe measurement gave contact resistance values in the 700-1000 kΩgange (as reported in figure 5 right side).

'TiC' CNTFETs exhibit a drastic improvement of performances. According to several authors TiC decrease contact resistance for the injection of both p- and ntype carriers [14]. Indeed, our 'TiC' CNTFETs are ambipolar i.e., they

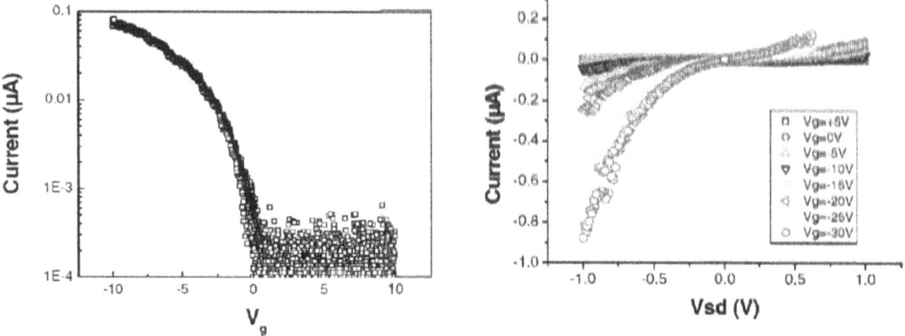

Figure 5. Transport characteristics of the 'classical' device. Left side: Transfer characteristics at room temperature with source to drain voltage V_{sd} = 200mV; dI_d/dV_g = 25nA/V; The ON/OFF current ratio is 10^4. Right side: I-V curves taken for gate voltage V_g in range [30V;+5V]; 2-probe resistance is 700 kΩ.

carried a strong current at both negative and positive values of Vg. dI_d/dV_g increase in the 10^{-7} A/V range and the current modulation occurred through 6-7 orders of magnitude which is two orders of magnitude better than 'classical' devices (see figure 6 left side). I-V curves obtained in 2 probes measurement gave contact resistance values in the 200-300 kΩgrange (as reported in figure 6 right side).

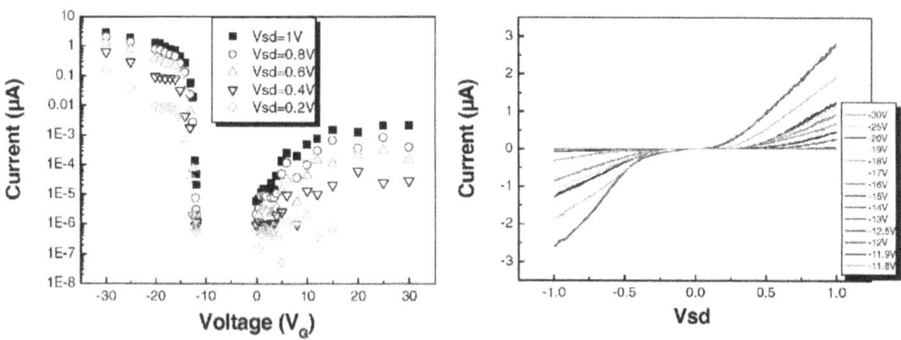

Figure 6. Transport characteristics of the 'TiC' device. Left side: Transfer characteristics with source to drain voltage V_{sd} = [0.2V: 1V]; dI_d/dV_g=0.17μA/V. The ON/OFF current ratio is 10^7. Right side: I-V curves taken for gate voltage V_g in range [30V: -11.8V]; 2 probes resistance is 230 kΩ.

The characteristic values of our 'TiC' devices are close to those obtained on similar back-gated device fabricated by random deposition directly on SiO2

[14]. In addition the behaviour of the transport characteristics of the 'TiC' and classical device with temperature (not shown) is quite similar to the p-type device of ref. 14.

The results clearly indicate that the nanotube/metal contact interfaces and their response to the applied electric fields determine the electrical performance of our CNTFETs. This obviously suggests that after RTA, the SAM used to direct the assembly of the nanotube does not perturb the transport characteristics of such fabricated CNTFETs. Thus, this work validates self-assembly approaches for large-scale production of nanotube-based electronics. Moreover, it should be noted that our process is fully compatible with the realization of top gate devices [15] and/or low thickness and high effective dielectric constants oxide films [16]. A controlled and systematic fabrication of CNTFETs with performances reaching actual state of the art of geometrically optimized CNTFET is therefore possible.

5. Conclusion

In this work we have demonstrated a high-density selective placement of individual or small bundles of SWNT via modification of the substrate properties (self-assembly). We showed that this technique is suitable to put a SWNT at a specific location on SiO2 (typically on 100 nm wide patterns line or crosses) where electrodes can be conveniently positioned. Electrical measurements with optimised contacts between semi-conducting SWNT and electrodes allowed us to obtain performances similar to comparable devices fabricated with randomly deposited SWNTs. The compatibility of our process with a top gate configuration with thin and high dielectric constant oxide should permit to reach state of the art device performance. The realisation of high performance devices via self-assembly techniques constitutes an important step for the realisation of large number of devices based on SWNT and their systematic study.

Acknowledgments

We acknowledge O. Jost of TU Dresden for supplying of SWNTs, R. Tsui, R. Martel and V. Derycke for useful discussions and E. Dujardin for pointing out the interest of the EDA. This work has been supported by the European project IST-1999-10593 SATURN.

References

[1] S. Iijima: *Nature* 354 (1991) 56

[2] H. W. Ch. Postma, T. Teepen, Z. Yao, M. Grifoni, and C. Dekker Science 293, 76-79 (2001)

[3] V. Derycke, R. Martel, J. Appenzeller, Ph. Avouris, Nano Lett. 1(9), 453 (2001)

[4] A.Bachtold, P.Hadley, Takeshi Nakanoshi, Cees Dekker Science **294**, 1317(2001)

[5] J. Liu, M.J. Casavant, M. Cox, D.A. Walters, P. Boul,W. Lu, A.J. Rimberg, K.A. Smith, D.T.Colbert, R.E. Smalley: Chem. Phys. Lett. **303**, 125 (1999)

[6] K.H. Choi ; J.P. Bourgoin ; S. Auvray ; D. Esteve ; G. S. Duesberg , S. Roth , M. Burghard, Surf. Sci. **462**, 195 (2000)

[7] S. M. Kanan,W. T. Y. Tze, C. P. Tripp, Langmuir **18,** 6623 (2002).

[8] K.D. Ausman, R. Piner, O. Lourie, R.S. Ruoff, J. Phys. Chem. B **104**, 8911 (2000)

[9] A. A. Gorbunov, R. Friedlein, O. Jost, M. S.; Golden, J.; Fink, WPompe, Appl. Phys. A,**69**, 593 (1999).

[10] L. Capes, E. Valentin, S. Esnouf, A. Ribayrol, O. Jost, A. Filoramo, J-N Patillon, 2002. IEEE-NANO 2002. Proceedings of the 2002 2nd IEEE Conference on Nanotechnology, pages 439-442 , 26-28 Aug. 2002.

[11] M. R. Diehl, S. N. Yaliraki, R. A. Beckman, M. Barahona and J. R. Heath Angew. Chem. Int. Ed. **41**, 353 (2002)

[12] J. Kong, H. Dai, J. Phys. Chem. B **105**, 2890 (2001)

[13] E. Valentin, S. Auvray, J. Goethals, J. Lewenstein, L. Capes, A. Filoramo, A. Ribayrol, R. Tsui, J.P. Bourgoin, J.N. Patillon, MICROELECTRONIC ENGINEERING **61**, 491 (2002)

[14] R.Martel, V. Derycke, C. Lavoie, J. Appenzeller, K. K. Chan, J. Tersoff, Ph. Avouris, Phys. Rev. Lett., **87,** 256805 1-4 (2001).

[15] S. J. Wind, J. Appenzeller, R. Martel, V. Derycke, Ph. Avouris, Appl. Phys. Lett., **80**, 3817 (2002).

[16] F. Nihey, H. Hongo, M. Yudasaka, S. Ijima, Jpn. J. Appl. Phys. **41,** 1049 (2002).

TWO-CHANNEL KONDO EFFECT IN A MODIFIED SINGLE ELECTRON TRANSISTOR

Yuval Oreg*†
Department of Condensed Matter Physics,
Weizmann Institute of Science, Rehovot, 76100, ISRAEL
yuval.oreg@weizmann.ac.il

David Goldhaber-Gordon
Geballe Laboratory for Advanced Materials and Department of Physics,
Stanford University, Stanford, CA 94305, USA
goldhaber-gordon@stanford.edu

Abstract We suggest a simple system of two electron droplets which should display two-channel Kondo behavior at experimentally-accessible temperatures. Stabilization of the two-channel Kondo fixed point requires fine control of the electrochemical potential in each droplet, which can be achieved by adjusting voltages on nearby gate electrodes. We study the conditions for obtaining this type of two-channel Kondo behavior, discuss the experimentally-observable consequences, and explore the generalization to the multi-channel Kondo case.

Keywords: Non Fermi Liquid, Multi channel Kondo effect, Quantum dots.

1. Introduction

The single-channel Kondo (1CK) effect has been studied for decades in metals with magnetic impurities [1]. The same phenomenon has recently been observed in the novel context of semiconductor nanostructures containing no magnetic impurities: here, an electron droplet with a degenerate ground state assumes the role of a magnetic impurity, and nearby electron reservoirs act as the surrounding normal metal [2, 3, 4, 5, 6, 7]. These semiconductor systems are extremely flexible. The electron droplet's shape and size are determined

*This manuscript is a copy of Yuval Oreg and David Goldhaber-Gordon Phys. Rev. Lett. 90, 136602 (2003)
†Yuval Oreg is Incumbent of the Louis and Ida Rich Career Development Chair.

A.S. Alexandrov et al. (eds.), Molecular Nanowires and Other Quantum Objects, 67–76.

by lithographic patterning, and its occupancy, energy levels, and coupling to external reservoirs can be precisely measured, and even tuned *in-situ* using gate voltages. This unique tunability has enabled the first precision measurements of Kondo temperature as a function of system parameters, yielding an excellent match to theory [4, 8]. Experiments on semiconductor nanostructures have also accessed new regimes, notably the low-temperature unitary limit [5], Kondo effect out of equilibrium [9], and the single mixed-valence impurity [4]. These experiments have even introduced exotic varieties of Kondo effect never seen in bulk studies, such as magnetic field-induced Kondo [10, 11] and two-impurity Kondo [12]. As with conventional 1CK, each of these systems displays an interesting many-body resonance. However, at very low T we can describe each system simply as a Fermi liquid superimposed with a resonance [13]; *i.e.*, there is no non-Fermi liquid ground state.

Studying two-channel Kondo (2CK) effect [14, 15, 16, 17, 18] in semiconductor nanostructures could be even more intriguing. In 2CK, a twofold degenerate system such as a local spin is antiferromagnetically coupled to not one, but two independent electron reservoirs. Since the reservoirs do not communicate, each attempts to screen the local spin, resulting in overall overscreening. Unlike 1CK, this system exhibits fascinating low-energy non-Fermi-liquid behavior [19, 20]. Yet there have been no conclusive experimental observations of 2CK [14, 16, 21]. Indeed, in contrast to single-channel Kondo, 2CK effect is not likely to occur in ordinary metals with magnetic impurities, due to intrinsic channel anisotropy [22]. Ralph reported observation of 2CK, with local near-degeneracies associated with atomic tunnelling in a disordered metal rather than the traditional spin. The observed behavior is striking, but its physical origin has remained controversial[16, 21].

2. Results and Discussion

In this paper we argue that a simple configuration of two electron droplets (see Fig. 1) attached to conducting leads can exhibit 2CK correlations [19, 20], retaining non-Fermi-liquid (NFL) behavior at low temperature.

The relevant fixed point is stabilized at low temperature by fine tuning the voltage on just one gate electrode. Near a 2CK fixed point, quantities such as specific heat, entropy and spin susceptibility [20] behave differently than they would in a Fermi liquid. The conductance through our model system should exhibit an anomalous power-law dependence on temperature, deviating from its $T = 0$ value as $\sqrt{T}$[20] rather than T^2. The simplicity of the structure and the ability to tune system parameters offer hope for detailed study of the NFL realm, including non-thermodynamic quantities such as transmission phase [23], noise [24], pumping [25], and tunnelling density of states.

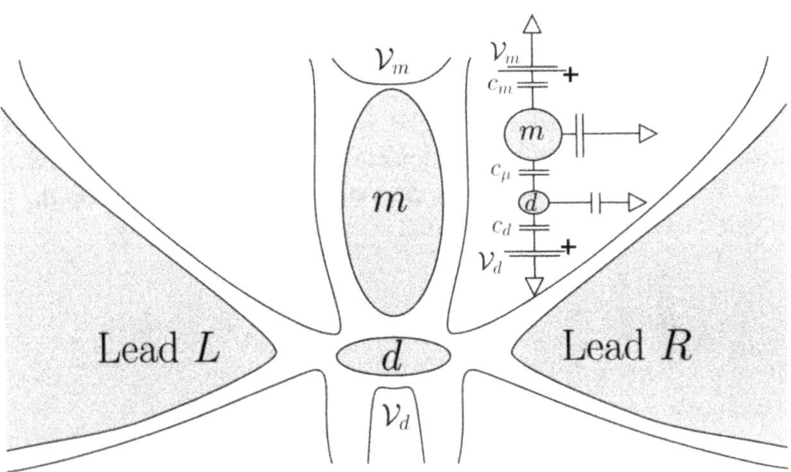

Figure 1. A proposed realization of the two-channel Kondo (2CK) model. Two non-interacting leads (L and R) and a large dot m are attached to a single-level small dot d. If dot d is occupied by a single electron, it can flip its spin by virtually hopping the electron onto either dot m or the leads, and then returning an electron with opposite spin to dot d. Dot m and the leads thus serve as the two distinct screening channels required to produce the 2CK effect. Crucially, when kT is smaller than the charging energy of dot m, Coulomb blockade blocks transfer of electrons between the leads and dot m. Fine tuning of the voltage $\mathcal{V}_m$ (and/or $\mathcal{V}_d$) can equalize the coupling to the two channels, stabilizing the 2CK fixed point.

In our model, a small central electron droplet (denoted by d) hosts a single level of energy ε_{ds}, which can be empty, or occupied by electrons of either or both spin directions $s = \uparrow, \downarrow$. Henceforth we refer to this droplet as *small dot d*. The spin of the (singly-occupied) small dot d serves as the local degeneracy needed for 2CK. Connected to small dot d by tunnelling are two conducting leads, plus an additional, much larger dot. In the large dot we neglect the discreteness of single-particle energy levels, while retaining a finite Coulomb energy. Thus, this dot behaves as a "Coulomb-interacting lead"; we refer to it as *large dot m*.

For fixed number of electrons: n_m on dot-m and n_d on dot-d, the electrostatic energy $E_{n_d}^{n_m} \equiv E_{n_d}^{n_m}(\mathcal{V}_m, \mathcal{V}_d)$ is:

$$E_{n_d}^{n_m} = U\left(n_d - \mathcal{N}_d\right)^2 + u_m\left(n_d + \alpha n_m - \mathcal{N}\right)^2, \tag{1}$$

where $U \equiv e^2/(2\tilde{C}_d) \gg u_m \equiv e^2/(2\tilde{C}_m - c_\mu^2/\tilde{C}_d)$, $|e|\mathcal{N}_d \equiv c_d\mathcal{V}_d$, $|e|\mathcal{N} \equiv c_m\mathcal{V}_m + c_d c_\mu/\tilde{C}_m\mathcal{V}_d$ and $\alpha \equiv c_\mu/\tilde{C}_d \lesssim 1$. Here $\tilde{C}_{m(d)}$ is the *total* capacitance of dot $m(d)$. See Fig. 1 for definitions of the other capacitances. Note that the parameter $\mathcal{N}_d$ controls the number of electrons on the small dot while $\mathcal{N}$

controls the total number of electrons on both dots combined. Since dot m is large we may assume that $\tilde{C}_m$ is much larger than all other capacitances.

To write down the full Hamiltonian H of the model system, it is useful to perform a transformation on the operators L_{ks} and R_{ks} for electrons in leads L and R, respectively [26]. We define $\psi_{ks} = \cos\theta\ L_{ks} + \sin\theta\ R_{ks}$, $\phi_{ks} = \cos\theta\ R_{ks} - \sin\theta\ L_{ks}$, $\tan\theta = V_R/V_L$, $V_\psi = \sqrt{|V_L|^2 + |V_R|^2}$. Without loss of generality we take the coupling constants V_i, $i = L, R, m$, to be real. With these definitions, the new effective lead ψ couples to the small dot d, but the effective lead ϕ does not couple:

$$
\begin{aligned}
H &= \sum_{i=\phi,\psi,m;ks} \varepsilon_{iks} i^\dagger_{ks} i_{ks} + \sum_s \varepsilon_{ds} d^\dagger_s d_s + E^{n_d}_{n_m} \\
&\quad + V_m \sum_{ks} m^\dagger_{ks} d_s + V_\psi \sum_{ks} \psi^\dagger_{ks} d_s + \text{h.c.},
\end{aligned} \tag{2}
$$

To obtain a 2CK fixed point we assume that $\mathcal{V}_d$ is tuned to make the average occupancy of the small dot $n_d = 1$, creating a local spin-$\frac{1}{2}$. We further assume that $D, U \gg u_m$, with D a cutoff of order the Fermi energy [27]. With decreasing temperature the system evolves through several stages. Formally, we integrate out the fast variables progressively in the renormalization group (RG) sense. Details of this calculation will be published elsewhere.

For $kT > U$, charge fluctuations on both the small and large dots are possible. Haldane showed [8] that in this regime only the level energy ε_d is renormalized, while the couplings to the leads remain the same. For $kT \lesssim U$ we may perform the Schrieffer-Wolff transformation. In this transformation charge fluctuations on the small dot are eliminated, and the effect of virtual electron hopping is simply to flip the spin on the small dot. Our Anderson-like Hamiltonian is mapped onto a 1CK Hamiltonian. In the present case we have four possible spin flip events. Two are diagonal processes in which an electron hops onto dot d from lead ψ (large dot m) and then an electron with opposite spin hops off to the same lead ψ (large dot m). And two are off-diagonal processes in which an electron hops onto dot d from lead ψ (large dot m) and then an electron with opposite spin hops off to large dot m (lead ψ). (Four "hole" processes, in which an electron first hops *from* the dot to lead ψ or large dot m, are also possible). As T decreases further, so long as $kT > u_m$ charge fluctuations on large dot m are allowed and the system flows according to the single channel Kondo RG laws [8]. However, for $kT < u_m$ charge fluctuations on the large dot are not possible and off-diagonal hopping is suppressed. Diagonal spin flip events remain possible. In this regime we obtain the standard two-channel Kondo model [20], with an additional free channel ϕ which decouples from the rest of the system [see also Eq. (6)]. The diagonal exchange coupling constants

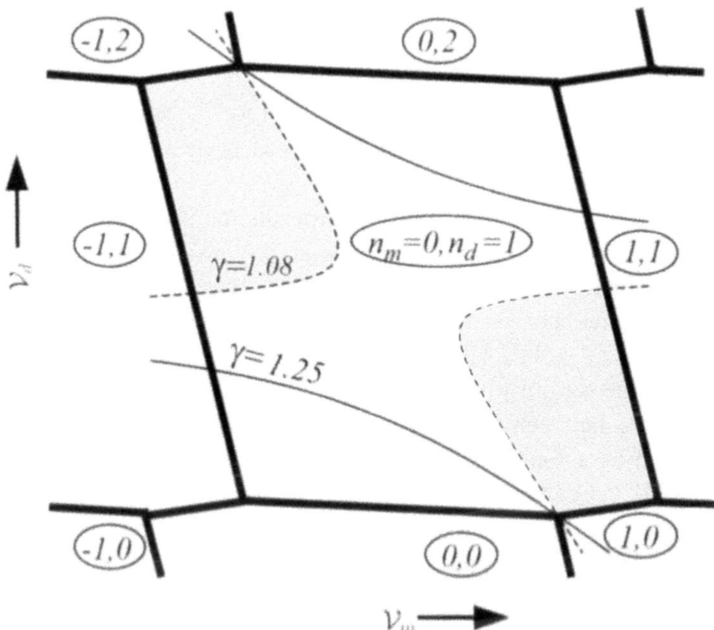

Figure 2. The number of electrons on dots m and d are functions of the gate voltages $\mathcal{V}_m$ and $\mathcal{V}_d$. Within the central hexagon, dot d is singly-occupied, a prerequisite for observation of Kondo effect. Curves superimposed on this hexagon ("2CK lines") map where in the $\mathcal{V}_m, \mathcal{V}_d$ plane the two-channel Kondo (2CK) effect is realized for two different values of the coupling ratio $\gamma \equiv \Gamma_m/\Gamma_\psi$ — each value gives rise to a pair of disjoint curves. As illustrated for $\gamma = 1.08$ (dashed) these two curves divide the hexagon into three regions with distinct low-temperature fixed points. On the curves, the 2CK effect is realized and the deviation of the inter-lead differential conductance from its $T \to 0, \mathcal{V}_{LR} \to 0$ limit $G(0,0)$ is $\propto \sqrt{\max(T, \mathcal{V}_{LR})}$[Eq. (4a)]. In the shaded regions at top left and bottom right, dot m "wins", forming an exclusive 1CK resonance with dot d, and driving $G(T, \mathcal{V}_{LR})$ close to zero [Eq. (4b)]. By contrast, in the large unshaded region, leads L and R "win" giving rise to familiar Fermi-liquid behavior $G(0,0) - G(T, \mathcal{V}_{LR}) \propto [\max(T, \mathcal{V}_{LR})]^2$, [Eq. (4c)]. With increasing γ, the regions where dot m wins grow and merge, while the region where the leads win shrinks and splits.

(at scale U) are:

$$\begin{aligned}\tilde{J}_{mm}(\mathcal{V}_m, \mathcal{V}_d) &\equiv \Gamma_m \left[\frac{1}{E_0^1 - E_1^0} + \frac{1}{E_2^{-1} - E_1^0}\right];\\ \tilde{J}_{\psi\psi}(\mathcal{V}_m, \mathcal{V}_d) &\equiv \Gamma_\psi \left[\frac{1}{E_2^0 - E_1^0} + \frac{1}{E_0^0 - E_1^0}\right].\end{aligned} \tag{3}$$

Here $\Gamma_{m(\psi)} = |V_{m(\psi)}|^2 \nu_{m(\psi)}$ is the rate of tunnelling between dot d and dot m (effective lead ψ); and $E_{n_d}^{n_m}$ are defined in Eq. (1). To obtain a 2CK fixed point we tune $\mathcal{V}_m$ and $\mathcal{V}_d$ to make $\tilde{J}_{mm}$, the antiferromagnetic coupling of dot d to dot m, equal to $\tilde{J}_{\psi\psi}$, the antiferromagnetic coupling of dot d to the leads. Eq. (3) and the ratio $\gamma \equiv \Gamma_m/\Gamma_\psi$ define a curve in the $\mathcal{V}_m, \mathcal{V}_d$ plane. In Fig. 2 we show these "2CK lines" for two different values of γ, both of order 1. On these lines 2CK physics should be realized at low T. We did not consider in Fig. 2 the renormalization of the parameters at scales below U, which may modify the detailed shape of the curves.

The 2CK fixed point can be reached experimentally by a three-step procedure: First, fix $\mathcal{V}_d$ to give one electron (or an odd number of electrons) in the small dot. Second, tune γ to roughly 1 by adjusting the individual tunnelling rates. No great precision is required in this step. Finally, fine-tune $\mathcal{V}_m$ so that $\tilde{J}_{mm}(\mathcal{V}_m, \mathcal{V}_d) = \tilde{J}_{\psi\psi}(\mathcal{V}_m, \mathcal{V}_d)$. In Fig. 2 this corresponds to tuning $\mathcal{V}_m$ until we hit a 2CK line for our given value of γ.

The current I between the left and right leads can be measured as a function of T, and as a function of $\mathcal{V}_{LR}$, the bias applied between leads L and R (see Fig. 1). Drawing on the extensive literature of 2CK physics [14] we can predict the qualitative behavior of the I-$\mathcal{V}_{LR}$ curve through the dot, for different values of the gate voltages $\mathcal{V}_m$, $\mathcal{V}_d$ that scan the hexagon of Fig. 2. On the 2CK lines (see Fig. 2) in the unitary limit — T,$\mathcal{V}_{RL} \ll$ Kondo temperature $T_K \cong \sqrt{U\Gamma_m}/2\, e^{-1/\tilde{J}_{mm}(\mathcal{V}_m,\mathcal{V}_d)}$ — the differential conductance $G(T, \mathcal{V}_{LR}) \equiv dI/d\mathcal{V}_{LR}$ should approaches its limiting value $G(0,0)$ as

$$2CK: \quad G(0,0) - G(T, \mathcal{V}_{LR}) \propto \sqrt{\max\left(\mathcal{V}_{LR}, kT\right)}. \tag{4a}$$

In the symmetric case $V_L = V_R$, we get $G(0,0) = G_K \equiv e^2/h$, half the maximal value of $G(0,0)$ in the 1CK effect [5, 7].

In the part of the hexagon where $\tilde{J}_{mm} > \tilde{J}_{\psi\psi}$ (shaded, for $\gamma = 1.08$), at low T the electrons in dot m screen the spin of dot d, while the leads are decoupled. In the RG sense $\tilde{J}_{\psi\psi}$ flows to zero, so that dot m "wins" over the leads and forms a 1CK state with dot d. In this case $dI/d\mathcal{V}_{LR}$ is small and given by

$$\text{Large dot wins:} \quad G(T, \mathcal{V}_{LR}) \propto \left[\max\left(\mathcal{V}_{LR}, kT\right)\right]^2. \tag{4b}$$

In contrast, in the unshaded part of the hexagon in Fig. 2, where $\tilde{J}_{mm} < \tilde{J}_{\psi\psi}$, dot m decouples from dot d at low T, leaving the leads to form a 1CK resonance with dot d and

$$\text{Leads win:}\ G(0,0) - G(T, \mathcal{V}_{LR}) \propto \left[\max\left(\mathcal{V}_{LR}, kT\right)\right]^2, \tag{4c}$$

where $G(0,0) = 2G_K$ for $V_L = V_R$.

At sufficiently low temperature the finite level spacing Δ_m in dot m will cut off the RG flow of the coupling constants [28]. We cannot make Δ_m infinitesimal as we must retain a finite Coulomb blockade energy [29] $u_m > kT$.

However, the ratio between charging energy and level spacing can be made large, allowing 2CK behavior to be observed over an order of magnitude in temperature before the system finally flows to the 1CK fixed point.

The above discussion can be generalized to include $M-1$ large dots, resulting in an M-channel Kondo (MCK) model, which may be possible (though challenging) to realize experimentally for $M > 2$.

To describe this system we use the model Hamiltonian:

$$\begin{aligned} H &= \sum_{aks} \varepsilon_{aks} a^\dagger_{ks} a_{ks} + \sum_a u_a (n_a - \mathcal{N}_a)^2 \\ &+ \sum_s \varepsilon_d d^\dagger_s d_s + U n_{d\uparrow} n_{d\downarrow} + \sum_{ks} V^*_{ak} a^\dagger_{ks} d_s + \text{h.c.} \end{aligned} \quad (5)$$

Here a_{ks} is the annihilation operator of an electron in state k, with spin s and energy ε_{aks} on large dot a. $a = 1, \ldots, M$, $n_a = \sum_{ks} a^\dagger_{ks} a_{ks}$, and the parameter $\mathcal{N}_a$ sets dot a's equilibrium occupancy.

The physical role of the u_a terms is clear: at low temperatures they forbid processes in which charge is ultimately transferred from one large dot to another. Spin flip events — *e.g.* when an electron hops onto the small dot and then an electron with opposite spin hops off the small dot to the same large dot — remain possible and under appropriate conditions lead to a MCK fixed point.

After integrating out energies larger then U [8], we perform the Schrieffer-Wolff transformation [30] and find that the second line of Eq. (5) is transformed to

$$\sum_{ab,kq} J^{kq}_{ab} \left[S^+ s^{-kq}_{ab} + S^- s^{+kq}_{ab} + 2S^z s^{zkq}_{ab} \right], \quad (6)$$

$S^\pm = d^\dagger_{\uparrow(\downarrow)} d_{\downarrow(\uparrow)}$, $2S^z = d^\dagger_\uparrow d_\uparrow - d^\dagger_\downarrow d_\downarrow$, $s^{\pm kq}_{ab} = a^\dagger_{k\uparrow(\downarrow)} b_{q\downarrow(\uparrow)}$ and $2s^{zkq}_{ab} = \frac{1}{2} a^\dagger_{k\uparrow} b_{q\uparrow} - a^\dagger_{k\downarrow} b_{q\downarrow}$. With $V_{aq} = V_{bk} \equiv V$, and assuming that $\varepsilon_{a(b)qs} \approx \varepsilon_{a(b)F}$, where $\varepsilon_{a(b)F}$ is the last empty (occupied) level in dot a (b), we find:

$$J^{kq}_{ab} = J_{ab} = |V|^2 \frac{U + u^-_b + u^+_a}{\left[U + \varepsilon_d + u^-_b\right]\left[u^+_a - \varepsilon_d\right]}, \quad (7)$$

where $u^\pm_p = u_p \left[1 \pm 2\left(n_p - \mathcal{N}_p\right)\right] \pm \varepsilon_{pF}$, $p = a, b$. For $-1/2 < \mathcal{N}_p - n_p + (\mu - \varepsilon_{pF})/(2u_p) < 1/2$ there are n_p electrons in dot p, where μ is the electrochemical potential of a reference reservoir. Particle-hole symmetry may be absent in the large dots, so in general $J_{ab} \neq J_{ba}$.

At $kT > \max\{u_{ab}\}$ the system evolves according to the 1CK RG flow, where $u_{ab} = \left(u^+_a + u^-_b\right)(1 - \delta_{ab})$. At $kT < \min\{u_{ab}\}, a \neq b$, the off-diagonal processes describing transfer of an electron from dot b to dot a are exponentially suppressed as $J_{ab} = J^0_{ab} e^{u_{ab}/(4kT)}$. Notice that $u_{aa} = 0$, since the charge on dot a is not changed when an electron hops from dot a onto dot d and then back to the same dot a. At $kT < \min\{u_{ab}\}$ only the diagonal terms of J_{ab} do not flow to zero. Assuming that we are not at a degeneracy point where $u_{ab} = 0$, an easy condition to avoid, the RG equations are identical to the MCK RG equations [22]. As in the case of classic MCK, our NFL fixed point is unstable to the introduction of channel anisotropy. If one of the coupling constants is larger than the others, the corresponding channel alone screens the local spin and forms a Kondo resonance while the other channels are decoupled from the local spin. In our model we can tune all the $\mathcal{N}_a$ to achieve $J_{aa} = J$ for all a.

Gate voltages capacitively control the energy of the last occupied level in each large dot, so excitations in each large dot will be around a different Fermi energy. This does not modify the RG equations, but does affect certain physical properties such as the small dot density of states at finite energies. A similar situation occurs in the discussion of 2CK in a dot out of equilibrium [15, 18].

3. Conclusions

In conclusion, if a small dot is coupled to two (or more) electron reservoirs, Coulomb blockade can suppress inter-reservoir charge transfer at low temperatures. Electrostatic gates provide the tunability needed to stabilize a 2CK fixed point, resulting in observable NFL behavior. Softer suppressions of inter-reservoir tunneling could also work in place of Coulomb blockade. For example, the reservoirs could be conductors with large impedance [31], one-dimensional Luttinger liquids [32] or conductors with strongly-interacting charge carriers. Finally, while the channel asymmetry parameter is relevant in the RG sense, for realistically well-matched channel couplings we expect that the system will remain near the 2CK fixed point, and will show NFL behavior, over a wide range of temperatures.

Acknowledgments

It is our pleasure to acknowledge useful discussions with Natan Andrei, Michel Devoret, Leonid Glazman, Gabi Kotliar, Leo Kouwenhoven, Yigal Meir, Ron Potok, Arcadi Shehter, Alex Silva, Ned Wingreen, and especially Avi Schiller. This work was supported by DIP grant c-7.1, by ISF grant 160/01-1, and by Stanford University.

References

[1] A. C. Hewson, *The Kondo Problem to Heavy Fermions* (Campridge University Press, Cambridge, 1993).

[2] D. Goldhaber-Gordon, H. Shtrikman, D. Mahalu, D. Abusch-Magder, U. Meirav, and M. A. Kastner, Nature **391**, 156 (1998).

[3] S. M. Cronenwett, T. H. Oosterkamp, and L. P. Kouwenhoven, Science **540**, 281 (1998).

[4] D. Goldhaber-Gordon, J. Göres, M. A. Kastner, H. Shtrikman, D. Mahalu, and U. Meirav, Phys. Rev. Lett. **81**, 5225 (1998).

[5] W. G. van der Wiel *et al.*, Science **289**, 2105 (2000).

[6] J. Nygard, D. H. Cobden, and P. E. Lindelof, Nature **408**, 342 (2000).

[7] W. Liang, M. Bockrath, and H. Park, Phys. Rev. Lett. **88**, 126801 (2002).

[8] F. D. M. Haldane, Phys. Rev. Lett. **40**, 416 (1978).

[9] S. De Franceschi *et al.* cond-mat/0203146 (2002).

[10] S. Sasaki *et al.*, Nature **764**, 405 (2000).

[11] M. Pustilnik, Y. Avishai, and K. Kikoin, Phys. Rev. Lett. **84**, 1756 (2000), and M. Pustilnik and L. I. Glazman Phys. Rev. B **64** 045328 (2001).

[12] H. Jeong, A. M. Chang, and M. R. Melloch, Science **293**, 2221 (2001).

[13] P. Noziéres, J. Low Temp. Phys. **17**, 31 (1974).

[14] D. Cox and A. Zawadowski, Advances in Physics **47**, 599 (1998), cond-mat/9704103.

[15] We emphasize that our proposal differs from other suggested experimental realizations of the 2CK effect [16, 17], in that our local degeneracy comes from spin rather than charge or some other two-level system. Our mapping differs from that of [17] in several essential ways: we consider a system with fixed electron number in what would otherwise be a Coulomb valley, not a charge degeneracy point; our Kondo temperature is set by the charging energy of a small dot; and we do not require single-mode coupling between the dot and leads. Our 2CK effect is realized at equilibrium, avoiding decoherence and suppression of 2CK by voltage-driven current [18].

[16] D. C. Ralph and R. A. Buhrman, Phys. Rev. Lett. **69**, 2118 (1992), and Phys. Rev. B 51, 3554 (1995); and D. C. Ralph *et al.* Phys. Rev. Lett. 72, 1064 (1994).

[17] K. A. Matveev, Phys. Rev. B **51**, 1743 (1995).

[18] A. Rosch, J. Kroha, and P. Wölfle, Phys. Rev. Lett. **87**, 156802 (2001).

[19] N. Andrei and C. Destri Phys. Rev. Lett. **52**, 364 (1984); P. B. Wiegmann and A. M. Tsvelick Z. Phys. B **54**, 201 (1985)

[20] I. Affleck and A. W. W. Ludwig, Phys. Rev. B **48**, 7297 (1993).

[21] N. S. Wingreen, B. L. Altshuler, and Y. Meir, Phys. Rev. Lett. **75**, 769 (1995), see also Ralph *et al.* Reply ibid 770 and Wingreen *et al.* Erratum ibid 81, 4280 (1998).

[22] P. Nozieres and A. Blandin, J. Physique **41**, 193 (1980).

[23] U. Gerland *et al.* Phys. Rev. Lett. **84**, 3710 (2000), and Y. Ji *et al.*, Science **290**, 779 (2000).

[24] R. de Picciotto *et al.* Nature **389**, 162 (1997).

[25] M. Switkes *et al.* Science **283**, 1907 (1999).

[26] L. I. Glazman and M. E. Raikh, Sov.Phys. Lett. **47**, 452 (1988), pis'ma Zh. Eksp. Teor. Fiz. **47**, 378-380 (1988).

[27] In semiconductor structures, generally $D > U$. We will assume this here, but relaxing this assumption has no significant impact on our conclusions.

[28] W. B. Thimm, J. Kroha, and J. von Delft, Phys. Rev. Lett. **82**, 2143 (1999).

[29] Our Kondo temperature is set by parameters of the small dot. For small (large) dot of diameter $100nm(2\mu m)$ in a GaAs/AlGaAs heterostructure, it is possible to achieve $u_m \sim 3K > T_K \sim 1K \gg \Delta_m \sim 50mK$.

[30] J. R. Schrieffer and P. A. Wolff, Phys. Rev. **149**, 491 (1966).

[31] M. H. Devoret *et al.*, Phys. Rev. Lett. **64**, 1824 (1990).

[32] M. Fabrizio and A. O. Gogolin, Phys. Rev. B. **51**, 17827 (1995).

SYNTHESIS AND STRUCTURAL CHARACTERISATION OF SINGLE WALL CARBON NANOTUBES FILLED WITH IONIC AND COVALENT MATERIAS

Jeremy Sloan[1,2], Angus I. Kirkland[2], John L. Hutchison[2], Steffi Friedrichs[1], and Malcolm L.H. Green[1]

[1] *Inorganic Chemistry Laboratory, University of Oxford, South Parks Road, Oxford, OX1 3QR, U.K.*

[2] *Department of Materials, University of Oxford, Parks Road, Oxford, OX1 3PH, U.K.*

malcolm.green@chem.ox.ac.uk

Abstract Carbon nanotubes have hollow interiors with internal diameters ranging from 0.5 – 20 nm although more typically 1 – 4 nm. It is now well established that these internal cavities can be filled with crystalline and molecular materials. Filling of single walled carbon nanotubes (SWNTs) are normally filled by direct mixing of the molten filling material and the nanotubes at elevated temperature. Characterisation of filled carbon nanotubes relies mainly on high-resolution transmission electron microscopy (HRTEM) and extended resolution HRTEM techniques. The crystals formed within the narrower SWNTs are often only a few atoms in cross section with a large proportion of atoms lying on the surface of the encapsulated crystals. An extreme example is in the case of a 2×2 crystal of potassium iodide which is effectively an 'all surface' crystal. The structures of the nanocrystals encapsulated inside SWNTs show other significant differences compared to the corresponding bulk materials. For example, inter-atomic differences can be substantially greater (i.e. by as much as $\sim$20%) for SWNT encapsulated materials. Several examples of filled SWNTs are described including those filled with simple metal halide salts, covalent metal halides and molecular species as in *o*-carborane.

Keywords: Single wall carbon nanotubes; 1D crystals; Transmission Electron Microscopy

1. Introduction

In his famous address *There's plenty of room at the bottom* in 1959 [1], Richard Feynman stated: "What could we do with layered structures with just the right layers ? What would the properties of materials be if we could really arrange the atoms the way we want them? They would be very interesting

A.S. Alexandrov et al. (eds.), Molecular Nanowires and Other Quantum Objects, 77–88.

to investigate theoretically. I can't see exactly what would happen, but I can hardly doubt that when we have some *control* of the arrangement of things on a small scale we will get an enormously greater range of possible properties that substances can have, and of different things that we can do".

In Oxford we have studied the synthesis and characterisation of nanoscale low dimensional materials within the hollow interior of single wall carbon nanotubes (SWNT). We have demonstrated the synthesis and *complete* structural characterisation of the atomic positions of the component atoms in crystals only a few atomic layers in cross section. This has been achieved by the development and application of advanced high resolution transmission electron microscopy (HRTEM) based imaging techniques to a precision directly comparable to that obtained in single crystal X-Ray structure determination. This work has established beyond doubt that these quantum sized nano-crystals have unprecedented 'Feynman'-type structures [2] (i.e. crystal of effectively integral numbers of atomic layers in cross-section) which are substantially different from those of the corresponding bulk materials both in co-ordination environments and interatomic distances. Importantly we have also demonstrated that an extended crystalline lattice is not required for the determination of atom positions and have also shown that discrete molecules and clusters of non-fullerenic molecules such as *o*-carborane (Figure 1(a) to (h)) can be detected within single wall nanotubes albeit not in this case at atomic resolution [3].

In this brief review we outline the synthetic and characterisation methodologies that permit the fabrication and study of these unique low dimensional systems and describe some of the remarkable new structural chemistry that they exhibit.

Synthetic aspects

Single walled carbon nanotubes (SWNTs)[4,5] are composed of rolled sheets of sp^2 graphene carbon terminated at one end by fullerenic carbon hemispheres, form well-defined cylindrical cavities (Figure 2(a)) within a strictly limited diameter range (typically 0.7-2.5 nm) when prepared by either catalytic carbon arc vapourisation [6] (i.e. Krätschmer-Huffman synthesis) or laser ablation of metal doped carbon rods [7]. Multi-gram quantities of such nanotubes can now be produced by a variety of catalyst-assisted decomposition techniques, most notably by pyrolysis of carbon monoxide in the presence of iron [8].

Multiwall carbon nanotubes MWNT filled with a variety of materials including metal oxides, metals, metal carbides, metal sulphides and proteins were already well known before the first successful filling of single wall carbon nanotubes (SWNTs) [9-16]. The first reported filled SWNTs capillaries were clusters and 1D nanowires of Ru metal [17]. Subsequently, SWNTs have been filled in high yield either from the vapour phase or by capillary wetting [18-21].

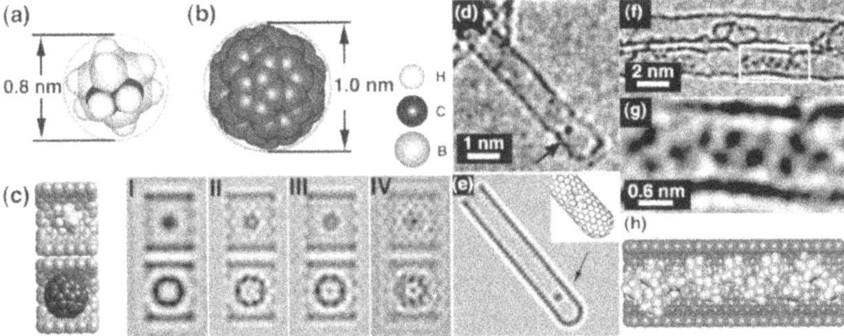

Figure 1. Structure models comparing outside dimensions of an o-carborane molecule (a) and a C_{60} molecule (b). (c) Packing models and image simulations comparing imaging properties (i.e. via conventional HRTEM) of a discrete o-carborane molecule (top) and a C_{60} molecule (bottom) inside SWNTs. (d) and (e) experimental HRTEM image and simulation of a discrete o-carborane molecule formed inside a SWNT tip. (f)-(g) conventional HRTEM image, detail and packing diagram of a cluster of ocarborane molecules formed inside a SWNT.

In the case of capillary wetting, SWNTs are prepared by heating the as-made or purified SWNTs with the chosen material at a temperature about 50-100°C above the melting point of the material [18]. Greater filling yield can be obtained by thermally cycling the SWNTs and the molten filling material [19]. Even though the SWNTs are initially have carbon caps the molten media in many cases, enters the tube and then the capilliarity wetting leads to the formation of continuous lengths of the filling material inside the SWNTs. The ability of a molten material to wet and fill SWNTs depends on the surface tension criterion determined by Ebbesen [22]. After cooling, examination of the a solidified material by HRTEM often shows the presence of regular arrays of dark spots. These are characteristic for the formation of a nano-crystal. When the filling material is a molten eutectic mixture, for example with molten mixtures of the silver halides AgCl and AgI then the apparently random array of dark spots can be interpreted in terms of essentially ordered silver ions surrounded tetrahedrally in a wurtzite-type configuration by randomly arranged Cl^- and I^- atoms [23]. Additionally exposing wide (i.e. $>$ 2 nm diameter) SWNTs filled with silver halides to light results in the formation of aligned 1D *fcc* silver nanowires within the capillaries [20,23]. SWNTs can now be filled with a wide variety of materials including metals and metal salts [17,20,21,23-28], oxides [29,31], helical iodine chains [32], and 1D chains of fullerene [18,33] endofullerene [34,35] or *o*−carborane molecules [3].

Structural aspects

With respect to the study of encapsulated halides within SWNTs, we have described a systematic survey of their structural chemistry, as elucidated by HRTEM, which is detailed in Tables 1 and 2 [2]. One of the earliest detailed structural studies was the filling with a 2 × 2 crystal of potassium iodide (see Fig. 2(b) to(f)), formed by heating SWNTs with molten KI [24]. This Figure shows that close correspondence between the arrangement of the same intensity dark spots observed experimentally in a two dimensional TEM projection with the calculated TEM image based on the model shown in Fig. 2(e) and given in Fig. 2(f). This 2 × 2 crystal of potassium iodide was the first example of an "all surface" crystal described and, as such, has no body ions. The distances between the I-K and K-I atom columns are 14% longer than the corresponding distances in bulk KI.

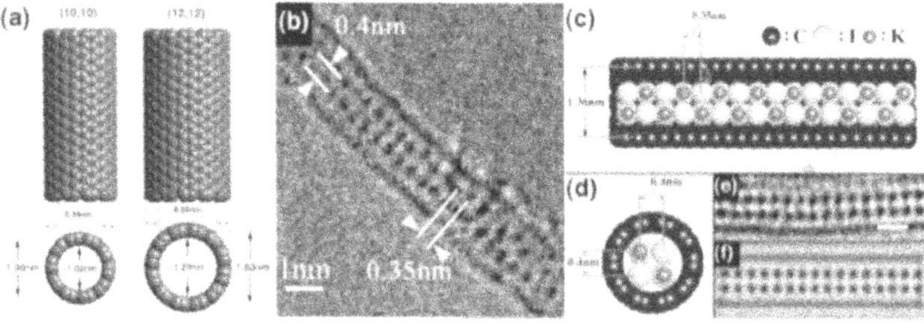

Figure 2. (a) Schematic representation of the van der Waals surface of (10,10) and (12,12) SWNTs with diameters of 1.4 and 1.6 nm, respectively, and internal diameters of ca. 1.0 and 1.3 nm respectively. (b) Conventional HRTEM image of a 2 × 2 KI crystal formed within a 1.4 nm diameter SWNT. (c) and (d) cutaway and end-on representations of a 2 × 2 KI crystal formed within a (10,10) SWNT. (e) and (f) HRTEM image and corresponding Scherzer defocus simulation of 2 × 2 KI crystal formed within a 1.4 nm SWNT.

We have also reported the formation of a 3 × 3 KI crystal in a larger diameter (1.6 nm) SWNT (Figure 3(a)-(c)) [25]. In this case, the crystal was imaged along the cell diagonal (i.e. <110> relative to bulk KI) which has the advantage that, in this projection, it is possible to visualise the K^+ and I^- sublattices as pure element columns (Figure 3(c)). This however has presented some difficulties for conventional HRTEM imaging as it was not possible to observe the weaker scattering K^+ sublattice directly. However, using a modified through focal series restoration approach [35], we have been able to restore the unaberrated phase (Figure 5(a)) of the exit surface wavefunction of the specimen [25]. Using this approach, the restored phase is obtained at close to the information limit of our HRTEM (*ca.* 0.1 nm) as opposed to the point resolution (*ca.* 1.6 nm) in conventional axial imaging. This additional information has made it possible to image the K^+ atom columns in addition to the more strongly scattering I^-

Stoichiometry	**C.N. of M**	3D complex	**C.N. of M**	Layer	**C.N. of M**	Chain	**C.N. of M**	**Molecular**
MX	4 6 8	Zn Blende **NaCl, AgX** **CsX**	5+2	TlI (yel.)	2	AuI		
MX_2	4 4 6 6 9 7+2 8	Silica-like ZnI_2 Rutile $CaCl_2$ **SrI_2,** **BaI_2,** EuI_2 $PbCl_2$ Fluorite	4 6 8	**HgI_2 (red)** **PbI_2** **CdI_2** **$CdCl_2$** ThI_2	3 3 4 4	$SnCl_2$ GeF_2 $BeCl_2$ $PdCl_2$	2 4	$HgCl_2$ Pt_6Cl_{12}
MX_3	6 7+2 8+1 9	ReO_3-type LaF_3 YF_3 UCl_3 **$LnCl_3$**	6 8+1	**$LnCl_3$** BiI_3 $PuBr_3$	4 6	AuF_3 ZrI_3	3 4 4	SbF_3 **Al_2Cl_6** Au_2Cl_6
MX_4	6 8 8	IrF_4 ZrF_4 **UCl_4,** **$ThCl_4$** **TeI_4**	6 8	PbF_4 ThI_4	5 6 6 6	TeF_4 $\alpha-NbI_4$ **$ZrCl_4$** **$HfCl_4$**	4 4	$SnBr_4$ **SnI_4**
MX_5	7	$\beta-UF_5$			6 6 7	BiF_5 CrF_5 $PaCl_5$	5 6 6	$SbCl_5$ Nb_2Cl_{10} Mo_4F_{20}
MX_6							6	**WCl_6** IrF_6

Table 1. General structure types of crystalline binary metal halides. The halides indicated in bold have all be encapsulated into SWNTs (C.N. = coordination number).

atom columns (Figures 3(a) and (b)). Furthermore, the phase gives a more faithful representation of the projected electron density of the specimen which has enabled us to measure individual atom column displacements (Figure 3(b)) within the incorporated 3×3 crystal . This in turn has allowed us to construct a more detailed structural model (Figures 3(c)) of the KI/SWNT composite than would have been possible via conventional HRTEM.

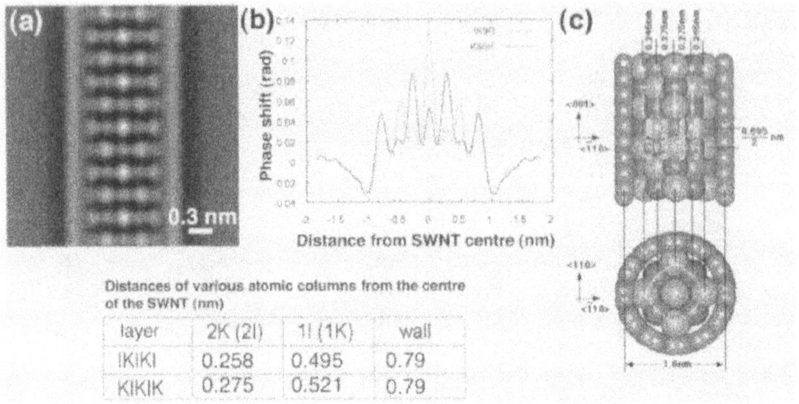

layer	2K (2I)	1I (1K)	wall
IKIKI	0.258	0.495	0.79
KIKIK	0.275	0.521	0.79

Figure 3. a) Reconstructed phase of the <110> projection of a 3 × 3 KI crystal in a 1.6 nm diameter SWNT. Note that the contrast in this image is reversed so that regions of high electron density appear bright [cf. Fig. 2(b)]. (b) Plot derived from alternating I-2K-3I-2K-I and K-2I-3K-2I-K layers in (a) showing relative displacements of the atom columns (measurements are in inset Table). (c) structure model derived from (a).

Other, non-rocksalt-type halides tend in general to form much more complex 1D crystal structures within SWNTs (see Tables 1(a) and (b)). BaI_2 for example is a polyhedral framework binary halide structure which, in the bulk, forms three different structural variants according to ambient pressure. All of these structures can be described in terms of 3D networks of face-sharing coordination polyhedra exhibiting varying iodine dispositions around a common tricapped trigonal prism motif (Figures 4(a) to (c)). Whereas the lower pressure forms exhibit 9-coordination (Figure 4(a) and (b)), the highest pressure version (Figure 4(c)) exhibits 10-coordination. However, within a 1.6 nm diameter SWNT, this material exhibited a completely different structure to that exhibited by any of the bulk forms [29].

Figure 4(d) and (e) show an HRTEM image and corresponding detail of a BaI_2 filled SWNT. The microstructure of the filling material is clearly visible as a 2D array of dark spots which show a lower intensity close to the walls of the SWNT and higher towards the centre. Since the electron scattering from Ba^{2+} is similar to that from I^- (*cf.* Ba (Z = 56) and I (Z = 53)) we propose that the dark spots correspond to 1-3 atom thick columns in projection each of which may contain some combination of Ba^{2+} and I^- ions.

In order to construct a model for this crystal, a 2D peak mapping program was used to determine the relative positions of the atomic columns in Figure 4(e), thus producing the atom map reproduced in Figure 4(f). In order to complete the crystal, this planar fragment was expanded vertically to give the end-on representation shown on the right of Figure 4(f) in which the fragment may be considered as being 1-2-3-2-1 atomic layers in thickness. Atomic positions

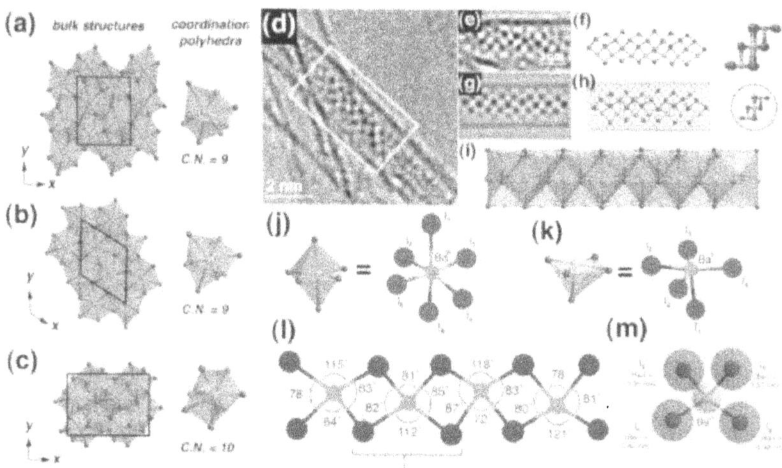

Figure 4. (a) Cotunnite form of BaI_2 formed under ambient conditions; (b) hexagonal P-62m modification of BaI_2 formed under 3 G Pa pressure; (c) $P112_1/a$ form of BaI_2 formed between 5 and 15 GPa pressure. (d) and (e) HRTEM image and detail of a 1D BaI_2 chain formed within a SWNT; (f) side-on and end-on views of atom positions derived from the lattice image; (g) HRTEM simulation of derived composite; (h) side-on and end-on views of the composite; (i) coordination model; (j) and (k) 6- and 5- coordination polyhedra observed within the 1D chain; (l) measured equatorial bond angles within the central region of the BaI_2 1D chain; (m) representative equatorial bond distances for I in (l).

were assigned based on a coordination scheme for Ba and I that would retain Ba in a +2 oxidation state. The vertical expansion was scaled according to bond distances anticipated from the ionic radii of Ba^{2+}_{VI} (0.136 nm) and I_{VI} (0.22 nm). A simulated HRTEM image (Figure 4(g)) calculated from this composite crystal and section of a (12,12) SWNT and the derived crystal fragment (Figure 4(h)) provides good agreement with the experimental image. The coordination polyhedra model derived from the 1D chain (Figure 4(i)) results in octahedral coordination for the Ba^{2+} ions along the centre of the chain (Figure 4(j)) and pentagonal coordination for the Ba^{2+} ions on the edges of the chain (Figure 4(k)).

A further striking example of the use of image enhancement techniques to study the structure of filled SWNTs is shown by the image of a SWNT filled with CoI_2 in Figure 5 (a) [36]. It shows the reconstructed phase of two sections of a discrete SWNT continuously filled with a 1D crystal of CoI_2 which have been Fourier filtered to remove spatial frequencies arising from the top-and-bottom surfaces of the SWNT. Detailed analysis of the structure of these crystal fragments shows that they differ fundamentally from the bulk form of the iodide which can be described as a hexagonal lattice with one or three $Co(hal)_2$ units

Halide Filling	Common Bulk Structure Types	**C.N. in bulk**	Structure **inside SWNTs**	**C.N.s observed or predicted within SWNTs**
KI	3D rocksalt	6	rocksalt	**4,5,6**
AgX	3D rocksalt,	6	"	4,5,6
CsX	wurzite	4	wurzite	**3,4**
SrI_2	3D rocksalt	6	rocksalt	4,5,6
BaI_2	*bcc*	8	*bcc*	4,6,8
UCl_4	3D network	7	1D PHC	**4,6**
$ThCl_4$	"	7+2	"	**5,6**
TeI_4	"	8	"	6,8
LnX_3	"	8	"	6,8
HgI_2 (red)	"	8	"	6,8
PbI_2	"	8+1	"	6,8
CdI_2	2D layered	3	"	2,3
$CdCl_2$	"	6	"	**5,6**
$LnCl_3$	"	6	"	5,6
$ZrCl_4$	"	6	"	5,6
$HfCl_4$	"	6	"	**5,6**
Al_2Cl_6	1D chain	6	"	as bulk
SnI_4	"	6	"	"
WCl_6	molecular	4	packed molecu-	"
	"	4	lar	"
	"	6	units	"

Table 2. Packing behaviour observed in SWNTs as a function of bulk structure type (C.N. = coordination number. 1D PDC = 1D polyhedral chain.).

per unit cell. The latter is similar to that of $Co(OH)_2$, consisting of stacked 2D layers of edge-sharing CoI_6 octahedra in which the anions form a *hcp* lattice.

In Fig. 5(a) the reconstructed phase shows a chain of pairs of atom columns, one pair separated by 0.54nm and the other by 0.26 nm, respectively perpendicular to the SWNT axis and by 0.21nm parallel to this axis. Single pixel width line profiles (Fig. 5(b)) (centred on the atomic column positions) along **I** and **II** (Fig. 5(a)) show higher projected potential at the column sites along **I**, consistent with a structural model in which these columns consist of a superimposed Co and I atom in projection with columns along **II** corresponding to single I atoms. The overall microstructure in this region therefore consists of a double chain of Co_2I_4 units in which each Co^{2+} is tetrahedrally coordinated to four bridging I^- atoms. In the second subregion (Fig. 5(c)) the reconstructed phase shows pairs of columns separated by 0.31nm (perpendicular) and again 0.21nm (parallel) to the tube axis and single pixel width line profiles (Fig. 5(d)) along **III** (Fig. 5(c)) show higher projected potentials at alternate column sites along **III**. This is entirely consistent with the same microstructure as determined from

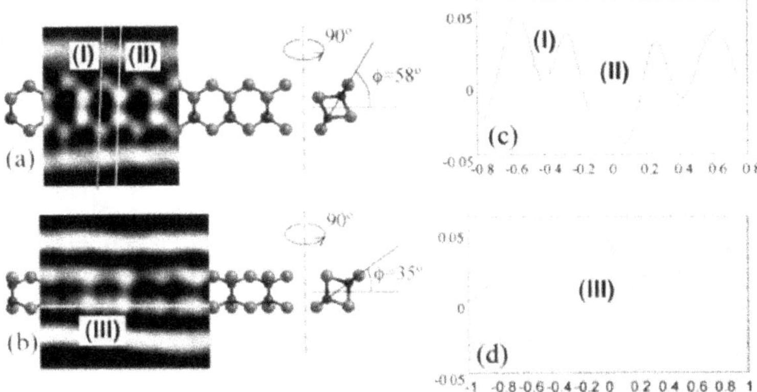

Figure 5. (a) and (b) reconstructed phase images and structure model for two sections of a CoI_2 1D crystal formed within a 1.6 nm diameter SWNT. (c) and (d) associated line profiles produced from the indicated regions in (a) and (b) respectively.

the first subregion but with a rotation of the crystal by 23° to give alternating columns of superimposed Co-I atoms and single I atoms in a rectangular arrangement.

Close inspection of the LHS of the whole image in Fig. 6(a) shows there are two central white spots and the separation between these two spots gradually decreases in the of the corresponding pairs on moving towards the left until they coincide. The apparent coalescence of these two spot in the two dimensional image can be interpreted in terms of a slow twisting of the crystal and image can be fully interpreted in terms of the model shown in Figure 6 (c). The tetrahedral structure of the CoI_2 chain is unprecedented and is not observed in the bulk material. Careful measurement of the diameter of the SWNT shown in Fig. 6(b) shows that is changes from a maximum of 1.4 nm at **I** to a minimum of 1.1 nm at **II**, which is strongly correlated to the measured crystal rotation. This variation arises as a consequence of an elliptical distortion in the SWNT (with an aspect ratio of 1.3) that rotates in sympathy with the helical crystal giving rise to different projected tube diameters (i.e. Fig. 6(e) and (f)).

These results provide the first detailed experimental evidence of a strong steric interaction between the crystal and encapsulating SWNT. In addition, they show that filled SWNTs can distort in ways that unfilled and unstrained SWNTs do not and such deformations have been predicted theoretically to alter their π band structures and therefore their conduction properties [37]. The determination of the complete crystal structure of the encapsulate as reported here, should therefore enable the computation of the combined influence of the crystal structure and tube distortion on the composite band structure.

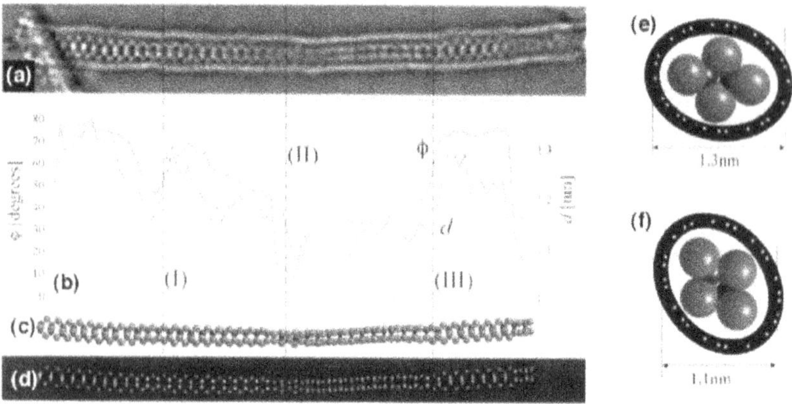

Figure 6. (a) reconstructed phase images of entire filled SWNT. (b) measured twist angle of 1D CoI_2 crystal (ϕ) and corresponding diameter ("d") measured along SWNT. (c) derived structure model. (d) multislice image simulation produced from (d). (d) and (f) end-on views of CoI_2 crystal fragments for minimum and near maximum measured "d" respectively.

2. Conclusions

The effect of the size of ultra-thin SWNT capillaries on the crystallisation of encapsulated molten binary or higher order species is to produce reduced or modified coordination structures relative to the corresponding bulk materials. The reason for this is that the proportion of surface atoms to body atoms within the bulk of the crystal is dramatically different in SWNT encapsulated crystals compared the same ratio for the corresponding bulk material. A simple illustration of this is given in Table 2 which shows the differences in ratios between body and surface atoms for crystals of increasing size. It follows from even such a simple analysis that for crystals such as the 2 ×g and 3 ×g3gKI crystals formed within SWNTs, surface properties are going to predominate over bulk properties as already indicated by the substantially expanded lattice spacings observed in these crystals.

Crystal size (atom layers or vol.)	No. of surface atoms (S)	No. of body atoms (B)	Ratio B/S
2× 2 × 2	8	0	∞
3 × 3× 3	26	1	26
4 × 4 × 4	56	8	7
5 × 5 × 5	98	27	3.62
1 × 1 × 1 mm^3 crystal	4.766×10^{13}	4.473×10^{19}	1.06×10^{-6}

Table 3. Ratio of surface atoms to body atoms.

The next stage for future studies of filled carbon SWNTs will be to determine the physical properties such as transport, mechanical, optical, and magnetic properties. Research towards these objectives is now in progress. The potential for application for filled SWNTs in the microelectronics industry is difficult to guage but should not be discounted. SWNT are indeed remarkable in many respects and their ability to encapsulate a *vast* number and variety of other materials is already certain. The next phase of the physical characterization of filled SWNTs may provide surprising, novel and potentially useful behaviour.

Acknowledgments

We are indebted to a large number of collaborators who have contributed to this project, including, in Oxford: David Wright, Miles Novotny, Neal Hanson, Sara Grosvenor, Eilidh Philp, Dr. Gareth Brown, Dr. Sam Bailey, Dr. Cigang Xu, Dr. Jens Hammer, Dr. Karl Coleman, Prof. Hee-Gweon Woo, Dr. Emmanuel Flahaut, Dr. Rüdiger Meyer and, in Cambridge: Dr. Rafal Dunin-Borowski and Dr. Owen Saxton. We acknowledge financial support from the EPSRC (Grant Nos. GR/L59238 and GR/L22324) and the Leverhulme Trust. J.S. is indebted to the Royal Society for a University Research Fellowship. S.F. is indebted to Hertford College for a Career Development Fellowship.

References

[1] Reproduced in R.P. Feynman, *Engin. Science*, 1960, **23**, 22.

[2] J. Sloan, A.I. Kirkland, J.L. Hutchison and M.L.H. Green, *J. Chem. Soc., Chem. Commun.*, 2002, 1319.

[3] D.A. Morgan, J. Sloan and M.L.H. Green, *J. Chem. Soc., Chem. Commun.*, 2002, 2442.

[4] S. Iijima and T. Ichihashi, *Nature*, 1993, **363**, 603.

[5] D.S. Bethune, C.H. Kiang, M.S. de Vries, G. Gorman, R. Savoy, J. Vazquez and R. Beyers, *Nature*, 1993, **363**, 605.

[6] P.M. Ajayan, J.M. Lambert, P. Bernier, L. Barbedette, C. Colliex and J.M. Planeix, *Chem. Phys. Lett.*, 1993, **215**, 509.

[7] T. Guo, P. Nikolaev, A. Thess, D.T. Colbert, R.E. Smalley, *Chem. Phys. Lett.*, 1995, **243**, 49.

[8] P. Nikolaev, M.J. Bronikowski, R. K. Bradley, F. Rohmund, D.T. Colbert, K.A. Smith, R.E. Smalley, *Chem. Phys. Lett.*, 1999, **313**, 91.

[9] P.M. Ajayan and S. Iijima, *Nature*, 1993, **361**, 6410.

[10] S.C. Tsang, Y.K. Chen, P.J.F. Harris and M.L.H. Green, *Nature*, 1994, **372**, 159.

[11] E.G. Bithell, A. Rawcliffe, S.C. Tsang, M.J. Goringe and M.L.H. Green, *Inst. Phys. Conf. Ser.*, 1995, **147**, 361.

[12] A. Chu, J. Cook, R.J.R. Heesom, J.L. Hutchison, M.L.H. Green and J. Sloan, *Chem. Mater.*, 1996, **8**, 2751.

[13] P.M. Ajayan, O. Stephan, P. Redlich and C. Colliex, *Nature*, 1995, **375**, 564.

[14] C. Guerret-Piécourt, Y. Le Bouar, A. Loiseau and H. Pascard, *Nature*, 1994, **372**, 761.

[15] J.J. Davis, R.J. Coles and H.A.O. Hill, *J. Electroanal. Chem.*, 1997, **279**, 440.

[16] J.J. Davis, M.L.H. Green, H.A.O. Hill, Y.C. Leung, P.J. Sadler, J. Sloan, A.V. Xavier and S.C. Tsang, *Inorg. Chim. Acta*, 1998, **272**, 261.

[17] J. Sloan, J. Hammer, M. Zweifka-Sibley and M.L.H. Green, *J. Chem. Soc., Chem. Commun.*, 1998, 347.

[18] B.W. Smith, D.E. Luzzi and Y. Achiba, *Chem. Phys. Lett.*, 2000, **331**, 137.

[19] K. Suenaga, M. Tence, C. Mory, C. Colliex, H. Kato, T. Okazaki, H. Shinohara, K. Hirahara, S. Bandow and S. Iijima, *Science*, 2000, **290**, 2280.

[20] J. Sloan, D.M. Wright, H.G. Woo, S. Bailey, G. Brown, A.P.E. York, K.S. Coleman, J.L. Hutchison and M.L.H. Green, *J. Chem. Soc.,Chem. Commun.*, 1999, 699.

[21] G. Brown, S.R. Bailey, M. Novotny, R. Carter, E. Flahaut, K.S. Coleman, J.L. Hutchison, M.L.H. Green and J. Sloan, *Appl. Phys. A*, 2003, **76**, 1-6.

[22] T.W. Ebbesen, *J. Phys. Chem. Solids*, 1996, **57**, 951.

[23] J. Sloan, M. Terrones, S. Nufer, S. Friedrichs, S.R. Bailey, H.G. Woo, M. Rühle, J.L. Hutchison and M.L.H. Green, *J. Am. Chem. Soc.*, 2002, **124**, 2116.

[24] J. Sloan, M.C. Novotny, S.R. Bailey, G. Brown, C. Xu, V.C. Williams, S. Friedrichs, E. Flahaut, R.L. Callendar, A.P.E. York, K.S. Coleman, M.L.H. Green, R.E. Dunin-Borkowski and J.L. Hutchison, *Chem. Phys. Lett.*, 2000, **329**, 61.

[25] R.R. Meyer, J. Sloan, R.E. Dunin-Borkowski, A.I. Kirkland, M.C. Novotny, S.R. Bailey, J.L. Hutchison and M.L.H. Green, *Science*, 2000, **289**, 1324.

[26] C. Xu, J. Sloan, G. Brown, S.R. Bailey, V.C. Williams, S. Friedrichs, K.S. Coleman, E. Flahaut, J.L. Hutchison, R.E. Dunin-Borkowski, M.L.H. Green, *J. Chem. Soc., Chem. Commun.*, 2000, 2427.

[27] J. Sloan, S. Grosvenor, S.R. Friedrichs, A.I. Kirkland, J.L. Hutchison, M.L.H. Green, *Angew. Chem. Int. Ed.*, 2002, 41, 1156.

[28] G. Brown, S.R. Bailey, J. Sloan, C. Xu, S. Friedrichs, E. Flahaut, K.S. Coleman, M.L.H. Green, J.L. Hutchison and R.E. Dunin-Borkowski, *J. Chem. Soc.,Chem. Commun.*, 2001, 845.

[29] J. Mittal, M. Monthioux, H. Allouche and O. Stephan, *Chem. Phys. Lett.*, 2001**, 339**, 311.

[30] S. Friedrichs, R.R. Meyer, J. Sloan, A.I. Kirkland, J.L. Hutchison and Malcolm L.H. Green, *J. Chem. Soc., Chem. Commun.*, 2001, 929.

[31] S. Friedrichs, J. Sloan, M.L.H. Green, J.L. Hutchison, R.R. Meyer and A.I. Kirkland, *Phys. Rev. B*, 2001, **64**, 045406/1.

[32] X. Fan, E.C. Dickey, P.C. Eklund, K.A. Williams, L. Grigorian, R. Buczko, S.T. Pantelides and S.J. Pennycook, *Phys. Rev. Lett.*, 2000, **84**, 4621.

[33] K. Suenaga, M. Tence, C. Mory, C. Colliex, H. Kato, T. Okazaki, H. Shinohara, K. Hirahara, S. Bandow and S. Iijima, *Science*, 2000, **290**, 2280.

[34] K. Hirahara, K. Suenaga, S. Bandow, H. Kato, T. Okazaki, H. Shinohara and S. Iijima, *Phys. Rev. Lett.*, 2000, **85**, 5384.

[35] R. R. Meyer, A. I. Kirkland and W. O. Saxton, *Ultramicroscopy*, 2003, **92**, 89.

[36] E. Philp, J. Sloan, A.I. Kirkland, R.R. Meyer, S. Friedrichs, J.L. Hutchison, M.L.H. Green, *Nature Mater.*, 2003, in press.

[37] C.L. Kane and E.J. Mele, *Phys. Rev. Lett.* 1997, **78**, 1932.

ELECTRON TRANSPORT IN CARBON NANOTUBE SHUTTLES AND TELESCOPES

I. M. Grace, S. W. Bailey and C. J. Lambert
Department of Physics, Lancaster University,
Lancaster, United Kingdom
c.lambert@lancaster.ac.uk

Abstract Multiwall carbon nanotubes (MWNT) are coaxial cylinders with low translational and rotational energy barriers allowing the inner tubes to easily slide with respect to the outer tubes. Using a tight binding, scattering technique, a computational study of the electron transport properties of MWNT's is presented. Two types of systems are investigated. The first (a shuttle system) consists of an outer single-wall nanotube (SWNT) connected to external reservoirs, inside of which is located a finite length SWNT of smaller diameter. The second (a telescoping system) consists of a SWNT inserted a finite distance into a larger-diameter SWNT, with the wider tube connected to one reservoir and the smaller-diameter tube connected to a second reservoir. For the first time, these structures are studied using a material-specific Hamiltonian derived from ab initio calculations based on density functional theory. Various commensurate systems are studied, including the (6,6)@(11,11) double wall nanotube.

Keywords: Carbon nanotubes, Multiwall carbon nanotubes, transport

1. Introduction

Carbon nantotubes[1] are narrow graphitic cylinders which possess remarkable electronic and mechanical properties[2]. These make them promising candidates for building nanoscale electronic circuits. Multiwall carbon nanotubes (MWNT) consist of concentric shells nested inside one another with an interwall spacing of 3.4Å, which is similar to the inter-layer spacing in graphite. Due to a weak interaction between the walls, these multi-wall tubes exhibit a rich variety of electronic properties.

The electronic properties of a single wall nanotube (SWNT) are primarily determined by its chirality, which can cause it to be either metallic or semiconducting[3]. The properties of a MWNT are additionally determined by the inter-wall interaction which can significantly alter the electronic structure around the Fermi energy, causing the formation of pseudogaps[4]. This inter-

A.S. Alexandrov et al. (eds.), Molecular Nanowires and Other Quantum Objects, 89–94.

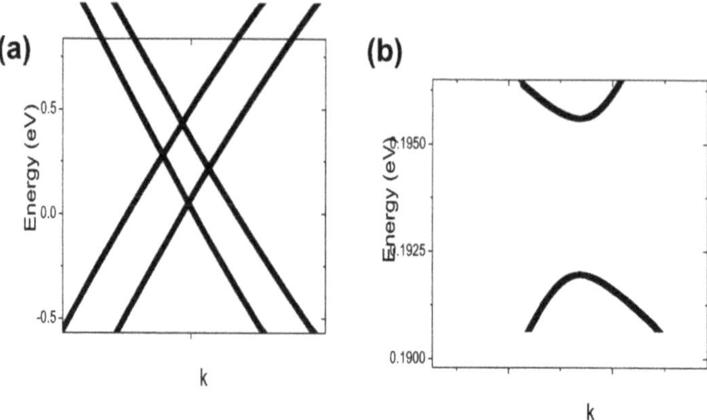

Figure 1. (a) (6,6)@(11,11) bandstructure close to the Fermi Energy (0eV). (b) Gap opening due to interwall interaction

action between walls causes similar effects in bundles of single wall tubes [5]. Conductance measurements on multiwall tubes show a range of behaviour for transport properties. Some show ballistic transport with the suggestion that current is carried predominately by the outer walls[6][7]. Others show that current passes between the walls[8], while another shows diffusive transport and Luttinger behaviour[9]. Recent experiments have shown that it is possible to slide the inner-walls of a MWNT in a 'telescoping' like motion[10]. This is made possible by the low energy barriers for sliding the inner-walls, and may allow for new types of nanomechanical systems such as a novel gigahertz oscillator [11]

In this paper we calculate the transport properties of a double wall structure. The systems studied consist of a finite MWNT contacted to two semi-infinite single wall leads. The focus here is on metallic armchair nanotubes, primarily the (6,6)@(11,11), this has the preferred interwall separation of 3.4Å. This system is representative of most armchair MWNTs, as it does not possess axial symmetry. An infinite single wall armchair nanotube is predicted to be metallic and has two bands crossing near the Fermi energy, this leads to a conduction of $2G_0$ ($G_0=2e^2/h$)[12]. Multiwall nanotubes behave as conductors in parallel and in the absence of inter-wall interactions the conductance is simply a sum of the conductance of each of the SWNTs. Therefore in the absence of inter-wall coupling, the conductance close to the Fermi energy of a clean double wall structure would be $4G_0$.

To calculate the transport properties in finite MWNT we use a previously developed recursive Greens function scattering technique[13], together with a tight binding Hamiltonian to model the nanotube. The Hamiltonian is generated

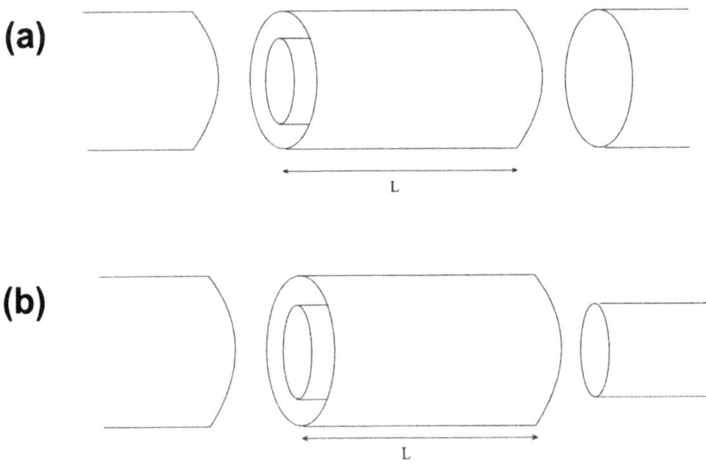

Figure 2. (a) Finite Multiwall region contacting the outer walls to two single wall nanotubes. (b) One side of scattering region contacted by outer wall to lead, the other to the inner wall.

using the first principles density functional theory code, SIESTA[14]. In what follows, we use the local density approximation parameterized by Perdue and Zunger[15] and nonlocal norm-conserving pseudopotentials[16]. The valence electrons were described by a single-ζ basis set and the cutoff radius for the s and p orbital was chosen to be 4.1 a.u. Using this Hamiltonian, the calculated bandstructure of an infinite (6,6)@(11,11) MWNT is shown in Figure 1. This shows that close to the Fermi energy there are four bands, which have small pseudogaps of the order 2meV, opening at the band crossings. The size of these gaps are much smaller than those seen in the (5,5)@(10,10)[4], in agreement with predictions[17] for double wall structures with no axial symmetry.

In what follows two different systems are investigated. For the finite system referred to as a 'shuttle', single wall leads are attached to both sides of the outer wall of the finite region shown in Figure 2(a), which contains an inner tube of finite length L. In the second system, called a 'telescope', the left lead is attached to the outer wall and the right lead is attached to inner tube (figure 2b).

2. Telescoping Nanotubes

First we look at the telescoping (6,6)@(11,11) double wall nanotube. In this case the only connection between the two leads is via the interwall interaction in the scattering region. The total transmission T(E)=G(E)/($2e^2$/h) is plotted for different length scattering regions and are shown in Figure 3. For each of these

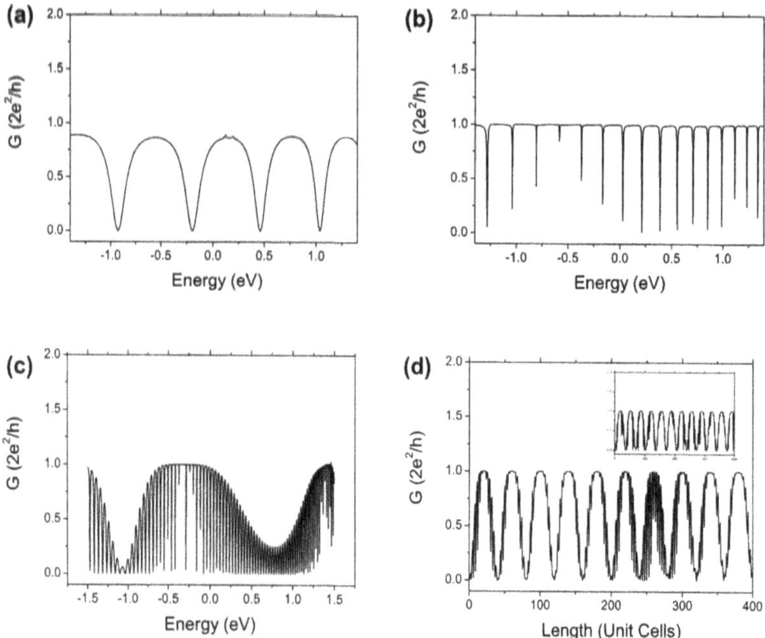

Figure 3. Conductance of the telescoping (6,6)@(11,11) nanotube as a function of energy for scattering length 10 unit cells(a), 50 unit cells (b) and 250 unit cells (c).(d) conductance as a function of length for varying strength interwall interactions

plots, the conductance varies with length and has a maximum value of $1G_0$ in the region around the Fermi energy. Since a single wall nanotube has two open channels, figure 3 suggests that one of the channels is closed between the walls (this suppression of one of the transmission channels is possibly because the (6,6)@(11,11) has no axial symmetry). For the short scattering region (figure 3(a)) with a length of 10 unit cells, we find an almost periodic sequence of antiresonance dips. As the length of the scattering region is increased the period of these antiresonances decreases (figures 3 (b) and (c)). Antiresonances have also been observed in previous calculations on the telescopic nanotubes [18] and nanotubes enclosing fullerenes[19]. For much larger overlap regions the transmissions become modulated with a longer period in agreement with previous calculations carried out on telescoping nanotubes[18].

The transmission values at the Fermi energy were also calculated with respect to the number of slices of nanotube in the overlap region. Here we can see (Figure 3(d)) that the conductance oscillates smoothly between 0 and $1G_0$. As shown in the inset the period of this oscillation depends on the strength of the interwall interaction, which can be increased by changing the cutoff radius of the orbitals. In this case the cutoff has been increased to 5.1 a.u. for both the s and p orbital when generating the Hamiltonian. This stronger interaction leads to a shorter period of oscillation.

3. Shuttle Nanotubes

We now compare the above behaviour with that of a shuttle. Figure 4 shows that in this case the transmission varies between $2G_0$ and $1G_0$ due to the fact there are two open channels in the leads, and the transport takes place mainly via the outer wall of the system. There is again one channel closed for transport between the walls as was the case in the telescopic case. The conductance properties show similar results to the telescoping system, i.e. there are anti-resonances whose period become shorter as the length of the scattering region increases and again shows beating behaviour for the longer MWNTs.

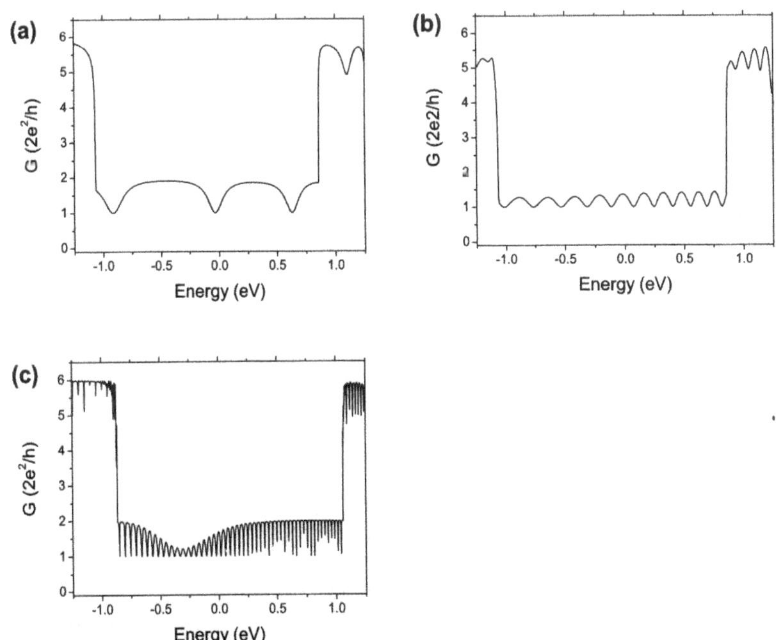

Figure 4. Conductance of the Shuttle (6,6)@(11,11) nanotube. (a) 10 unit cells. (b) 50 unit cells. (c) 250 unit cells.

4. Conclusions

We have calculated the conductance in both shuttle and telescoping finite MWNT. For the first time, these structures have been studied using a material-specific Hamiltonian derived from ab initio calculations based on density functional theory.

References

[1] S. Iijima, Nature 354, 56 (1991)

[2] R. Saito,G. Dresselhaus, M.S. Dresselhaus, Physical Properties of Carbon Nanotubes. Imperial College London, 1998.

[3] J. W. Mintmire, B. I. Dunlap, and C. T. White, Phys. Rev. Lett. 68, 631 (1992)

[4] Y. Kwon, and D. Tomanek, Phys. Rev B, 58, R16001 (1998)

[5] Y. Kwon, S. Saito, and D. Tomanek, Phys. Rev B, 58,R13314 (1998)

[6] S. Frank, P. Poncharal, Z. I. Wang, and W.A. de Heer, Science 280, 1744 (1998)

[7] A. Urbina, et al, Phys. Rev. Lett. 90, 106603 (2003)

[8] P. G. Collins, M. S. Arnold, and P. Avouris, Science, 292, 706 (2001)

[9] C. Schonenberger, A.Bachtold, C. Strunk, J. P. Salvelat, L. Forro, Appl. Phys. A 69, 283 (1999)

[10] J. Cummings and A. Zettl, Science 289, 602 (2000)

[11] Q. Zheng, Q. Jiang, Phys. Rev. Lett. 88, 045503 (2002)

[12] L. Chico, L. X. Benedict, S. G. Louie, and M. L. Cohen, Phys. Rev. B 54, 2600 (1996)

[13] S. Sanvito, C. J. Lambert, J. Jefferson, and A. Bratkovsky, Phys. Rev. B, 59, 11936 (1999)

[14] J. M. Soler, E. Artacho, J. D. Gale, A. García, J. Junquera, P. Ordejón and D. Sánchez-Portal, J. Phys.: Condens. Matter 14, 2745-2779 (2002)

[15] J.P. Perdew and A. Zunger, Phys. Rev. B 23, 5048 (1981)

[16] N. Troullier and J.L. Martin, Phys. Rev. B 43, 1993 (1991)

[17] Ph. Lambin, V. Meunier, and A. Rubio, Phys. Rev. B 62, 5129 (2000)

[18] D.-H. Kim, and K.J. Chang, Phys. Rev. B 66, 155402 (2002)

[19] D.-H. Kim, H.-S. Sim, and K.J. Chang, Phys. Rev. B 64, 115409 (2001)

ARGUMENTS FOR QUASI-ONE-DIMENSIONAL ROOM TEMPERATURE SUPERCONDUCTIVITY IN CARBON NANOTUBES

Guo-meng Zhao

Department of Physics and Astronomy, California State University at Los Angeles, Los Angeles, CA 90032, USA

gzhao2@calstatela.edu

Abstract In this article and references herein, I provide over twenty arguments for room temperature superconductivity in carbon nanotubes. The one-dimensionality of the nanotubes complicates the right-of-passage for prospective quasi-one-dimensional superconductors. The Meissner effect for individual tubes is less visible because the diameters of the tubes are much smaller than the penetration depth. Zero resistance is less obvious because of the quantum contact resistance and significant quantum phase slip, both of which are associated with a finite number of transverse conduction channels. Nonetheless, on-tube resistance at room temperature has been found to be indistinguishable from zero for many individual multi-walled nanotubes and a large Meissner effect has been observed in large bundles of multi-walled nanotubes at room temperature. On the basis of these arguments, I suggest that carbon nanotubes deserve to be classified as room temperature superconductors.

Keywords: room-temperature superconductivity, carbon nanotubes

1. Introduction

Finding room temperature (RT) superconductors is one of the most challenging problems in science. In 1999, Tsebro et al. [1] observed a substantial remnant magnetization in multi-walled carbon nanotubes (MWNTs) up to room temperature. Such a result is consistent with either RT superconductivity or ferromagnetism. Without ruling out or even discussing ferromagnetism, Tsebro et al. speculated that the data are consistent with RT superconductivity. In November 2001, Zhao and Wang [2] provided subtle experimental evidence for RT superconductivity (SC) in MWNTs. Since then, Zhao [3, 4, 5, 6] has provided over twenty arguments for RTSC in both single-walled and multi-walled carbon nanotubes. Very recently, Kopelevich and coworkers [7] have given evidence for RTSC in graphite-sulfur composites.

A.S. Alexandrov et al. (eds.), Molecular Nanowires and Other Quantum Objects, 95–106.

It is generally believed that the superconducting transition temperature T_c cannot be higher than 30 K within the conventional phonon-mediated mechanism although there is no theoretical justification for the T_c limit. Alexandrov and Mott have demonstrated that Bose-Einstein condensation of bipolarons in polar materials can lead to a T_c higher than 100 K [8]. Ginzburg [9] and Little [10] have proposed that high-temperature superconductivity could be realized by exchanging high-energy electronic excitations such as excitons and plasmons. Lee and Mendoza have shown that superconductivity as high as 500 K can be reached through a pairing interaction mediated by undamped acoustic plasmon modes in a quasi-one-dimensional (1D) electronic system [11]. Moreover, high-temperature superconductivity can occur in a multi-layer electronic system due to an attraction of charge carriers in the same conducting layer via exchange of virtual plasmons in neighboring layers [12]. If these theoretical studies are relevant, one should be able to find high-temperature superconductivity in quasi-one-dimensional and/or multi-layer systems such as cuprates, carbon nanotubes and graphite-sulfur composites. In this article, I will review some arguments for RTSC especially in single-walled nanotubes (SWNTs). Over twenty detailed arguments for RTSC in both SWNTs and MWNTs, and alternative explanations to some data have been given in Refs. [3, 4, 5, 6].

2. Electrical transport and tunneling spectra in SWNTs

It is well known that, for quasi-1D systems disorder has extremely strong effects on electrical transport [13, 14]. For a noninteracting 1D system, all states get localized in the presence of an infinitesimal random potential [15]. Interactions can modify this picture, as shown clearly from theoretical works on one-chain and two-chain systems on the basis of bosonization and renormalization-group techniques [13, 14]. It is shown that metal-like conductivity occurs only for strongly attractive interactions that lead to quasi-1D s-wave high-temperature superconductivity [13, 14]. In contrast, for repulsive interactions d-wave superconductivity would occur for the pure system and, in the presence of disorder and/or impurities, the resistivity decreases monotonically with increasing temperature (semiconductor-like behavior) [14]. Thus, quasi-1D d-wave superconductors due to repulsive interactions behave like insulators at low temperatures [14]. Alternatively, the metal-like conductivity below a mean-field superconducting transition temperature T_{c0} in quasi-1D s-wave superconductors can be understood in terms of quantum phase slips (QPSs). Indeed, QPS theories [16, 17] can well explain a number of experiments that show a metal-like temperature dependence of the resistance well below T_{c0} in thin superconducting wires [16, 18, 19].

Within a QPS theory [17], the intrinsic on-wire (on-tube) resistance $R_i \propto T^{2\mu-3}$ for $\Phi_o I/ck_B << T < 0.5T_{c0}$, and $R_i \propto I^{2\mu-3}$ (independent of temperature) for $T << \Phi_\circ I/ck_B$. Here $\Phi_\circ$ is the quantum flux, c is the speed of light, I is the measuring current, and μ is a quantity that characterizes the ground state. The zero temperature resistance can approach zero when $\mu > 2$ and the current approaches zero, but is finite when $\mu \leq 2$. When μ <1.5, R_i increases with decreasing temperature (insulating behavior), while for $\mu > 1.5$, R_i decreases with decreasing temperature (metallic behavior). Only if the QPSs are strongly suppressed, can zero on-tube resistance be approached below T_{c0}.

Considering the contact resistance and the on-tube resistance due to QPSs, we obtain the two-probe resistance of a superconducting tube:

$$R(T) = R_0 + aT^p. \tag{1}$$

Here $p = 2\mu - 3$, $R_0 = R_Q/tN_{ch} + 2R_c$, t is the transmission coefficient ($t \leq 1$), $R_Q = h/2e^2$ =12.9 kΩ is the resistance quantum, and R_c is the contact resistance. The coefficient a normalized by the tube length L is given by [17]

$$a/L \propto \exp(-2S_{core}), \tag{2}$$

where

$$S_{core} \propto [N(E_F)]^{1/3}\sigma^{2/3}\Delta^{2/3}, \tag{3}$$

where $N(E_F)$, σ and Δ are the density of states at the Fermi level, the normal-state conductivity and the superconducting gap, respectively. It is apparent that the quantity a/L is very sensitive to S_{core} and thus to σ when S_{core} is significant. Thus, a/L should strongly depend on disorder of tubes.

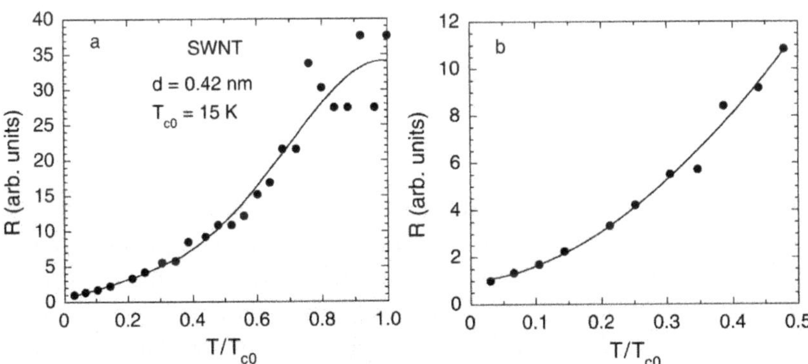

Figure 1. a) The resistance as a function of T/T_{c0} for the smallest diameter SWNTs with d = 0.42 nm. The data are extracted from Ref. [20]. b) The temperature dependence of the resistance below $0.5T_{c0}$.

Now we compare the QPS theory with the resistance data of SWNTs that are known to exhibit quasi-1D superconductivity. For the smallest diameter

SWNTs with $d = 0.42$ nm, T_{c0} was found to be about 15 K from both magnetic and electrical measurements [20]. In Fig. 1a, we plot the two-probe resistance as a function of T/T_{c0} for this SWNT. It is apparent that the resistance increases more rapidly above $0.5T_{c0}$ and flattens out towards T_{c0}. Above T_{c0}, the resistance decreases with increasing temperature [20]. The broad transition and finite resistance at any finite temperatures below T_{c0} are caused by a large superconducting fluctuation in quasi-1D systems [20]. Zero on-tube resistance state is only possible at zero temperature [20, 17].

As demonstrated in Fig. 1b, the temperature dependence of the resistance below $0.5T_{c0}$ can be well fit by Eq. 1. From the fit, we find that $p = 1.77 \pm 0.18$. This suggests that the QPS theory can indeed explain the quasi-1D superconductivity in this SWNT. Using $p = 1.77$ and $p = 2\mu - 3$, we obtain $\mu = 2.4 >$

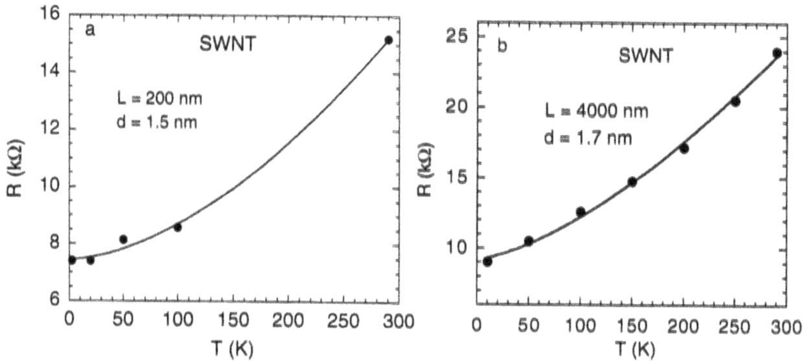

Figure 2. The temperature dependences of the resistance at zero gate voltage for two SWNTs with $L = 200$ and 4000 nm. The data are extracted from Refs. [21, 22].

In Fig. 2, we show the temperature dependences of the resistance at zero gate voltage for two metallic chirality SWNTs with different lengths ($L = 200$ and 4000 nm, respectively) and similar diameters ($d = 1.5$ and 1.7 nm, respectively). The contacts to the nanotubes are nearly ideal with the transmission probability close to 1 (Ref. [21, 22]). In this case, the two-probe resistance approaches 6.45 kΩ if the on-tube resistance approaches zero. It is remarkable that the temperature dependences of the resistance can be well fit by Eq. 1 with $p = 1.71 \pm 0.23$ for $L = 200$ nm and $p = 1.48 \pm 0.11$ for $L = 4000$nm. The exponents are very close to the one ($p = 1.77 \pm 0.18$) found for the smallest SWNTs with $T_{c0} = 15$ K. If we use the fixed parameter $p = 1.48$ for both $L = 200$ and 4000 nm tubes, the best fits show that the coefficient a/L for $L = 200$ nm tube is about 10 times larger than that for $L = 4000$ nm. Such a large difference in the a/L values of the two tubes can be only explained by Eq. 2 and Eq. 3 that are predicted from QPS theory for quasi-1D s-wave superconductivity. This is because a small difference in the normal-state conductivity can lead to a huge difference in the a/L values according to Eq. 2 and Eq. 3. On the other hand, due to phonon backscattering and/or electron umklapp scattering in a Luttinger liquid (LL) with repulsive interactions, the on-tube resistivity is theoretically shown to exhibit a semiconductor-like behavior at low temperatures and a power-law behavior with $p \simeq 0.6$ at high temperatures [23, 24]. None of the theoretically predicted features is consistent with the data in Fig. 2. Further, the inelastic mean free path at room temperature for L = 200 nm tube would be about 160 nm if the tube were not a quasi-1D s-wave RT superconductor [6]. Such a short inelastic mean free path at RT for L= 200 nm tube and a huge difference in the a/L values of the two similar tubes are not consistent with any inelastic scattering processes in carbon nanotubes.

For another tube with L = 800 nm, the resistance at zero gate voltage is independent of temperature below 270 K (Ref. [21]), i.e., $p = 2\mu - 3 \simeq 0$. It was also shown that [21] underdoping leads to $p < 0$ while overdoping gives rise to $p > 0$. This doping dependence of p is only consistent with the QPS theory. Within the QPS theory [17], $\mu \simeq 50r/\lambda_L \propto \sqrt{n}$ (where λ_L is the London penetration depth and n is the carrier density). It is clear that μ increases with n such that p can cross over from a negative value at low doping to a positive one at high doping, in agreement with experiment [21].

Now we discuss tunneling spectra. From a single-particle tunneling spectrum obtained through two high-resistance contacts (see Fig. 1b of Ref. [25]), we can clearly see a pseudo-gap feature that appears at an energy of about 220 meV. Having ruled out the LL behavior (see the above discussion), we could relate the pseudo-gap feature to the superconducting gap smeared by QPSs. Considering the broadening of the gap feature due to QPSs and the double tunneling junctions in series, we estimate the superconducting gap $\Delta(0)$ to be about 100 meV. The scanning tunneling microscopy and spectroscopy [26] on individual single-walled nanotubes also show pseudo-gap features with $\Delta(0) > 100$ meV in doped metallic SWNTs. Using $k_B T_{c0} = \Delta(0)/1.76$, we find $T_{c0} \simeq 660$ K.

This gap value deduced from the tunneling spectrum is in good agreement with electrical transport data for a SWNT mat (see Fig. 3). Below 200 K the resistivity is nearly temperature independent while above 200 K the resistivity increases suddenly and starts to flatten out above 550 K. Such a temperature dependence is expected for granular superconductors [28] where superconducting islands are separated by non-superconducting materials and/or voids. For the SWNT mat, the nonsuperconducting materials are semiconducting chirality tubes. If the resistivity jump above 200 K were due to inelastic scattering, then the inelastic mean free path at 600 K would be less than 2 nm according to the resistivity difference between 600 K and 200 K. Such a short inelastic mean

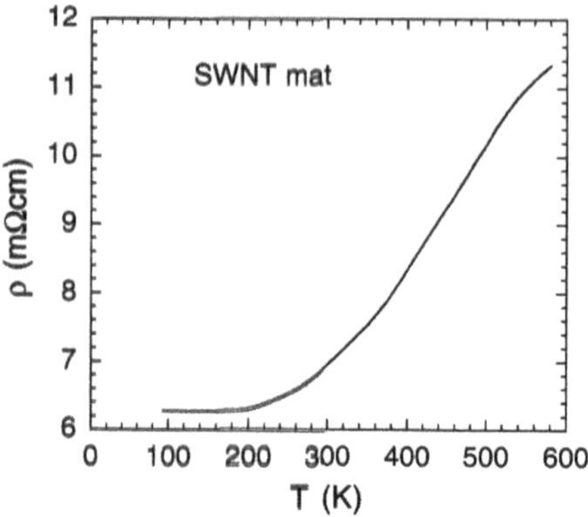

Figure 3. Temperature dependence of the resistivity for a SWNT mat with $T_{c0} \simeq 640$ K. The data are extracted from Ref. [27].

free path at 600 K is not compatible with any inelastic scattering processes in carbon nanotubes. Alternatively, by comparing Fig. 3 with Fig. 1a, we find that this temperature dependence of the resistance agrees with quasi-1D superconductivity with a $T_{c0} \simeq 640$ K.

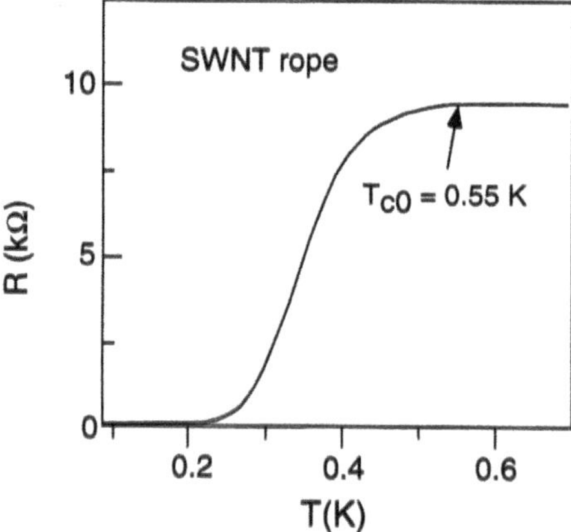

Figure 4. The temperature dependence of the resistance for a SWNT rope with $T_{c0} = 0.55$ K. The figure is reproduced from Ref. [29].

The electrical transport and single-particle tunneling spectra thus consistently suggest that T_{c0} in SWNTs can be higher than 500 K. Nevertheless, a much lower T_{c0} ($\sim$0.55 K) was observed in a SWNT rope [29]. The rope consists of about 400 individual SWNTs with $d \simeq 1.5$ nm. Fig. 4 shows the resistance data for the SWNT rope. One can see that the resistance starts to drop below about 0.55 K. It is interesting to note that the resistivity of the rope decreases with increasing temperature up to 300 K in the normal state, and that the elastic mean free path just above T_{c0} is very short (8 nm) [29], similar to the case of the smallest SWNTs [20]. This suggests that these nanotubes behave like semiconductors in the normal state and like metals in the superconducting state. Very low superconductivity in this SWNT rope may be due to the fact that the tubes are very lightly doped. This is in agreement with the tunneling spectrum, which shows no pseudo-gap feature at 5 K in an undoped or very lightly doped armchair tube [30].

3. Raman scattering in a SWNT rope

It is known that Raman scattering has provided essential information about the electron-phonon coupling and the electronic pair excitation energy in the high-T_c cuprate superconductors [32, 33, 34]. The anomalous temperature-dependent broadening of the Raman active B_{1g}-like mode of 90 K superconductors $RBa_2Cu_3O_{7-y}$ (R is a rare-earth element) allows one to precisely determine the superconducting gap [33]. The pronounced softening observed only for the B_{1g} mode is due to the fact that the phonon energy of the B_{1g} mode is

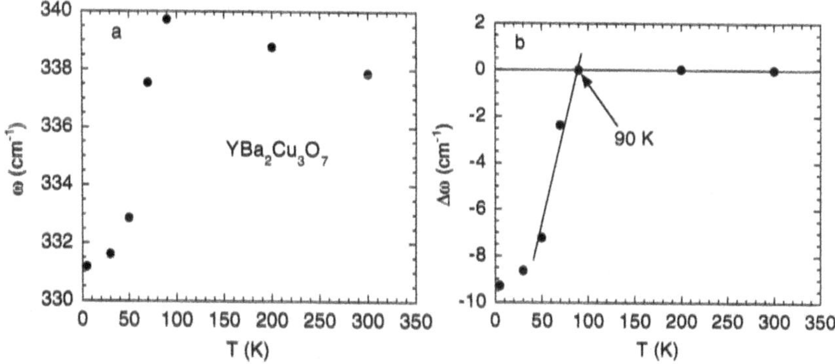

Figure 5. a) Temperature dependence of the frequency for the Raman-active B_{1g} mode of a 90 K superconductor $YBa_2Cu_3O_{7-y}$. The data are extracted from Ref.[32]. b) The difference between the measured frequency and the linearly fitted curve above T_c.

The temperature dependence of the frequency for the Raman-active B_{1g} mode of a 90 K superconductor $YBa_2Cu_3O_{7-y}$ is shown in Fig. 5a. It is apparent that the frequency decreases linearly with increasing temperature above T_c and that the mode starts to soften below about $0.95T_c$. The temperature dependence of the frequency above T_c is caused by thermal expansion. The temperature dependence of the frequency will become more pronounced at higher temperatures because the magnitude of the slope $-d\ln\omega/dT$ is essentially proportional to the lattice heat capacity that increases monotonically with temperature. The significant softening of the mode below T_c occurs only if the energy of the Raman mode is very close to $2\Delta(0)$ and the electron-phonon coupling is substantial [35], as it is the case in the 90 K superconductor $YBa_2Cu_3O_{7-y}$ [32, 33, 34]. In order to see more clearly the softening of the mode, we show in Fig. 5b the difference of the measured frequency and the

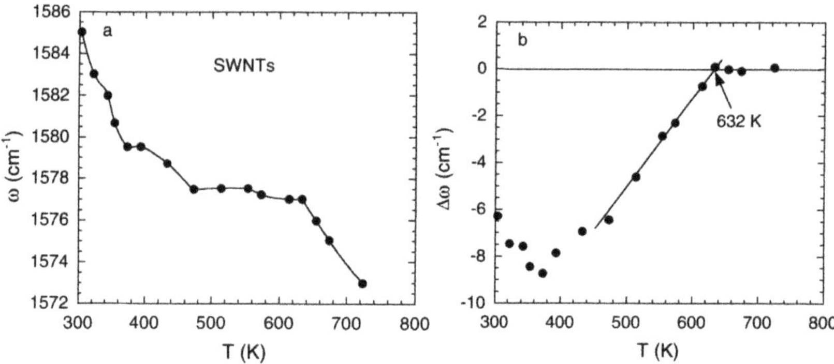

Figure 6. a) Temperature dependence of the frequency for the Raman active G-band of a SWNT rope. The data are from Ref. [31]. b) The difference of the measured frequency and the linearly fitted curve above the kink temperatures (see text).

Fig. 6a shows the temperature dependence of the frequency for the Raman active G-band of a SWNT rope. It is striking that the frequency data show a clear tendency of softening below about 630 K. Above 630 K, the frequency decreases linearly with increasing temperature with a slope much larger than that in $YBa_2Cu_3O_{7-y}$. This is due to a much larger thermal expansion at such high temperatures in SWNTs. The slope will decrease with decreasing temperature and eventually level off at low temperatures. The leveling off between 450-630 K must be due to the competition between the mode hardening due to the normal thermal effect and the mode softening possibly due to a superconducting transition or a transition to charge/spin density wave. The transition to charge/spin density wave will lead to a metal-insulator transition upon cooling, in contrast to the transport data in Fig. 3.

In order to see more clearly the softening of the mode, we show in Fig. 6b the difference between the measured frequency and the linearly fitted curve above the kink temperature (~630 K). One can see that the result shown in Fig. 6b is similar to that shown in Fig. 5b. This suggests that the softening of the Raman active G-band in the SWNTs may have the same microscopic origin as the softening of the Raman active B_{1g} mode in $YBa_2Cu_3O_{7-y}$. This explanation is plausible only if the phonon energy of the G-band is very close to

$2\Delta(0)$. We are fortunate that the phonon energy of the G-band is 197 meV, very close to $2\Delta(0) \simeq 200$ meV deduced above from the tunneling spectrum and the electrical transport experiment. Therefore, it is very likely that the softening of the Raman mode in the SWNTs is related to a superconducting phase transition.

From Fig. 6, we can clearly see that the softening starts at about 632 K. Using the fact that the softening starts at $0.95T_{c0}$ (Ref. [35]), we can assign the mean-field transition temperature $T_{c0} = 665$ K. It is interesting to note that there is a clear minimum at $T^* = 370$ K $= 0.57T_{c0}$ in Fig. 6b. It is remarkable that such a minimum is also seen at about $0.6T_{c0}$ in a calculated curve for a superconductor (see Fig. 8 of Ref. [35]). This shallow minimum in the frequency shift is related to a sharp minimum in the real part of the polarization $\Pi(\omega, T)$, which occurs at $\hbar\omega/2\Delta(T^*) = 1$ for weak coupling [35, 36]. From the temperature dependence of the BCS gap [37], we find that $\Delta(T^*) = \Delta(0.57T_{c0}) = 0.93\Delta(0)$. With $\hbar\omega/2\Delta(T^*) = 1$ and $\hbar\omega = 197$ meV, we finally obtain $\Delta(0) = 105$ meV, which is in excellent agreement with the value deduced from the tunneling spectrum [4]. Then we calculate $2\Delta(0)/k_BT_{c0} = 3.66$, which is in excellent agreement with the prediction based on the plasmon-mediated pairing mechanism [11]. Moreover, the electron-phonon coupling constant of the hardest optic mode deduced from the Raman data is in quantitative agreement with an independent theoretical prediction [6].

4. Meissner effect and Remnant Magnetization in MWNTs

It is well known that the diamagnetic susceptibility of graphite is very small when the magnetic field is along the plane. It is also shown [38] that the diamagnetic susceptibility is small ($\sim 10^{-6}$ emu/g) when the field is along the tube axis direction. This theoretically predicted value is about one order of magnitude smaller than the measured one (-1.1×10^{-5} emu/g) [39]. This large discrepancy can be naturally resolved if the MWNTs are superconductors that contribute to diamagnetism due to the Meissner effect.

For tubes of radius r with magnetic field parallel to the tube axis direction, the zero-temperature diamagnetic susceptibility due to the Meissner effect is given by

$$\chi_{\parallel}(0) = -\frac{\bar{r^2}}{32\pi\lambda_\theta^2(0)}. \tag{4}$$

Here r is the radius of tubes, $\bar{r^2}$ is the average value of r^2, and $\lambda_\theta(0)$ is the penetration depth when the screening current is along the circumferential direction. The above equation is valid only if $\lambda_\theta(0)$ is larger than the maximum radius of tubes. If we assume that the radii of tubes are equally distributed from 0 to 100 Å, we find $\bar{r^2}$= 6000 Å^2. With the weight density of 2.17 g/cm^3

[40] and $\chi_{\parallel}(0) = -0.93 \times 10^{-5}$ emu/g [39] (we have subtracted the normal-state diamagnetic susceptibility), we calculate $\lambda_\theta(0) \simeq 1724$ Å. This value of the penetration depth corresponds to n/m_θ^* = $0.96\times10^{21}/\mathrm{cm}^3 m_e$, where m_θ^* is the effective mass of carriers along the circumferential direction. With n = $1.6\times10^{19}/\mathrm{cm}^3$ (Ref. [41]), we find that $m_\theta^* \simeq 0.017\ m_e$, which is larger than the in-plane effective mass of 0.012 m_e for graphites [42]. Since the effective mass along the tube axis direction is nearly zero [43], the value of m_θ^* indicates

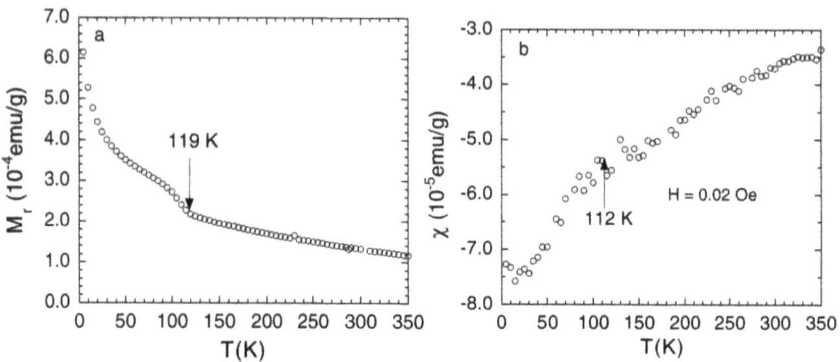

Figure 7. a) Temperature dependence of the remnant magnetization for multi-walled nanotube mat. b) The field-cooled susceptibility as a function of temperature in a field of 0.020 Oe. After Ref. [2].

To further show that MWNTs are room-temperature superconductors, we show in Fig. 7 the temperature dependencies of the remnant magnetization M_r and the diamagnetic susceptibility for our MWNT ropes. It is apparent that the temperature dependence of M_r (Fig. 7a) is similar to that of the diamagnetic susceptibility (Fig. 7b) except for the opposite signs. This behavior is expected for a superconductor. Although the M_r was also observed by Tsebro *et al.* up to 300 K [1], the observation of M_r alone does not give unambiguous evidence for RT superconductivity since such a M_r could be caused by ferromagnetic impurities.

We now rule out the existence of ferromagnetic impurities. If there were ferromagnetic impurities, the total susceptibility would tend to turn up below 120 K where the M_r increases suddenly. This is because paramagnetic susceptibility and M_r should increase simultaneously for ferromagnetic impurities. In contrast, the susceptibility suddenly turns down rather than turns up below 120 K (Fig. 7b). This provides strong evidence that the observed M_r in our MWNTs is not associated with the presence of random ferromagnetic impurities but with superconductivity. Moreover, the anomalies at about 120 K in the diamagnetic susceptibility and remnant magnetization correspond to the anomalies in both the conductance and Hall coefficient [3]. All these results can be consistently explained by RTSC [3].

5. Concluding remarks

Although the present article and Refs. [2, 3, 4, 5, 6] have provided over twenty arguments for RTSC in carbon nanotubes, and we cannot find any published data that are contradictory with the RTSC explanation, the ultimate proof of RTSC in carbon nanotubes should be the observations of a large Meissner effect and zero resistance at RT. As discussed above, the Meissner effect for isolated tubes is intrinsically small due to the fact that the penetration depth is much larger than the tube diameters. For Josephson-coupled large bundles of superconducting MWNTs, there should be a substantial Meissner effect. Indeed, I was told in this ARW conference that a Russian group has observed a large Meissner effect at RT in large bundles of MWNTs. Further, a negligible on-tube resistance ($<86\ \Omega/\mu$m) has been observed at RT in the majority of individual MWNTs [44, 45]. Another independent group [46] has also observed nearly zero on-tube resistance over a 4 μm long MWNT at RT ($<1.0\ \Omega/\mu$m). These results cannot be explained by ballistic transport at RT [6], but are consistent with quasi-1D RTSC.

Acknowledgments

I am grateful to Dr. R. Walter *et al.* for sending me their unpublished data. I thank Dr. Pieder Beeli for stimulating discussions.

References

[1] V. I. Tsebro, O. E. Omelyanovskii, and A. P. Moravskii, JETP Lett. **70**, 462 (1999).

[2] G. M. Zhao and Y. S. Wang, cond-mat/0111268.

[3] G. M. Zhao, cond-mat/0208197.

[4] G. M. Zhao, cond-mat/0208198.

[5] G. M. Zhao, cond-mat/0208200.

[6] G. M. Zhao, cond-mat/0208201.

[7] Y. Kopelevich et al., Physica C (in press).

[8] A. S. Alexandrov and N. F. Mott, *Polarons and Bipolarons* (World Scientific, Singapore, 1995).

[9] V.L. Ginzburg, in: V.L. Ginzburg, D. A. Kirzhnits Eds., *High-Temperature Superconductivity*, Consultants Bureau, New York, 1982.

[10] W. A. Little, Phys. Rev. **164**, A1416 (1964).

[11] Y.C. Lee and B. S. Mendoza, Phys. Rev. B **39**, 4776 (1989).

[12] S. M. Cui and C. H. Tsai, Phys. Rev. B **44**, 12500 (1991).

[13] T. Giamarchi and H. J. Schulz, Phys. Rev. B **37**, 325 (1988).

[14] E. Orignac and T. Giamarchi, Phys. Rev. B **56**, 7167 (1997).

[15] A. A. Abrikosov and J. A. Rhyzkin, Adv. Phys. **27**, 147 (1978).

[16] N. Giordano, Phys. Rev. B **41**, 6350 (1990).

[17] A. D. Zaikin *et al.*, Phys. Rev. Lett. **78**, 1552 (1997).

[18] N. Giordano and E. R. Schuler, Phys. Rev. Lett. **63**, 2417 (1989).

[19] N. Giordano, Phys. Rev. B **43**, 160 (1991).

[20] Z. K. Tang *et al.*, Science, **292**, 2462 (2001).

[21] J. Kong *et al.*, Phys. Rev. Lett. **87**, 106801 (2001).

[22] D. Mann et al., cond-mat/0309044.

[23] A. Komnik and R. Egger, cond-mat/9906150 (1999).

[24] C. Kane, L. Balents, and M. P. A. Fisher, Phys. Rev. Lett. **79**, 5086 (1997).

[25] Z. Yao, C. L. Kane, and C. Dekker, Phys. Rev. Lett. **84**, 2941 (2000).

[26] J. W. G. Wildoer *et al.*, Nature (London) **391**, 59 (1998).

[27] R. S. Lee *et al.*, Nature (London) **388**, 255 (1997).

[28] L. Merchant *et al.*, Phys. Rev. B **63**, 134508 (2001).

[29] M. Kociak *et al.*, Phys. Rev. Lett. **86**, 2416 (2001).

[30] M. Ouyang *et al.*, Science **292**, 702 (2001).

[31] R. Walter *et al.*, Bull Am. Phys. Soc. **47**, 361 (2002); The Raman data were sent by Dr. R. Walter from the University of North Carolina.

[32] M. Krantz *et al.*, Phys. Rev. B **38**, 4992 (1988).

[33] B. Friel, C. Thomsen, and M. Cardona, Phys. Rev. Lett. **65**, 915 (1990).

[34] K. M. Ham, *et al.*, Phys. Rev. B **47**, 11 439 (1993).

[35] R. Zeyher and G. Zwicknagl, Z. Phys. B **78**, 175 (1990).

[36] R. Zeyher and G. Zwicknagl, Solid State Commun. **66**, 617 (1988).

[37] M. Tinkham, *Introduction to Superconductivity* (McGraw-Hill, 1996).

[38] J. P. Lu, Phys. Rev. Lett. **74**, 1153 (1995).

[39] O. Chauvet *et al.*, Phys. Rev. B **52**, R6963 (1995). The aligned nanotube films were produced by a process in which the tubes are ultrasonically separated.

[40] D. Qian *et al.*, Appl. Phys. Lett. **76**, 2828 (2000).

[41] G. Baumgartner *et al.*, Phys. Rev. B**55**, 6704 (1997).

[42] V. Bayot *et al.*, Phys. Rev. B **40**, 3514 (1989).

[43] P. L. McEuen *et al.*, Phys. Rev. Lett. **83**, 5098 (1999).

[44] S. Frank *et al.*, Science **280**, 1744 (1998).

[45] P. Poncharal *et al.*, J. Phys. Chem. B 106, 12104 (2002).

[46] P. J. de Pablo *et al.*, Appl. Phys. Lett. **74**, 323 (1999).

THERMODYNAMIC INEQUALITIES IN SUPERFLUID AND CRITICAL VELOCITIES IN NARROW ORIFICES

A.F. Andreev and L.A. Melnikovsky
Kapitza Institute, Moscow
andreev@kapitza.ras.ru
leva@kapitza.ras.ru

Abstract We investigate general thermodynamic stability conditions for the superfluid. This analysis is performed in an extended space of thermodynamic variables containing (along with the usual thermodynamic coordinates such as pressure and temperature) superfluid velocity and momentum density. The stability conditions lead to *thermodynamic inequalities* which replace the Landau superfluidity criterion at finite temperatures.

Keywords: superfluid stability, thermodynamic inequalities, critical velocities, liquid helium

1. Introduction

Usually in experiments the vortices destroy superfluidity at velocities far below the Landau critical velocity. This is why the superfluid hydrodynamics equations can be expanded in powers of low velocities and one safely uses the first nontrivial terms of this expansion.

Nevertheless, there is a number of experiments (see [1]) where the superfluid flow is investigated in small orifices. It has been shown that in these circumstances the maximum velocity is a decreasing function of the orifice width and may reach the order of the Landau critical velocity if the aperture is small enough. This means that all thermodynamic quantities of the superfluid become nontrivial functions of the not small superfluid velocity (*i.e.*, they depend not only on the usual thermodynamic coordinates such as pressure and temperature). The only assumption one can make (and we do it) is that the fluid at rest is isotropic. This quite general statement of the problem is used in the paper; we find the complete set of thermodynamic inequalities in this light, *i.e.*, the conditions imposed on thermodynamic functions for the superfluid to remain stable.

A.S. Alexandrov et al. (eds.), Molecular Nanowires and Other Quantum Objects, 107–115.

Finally we employ the Landau phonon-roton model to calculate the highest velocity compatible with obtained thermodynamic inequalities and show that it can be interpreted as a critical velocity. This thermodynamic scenario supposedly explains the superfluidity break-up in small orifices.

2. Stability

When deriving general superfluid hydrodynamic equations it is usually supposed [2] that each infinitesimal volume of the liquid is (locally) in equilibrium and this equilibrium is stable. For the state of the liquid to be stable, it should provide an entropy maximum (at least local) for an isolated system. Instead of investigating the condition of the entropy maximality, it is convenient [3] to use another, equivalent to the first one, condition, that is the condition of the energy minimality under constant entropy and additive integrals of motion. Thus, to examine if the state is stable or not, one must investigate the second variation of the energy. Such analysis will provide sufficient conditions for the energy minimality.

Total energy of the superfluid E_{tot} is an integral of the energy density E over the entire volume

$$E_{tot} = \int E \, \mathrm{d}\mathbf{r}. \tag{1}$$

The energy density can be obtained via a Galilean transformation

$$E = \frac{\rho v_s^2}{2} + \mathbf{v}_s \mathbf{j}_0 + E_0. \tag{2}$$

Here $\mathbf{v}_s$ is the superfluid velocity, ρ is the mass density and subscript 0 denotes quantities measured in the frame of reference of the superfluid component (that is the frame where the superfluid velocity is zero). Namely, E_0 and $\mathbf{j}_0$ are the energy density and the momentum density (or, equally, the mass flux) with respect to the superfluid component. The former is a function of ρ, $\mathbf{j}_0$, and the entropy density S. Its differential can be written as

$$\mathrm{d}E_0 = T \, \mathrm{d}S + \mu \, \mathrm{d}\rho + \mathbf{w} \, \mathrm{d}\mathbf{j}_0, \tag{3}$$

where Lagrange multipliers T, μ, and $\mathbf{w}$ are the temperature, the chemical potential, and the so-called relative velocity of normal and superfluid components. The liquid is isotropic and, consequently, the velocity $\mathbf{w}$ and the momentum density $\mathbf{j}_0$ are parallel to each other, as expressed by

$$\mathbf{j}_0 = j_0(T, \rho, w) \frac{\mathbf{w}}{w}.$$

This leads to a useful identity for the partial derivatives of $\mathbf{j}_0$ with respect to $\mathbf{w}$:

$$\left(\frac{\partial j_0^k}{\partial w^l} \right)_{T,\rho} = \frac{w^k w^l}{w^2} \left(\frac{\partial j_0}{\partial w} \right)_{T,\rho} + \left(\frac{\delta^{kl}}{w} - \frac{w^k w^l}{w^3} \right) j_0. \tag{4}$$

Further transforming (2), we can rewrite it with the help of (3) in the form

$$\mathrm{d}E = T\,\mathrm{d}S + \left(\mu + \frac{v_s^2}{2} - \mathbf{v}_s\mathbf{v}_n\right)\mathrm{d}\rho + (\mathbf{j} - \rho\mathbf{v}_n)\,\mathrm{d}\mathbf{v}_s + \mathbf{v}_n\,\mathrm{d}\mathbf{j}, \qquad (5)$$

where we denoted the total momentum density $\mathbf{j} = \rho\mathbf{v}_s + \mathbf{j}_0$ and the normal velocity $\mathbf{v}_n = \mathbf{v}_s + \mathbf{w}$.

As usual, stability implies that each "allowed" fluctuation increases the total energy of the system E_{tot}. Allowed are the fluctuations leaving conserved quantities unchanged. This means that the minimality of E_{tot} must be investigated under fixed entropy and all additive integrals of motion: mass, momentum, and superfluid velocity. While the conservation of mass and momentum is well-known, conservation of the superfluid velocity worths a special comment. Really, since the superfluid flow is irrotational, the velocity $\mathbf{v}_s$ is a gradient of a scalar: $\mathbf{v}_s = \nabla\phi$. The same is true for the time derivative $\dot{\mathbf{v}}_s = \nabla\dot{\phi}$. This formula expresses the conservation of all three components of the vector

$$\mathbf{V}_s = \int \mathbf{v}_s\,\mathrm{d}\mathbf{r}. \qquad (6)$$

Consider a macroscopic fluctuation of all the variables δS, $\delta\rho$, $\delta\mathbf{v}_s$, and $\delta\mathbf{j}$. They are conserved and this ensures that the first variation of the total energy for a uniform system is identically zero

$$\delta E_{tot} = \int \left(\left(\frac{\partial E}{\partial S}\right)_{\rho,\mathbf{v}_s,\mathbf{j}} \delta S + \left(\frac{\partial E}{\partial \rho}\right)_{S,\mathbf{v}_s,\mathbf{j}} \delta\rho + \left(\frac{\partial E}{\partial \mathbf{v}_s}\right)_{S,\rho,\mathbf{j}} \delta\mathbf{v}_s + \left(\frac{\partial E}{\partial \mathbf{j}}\right)_{S,\rho,\mathbf{v}_s} \delta\mathbf{j} \right) \mathrm{d}\mathbf{r} \equiv 0. \qquad (7)$$

The minimality criterion must be obtained as the condition of the positive definiteness of the second differential quadratic form. The matrix of this quadratic form is a Jacobian matrix 8×8:

$$Q = \left\| \frac{\partial(T,\ \mathbf{v}_n,\ \mu + v_s^2/2 - \mathbf{v}_s\mathbf{v}_n,\ \mathbf{j} - \rho\mathbf{v}_n)}{\partial(S,\ \mathbf{j},\ \rho,\ \mathbf{v}_s)} \right\|. \qquad (8)$$

Common rule states that it is positive definite if all principal minors $M_1, M_2, \ldots M_8$ in the top-left corner are positive. We recursively test these minors:

- The first positivity condition

$$M_1 = \frac{\partial(T, \mathbf{j}, \rho, \mathbf{v}_s)}{\partial(S, \mathbf{j}, \rho, \mathbf{v}_s)} = \frac{\partial(T,\ \mathbf{j},\ \rho, \mathbf{v}_s)}{\partial(T, \mathbf{v}_n, \rho, \mathbf{v}_s)} \frac{\partial(T, \mathbf{v}_n, \rho, \mathbf{v}_s)}{\partial(S,\ \mathbf{j},\ \rho, \mathbf{v}_s)} = \left(\frac{\partial j_0}{\partial w}\right)_{T,\rho} \left(\left(\frac{\partial S}{\partial T}\right)_{\rho,w} \left(\frac{\partial j_0}{\partial w}\right)_{T,\rho} - \left(\frac{\partial j_0}{\partial T}\right)^2_{\rho,w} \right)^{-1} > 0$$

corresponds to the usual requirement of the heat capacity positivity. It is shown below that $(\partial j_0/\partial w)_{T,\rho} > 0$, hence the last inequality eventually becomes

$$\left(\frac{\partial S}{\partial T}\right)_{\rho,w}\left(\frac{\partial j_0}{\partial w}\right)_{T,\rho} - \left(\frac{\partial j_0}{\partial T}\right)^2_{\rho,w} > 0. \tag{9}$$

- Positivity of the next group of minors is easily verified with the following transformation

$$Q' = \left\| \frac{\partial(T, \mathbf{v}_n, \rho, \mathbf{v}_s)}{\partial(S, \ \mathbf{j}, \ \rho, \mathbf{v}_s)} \right\| = \left\| \frac{\partial(T, \mathbf{j}, \rho, \mathbf{v}_s)}{\partial(S, \mathbf{j}, \rho, \mathbf{v}_s)} \right\| \left\| \frac{\partial(T, \mathbf{v}_n, \rho, \mathbf{v}_s)}{\partial(T, \ \mathbf{j}, \ \rho, \mathbf{v}_s)} \right\|. \tag{10}$$

Whether the minors M_2, M_3, M_4 are positive is determined by the second multiplier in (10). Required condition is therefore equivalent to the positive definiteness of the matrix

$$\left\| \left(\frac{\partial \mathbf{j}}{\partial \mathbf{v}_n}\right)_{T,\rho,\mathbf{v}_s} \right\|^{-1} = \left\| \left(\frac{\partial \mathbf{j}_0}{\partial \mathbf{w}}\right)_{T,\rho} \right\|^{-1} = \left\| \begin{matrix} (\partial j_0/\partial w)_{T,\rho} & 0 & 0 \\ 0 & j_0/w & 0 \\ 0 & 0 & j_0/w \end{matrix} \right\|^{-1}.$$

Here we used (4) and chosen the direction of the $\mathbf{w}$ vector as the first coordinate. This adds to our collection two more inequalities

$$\mathbf{j}_0\mathbf{w} \geq 0, \tag{11}$$

$$\left(\frac{\partial j_0}{\partial w}\right)_{T,\rho} > 0. \tag{12}$$

- The same transformation applied to the biggest minors gives:

$$Q = Q' \left\| \frac{\partial(T, \mathbf{v}_n, \mu + v_s^2/2 - \mathbf{v}_s\mathbf{v}_n, \mathbf{j} - \rho\mathbf{v}_n)}{\partial(T, \mathbf{v}_n, \qquad \rho, \qquad \mathbf{v}_s \quad)} \right\| = Q'Q''.$$

Again, the minors M_5, M_6, M_7, M_8 correspond to nontrivial principal minors of Q''. We use the thermodynamic identity to relate the chemical potential μ and the conventional pressure p

$$\mathrm{d}\mu = \frac{\mathrm{d}p}{\rho} - \frac{S}{\rho}\mathrm{d}T - \frac{\mathbf{j}_0}{\rho}\mathrm{d}\mathbf{w}.$$

This gives

$$\left(\frac{\partial\left(\mu + v_s^2/2 - \mathbf{v}_s\mathbf{v}_n\right)}{\partial\rho}\right)_{T,\mathbf{v}_n,\mathbf{v}_s} = \left(\frac{\partial\mu}{\partial\rho}\right)_{T,\mathbf{w}} = \frac{1}{\rho}\left(\frac{\partial p}{\partial\rho}\right)_{T,w}.$$

The following is an explicit representation of Q'' sub-matrix corresponding to a four-dimensional space ρ, v_s^x, v_s^y, v_s^z; as before we let the x-axis run along $\mathbf{w}$ direction. Using (4) we obtain

$$\begin{Vmatrix} (\partial p/\partial \rho)_{T,w}/\rho & (\partial j_0/\partial \rho)_{T,w} - w & 0 & 0 \\ (\partial j_0/\partial \rho)_{T,w} - w & \rho - (\partial j_0/\partial w)_{T,\rho} & 0 & 0 \\ 0 & 0 & \rho - j_0/w & 0 \\ 0 & 0 & 0 & \rho - j_0/w \end{Vmatrix}.$$

Appropriate inequalities are:

$$\left(\frac{\partial p}{\partial \rho}\right)_{T,w} > 0, \tag{13}$$

which is literally a generalized (to a non-zero inter-component velocity w) positive compressibility requirement,

$$j_0 < w\rho, \tag{14}$$

and

$$\left(\frac{\partial p}{\partial \rho}\right)_{T,w} \left(\rho - \left(\frac{\partial j_0}{\partial w}\right)_{T,\rho}\right) - \rho\left(\left(\frac{\partial j_0}{\partial \rho}\right)_{T,w} - w\right)^2 > 0. \tag{15}$$

Inequalities (9), (11), (12), (13), (14), and (15) are sufficient conditions for the thermodynamic stability.

3. Discussion

In a "stopped-normal-component" arrangement, the mass flux $\mathbf{f}$ with respect to the normal component may become more convenient than $\mathbf{j}_0$—the mass flux relative to the superfluid one. The obvious relation between them $f = \rho w - j_0$ leads to the following reformulation of the inequalities:

$$\mathbf{fw} < 0, \quad f < w\rho, \tag{16}$$

$$0 < \left(\frac{\partial f}{\partial w}\right)_{\rho,T} < \rho \tag{17}$$

$$\left(\frac{\partial S}{\partial T}\right)_{\rho,w} \left(\rho - \left(\frac{\partial f}{\partial w}\right)_{T,\rho}\right) > \left(\frac{\partial f}{\partial T}\right)^2_{\rho,w}, \tag{18}$$

$$\left(\frac{\partial p}{\partial \rho}\right)_{T,w} \left(\frac{\partial f}{\partial w}\right)_{\rho,T} > \rho\left(\frac{\partial f}{\partial \rho}\right)^2_{w,T} \tag{19}$$

As a simple application of the derived inequalities, consider them at $w = 0$. From (16), (17), (18), and (19) we get

$$\left(\frac{\partial S}{\partial T}\right)_{\rho,w} > 0, \quad \left(\frac{\partial p}{\partial \rho}\right)_{T,w} > 0, \tag{20}$$

$$\rho > \left(\frac{\partial j_0}{\partial w}\right)_{T,\rho} > 0. \tag{21}$$

Using conventional notation, last inequality reads in the limit $w \to 0$

$$\rho_{\mathrm{s}} > 0, \quad \rho_{\mathrm{n}} > 0. \tag{22}$$

4. Phonon-Roton model

Here we provide a usage example of the stability criteria for real superfluid ^{4}He. To calculate derivatives involved in the inequalities one take refuge in the microscopic approach. Simple and clear Landau phonon-roton model works pretty well in wide temperature and velocity ranges. We use this model to calculate the contribution of these quasiparticles to the "modified" free energy in the frame of reference of the superfluid component:

$$\tilde{\mathcal{F}}_0 = \mathcal{F}_0 - \mathbf{w}\mathbf{j}_0. \tag{23}$$

Differential of this potential is given by

$$\mathrm{d}\tilde{\mathcal{F}}_0 = -S\,\mathrm{d}T - \mathbf{j}_0\,\mathrm{d}\mathbf{w}. \tag{24}$$

The modified free energy is obtained from the excitation spectrum with a conventional formula

$$\tilde{\mathcal{F}}_0 = T\int \ln\left(1 - \exp\left(\frac{\mathbf{p}\mathbf{w} - \epsilon(p)}{T}\right)\right)\frac{\mathrm{d}\mathbf{p}}{(2\pi\hbar)^3}. \tag{25}$$

We denoted the excitation energy $\epsilon(p)$, which is given for two branches by the expressions

$$\epsilon_{\mathrm{ph}}(p) = cp, \quad \epsilon_{\mathrm{r}}(p) = \Delta + \frac{(p - p_0)^2}{2m}. \tag{26}$$

Here and below, subscripts distinguish the quantities related to phonons and rotons, c is the sound velocity, Δ is the roton energy gap, m is the effective mass, and p_0 is the momentum at the roton minimum[1]. A small dimensionless parameter $m\Delta/p_0^2 \sim 0.03 \ll 1$ ensures, *e.g.*, that the Landau critical velocity is determined by $v_{\mathrm{L}} = \Delta/p_0$.

When integrated, these dispersion laws give the following contributions to the free energy:

$$\tilde{\mathcal{F}}_{0,\mathrm{ph}} = -\frac{T^4\pi^2}{90\hbar^3 c^3}\left(1-\frac{w^2}{c^2}\right)^{-2}, \tag{27}$$

$$\tilde{\mathcal{F}}_{0,\mathrm{r}} = -\frac{T^{5/2}m^{1/2}}{2^{1/2}\pi^{3/2}\hbar^3}\frac{p_0}{w}\sinh\frac{wp_0}{T}\exp\left(-\frac{\Delta}{T}\right). \tag{28}$$

One can obtain all[2] thermodynamic variables by differentiating this potential. Namely

$$S = -\left(\frac{\partial\tilde{\mathcal{F}}_0}{\partial T}\right)_{w,\rho},$$

$$j_0 = -\left(\frac{\partial\tilde{\mathcal{F}}_0}{\partial w}\right)_{T,\rho}.$$

Inequality (15) is the first to become invalid. Appropriate validity region is plotted in Fig. 1. The liquid is unstable above the curve.

At zero temperature the critical velocity becomes the Landau critical velocity v_{L}. It should also be noted that for systems where all quasiparticles can be described hydrodynamically (in other words, systems lacking roton branch) inequality (15) at zero temperature includes $(\partial p/\partial\rho)_{T,w} - w^2 > 0$ *i.e.*, $w < c$.

5. Conclusion

Experimentally, the superfluidity break-up in small orifices is believed to have the following nature (see [1]). Until the aperture size is too small the critical velocity does not depend on the temperature and increases as the size decreases. This is the very behaviour that is specific to the vortex-related critical velocity.

When the orifice width is narrow enough the vortex-related critical velocity becomes so high, that the break-up scenario and its features change. The critical velocity does not depend on the aperture any more but decreases when the temperature increases. This behaviour is commonly associated (see [1]) with the Iordanski-Langer-Fisher mechanism (see [6]). Nevertheless, this association lacks numerical comparison because no reliable information about the actual orifice shape is available.

On the other side experimentally observed behaviour of the critical velocity can be attributed to the suggested stability criterion. In other words we provide an alternative explanation of experimental results based on an assumption that in narrow orifices the thermodynamic limit of w_{c} is reached.

We should also note that our approach to the critical velocity as a stability limit is similar to that used by Kramer[7]. Actually the inequality he employed

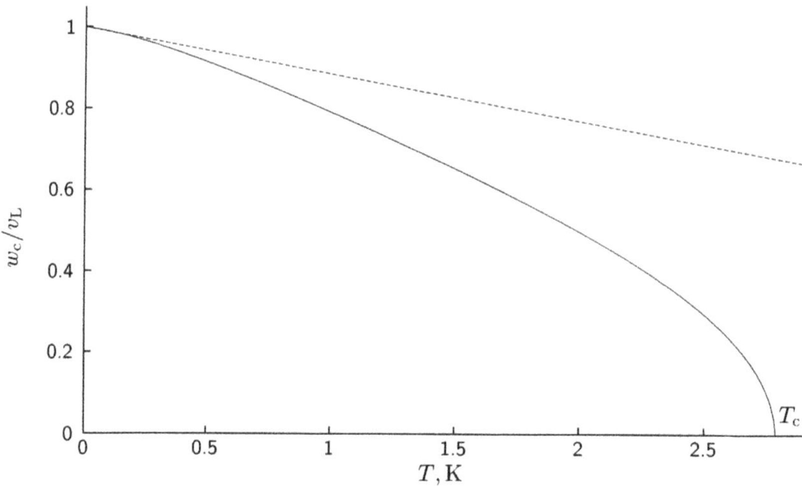

Figure 1. Critical velocity w_c versus temperature T at normal pressure. Dashed line corresponds to the equation $T = \Delta - p_0 w$. Note that the condition $T < \Delta - p_0 w$ holds true over entire stability domain. The "stability" critical velocity w_c coincides with the Landau critical velocity v_L at zero temperature and vanishes completely (22) at the critical temperature T_c (the λ-point). In the phonon-roton model the critical temperature is $T_c \approx 2.8$ K.

is not a thermodynamic one. Moreover, generally speaking it is wrong. But numerical results for the critical velocity he obtained using the phonon-roton model do not deviate much from those plotted in Fig.1.

Acknowledgments

It is a pleasure for us to thank I.A.Fomin for useful discussions. This work was supported in parts by INTAS grant 01-686, CRDF grant RP1-2411-MO-02, RFBR grant 03-02-16401, and RF president program.

Notes

1. Data taken from [4, 5]: $\rho = 0.145\,\mathrm{g/cm^3}$, $\Delta = 8.7\,\mathrm{K}$, $m = 0.16 m_{\mathrm{He}}$, $p_0 = 3.673\,10^8\,\mathrm{g}^{-1/3}\rho^{1/3}\hbar$, $c = 23800\,\mathrm{cm/s}$, $\partial\Delta/\partial\rho = -0.47\,10^{-14}\,\mathrm{cm^5 s^{-2}}$, $\partial m/\partial\rho = -0.45\,10^{-23}\,\mathrm{cm^3}$, $\partial c/\partial\rho = 467\,10^3\,\mathrm{cm^4 s^{-1} g^{-1}}$.

2. We neglect the quasiparticle contribution to the pressure derivative because it is just a small correction to the speed of sound.

References

[1] E.Varoquaux, W.Zimmermann, and O.Avnel, in *Excitations in Two-Dimensional and Three-Dimensional Quantum Fluids*, edited by A.F.G.Wyatt and H.J.Lauter (NATO ASI Series, Plenum Press, New York-London, 1991), p.343.

[2] I.M.Khalatnikov, *An Introduction to the theory of superfluidity*. (W.A.Benjamin, New York-Amsterdam 1965).

[3] L.D.Landau, E.M.Lifshitz, *Statistical Physics*, part 1 (Pergamon Press, Oxford, 1980).

[4] R.J.Donnelly, P.H.Roberts, *Journal of Low Temperature Physics*, **27**, 687 (1977).

[5] R.J.Donnelly, J.A.Donnelly, R.N.Hills, *Journal of Low Temperature Physics*, **44**, 471 (1981).

[6] J.S.Langer, J.D.Reppy, *Prog. Low. Temp. Phys.*, Vol. VI, ed. C.J.Gorter (North-Holland, Amsterdam, 1970) Ch.1.

[7] L.Kramer, *Phys. Rev.*, **179**, 149 (1969).

SHOT NOISE IN MESOSCOPIC DIFFUSIVE ANDREEV WIRES

Wolfgang Belzig

Department of Physics and Astronomy, University of Basel,
Klingelbergstr. 82, CH-4056 Basel, Switzerland
Wolfgang.Belzig@unibas.ch

Abstract We study shot noise in mesoscopic diffusive wires between a normal and a superconducting terminal. We particularly focus on the regime, in which the proximity-induced reentrance effect is important. We will examine the difference between a simple Boltzmann-Langevin description, which neglects induced correlations beyond the simple conductivity correction, and a full quantum calculation. In the latter approach, it turns out that two Andreev pairs propagating coherently into the normal metal are anti-correlated for $E \lesssim E_c$, where $E_c = \hbar D/L^2$ is the Thouless energy. In a fork geometry the flux-sensitive suppression of the effective charge was confirmed experimentally.

Keywords: Shot noise, counting statistics, Andreev reflection, proximity effect

1. Introduction

Fluctuations of the current in mesoscopic conductors originate from the quantum scattering of the charge carriers, and are sensitive to their interference, statistics and interaction. This makes the theoretical and experimental study of noise in small electronic circuits interesting and challenging (for recent reviews, see Refs. [1, 2]).

Statistical correlations in the transport of fermions have led to a number of interesting predictions. For example, the noise of a single-channel quantum contact of transparency T at zero temperature has the form $S_I = 2|eI|(1-T)$ [3, 4, 5, 6]. The noise is thus suppressed below the Schottky value $2|eI|$ of uncorrelated charge transfer. The suppression is a direct consequence of the Pauli principle, which prevents two electrons from tunneling together. It is convenient to measure the deviation from the Schottky result by the so-called Fano factor $F = S_I/2|eI|$. For a number of generic conductors, it turns out that the suppression of the Fano factor is universal, i. e. it does not depend on details of the conductor like geometry or impurity concentration. In particular,

A.S. Alexandrov et al. (eds.), Molecular Nanowires and Other Quantum Objects, 117–128.

a diffusive metal with elastic scattering leads to $F_{diff}^{N} = 1/3$ [7, 8], which is independent on the concrete shape of the conductor [9] and has been confirmed experimentally [10, 11].

If superconductivity comes into play the fundamental charge transport mechanism at energies below the superconducting gap is Andreev reflection. Two electrons enter the superconductor simultaneously and form a Cooper pair, which can propagate in the superconductor. Thus, in this process a charge transfer of $2e$ occurs, but with a reduced probability, since two particles have to tunnel. The shot noise is proportional to the charge of the elementary processes, and one thus naivly expects a doubling of the shot noise, which was indeed found theoretically [12, 13] and experimentally [14, 15] for diffusive conductors. It is remarkable that this doubling occurs for diffusive conductors, whereas it is not found for other conductors like, e. g., single-channel contacts [3, 12, 16, 17, 18], double tunnel junctions [19, 20, 21, 22], or diffusive junctions with a tunneling barrier [23, 24]. A doubling of the full Schottky noise was recently observed experimentally [25].

In this article we address the counting statistics and the noise in diffusive structures with normal and/or superconducting terminals. In particular we concentrate on the energy- and phase-dependent shot noise in an Andreev interferometer. First, we briefly review the theory of full counting statistics using the Keldysh Green's function approach. We derive the counting statistics of diffusive conductors for various limits. In Sec. 3 we obtain generic results for the shot noise in diffusive conductors in some limiting cases. Finally, in Secs. 4 and 5 we discuss the information contained in the energy- and phase-dependence of the shot noise in diffusive wires and Andreev interferometers. A good qualitative agreement of experimental results and our full quantum calculation is demonstrated.

2. Current Statistics

Consider the following Gedanken experiment. A constant voltage bias is applied to a mesoscopic conductor for a certain time interval t_0. During this time interval we count the number of charges N passing the conductor. Due to quantum-mechanical uncertainty the outcome of the experiment is described by a probability $P_{t_0}(N)$ that N charges have passed the conductor. This is the so-called *full counting statistics* (FCS), which is the most natural description of quantum transport. Alternatively we may study the *characteristic function* $\Phi(\chi) = \sum_N P_{t_0}(N)e^{iN\chi}$. The kth coefficient in the expansion of the characteristic function in powers of χ yields the moments of the counted charge $\langle N^k \rangle$. An equivalent description is to obtain the *cumulant generating function* (CGF) $S(\chi) = \ln \Phi(\chi)$, which gives the cumulants directly. This will be the quantity which we obtain below. The cumulants are directly connected to the (measureable) zero-frequency current correlations functions.

One route to counting statistics is the Keldysh-Green's function approach in combination with the circuit theory of mesoscopic transport developed by Nazarov [26, 27, 28]. For details, we refer to several articles in [2]. In this approach, terminals are described by 4×4 Green's function matrices. For a normal terminal (N) with occupation f we introduce the 2×2 matrix $\hat{f}$ =diag$(f(E), f(-E))$ and have

$$\check{G}_N(\chi) = \begin{pmatrix} \hat{\tau}_3(1-2\hat{f}) & -2\hat{\tau}_3 e^{i\chi\hat{\tau}_3}\hat{f} \\ -2\hat{\tau}_3 e^{-i\chi\hat{\tau}_3}\hat{f} & \hat{\tau}_3(2\hat{f}-1) \end{pmatrix} . \tag{1}$$

For a superconducting terminal (S) at equilibrium and for $E \ll \Delta$ we have

$$\check{G}_S = \begin{pmatrix} \hat{\tau}_1 & 0 \\ 0 & \hat{\tau}_1 \end{pmatrix} . \tag{2}$$

For a general contact described by a scattering matrix with corresponding transmission eigenvaluese $\{T_n\}$ the counting statistics is obtained as [29]

$$S(\chi) = \frac{t}{h}\int dE \sum_n \mathrm{Tr}\ln\left[1 + \frac{T_n}{4}\left(\{\check{G}_L, \check{G}_R\} - 2\right)\right] . \tag{3}$$

To give a simple example, we evaluate the counting statistics of a quantum contact between two normal terminals and obtain in agreement with the scattering matrix approach [30, 31] (see Appendix for a short summary)

$$S(\chi) = \frac{t}{h}\int dE \sum_n \ln\Big[1 + T_n f_L(1-f_R)(e^{i\chi}-1) + T_n f_R(1-f_L)(e^{-i\chi}-1)\Big] . \tag{4}$$

Thus, the statistics is a simple multinomial form of the two possible events of left and right transfer of charges. The formula (3) includes in addition the statistics of SN- and SS-contacts [16, 29].

In diffusive conductors we have to find the general counting statistics by a different method. The quantum kinetic equation for the diffusive wire is the so-called Usadel equation [32]

$$\hbar\frac{\partial}{\partial x}D(x)\check{G}(x)\frac{\partial}{\partial x}\check{G}(x) = -i\left[E\hat{\tau}_3, \check{G}(x)\right] , \tag{5}$$

with continuous boundary conditions at both ends of the conductor. The right-hand side of Eq. (5) accounts for the decoherence of electrons and holes during the propagation in the normal metal at finite energies E. A full solution has so far been only obtained numerically, and we will discuss the implications of the decoherence on the shot noise later.

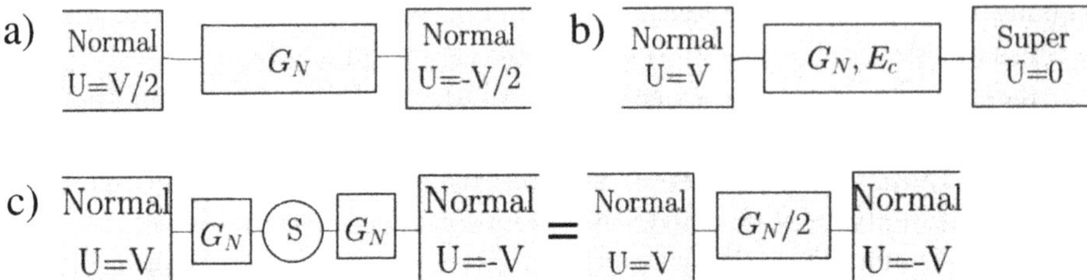

Figure 1. a) diffusive wire with conductance G_N between two normal terminals. b) diffusive wire betweena normal and a superconducting terminal. The wire is characterized by the characteristic energy $E_c = \hbar D/L^2$. c) in the incoherent regime the wire b) is mapped onto a series of two wires between normal terminals. For diffusive connectors the system is equivalent to one wire with conductance $G_N/2$.

For diffusive wires between normal terminals or with one superconducting terminal (see Fig. 1a and Fig. 1b for smalle energies $eV, k_BT \ll E_c$) the right hand side of Eq. (5) can be neglected. We obtain the general solution [27, 33, 34]

$$S(\chi) = \frac{t_0 G_N}{8h} \int dE \, \mathrm{Tr} \, \mathrm{acosh}^2 \left(\frac{1}{2} \{ \check{G}_L, \check{G}_R \} \right) . \tag{6}$$

The same result can be obtained by averaging Eq. (3) over the transmission eigenvalue distribution of a diffusive scatterer, i.e. the bimodal distribution [35]

$$\rho(T) = \frac{G_N}{2G_Q} \frac{1}{T\sqrt{1-T}} . \tag{7}$$

Another drastic simplification can be made if the proximity effect is negligible , i. e. the right hand side of Eq. (5) is dominant. In Ref. [36] it was shown, that the diffusive wire can be mapped onto a series of two diffusive wires contacted by normal terminals, which constitute the electron and hole propagation (see Fig. 1c). For the special case of diffusive connectors the counting statistics is independent of the geometry. As a consequence, the counting statistics is exactly given by that of a *normal* conductor, with *halved* conductance and a *negative* counting field for the hole terminal.

The energy dependence resulting from the right hand side of Eq. (5) leads to interesting effects related to the quantum propagation of electron-hole pairs [37]. Below, we will address in detail the dependence of the shot noise on voltage and temperature, and compare our theoretical predictions, based on (5) with available experimental results.

3. Shot Noise in Diffusive Conductors

To obtain the shot noise from the counting statistics we calculate

$$S_I = -\frac{2e^2}{t_0}\frac{\partial^2}{\partial\chi^2}S(\chi)\Big|_{\chi=0} . \tag{8}$$

From Eq. (3) we obtain for a general scatterer between normal terminals

$$\begin{aligned} S_I^{NN} &= \frac{2e^2}{h}\sum_n \int dE \Big[T_n(1-T_n)\left(f_L(E)-f_R(E)\right)^2 \\ &\quad + T_n\big[f_L(E)(1-f_L(E)) + f_R(E)(1-f_R(E))\big]\Big] . \end{aligned} \tag{9}$$

Averaging (9) over the transmission eigenvalue distribution (7) we find

$$\begin{aligned} S_I^{NN} = 2G_N \int dE \; \Big[& f_L(E)\,(1-f_L(E)) + f_R(E)\,(1-f_R(E)) \\ & + \frac{1}{3}\left(f_L(E)-f_R(E)\right)^2 \Big] . \end{aligned} \tag{10}$$

The energy integration can be done using the formulas in the appendix. As result we obtain

$$S_I^{NN}(T,V) = \frac{4}{3}G_N k_B T + \frac{2}{3}eG_N V \coth\left(\frac{eV}{2k_B T}\right) , \tag{11}$$

where we introduced the conductance $G_N = \frac{e^2}{h}\sum_n T_n$.

Using the result (6) we can obtain at $E = 0$ the general formula for the noise

$$S_I = -\frac{G}{4h}\int dE \, \mathrm{Tr}\left[\frac{\partial^2}{\partial\chi^2}\{\check{G}_L,\check{G}_R\} - \frac{2}{3}\left(\frac{\partial}{\partial\chi}\{\check{G}_L,\check{G}_R\}\right)^2\right]_{\chi\to 0} . \tag{12}$$

For a diffusive wire between a normal and a superconducting terminal at $eV, k_B T \ll E_c$ we obtain

$$S_I^{NS}(V) = \frac{2G_N}{3}\int dE \, [f(E)+f(-E)]\,[2-f(E)-f(-E)] . \tag{13}$$

Evaluating the energy integration we find

$$S_I^{NS}(V) = \frac{8}{3}G_N k_B T + \frac{4}{3}eG_N V \coth\left(\frac{eV}{k_B T}\right) . \tag{14}$$

This result can, indeed, be infered from the normal-state result Eq. (12) by the replacement $G_N \to 2G_N$ and $e \to 2e$.

While the derivations presented previously are valid in the limit $eV, k_BT \ll E_c$, we can also obtain the noise in the limit $eV \gg E_c$ or $k_BT \gg E_c$. Here we employ the incoherent Andreev circuit theory approach [36]. According to the mapping rules we obtain the noise from the normal result by the replacement $G_N \to G_N/2$, $V \to 2V$ and $S_I \to 4S_I$. By applying these substitutions to Eq. (11) we again obtain the result (14).

We see that the shot noise (as well as the full counting statistics) in the discussed regimes is universal. First, at low energies $eV, k_BT \ll E_c$ universality means, that the noise depends only on the normal-state conductance G_N and is independent of the detailed geometry. Furthermore, it turns out that in the incoherent regime the full counting statistics and, therefore the current noise is also the same. The universality of the noise is quite surprising, since the transport mechanisms differ quite drastically in both limits. This remarkable coincidence holds, however, only for diffusive conductors. For double tunnel junctions, chaotic cavities, or other combinations of scatterers the transport properties differ (see Ref. [34] and references therein).

4. Energy-dependent current noise

The full quantum-mechanical description requires the solution of the Usadel equation (5) to first order in the counting field. We will discuss these results later. Let us first note, that one can obtain an approximate expression for the energy dependence of the shot noise by a generalized Boltzmann-Langevin approach. We recall, that the kinetic equation for the average distribution function has the form

$$\frac{\partial}{\partial x}\sigma(E,x)\frac{\partial}{\partial x}\left(1-f(E,x)-f(-E,x)\right)=0\,. \tag{15}$$

The local energy-dependent conductivity incorporates the effect of the proximity induced coherence and is given by $\sigma(E,x) = \sigma_N \cosh(\mathrm{Re}\theta(E,x))$, where the spectral angle θ obeys $\hbar D(\partial^2/\partial x^2)\theta(E,x) = -iE\sin(\theta(E,x))$ with appropriate boundary conditions (see Ref. [38] for details). Solving Eq. (15) the current (per unit area of the contact) is

$$I(V,T)=\frac{1}{e}\int dEG(E)\left[1-f_N(E)-f_N(-E)\right]. \tag{16}$$

Here we have defined the spectral conductance

$$\frac{1}{G(E)}=\int_0^L\frac{dx}{L}\frac{1}{\sigma(E,x)}, \tag{17}$$

which have written here for a one-dimensional wire of length L with a uniform cross section. For an arbitrary geometry, the spectral conductance has to be found from the solution of a diffusion equation.

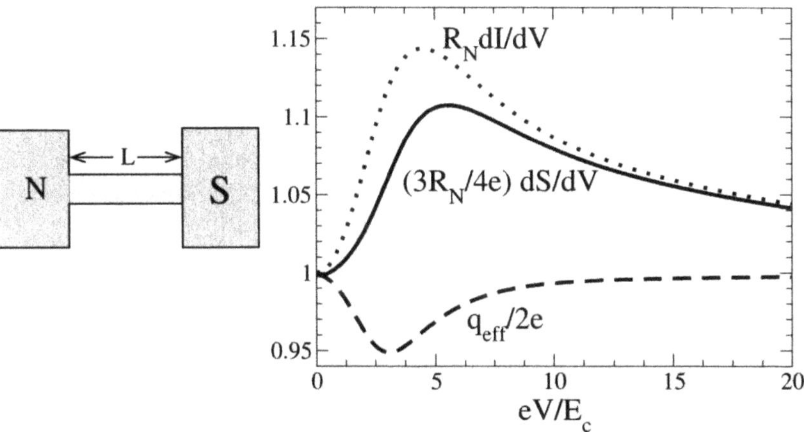

Figure 2. Shot noise of a diffusive proximity wire. Both the differential conductance and the noise show a reentrant behaviour. The effective charge reveals that the correlated Andreev pair transport suppresses the noise below the uncorrelated Boltzmann-Langevin result.

Guided by the kinetic equation for the average transport, we may try to find the current noise by generalizing the Boltzmann-Langevin approach [13] to include the energy- and space-dependent conductivity. In fact, this task has already been performed. In Ref. [27] it was shown, that the counting statistics of a diffusive normal wire only depends on the conductance, defined in exactly the same way as (17). In this proof an energy-independent conductivity $\sigma(x)$, but an arbitrary form of the diffusive metal was taken into account. However, it is obvious, that the same result is obtained if a spectral conductivity $\sigma(E,x)$ is assumed. As a result the current would be given by Eq. (16).

Next, we combine this observation with the results for incoherent Andreev transport obtained in Ref. [36]. In this work it was shown that the counting statistics of a diffusive wire between a normal and a superconducting terminal is mapped onto a series of two diffusive wires. Again, this holds equally well if we assume *ad hoc* an energy-dependent conductance of both diffusive wires. Due to electron-hole symmetry the spectral conductance of both wires and therefore of the total wire is the same.

Collecting these observations, we obtain the current noise from Eq. (10) by inserting the spectral conductance (17) *inside* the energy integral and multiplying by a factor of 2. In this way we obtain the result of the modified Boltzmann-Langevin approach

$$S_I^{BL} = 4\int dEG(E)\,\Big[f(E)(1-f(E)) + f(-E)(1-f(-E)) \\ + \tfrac{1}{3}(f(E)-f(-E))^2\Big] . \qquad (18)$$

Evaluating the energy integral with the help of the formulas in the appendix we find

$$S_I^{BL} = \frac{8}{3}G(V,T)k_BT + \frac{4}{3}eI(V,T)\coth\left(\frac{eV}{k_BT}\right) . \tag{19}$$

Here we introduced the temperature-dependent differential conductance $G(V,T) = dI(V,T)/dV$. The same result was obtained by a different method recently in Ref. [39]. At zero temperature, we obtain the result [40]

$$S_I^{BL}(V) = \frac{4}{3}eI(V) . \tag{20}$$

This result represents the starting point for our further considerations. It was derived neglecting correlations of scattering events *between electrons and holes* in the normal metal. Thus, in the following we will be specifically interested in the deviations of the noise from the simple Boltzmann-Langevin result (20) and introduce the *effective charge* $q_{eff}(V) = (3/2)\partial S_I/\partial I$ [41].

The full quantum-mechanical calculation of the energy-dependent shot noise was performed in Ref. [37] and the results are shown in Fig. 2. A direct comparison of the differential shot noise and the differential conductance (for zero temperature) shows the difference in the energy dependence. The effective charge defined above displays the clear deviation of the quantum noise from the Boltzmann-Langevin result of $2e$. At energies below the Thouless energy $E_c = \hbar D/L^2$ the effective charge is suppressed below $2e$. This shows that the correlated Andreev pair transport suppresses the noise below the uncorrelated Boltzmann-Langevin result.

5. Phase-dependent shot noise

To experimentally probe the pair correlations in diffusive superconductor-normal metal-heterostructures it is most convenient to use an Andreev interferometer. An example is shown in the left part of Fig. 3. A diffusive wire connected to a normal terminal is split into two parts, which are connected to two different point of a superconducting terminal. By passing a magnetic flux through the loop one can effectively vary the phase difference between the two connections to the superconductor. Such a structure has been experimentally realized by the Yale group [41]. In Fig. 3 we present a direct comparison between our theoretical predictions and the experimentally obtained effective charge. Note that we have included the experimental temperature in the theoretical modelling. The finite temperature explains the strong decrease of the effective charge in the regime $|eV| \leq k_BT$, where the noise is fixed by the fluctuation-dissipation theorem. The disagreement between theory and experiment in this regime stems solely from differences in the measured temperature-dependent conductance from the theoretical prediction. We attribute this to heating effects.

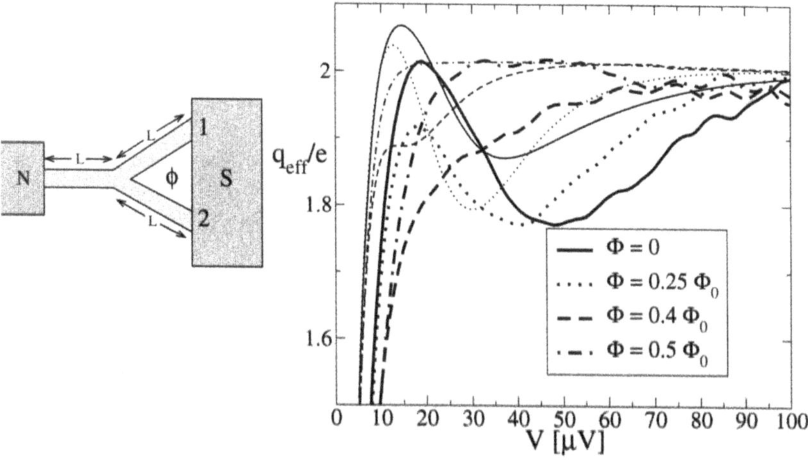

Figure 3. Effective charge of the Andreev interferometer shown in the left (realized experimentally in Ref. [41]). The right panel shows the experimental (thick lines) and theoretical (thin lines) results for the effective charge are shown in the main plot for different magnetic fluxes. The theoretical results contain no fitting parameter, since the only relevant energy scale $E_c = 30\mu$eV was extracted from the sample geometry. Therefore, we believe the deviations between experimental and theoretical results comes from possible heating effects in the experiment, which are not accounted for in the theoretical calculation.

The qualitative agreement in the shot-noise regime $|eV| \geq k_B T$ is satisfactory, if one takes into account, that we have no free parameters for the theoretical calculation. Both, experiment and theory show a suppression of the effective charge for some finite energy, which is of the order of the Thouless energy and depends on flux in a qualitative similar manner. Remarkably for half-integer flux the effective charge is completely flat, in contrast to what one would expect from circuit arguments based on the conductance distribution in the fork geometry. Currently we have no explanation for this behaviour, and therefore more work is needed in this direction.

6. Conclusions

Shot noise in diffusive heterostructures between normal and superconducting terminal provides valuable information on the correlated Andreev pair transport. We have examined the difference between a simple Boltzmann-Langevin description, which neglects these correlations, and a full quantum calculation. Examining the effective charge we have shown that two Andreev pairs propagating coherently into the normal metal are anti-correlated. The phase sensitivity of the suppression of the effective charge was confirmed experimentally.

Acknowledgments

We thank C. Bruder, Yu. V. Nazarov, D. Prober, B. Reulet, and P. Samuelsson for discussions. This work was supported by the Swiss National Science Foundation and the NCCR Nanoscience.

Appendix: Scattering Approach

One of the most general approaches to quantum transport is the scattering matrix approach. Following Levitov and Lesovik [30, 31] the counting statistics can be found from the a slight modification of the standard scattering matrix approach. For a two terminal structure we label the scattering states in the leads by $\psi = (c_1^L, \ldots, c_{n_L}^L, c_1^R, \ldots, c_{n_R}^R)$. The modified scattering matrix and the occupation matrix read

$$S_\chi = \begin{pmatrix} r & te^{i\chi/2} \\ t'e^{-i\chi/2} & r' \end{pmatrix} \quad ; \quad f = \begin{pmatrix} f_L & 0 \\ 0 & f_R \end{pmatrix} . \tag{A.1}$$

The counting statistics is then given by

$$\Phi_E(\chi) = \det\left(1 - f + fS_\chi S_{-\chi}^\dagger\right) , \; S(\chi) \equiv \ln \Phi(\chi) = \frac{t_0}{h}\int dE \ln \Phi_E(\chi) . \tag{A.2}$$

Using the standard polar decomposition for the scattering matrix we obtain the counting statistics (12).

Appendix: Integrals

In the derivations in this article we encounter integral expressions of the forms

$$I_1(U) = \int dE G(E) \left[f_1(E)(1 - f_1(E)) + f_2(E)(1 - f_2(E))\right] , \tag{B.1}$$

$$I_2(U) = \int dE G(E) \left[f_1(E)(1 - f_2(E)) + f_2(E)(1 - f_1(E))\right] . \tag{B.2}$$

Here we introduced $f_1(E) = f_D(E+U)$ and $f_2 = f_D(E-U)$, where $f_D(E) = (\exp(E/k_B T)+1)^{-1}$ is the Fermi-Dirac distribution. The goal is to reduce the integrals to expressions related to the current $I(U) = \int dE G(E)(f_1(E) - f_2(E))$. The first integral is solved by noting that $-k_B T(\partial/\partial U) f_{1(2)}(E) = \pm f_{1(2)}(E)(1 - f_{1(2)}(E))$ and we find for the first integral

$$I_1(U) = k_B T \frac{\partial I(U)}{\partial U} . \tag{B.3}$$

With the help of the identity

$$f_1(E)(1 - f_2(E)) + f_2(E)(1 - f_1(E)) = (f_1(E) - f_2(E)) \coth\left(\frac{U}{k_B T}\right), \tag{B.4}$$

we find for the second integral

$$I_2(U) = I(U) \coth\left(\frac{U}{k_B T}\right). \tag{B.5}$$

References

[1] Ya. M. Blanter and M. Büttiker, Phys. Rep. **336**, 1 (2000).
[2] *Quantum Noise in Mesoscopic Physics*, edited by Yu. V. Nazarov (Kluwer, Dordrecht, 2003).
[3] V. A. Khlus, Sov. Phys. JETP **66**, 1243 (1987).
[4] G. B. Lesovik, Pis'ma Zh. Eksp. Teor. Fiz. **49**, 513 (1989) [JETP Lett. **49**, 592 (1989)].
[5] M. Büttiker, Phys. Rev. Lett. **65**, 2901 (1990).
[6] M. Büttiker, Phys. Rev. B **46**, 12485 (1992).
[7] C. W. J. Beenakker and M. Büttiker, Phys. Rev. B **46**, 1889 (1992).
[8] K. E. Nagaev, Phys. Lett. A **169**, 103 (1992); Phys. Rev. B **57**, 4628 (1998).
[9] Yu. V. Nazarov, Phys. Rev. Lett. **73**, 1420 (1994).
[10] P. Dieleman *et al.*, Phys. Rev. Lett. **79**, 3486 (1997);
[11] T. Hoss *et al.*, Phys. Rev. B **62**, 4079 (2000).
[12] M. J. M. de Jong and C. W. J. Beenakker, Phys. Rev. B **49**, 16070 (1994).
[13] K. E. Nagaev and M. Büttiker, Phys. Rev. B **63**, 081301(R) (2001).
[14] X. Jehl *et al.*, Phys. Rev. Lett. **83**, 1660 (1999); X. Jehl *et al.*, Nature **405**, 50 (2000).
[15] A. A. Kozhevnikov, R. J. Schoelkopf, and D. E. Prober, Phys. Rev. Lett **83**, 1660 (2000).
[16] B. A. Muzykantskii and D. E. Khmelnitzkii, Phys. Rev. B **50**, 3982 (1994).
[17] T. Martin, Phys. Lett. A **220**, 137 (1996).
[18] M. P. Anantram and S. Datta, Phys. Rev. B **53**, 16390 (1996).
[19] S. V. Naidenov and V. A. Khlus, Fiz. Nizk. Temp. **21**, 594 (1995) [Low Temp. Phys. **21**, 462 (1995)]
[20] A. L. Fauchere, G. B. Lesovik, and G. Blatter, Phys. Rev. B **58**, 11177 (1998).
[21] K. M. Schep and G. E. W. Bauer, Phys. Rev. B **56**, 15860 (1997).
[22] P. Samuelsson, Phys. Rev. B **67**, 054508 (2003).
[23] M. P. V. Stenberg and T T. Heikkilä, Phys. Rev. B **66**, 144504 (2002).
[24] F. Pistolesi, G. Bignon, and F. W. J. Hekking, cond-mat/0303165 (unpublished).
[25] F. Lefloch, C. Hoffmann, M. Sanquer, and D. Quirion, Phys. Rev. Lett. **90**, 067002 (2003).
[26] Yu. V. Nazarov, Phys. Rev. Lett. **73**, 134 (1994).
[27] Yu. V. Nazarov, Ann. Phys. (Leipzig) **8**, SI-193 (1999).
[28] Yu. V. Nazarov, Superlattices Microst. **25**, 1221 (1999).
[29] W. Belzig and Yu. V. Nazarov, Phys. Rev. Lett. **87**, 197006 (2001).
[30] L. S. Levitov and G. B. Lesovik, JETP Lett. **58**, 230 (1993).
[31] L. S. Levitov, H. W. Lee, and G. B. Lesovik, J. Math. Phys. **37**, 4845 (1996).
[32] K. D. Usadel, Phys. Rev. Lett. **25**, 507 (1970).
[33] H. Lee, L. S. Levitov, and A. Yu. Yakovets, Phys. Rev. B **51**, 4079 (1996).
[34] W. Belzig in [2].

[35] O. N. Dorokhov, Solid State Comm. **51**, 381 (1984).

[36] W. Belzig and P. Samuelsson, cond-mat/0305249, to appear in Europhys. Lett.

[37] W. Belzig and Yu. V. Nazarov, Phys. Rev. Lett. **87**, 067006 (2001).

[38] Yu. V. Nazarov and T. H. Stoof, Phys. Rev. Lett. **76**, 823 (1996).

[39] M. Houzet and V. P. Mineev, Phys. Rev. B **67**, 184524 (2003).

[40] B. Reulet, D.E. Prober, and W. Belzig, in [2].

[41] B. Reulet, A.A. Kozhevnikov, D.E. Prober, W. Belzig, and Yu.V. Nazarov, Phys. Rev. Lett. **90**, 066601 (2003).

PROXIMITY EFFECT IN SUPERCONDUCTOR/FERROMAGNET LAYERED STRUCTURES

Anatoli Sidorenko
Institute of Applied Physics, LISES, MD2028 Kishinev, Moldova
sidorenko@lises.asm.md

Abstract Superconducting properties of proximity coupled Nb/Ni bilayers were investigated. Distinct oscillations of the superconducting critical temperature upon increasing of the ferromagnet layer thickness, d_{Ni}, are observed. The results are discussed in terms of Larkin-Ovchinnikov-Fulde-Ferrell (LOFF) quasi-one-dimensional superconducting state.

Keywords: superconductivity, proximity effect

1. Introduction

The coexistence of superconductivity (SC) and ferromagnetism (FM) seems to be very doubtful, because SC requires Cooper-pairing between electrons with antiparallel spin, whereas FM favors a parallel alignment of the electron spins. Nevertheless, the coexistence of both phenomena was theoretically considered by A. Larkin, Yu. Ovchinnikov, and P. Fulde, R. Ferrell [1,2]. Under a strong restricted conditions for the exchange field,

$$0.71\Delta_0 < E_{exc} < 0.76\Delta_0 \quad (1)$$

where Δ_0 is the superconducting gap, they predicted the existence of a superconducting state in the ferromagnet with a spatially modulated pair amplitude and nonzero pairing momentum, $\delta \boldsymbol{p}$. This state, known as the „3D-LOFF state“, is very difficult to realize experimentally because the value of the exchange field for ferromagnets usually is much higher than Δ_0 ($E_{exc} \sim$ 10000K, $\Delta_0 \sim$ 10K).

A more realistic system for the observation of the LOFF-like state was considered in a planar geometry (quasi-1D LOFF-like state) by A. Buzdin, and Z. Radovich [3,4]. They proposed to separate superconductor and ferromagnet, then the strict limitation (1) for the exchange field will be not important. Such

A.S. Alexandrov et al. (eds.), Molecular Nanowires and Other Quantum Objects, 129–138.

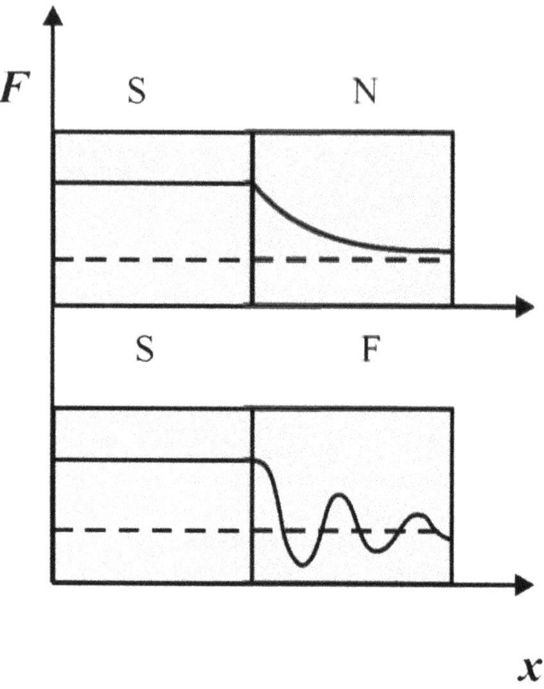

Figure 1. Proximity effect in a superconductor/normal metal (S/N) bilayer with an exponential decay of pairing function, $\boldsymbol{F}_F \sim \boldsymbol{e}_N^{-x/\xi}$ in the N-layer, and in a proximity coupled superconductor/ferromagnet (S/F) bilayer with oscillating $\boldsymbol{F}_F \sim \boldsymbol{e}_F^{-ix/\xi}$ within the F-layer.

"proximity-coupled" S/F system is shown in Fig.1. In the usual S/N- proximity coupled system an monotonous decrease of the paring function, $\boldsymbol{F}$, within the N-layer [5] leads to an monotonous decrease of the critical temperature of S/N bilayers when the N-layer thickness increases, as was confirmed in a lot of experimental works at the beginning of 70-es. In contrast to S/N systems, for the S/F coupled bilayers a very unusual phenomenon was predicted by A.Buzdin: an oscillating dependence of T_c as a function of magnetic layer thickness, d_F. Due to the proximity effect, superconducting electrons diffuse from the S-layer into the ferromagnetic layer and are subjected to the exchange field. A maximum of $T_c(d_F)$ occurs when the period of the Usadel function describing the pair amplitude, $\boldsymbol{F}$, matches the thickness of the ferromagnetic layer. Further theoretical investigations [6] predict rather different behavior of $T_c(d_F)$: from a monotonic suppression of T_c and its saturation at some d_F up to re-entrant superconductivity. In spite of the well developed theory of S/F layered systems, there exist only a few experimental investigations of $T_c(d_F)$ in which reliable oscillations of $T_c(d_F)$ were observed (see [7] and the References therein). Crucial difficulties arose in these experiments: high roughness of the S/F interface, which destroys the oscillating phenomena, and the problem of precise thickness control of the ferromagnet layer of some nanometer thickness. We will consider in the present work the experimental conditions necessary for reproducible observation of the LOFF-like 1D superconducting state with T_c–oscillations.

2. Sample preparation and characterization

Sample deposition

Among a different possible S/F – couples for sample preparation we have chosen Nb and Ni. The reason of this choice was a very low mutual solubility of this two metals, less than 4 at.% at ambient conditions [8], that leads to obtaining a well resolved clean S/F-interface. We did not take absolute immiscible metals, like Nb and Gd, which could give an island-like growth of the F-layer and as the result, a poor quality rough S/F interface. The Nb/Ni samples have been prepared by DC magnetron sputtering on a flame-polished glass substrate kept at room temperature. Pure Nb (99,99%, "Redmet") and Ni (99,99%, "Johnson Matthey") were used as sputtering targets. The glass substrate of 10x75 mm^2 was put into the chamber symmetrically with respect to the Nb target. An oscillation motion of the substrate was applied during the Nb layer deposition to obtain thickness homogeneity. Then, without breaking the vacuum, a wedge-shaped Ni layer was deposited by shifting the substrate away from the symmetry axis of the Ni target. The resulting wedge sample was cut across the wedge to 1.5 mm wide strips using a diamond cutter. Following this routine we obtained a set of Nb/Ni samples with variable Ni layer thickness at constant Nb layer thickness. Platinum 50-μm wires were attached by silver paste for four-contact resistance measurements. The optimal deposition rate, controlled during deposition by magnetron current stabilization, was 24 nm/min for the Nb and 0.6 nm/s for the Ni layers.

X-ray diffraction analysis

The X-ray small angle reflectivity scans ($2\theta < 10^0$, Cu K_α radiation) were used to determine the layers roughness. Figure 2 shows a typical reflectivity pattern measured for one of the Nb/Ni sample, demonstrates well resolved oscillations of the reflectivity as a function of the angle with more than 15 periods. The weakly damped oscillations, persisting to a relatively large angle of about 5^0 give evidence for the high quality of the layers with smooth surfaces and interfaces. For a quantitative analysis of the X-ray data we used a simulation program, based on the Parratt formalism [9]. Fitting of the X-ray data gives the value of the layer thickness and the roughness σ_{rms} as the mean square deviation of the layer thickness from an "ideally" smooth layer. The maximal value of σ_{rms} for the investigated samples was less than 0.3 nm.

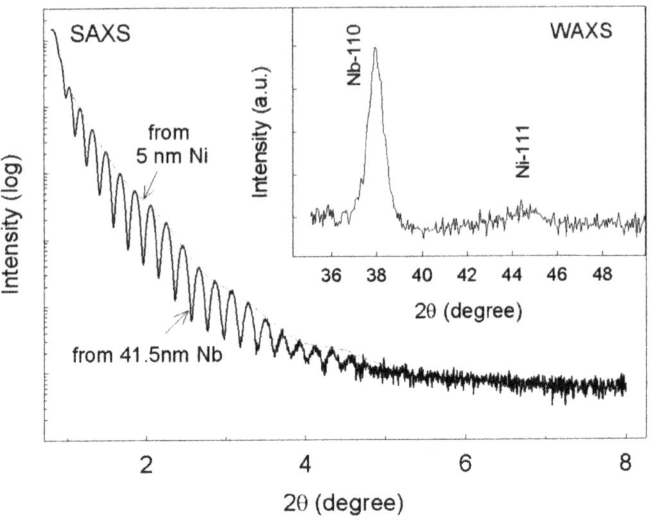

Figure 2. Small-angle X-ray spectra (SAXS) for a Nb/Ni bilayer on a glass substrate (d_{Nb} =41.5 nm, d_{Ni}=5 nm) demonstrating well defined oscillations of two periods: the short one arising from the total sample thickness (Nb+Ni) and the long one (dashed line) corresponding to the thin Ni layer. The solid line represents the fit according to the Parratt formalism [9]. The inset shows the wide-angle X-ray scattering (WAXS) pattern of the sample, indicating the (110)-textured structure of the Nb film.

RBS spectrometry

For the precise thickness measurements we used Rutherford backscattering spectrometry, giving the possibility to determine the absolute thickness of Ni layers at the level of 1 nm with an accuracy of 0.03 nm. Details of the RBS measurements are described in [7]. The results of thickness determination by RBS for one of investigated Nb/Ni-wedge samples are presented in Fig.3.

3. Results and discussion

Resistive transitions for one set of the Nb/Ni specimens are presented in Fig.4. The width of transition ($0.1R_N - 0.9R_N$ criteria) for all investigated samples

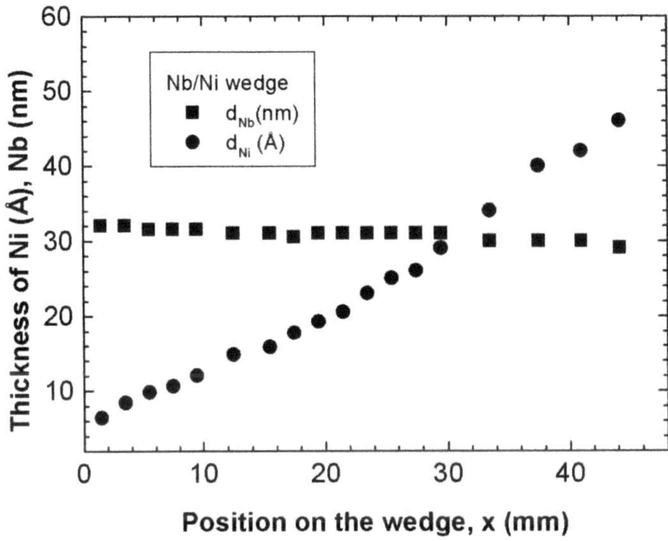

Figure 3. The results of RBS measurements for the thickness of Nb and Ni layers (samples set from Nb/Ni-wedge with $d_{Nb} = const = 31$ nm, d_{Ni} = variable).

was less than 0.1 K, which allows determination of the critical temperature from the midpoint of the resistive transitions with a proper accuracy. Figure 5 shows the dependence of the superconducting T_c versus the thickness of the ferromagnetic Ni layer for the samples with the thickness of Nb layer fixed at 31 nm. The critical temperature drops sharply when increasing d_{Ni} up to 1.2 nm, then it passes through a minimum and increases till the thickness d_{Ni} approaches $\sim$ 2.5 nm. With a further increase of the Ni layer thickness, T_c (d_{Ni}) passes through the maximum and decreases again showing an oscillatory behavior as a function of the Ni layer thickness. This oscillatory behavior is the evidence of the 1D-LOFF-like superconducting state existence in the investigated Nb/Ni samples.

As it was mentioned above, the proximity effect in S/F system has a crucial difference from that of the superconductor/normal metal (S/N) system. In a usual S/N system the pairing function exponentially relaxes deep into the N-layer on the scale of the coherence length $\xi_N = (\hbar D_N/2\pi k_B T)^{1/2}$, here D_N is the diffusion constant of the normal metal, $k_B$is the Bolzman constant. In S/F layered system, in contrast to S/N proximity coupled layers, the pairing function in a ferromagnet oscillates on a distance of the magnetic coherence length, $\xi_F = (\hbar v_F/E_{ex})$, and relaxes deep into the F-layer on the scale of

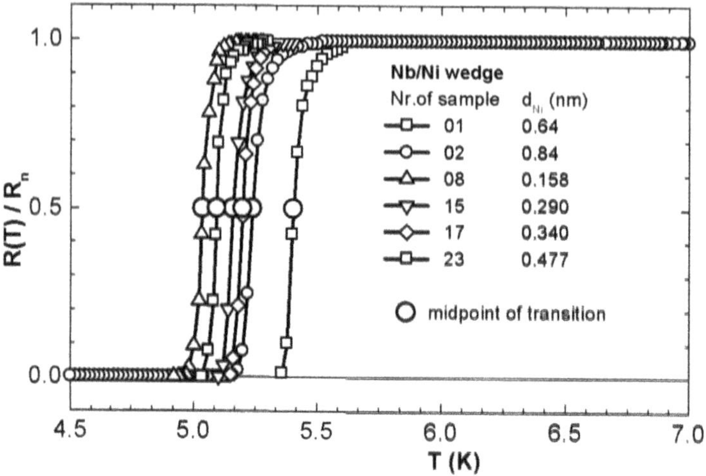

Figure 4. The resistive transitions $R(T)$ normalized to $R_n = R(T = 10$ K) for some of the measured samples from Nb/Ni-wedge with $d_{Nb} = const = 31$ nm, d_{Ni} = variable.

electron mean free path, l_F. In strong ferromagnets like Ni, Fe the oscillation period of the pairing function can be a few times shorter than the decay length l_F. If the thickness of the F-layer is smaller than or comparable to the l_F, the weakly damped pairing function wave incident on the S/F interface, will interfere with the wave reflected from the opposite surface of the F-layer. Using an analogy with light, the F-layer works like a Fabry-Perot interferometer, which can be highly reflective or almost transparent depending on the relation between the wavelength and the thickness of the interferometer. Thus, due to the interference, the pairing function flux will be modulated as a function of ferromagnet layer thickness d_F.

As a result, the coupling between S and F layers is modulated, and the superconducting T_c oscillates as a function of d_F. This type of oscillations we have observed in investigated Nb/Ni samples. For the quantitative comparison with the S/F-proximity theory we used the model developed by L. Tagirov [6] for bilayers with a finite S/F interface transparency T_F. In frame of the theory [6] superconducting transition temperature of the S/F-sandwich can be determine as a solution of equation

$$lnt_c + Re\psi(1/2 + \rho/t_c) - \psi(1/2) \tag{2}$$

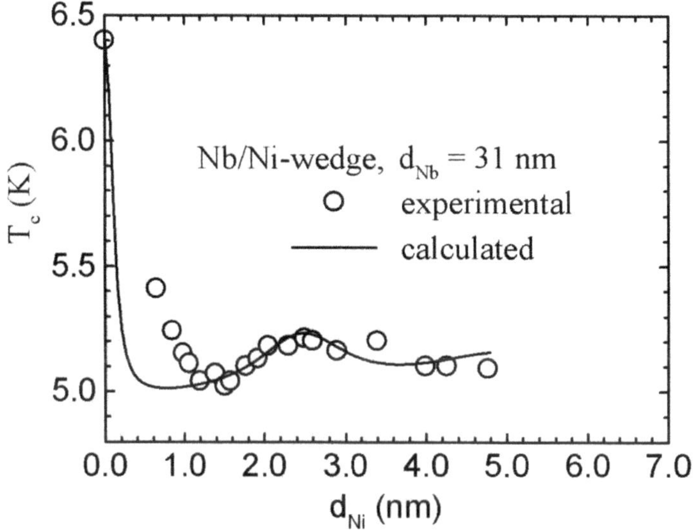

Figure 5. The dependence of the superconducting transition temperature T_c on the Ni layer thickness d_{Ni} for the sample set from the Nb/Ni-wedge with d_{Nb} =31 nm. The solid line demonstrates the best fit by the theory [6], the solid line is calculated using the parameters: $T_{c0} = 6.4$ K, $\xi_{BCS}(Nb) = 42$ nm, $\xi_S = 6.2$ nm, $(N_F v_F)/(N_S v_S) = 0.51$, $\xi_F/l_F = 0.5$, $T_F = 2.0$, $\xi_F = 0.88$ nm.

where $Re\psi$ is the real part of the digamma function, $t_c = T_c/T_{c0}$ is the reduced transition temperature,T_{c0} is the transition temperature of thick superconducting layer. Here the complex-valued pair-breaking parameter ρ is defined as $\rho = (k_S\xi_S)^2/2$, and k_S is the wave number of the pairing function propagation in the S-layer, which satisfy to the solution of equation:

$$(k_S d_S) tan(k_S d_S) = \frac{N_F v_F d_S \xi_{BCS}}{(N_S v_S \xi_S^2} [\frac{tanh(k_F d_F)}{(1 - i\xi_F/l_F)^{1/2}} + \frac{2}{T_F} tanh(k_F d_F)] \quad (3)$$

where N_F, N_S are the density of states at the Fermi level in F and S layers, respectively, v_F, v_S is the Fermi velocity, $\xi_{BCS} = (1.78\hbar v_S)/(\pi^2 k_B T_{c0})$, ξ_S =$(2/\pi)\xi_{GL}(0)$, Ginzburg-Landau coherence length $\xi_{GL}(0)$ is determined from the slope of the upper critical field, $\xi_{GL}(0) = [\frac{-dB_{c2}(T)}{dT}(2\pi T_c\phi_0)]^{-1/2}$, and T_F is the quantum mechanical interface transparency parameter, which equals to infinity for the perfectly transparent S/F interface, or is equal to zero for the completely reflecting interface. It takes into account the reflection of electrons from the interface caused by a mismatch of the Fermi momenta of the contacting metals.

According to the S/F proximity theory [6] three main regimes for T_c behavior can be observed in S/F bilayers depending on the thickness of the superconducting layer: a) for a large thickness, $d_S >> \xi_S$ the superconducting T_c can oscillate as a function of the ferromagnetic layer thickness d_F; b) for a thinner superconducting layer, when $T_c \sim (0.2\text{-}0.3)T_{c0}$ the regime of re-entrant superconductivity can be realized, i.e. the superconducting transition temperature drops to zero with the ferromagnetic layer thickness increasing, but with a further increase d_F the superconductivity appears again [10]; c) for the more thin superconducting layers the superconducting T_c falls down to zero upon increasing d_F.

One can conclude, that our experimental data belong to the regime a), demonstrating distinct oscillations of T_c (d_F). Fitting of the experimental results to the proximity effect theory [6], shown by solid cure in Fig.5 (with mediate value of the adjustable parameter of interface transparency T_F = 2.0), found a rather good agreement in the relevant range of Ni layer thickness above 1nm, but there is a deviation in the range of small thickness below 1 nm. The possible reason responsible for this deviation may be the existence of the magnetically "dead" layer at the S/F interface. As demonstrated in [11] the Nb/Fe couple always has an interdiffused layer at the S/F interface, which is magnetically "dead", i.e. the local magnetic moment disappears when increasing the number of Nb neighbors surrounding the Fe ion in the cluster. The same physical picture is realized also in V/Fe couples [12]. A magnetically dead-layer has also been observed at the V/Ni interface [13], where the vanishing of the Ni layer magnetic moment at d_{Ni} below 1 nm has been found. Finally, from a detailed study of Nb/Ni multilayers prepared by magnetron sputtering on a sapphire substrate [14] the authors concluded that below a Ni layer thickness of 1.4 nm the magnetic moment in the Ni layer disappears due to the interdiffusion of Nb - Ni and no ferromagnetic order is present. This thickness is close to the thickness $d_{Ni} \sim$ 1.0 nm in our study (see Fig.5), when the theory deviates from the experimental points. The suppression of superconductivity by a nonmagnetic dead-layer is weaker than by ferromagnetic Ni, and $T_c(d_{Ni})$ decreases slowly which meets our experimental observation.

4. Conclusions

We investigated the proximity effect in Nb/Ni bilayers prepared by DC magnetron sputtering on smooth glass substrates. The high quality of the films surface and the S/F interface was characterized by small-angle X-ray diffraction analysis. We observed distinct oscillations of the superconducting critical temperature for specimens with constant Nb layer thickness upon increasing the thickness of the Ni layer as an evidence of 1D-LOFF- like inhomogeneous superconducting pairing in the ferromagnetic Ni layer. We can now formulate the necessary conditions for experimental realization of the 1D-LOFF like state in S/F layered structures:

-The sputtering technique has advantages in comparison with MBE to fabricate S/F samples showing oscillating behaviour of $T_c(d_F)$;

-The S/F couples should not consist of immiscible metals, but of metals with restricted solubility and narrow composition ranges of intermetallic compounds formation for obtaining the smooth S/F interface;

-The substrate quality and deposition rate should ensure the S/F interface roughness less than the F-layer thickness.

S/F layered system prepared under those conditions will demonstrate 1D-LOFF-like superconducting state.

Acknowledgments

This work was partially supported through the INTAS Project Nr.9900585, and BMBF Project MDA02/002.

References

[1] Larkin, A.I., and Ovchinnikov, Yu.N. (1964) Nonuniform State of Superconductors, *Zh.Eksp.Teor.Fiz.* **47**,1136-1146.

[2] Fulde, P., and Ferrell, R. (1964) Superconductivity in a Strong Spin-Exchange Field, *Phys.Rev.* **135**, A550-A563.

[3] Buzdin, A.I., and Bulaevskii, L.N. (1982) Critical current oscillations as a function of the exchange field and thickness of the ferromagnetic metal (F) in an S-F-S Josephson junction, *Pis'ma Zh.Eksp.Teor.Fiz.* **35**, 147-148.

[4] Radovic, Z., Dobrosavljevic-Grujic, L., Buzdin, A.I., and Clem, J. (1988) Upper critical fields of superconductor-ferromagnet multilayers, *Phys.Rev.*B**38**,2388-2393.

[5] De Gennes, P.G. (1964) Boundary Effects in Superconductors, *Rev.Mod.Phys.***36**, 225-237.

[6] Tagirov, L.R. (1988) Proximity effect and superconducting transition temperature in superconductor/ferromagnet sandwiches, *Physica* C **307**, 145-163.

[7] Sidorenko, A.S., Zdravkov, V.I., Prepelitsa, A.A., Helbig, C., Luo, Y., Gsell, S., Schreck, M., Klimm, S., Horn, S., Tagirov, L.R., and Tidecks, R. (2003) Oscillations of the critical temperature in superconducting Nb/Ni bilayers, Ann.Phys.(Leipzig) **12**, 37-50.

[8] Massalski, Th.B. (ed.-chief) (1990) *Binary Alloy Phase Diagrams*, ASM International, Materials Park, Ohio.

[9] Parratt, L.G. (1954) Surface Studies of Solids by Total Reflection of X-Rays, *Phys.Rev.* **95**, 359-369.

[10] Tagirov, L.R., Garifullin, I.A., Garif'yanov N.N., Khlebnikov, S.Ya, Tikhonov, D.A., Westerholt, K., and Zabel, H. (2002) Re-entrant superconductivity in the V/Fe superconductor/ferromagnet layered system, *J. of Magnetism and Magnetic Materials* **240**, 577-579.

[11] Mühge, Th., Westerholt, K., Zabel, H., Garifyanov, N.N., Goryunov, Yu.V., Garifullin, I.A., and Khaliullin, G.G. (1997) Magnetism and superconductivity of Fe/Nb/Fe trilayers, *Phys. Rev.* B **55**, 8945-8954.

[12] Izquiero, J., Robles, R., Vega, A., Talanana, M., and Demangeat, C. (2001) Origin of dead magnetic Fe overlayers on V(110), *Phys. Rev.* B **64**, 060404-1-060404-4.

[13] Homma, H., Chun, C.S.L., Zheng, G.G., and I.K. Schuller, I.K. (1986) Interaction of superconductivity and itinerant-electron magnetism: Critical fields of Ni/V superlattices, *Phys. Rev.* B **33**, 3562-3565.

[14] Mattson, J.E., Osgood III, R.M., Potter, C.D., Sowers, C.H., and Bader, S.D. (1997) Properties of Ni/Nb magnetic/superconducting multilayers, *J. Vac. Sci. Technol.* A**15**, 1774-1779.

POLARONS IN SEMICONDUCTOR QUANTUM STRUCTURES

J. T. Devreese
Theoretische Fysica van de Vaste Stoffen (TFVS),
Universiteit Antwerpen, Universiteitsplein 1, B-2610 Antwerpen, Belgium;
also at: TU Eindhoven, P. O. Box 513, 5600 MB Eindhoven, The Netherlands
devreese@uia.ua.ac.be

Abstract In this presentation three recent contributions to the theory of continuum (or "Fröhlich"-) polarons are reviewed. (i) Using a generalization of the Jensen-Feynman variational principle within the path-integral formalism for identical particles, the ground-state energy of a confined N-polaron system is studied as a function of N and of the electron-phonon coupling strength. (ii) A theoretical investigation of the optical properties of stacked quantum dots is presented, which is based on the non-adiabatic approach. (iii) Cyclotron-resonance (CR) spectra of a gas of interacting polarons in a GaAs/AlAs quantum well are theoretically investigated taking into account the magnetoplasmon-phonon mixing and the band nonparabolicity. The theory explains that, for a high-density polaron gas, anticrossing of the CR spectra occurs near the GaAs TO-phonon frequency rather than near the GaAs LO-frequency in a good agreement with experimental data.

Keywords: polaron, quantum dot, cyclotron resonance, optical absorption, luminescence

1. Interacting polarons in a quantum dot

Thermodynamic and optical properties of interacting polarons have attracted increasing attention because of their possible relevance to physical phenomena in high-T_c superconductors (see, e. g., Ref. [1] and references therein). In this section, a system of N electrons with mutual Coulomb repulsion and interacting with the lattice vibrations is considered. A parabolic confinement potential, characterised by the frequency parameter Ω_0, is assumed. The total number of electrons is $N = \sum_\sigma N_\sigma$, where N_σ is the number of electrons with spin projection $\sigma = \pm 1/2$. A canonical ensemble is treated, where the numbers of electrons N_σ for each σ are fixed. The bulk phonons (characterized by wave vectors $\mathbf{q}$ and frequencies $\omega_\mathbf{q}$) are described by complex coordinates. The full set of electron coordinates is denoted by $\bar{\mathbf{x}} \equiv \{\mathbf{x}_{j,\sigma}\}$.

A.S. Alexandrov et al. (eds.), Molecular Nanowires and Other Quantum Objects, 139–150.

The partition function $Z(\{N_\sigma\},\beta)$ of the system can be expressed as a path integral over all electron and phonon coordinates. The path integral over the phonon variables in $Z(\{N_\sigma\},\beta)$ can be calculated analytically [2]. As a result, the partition function of the electron-phonon system factorises into a product of a free-phonon partition function with a partition function $Z_p\left(N_{1/2},N_{-1/2}|\beta\right)$ of interacting polarons, which is a path integral over the electron coordinates only:

$$Z(\{N_\sigma\},\beta)=Z_p(\{N_\sigma\},\beta)\prod_{\mathbf{q}}\frac{1}{2\sinh\left(\beta\hbar\omega_{\mathrm{LO}}/2\right)},\tag{1}$$

$$Z_p(\{N_\sigma\},\beta)=\sum_P\frac{(-1)^{\xi_P}}{N_{1/2}!N_{-1/2}!}\int d\bar{\mathbf{x}}\int_{\bar{\mathbf{x}}}^{P\bar{\mathbf{x}}}D\bar{\mathbf{x}}(\tau)\,e^{-S_p[\bar{\mathbf{x}}(\tau)]},\tag{2}$$

where $S_p[\bar{\mathbf{x}}(\tau)]$ results from the elimination of the phonon variables and contains the "influence phase" of the phonons. It describes the phonon-induced retarded interaction between the electrons, including the retarded self-interaction of each electron. The parameter $\beta\equiv 1/(k_BT)$ is inversely proportional to temperature T. In order to take the Fermi-Dirac statistics into account, the integral over the electron paths $\{\bar{\mathbf{x}}(\tau)\}$ contains a sum over all permutations P of the electrons with equal spin projections, with ξ_P denoting the parity of a permutation P. The free energy of a system of interacting polarons $F_p(\{N_\sigma\},\beta)$ is related to the partition function (2) by the relation:

$$F_p(\{N_\sigma\},\beta)=-\frac{1}{\beta}\ln Z_p(\{N_\sigma\},\beta).\tag{3}$$

At present no method is known to calculate the non-Gaussian path integral (2) analytically. For *distinguishable* particles, the Jensen-Feynman variational principle [2] provides a convenient approximation technique. It yields a lower bound to the partition function, and consequently an upper bound to the free energy. A many-body extension of the Jensen-Feynman inequality was found, which can be used for interacting identical particles (Ref. [3], p. 4476). A more detailed analysis of this variational principle for both local and retarded interactions can be found in Ref. [4]. It is required that the potentials are *symmetric* with respect to all permutations of the particle positions, and that both the exact propagator and the model propagator are *antisymmetric* (for fermions) with respect to permutations of any two electrons at any point in time. This means that those propagators must be defined on the same configuration space. Keeping in mind these requirements, the variational inequality for identical particles has the same form as the standard Jensen-Feynman variational principle:

$$F_p\leq F_0+\frac{1}{\beta}\left\langle S_p-S_0\right\rangle_{S_0},\tag{4}$$

where S_0 is a model action with corresponding free energy F_0. S_0 must fulfil the properties, which were mentioned above.

In Ref. [5], we have chosen a model system consisting of N electrons with coordinates $\bar{\mathbf{x}} \equiv \{\mathbf{x}_{j,\sigma}\}$ coupled to N_f "fictitious" particles with coordinates $\bar{\mathbf{y}} \equiv \{\mathbf{y}_j\}$ in a harmonic confinement potential with elastic interparticle interactions as studied in Ref. [6]. Both the free energy and the correlation functions of the model system are calculated analytically using the generation-function technique [7]. Further on, the zero-temperature case is considered.

Confined few-electron systems without the electron-phonon interaction can exist in one of two phases: the spin-polarized state and a state obeying Hund's rule, depending on the confinement frequency (see, e. g., Ref. [8]). We found [5] that for interacting few-polaron systems, besides the above two phases, there may occur also a third phase — the state with minimal spin — in quantum dots of polar substances with sufficiently strong electron-phonon coupling $\alpha \geq 3$ (for instance, of high-T_c superconductors [9]).

To investigate the optical properties of the many-polaron system, in Ref. [5], the memory-function formalism of Ref. [10] is extended to the case of interacting polarons in a quantum dot. Within this technique, the optical conductivity for a system of interacting polarons in a parabolic confinement potential is given in terms of the memory function $\chi(\omega)$,

$$\mathrm{Re}\sigma(\omega) = -\frac{e^2}{m} \frac{\omega \mathrm{Im}\chi(\omega)}{\left[\omega^2 - \Omega_0^2 - \mathrm{Re}\chi(\omega)\right]^2 + \left[\mathrm{Im}\chi(\omega)\right]^2}, \tag{5}$$

where $\chi(\omega)$ is

$$\chi(\omega) = \sum_{\mathbf{q}} \frac{2\left|V_{\mathbf{q}}\right|^2 q^2}{3N\hbar\omega} \int_0^\infty \left(e^{i\omega t} - 1\right) \mathrm{Im}\left[T_{\omega_{\mathrm{LO}}}^*(t) \left\langle \rho_{\mathbf{q}}(t) \rho_{-\mathbf{q}}(0)\right\rangle_M\right] dt. \tag{6}$$

Here, $T_\omega(t) = \cos\left[\omega\left(t - i\hbar\beta/2\right)\right] / \sinh\left(\beta\hbar\omega/2\right)$ is the phonon Green's function, while $\left\langle \rho_{\mathbf{q}}(t) \rho_{-\mathbf{q}}(0)\right\rangle_M$ is the density-density correlation function calculated using the model system of electrons harmonically interacting with fictitious particles. For a translationally invariant system $\Omega_0 \to 0$ the weak-coupling limit of the optical conductivity (5) reproduces the "central peak" of the polaron optical conductivity (see Eq. (5) of Ref. [11]). In the zero-temperature limit, the memory function (6) has been found analytically [5]. The changes of the shell filling schemes, which occur when varying the confinement frequency, also manifest themselves in the spectra of the optical conductivity. In Fig. 1, optical conductivity spectra for $N = 20$ polarons are presented for a quantum dot with the parameters of CdSe: $\alpha = 0.46$, $\eta = 0.656$ [12] and with different values of the confinement energy $\hbar\Omega_0$. In this case, the spin-polarized ground state changes to the ground state satisfying Hund's rule with increasing $\hbar\Omega_0$ in the interval $0.0421H^* < \hbar\Omega_0 < 0.0422H^*$.

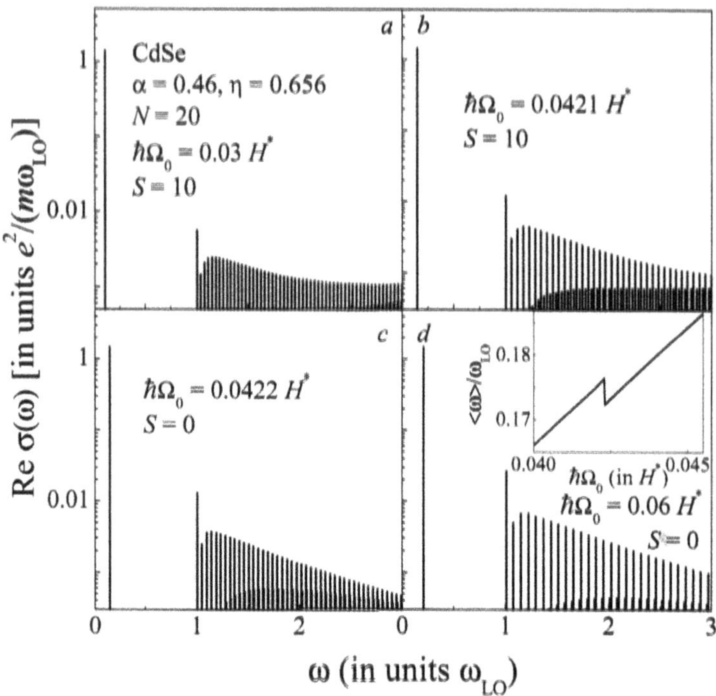

Figure 1. Optical conductivity spectra of $N = 20$ interacting polarons in quantum dots with $\alpha = 0.46$, $\eta = 0.656$ for different confinement energies close to the transition from a spin-polarized ground state to a ground state obeying Hund's rule. *Inset*: the first frequency moment $\langle\omega\rangle$ of the optical conductivity as a function of the confinement energy. (From Ref. [5].)

In the inset to Fig. 1, the first frequency moment of the optical conductivity as a function of $\hbar\Omega_0$ shows a *discontinuity*, at the value of the confinement energy corresponding to the change of the shell filling schemes from the spin-polarized ground state to the ground state obeying Hund's rule. This discontinuity should be observable in optical measurements.

The shell structure for a system of interacting polarons in a quantum dot is clearly revealed when analysing both the addition energy and the first frequency moment of the optical conductivity. The addition energy $\Delta(N)$ needed to put an extra electron into a quantum dot containing N electrons is defined as

$$\Delta(N) = E^0(N+1) - 2E^0(N) + E^0(N-1), \tag{7}$$

where $E^0(N)$ is the ground-state energy. In Fig. 2, we show both the function

$$\Theta(N) \equiv \langle\omega\rangle|_{N+1} - 2\langle\omega\rangle|_N + \langle\omega\rangle|_{N-1}, \tag{8}$$

and the addition energy $\Delta(N)$.

As seen from Fig. 2, distinct peaks appear in $\Theta(N)$ and $\Delta(N)$ at the "magic numbers" for closed-shell configurations at $N = 8, 20$ for the state obeying Hund's rule in panels a, b. In the case when the shell filling scheme is one and the same for different N (see panels a, b, where the filling obeys Hund's rule), each of the peaks of $\Theta(N)$ corresponds to a peak of the addition energy. In the case when the shell filling scheme changes with varying N (panels c, d), the function $\Theta(N)$ exhibits pronounced minima for N corresponding to the change of the filling scheme from the states, obeying Hund's rule, to the spin-polarized states.

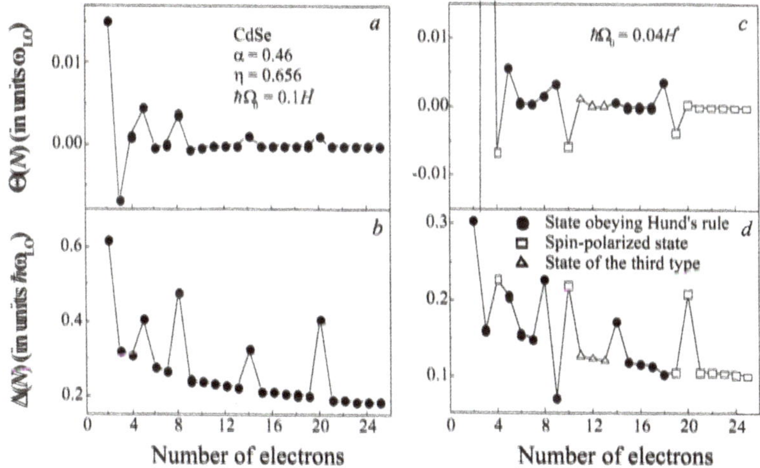

Figure 2. The function $\Theta(N)$ and the addition energy $\Delta(N)$ for systems of interacting polarons in CdSe quantum dots with $\alpha = 0.46$, $\eta = 0.656$. Open squares denote the spin-polarized ground state, full dots denote the ground state, obeying Hund's rule. The open triangles denote the ground state of the third type, with more than one partly filled shells, which is not totally spin-polarized. (From Ref. [5].)

It follows that measurements of the addition energy and the first frequency moment of the optical absorption as a function of the number of polarons in a quantum dot can reflect the difference between open-shell and closed-shell configurations. In particular, the closed-shell configurations may be revealed through peaks in the function $\Theta(N)$. The filling patterns for a many-polaron system in a quantum dot can be determined from the analysis of the first moment of the optical absorption for different numbers of polarons. The appearance of minima in the function $\Theta(N)$ will then indicate a transition from the states which are filled according to Hund's rule to the spin-polarized states.

2. Optical properties of polaronic excitons in quantum dots: Role of non-adiabaticity

In this section, I present a theoretical investigation of the optical properties of quantum dots, which is based on the non-adiabatic approach developed in Refs. [13, 14, 15]. The effects of non-adiabaticity are important to interpret the surprisingly high intensities of the phonon 'sidebands' observed in the optical absorption, the photoluminescence and the Raman spectra of quantum dots. Deviations of the phonon-peak intensities, observed in some experimental optical spectra, from the Franck-Condon progression, which is prescribed by the commonly used adiabatic approximation, find a natural explanation within our non-adiabatic approach [13, 14, 15, 16].

Here, the non-adiabatic approach is applied to the particular case of stacked self-assembled quantum dots. Recently, stacked quantum dots have gained an increasing attention (see, e. g., Refs. [17, 18, 19]) due to a possibility to finely control their energy spectra. A parallelepiped-shaped quantum dot is considered as a model for a self-assembled quantum dot in a stack. The exciton-phonon interaction is taken into account for all phonon modes specific for these quantum dots (bulk-like, half-space and interface phonons).

We calculate the optical absorption spectrum of stacked quantum dots using the Kubo formula. Within our non-adiabatic approach the following expression results for the linear coefficient of the optical absorption by the exciton-phonon system in a quantum-dot structure:

$$\alpha(\Omega) \propto \mathrm{Re} \sum_{\beta,\beta'} d_\beta^* d_{\beta'} \int_0^\infty dt\, e^{i(\Omega-\Omega_\beta+i\epsilon)t} \left\langle \beta \left| \bar{U}(t) \right| \beta' \right\rangle , \tag{9}$$

where Ω is the frequency of the incident light, while d_β and Ω_β are, respectively, the electric dipole matrix element and the Franck-Condon frequency of a transition between the exciton vacuum state and the one-exciton state $|\beta\rangle$. The evolution operator averaged over the phonon ensemble, $\bar{U}(t)$, is

$$\begin{aligned} \bar{U}(t) &= \mathrm{T} \exp\left\{ -\frac{1}{\hbar^2} \sum_\lambda \int_0^t dt_1 \int_0^{t_1} dt_2 \left[\bar{n}_\lambda \mathrm{e}^{i\omega_\lambda(t_1-t_2)} \gamma_\lambda^\dagger(t_1)\gamma_\lambda(t_2) \right.\right. \\ &+ \left.\left. (\bar{n}_\lambda + 1)\, \mathrm{e}^{-i\omega_\lambda(t_1-t_2)} \gamma_\lambda(t_1)\gamma_\lambda^\dagger(t_2) + \right] \right\}. \end{aligned} \tag{10}$$

In Eq. (10), T is the time ordering operator, λ labels the phonon modes specific for the quantum-dot structure under consideration, ω_λ are phonon frequencies, $\gamma_\lambda(t)$ are the exciton-phonon interaction amplitudes in the interaction representation, and $\bar{n}_\lambda = [\exp(\hbar\omega_\lambda/k_B T) - 1]^{-1}$.

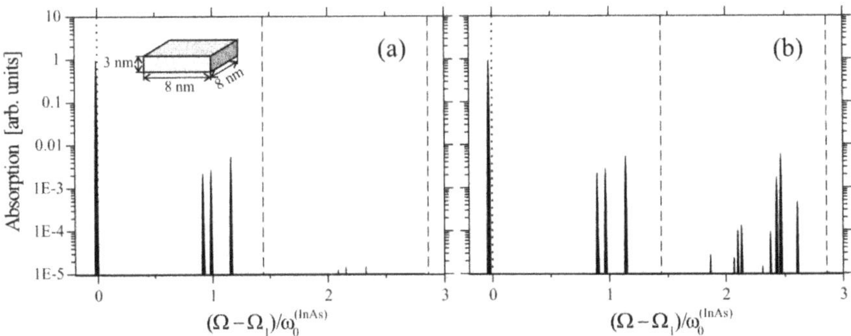

Figure 3. Absorption spectra, calculated for a single quantum dot with the adiabatic approximation [panel (a)] and with the non-adiabatic approach [panel (b)]. Optically active and non-active energy levels of a bare exciton are shown as dotted and dashed lines, respectively. Ω_1 is the transition frequency for the lowest state of a bare exciton. (From Ref. [22].)

Within the adiabatic approximation, which has been widely used to calculate the optical spectra of quantum dots, non-diagonal matrix elements of the exciton-phonon interaction are neglected when calculating $\alpha(\Omega)$ as given by Eq. (9) with Eq. (10). In the adiabatic approach [20, 21] one supposes that (i) both the initial and the final states of a quantum transition are non-degenerate, (ii) the energy differences between the exciton states are much larger than the phonon energies. It has been shown in Refs. [13, 14, 15, 16] that these conditions are often violated for optical transitions in quantum dots. In other words, the exciton-phonon system in a quantum dot can be essentially *non-adiabatic*.

In Ref. [13], a method was proposed to calculate the absorption spectrum given by Eqs. (9) and (10) for a spherical quantum dot taking into account non-adiabaticity of the exciton-phonon system. This approach has been further refined in Ref. [22]: for the matrix elements of the evolution operator a closed set of equations has been obtained using a diagrammatic technique. This set describes the effect of non-adiabaticity both on the intensities and on the positions of the absorption peaks.

In Figs. 3 and 4 the calculated optical absorption spectra are shown for a single quantum dot and for a system of two stacked quantum dots, respectively. From the comparison of the spectra obtained in the adiabatic approximation with those resulting from the non-adiabatic approach, the following effects of non-adiabaticity are revealed.

First, the *polaron shift* of the zero-phonon lines with respect to the bare-exciton levels is larger in the non-adiabatic approach than in the adiabatic approximation. Second, there is a strong *increase of the intensities of the phonon satellites* as compared to those given by the adiabatic approximation. This increase can be by more than two orders of magnitude. Third, in the optical ab-

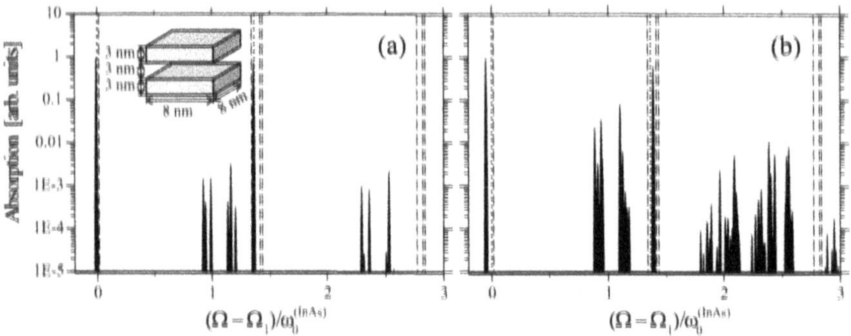

Figure 4. Absorption spectra, calculated for a system of two stacked quantum dots with the adiabatic approximation [panel (a)] and with the non-adiabatic approach [panel (b)]. Optically active and non-active energy levels of a bare exciton are shown as dotted and dashed lines, respectively. Ω_1 is the transition frequency for the lowest state of a bare exciton. (From Ref. [22].)

sorption spectra found within the non-adiabatic approach, there appear phonon satellites related to *non-active bare exciton states.*

Fourth, the optical-absorption spectra demonstrate the crucial role of *non-adiabatic mixing* of different exciton and phonon states in quantum dots. This results in a rich structure of the absorption spectrum of the exciton-phonon system [23, 14, 15]. For the stacked quantum dots, this effect is enhanced in the (quasi-) resonant case, when the exciton-level splitting, caused by the coupling between quantum dots, is close to a LO phonon energy [see panel (b) in Fig. 4].

Due to non-adiabaticity, multiple absorption peaks appear in spectral ranges characteristic for phonon satellites. From the states, which correspond to these peaks, the system can rapidly relax to the lowest emitting state. Therefore, in the photoluminescence excitation (PLE) spectra of quantum dots, pronounced peaks can be expected in spectral ranges characteristic for phonon satellites. New experimental evidence of the enhanced phonon-assisted absorption due to effects of non-adiabaticity has been recently provided by PLE measurements on single self-assembled InAs/GaAs [24] and InGaAs/GaAs [25] quantum dots.

3. Many-polaron cyclotron resonance in quantum wells

In the presence of a magnetic field, resonant magnetopolaron coupling (i. e., the anticrossing of zero-phonon and one-phonon states of the polaron system) can occur. The resonant magnetopolaron coupling manifests itself near the LO-phonon frequency for low electron densities (see, e. g., Refs. [26, 27, 28]). Cyclotron-resonance measurements performed on semiconductor quantum wells with high electron density [29] reveal anticrossing near the TO-phonon frequency rather than near the LO-phonon frequency. In Ref. [29], this effect is interpreted by invoking mixing between magnetoplasmons and phonons and in terms of a resonant coupling of the electrons with the mixed magneto-plasmon-phonon modes.

In Ref. [30], CR spectra for a polaron gas in a GaAs/AlAs quantum well are theoretically investigated, taking into account (i) the electron-electron interaction and the screening of the electron-phonon interaction, (ii) the magnetoplasmon-phonon mixing, (iii) the electron-phonon interaction with all the phonon modes specific for the quantum well under investigation. As a result of this mixing, different magnetoplasmon-phonon modes appear in the quantum well, which give contributions to the CR spectra.

In order to take into account the splitting of the CR peaks due to the non-parabolicity of the conduction band, the memory-function expression [31, 32] for the optical conductivity is generalized in the following way,

$$\mathrm{Re}\sigma\left(\omega\right)=-\frac{n_{S}e^{2}}{m_{b}}\sum_{n,l,\sigma}\mathrm{Im}\frac{\varkappa_{nl\sigma}}{\omega-\omega_{C}^{(nl\sigma)}-\chi\left(\omega,\omega_{C}^{(nl\sigma)}\right)/\omega+i0^{+}},\tag{11}$$

where n_S is the 2D electron density, $\chi\left(\omega,\omega_{C}^{(nl\sigma)}\right)$ is the memory function [30],

$$\omega_{C}^{(nl\sigma)}\equiv\frac{E_{n,l+1,\sigma}-E_{nl\sigma}}{\hbar}\tag{12}$$

is the transition frequency. The weight $\varkappa_{nl\sigma}$ is proportional to the number of open channels for the transitions $(l\rightarrow l+1)$. The normalized weights $\varkappa_{nl\sigma}$ are

$$\varkappa_{nl\sigma}=\frac{f_{nl\sigma}\left(1-f_{n,l+1,\sigma}\right)}{\sum_{n',l',\sigma'}f_{n'l'\sigma'}\left(1-f_{n',l'+1,\sigma'}\right)},\tag{13}$$

where $f_{nl\sigma}$ is the Fermi occupation number.

In Fig. 5, a set of the calculated magnetoabsorption spectra for a 10-nm GaAs/AlAs quantum well is plotted as a density map. This map shows the magnetoabsorption intensity as a function of both the magnetic field and the frequency. For comparison of these spectra with the experimental data of Ref. [29], the peak positions from Fig. 3 of Ref. [29] are shown in the same graph. Splitting of the peaks occurs because the electron Landau levels in a non-parabolic conduction band are non-equidistant. In Fig. 5, the higher-energy part of each split peak corresponds to the transitions between the Landau levels $(0\rightarrow 1)$, while the lower-energy part corresponds to the transitions $(1\rightarrow 2)$. With increasing magnetic-field strength, the upper filled level with $l=1$ becomes less populated, so that the relative intensity of the lower-energy peak diminishes. For the series of peaks indicated by crosses the splitting due to the band nonparabolicity is not experimentally resolved.

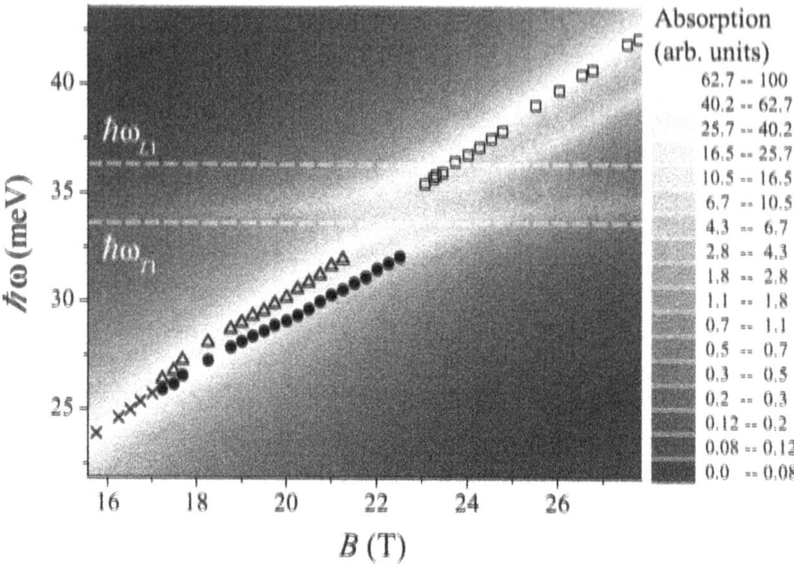

Figure 5. Density map of the magnetoabsorption spectra for a 10-nm GaAs/AlAs quantum well as calculated in Ref. [30]. Symbols indicate peak positions of the experimental spectra (which are taken from Fig. 3 of Ref. [29]). Dashed lines show LO- and TO-phonon energies in GaAs. (After Ref. [30].)

It is clearly seen from Fig. 5, that for the high-density polaron gas, anticrossing of the CR spectra occurs near the GaAs TO-phonon frequency ω_{T1} rather than near the GaAs LO-phonon frequency ω_{L1} for both the experimental and the calculated spectra. This effect is in contrast with the cyclotron resonance of a low-density polaron gas in a quantum well, where anticrossing occurs near the LO-phonon frequency. The appearance of the anticrossing frequency close to ω_{T1} instead of ω_{L1} is due to the screening of the electron-phonon interaction by the plasma vibrations. A similar effect appears also for the magnetophonon resonance: as shown in Ref. [33], the magnetoplasmon-phonon mixing leads to the shift of the resonant frequency of the magnetophonon resonance in quantum wells from $\omega \approx \omega_{L1}$ to $\omega \approx \omega_{T1}$.

Acknowledgments

I thank V. Fomin, F. Brosens, V. Gladilin, S. Klimin and J. Tempere for discussions during the preparation of this manuscript. This work has been supported by the GOA BOF UA 2000, I.U.A.P., F.W.O.-V. projects G.0274.01, G.0435.03, the W.O.G. WO.025.99 (Belgium) and the European Commission GROWTH Programme, NANOMAT project, contract No. G5RD-CT-2001-00545.

References

[1] Alexandrov, A. S., and Mott, N., *Polarons and bipolarons*, World Scientific, Singapore, 1996.

[2] Feynman, R. P., *Statistical Mechanics*, Benjamin, Massachusetts, 1972.

[3] Lemmens, L. F., Brosens, F., and Devreese, J. T., *Phys. Rev. E* **53**, 4467 (1996).

[4] Devreese, J. T., "Note of the path-integral variational approach in the many-body theory", in *Fluctuating Paths and Fields*, World Scientific, Singapore, 2001, pp. 289 - 304.

[5] Klimin, S. N., Fomin, V. M., Brosens, F., and Devreese, J. T., to be published.

[6] Devreese, J. T., Klimin, S. N., Fomin, V. M., and Brosens, F., *Solid State Commun.* **114**, 305 (2000).

[7] Brosens, F., Devreese, J. T., and Lemmens, L. F., *Phys. Rev. E* **55**, 227 (1997); **55**, 6795 (1997); **58**, 1634 (1998).

[8] Reimann, S. M., Koskinen, M., and Manninen, M., Phys. Rev. B **62**, 8108 (2000).

[9] Devreese, J. T., and Tempere, J., *Solid State Commun.* **106**, 309 (1998).

[10] Devreese, J. T., De Sitter, J., and Goovaerts., M., *Phys. Rev. B*, **5**, 2367 (1972).

[11] Devreese, J. T., Lemmens, L. F., and Van Royen, J., *Phys. Rev. B* **15**, 1212 (1977).

[12] Kartheuser, E., in: *Polarons in Ionic Crystals and Polar Semiconductors*, North-Holland, Amsterdam, 1972, pp. 717-733.

[13] Fomin, V. M., Gladilin, V. N., Devreese, J. T., Pokatilov, E. P., Balaban, S. N., and Klimin, S. N., *Phys. Rev. B* **57**, 2415 (1998).

[14] Devreese, J. T., Fomin, V. M., Gladilin, V. N., Pokatilov, E. P., and Klimin, S. N., *Nanotechnology* **13**, 163 (2002).

[15] Devreese, J. T., Fomin, V. M., Pokatilov, E. P., Gladilin, V. N., and Klimin, S. N., *Phys. Stat. Sol. (c)* **0**, 1189 (2003).

[16] Pokatilov, E. P., Klimin, S. N., Fomin, V. M., Devreese, J. T., and Wise, F. W., *Phys. Rev. B* **65**, 075316, 8 pages(2002).

[17] Anisimovas, E., and Peeters, F. M., Phys. Rev. B **65**, 233302 (2002).

[18] Holleitner, A. W., Blick, R. H., and Eberl, K., Appl. Phys. Letters **82**, 1887 (2003).

[19] Rebohle, L. Schrey, F. F., Hofer, S., Strasser, G., and Unterrainer, K., Appl. Phys. Letters **81**, 2079 (2002).

[20] Pekar, S. I., *Zh. Eksp. Teor. Fiz.* **20**, 267 (1950).

[21] Huang, K., and Rhys, A., *Proc. R. Soc. London, Ser. A* **204** , 406 (1950).

[22] Gladilin, V. N., Klimin, S. N., Fomin, V. M., and Devreese, J. T., to be published.

[23] Gladilin, V. N., Balaban, S. N., Fomin, V. M., and Devreese, J. T., in: *Proc. 25th Int. Conf. on the Physics of Semiconductors, Osaka, Japan, 2000*, Springer, Berlin, 2001, Part II, p. 1243-1244.

[24] Lemaître, A., Ashmore, A. D., Finley, J. J., Mowbray, D. J., Skolnick, M. S., Hopkinson, M., and Krauss, T. F., *Phys. Rev. B* **63**, 161309(R)(2001).

[25] Zrenner, A., Findeis, F., Baier, M., Bichler, M., and Abstreiter, G., *Physica B* **298**, 239 (2001).

[26] Chang, Y.-H., McCombe, B. D., Mercy, J.-M., Reeder, A. A., Ralston, J., and Wicks, G. A., *Phys. Rev. Lett.* **61** 1408 (1988).

[27] Vaughan, T.A., Nicholas, R.J., Langerak, C.J.G.M., Murdin, B.N., Pidgeon, C.R., Mason, N.J., and Walker, P.J., Phys. Rev. B **53**, 16481(1996).

[28] Wang, Y.J., Nickel, H.A., McCombe, B.D., Peeters, F.M., Shi, J.M., Hai, G.Q., Wu, X.-G., Eustis, T.J., and Schaff, W., Phys. Rev. Lett. **79**, 3226 (1997).

[29] Poulter, A. J. L., Zeman, J., Maude, D. K., Potemski, M., Martinez, G., Riedel, A., Hey, R., and Friedland, K. J., *Phys. Rev. Lett.* **86**, 336 (2001).

[30] Klimin, S. N. and Devreese, J. T., to be published.

[31] Wu, X.-G., Peeters, F. M., and Devreese, J. T., *Phys. Rev. B* **36**, 9760 (1987).

[32] Wu, X., Peeters, F. M., and Devreese, J. T., *Phys. Rev. B* **40**, 4090 (1989).

[33] Afonin, V. V., Gurevich, V. L., and Laiho, R., *Phys. Rev. B* **62**, 15913 (2000).

POLARONS IN COMPLEX OXIDES AND MOLECULAR NANOWIRES

A. S. Alexandrov
Department of Physics, Loughborough University, Loughborough, United Kingdom
a.s.alexandrov@lboro.ac.uk

Abstract There is a growing understanding that transport properties of complex oxides and individual molecules are dominated by polaron physics. In superconducting oxides the long-range Fröhlich and short-range Jahn-Teller electron-phonon interactions bind carriers into real space pairs - small bipolarons with surprisingly low mass but sufficient binding energy, while the long-range Coulomb repulsion keeps bipolarons apart preventing their clustering. The bipolaron theory numerically explains high Tc values without any fitting parameters and describes other key features of the cuprates. The same approach provides a new insite into the theory of transport through molecular nanowires and quantum dots (MQD). Attractive polaron-polaron correlations lead to a "switching" phenomenon in the current-voltage characteristics of MQD. The degenerate MQD with strong electron-vibron coupling has two stable current states (a volatile memory), which might be useful in molecular electronics.

Keywords: Polarons, bipolarons, superconductivity, molecular memory

1. Introduction

When the electron-phonon (e-ph) interaction energy E_p is larger than their kinetic energy, electrons in the Bloch band are "dressed" by phonons. If phonon frequencies are very low, the local lattice deformation traps the electron even in a perfect crystal lattice. This *self-trapping* phenomenon was predicted by Landau [1]. It has been studied in greater detail by Pekar [2], Fröhlich [3], Feynman [4], Devreese [5] and other authors in the effective mass approximation, which leads to the so-called *large* or *continuous* polaron. The large polaron propagates through the lattice as a free electron but with an enhanced effective mass.

In the strong-coupling regime, $\lambda = E_p/D > 1$, the finite bandwidth $2D$ becomes important, so that the effective mass approximation cannot be applied. The electron is called a *small* or *lattice* polaron in this regime. The self-trapping is never "complete", that is any polaron can tunnel through the lattice. Only

A.S. Alexandrov et al. (eds.), Molecular Nanowires and Other Quantum Objects, 151–166.

in the extreme *adiabatic* limit, when the phonon frequencies tend to zero, the self-trapping is complete, and the polaron motion is no longer translationally continuous. The main features of the small polaron were understood by Tjablikov [6], Yamashita and Kurosava [7], Sewell [8], Holstein [9] and his school [10, 11], Lang and Firsov [12], Eagles [13], and others and described in several review papers and textbooks [5, 14, 15, 16, 17, 18]. An exponential reduction of the bandwidth at large values of λ and phonon side-bands are among those features.

The lattice deformation also strongly affects the interaction between electrons. At large distances polarons repel each other in ionic crystals, but their Coulomb repulsion is substantially reduced due to the ion polarization. Nevertheless two *large* polarons can be bound into a *large* bipolaron by an exchange interaction even with no additional e-ph interaction but the Fröhlich one [5].

When a short-range deformation potential and molecular-type (i.e. Jahn-Teller [19]) e-ph interactions are taken into account together with the Fröhlich interaction [20], they can overcome the Coulomb repulsion. The resulting interaction becomes attractive at a short distance of about a lattice constant. Then two small polarons easily form a bound state, i.e. a *small* bipolaron, because their band is narrow. Consideration of particular lattice structures shows that small bipolarons are mobile even when the electron-phonon coupling is strong and the bipolaron binding energy is large [20]. Hence the polaronic Fermi liquid transforms into a Bose liquid of double-charged carriers in the strong-coupling regime. The Bose-liquid is stable because bipolarons repel each other [20]. Here we encounter a novel electronic state of matter, a charged Bose liquid, qualitatively different from the normal Fermi-liquid and from the BCS superfluid.

Experimental evidence for an exceptionally strong electron-phonon interaction in high temperature superconductors is now overwhelming. As we discussed in detail elsewhere [21], the extension of the BCS theory towards the strong interaction between electrons and ion vibrations describes the phenomenon naturally. High temperature superconductivity exists in the crossover region of the electron-phonon interaction strength from the BCS-like to bipolaronic superconductivity as was predicted before [22], and explored in greater detail by many authors after the discovery [23].

Small polarons with their phonon side-bands and attractive correlations are quite feasible also in molecular nanowires and quantum dots (MQD) used as the "transmission lines" [24, 25] and active molecular elements [26, 27] in molecular-scale electronics [26]. It has been experimentally demonstrated that the low-bias conductance of molecules is dominated by resonant tunneling through coupled electronic and vibration levels [28]. Conductance peaks due to electron-vibron interactions has been seen in C_{60} [29]. Different aspects of the electron-phonon/vibron (e-ph) interaction effect on the tunneling through molecules and quantum dots (QD) have been studied by several authors [30, 31, 32, 33, 34, 35, 36]. In particular, Glazman and Shekhter, and later Wingreen *et al.*[30] presented the exact resonant-tunneling transmission probability fully taking into account the e-ph interaction on a nondegenerate resonant site. Phonons produced transmission side-bands but did not affect the integral transmission probability. Li, Chen and Zhou [31] studied the conductance of a double degenerate (due to spin) quantum dot with Coulomb repulsion and the e-ph interaction. Their numerical results also showed the side-band peaks and the main peak related to the Coulomb repulsion, which was decreased by the e-ph interaction. Kang [32] studied the boson (vibron) assisted transport through a double-degenerate QD coupled to two *superconducting* leads and found multiple peaks in the I-V curves, which originated from the singular BCS density of states and the phonon side-bands.

While a correlated transport through mesoscopic systems with repulsive electron-electron interactions received considerable interest in the past, and continues to be the focus of intense investigations [37], much less has been known about a role of attractive correlations in MQD. Recently we have proposed a negative$-U$ Hubbard model of a d-fold degenerate quantum dot [38]. We argued that the *attractive* electron correlations caused by a strong electron-phonon (vibron) interaction in the molecule, and/or by the valence fluctuations provide a molecular switching effect, when the current-voltage (I-V) characteristics show two branches with high and low current for the same voltage. The effect was observed in a few experimental studies with complex [27] and simple molecules [39].

Here we review the analytical theory of a *correlated* transport through a degenerate molecule quantum dot (MQD) fully taking into account both Coulomb and e-ph interactions [40]. We show that the phonon side-bands significantly modify the switching behavior of the I-V curves in comparison with the negative-U Hubbard model [38]. Nevertheless, the switching effect is robust. It shows up when the effective interaction of polarons is attractive and the state of the dot is multiply degenerate, $d > 2$.

2. Attractive correlations of small polarons

Employing the canonical polaron formalism with a generic "Fröhlich-Coulomb" Hamiltonian, allows us explicitly calculate the effective attraction of small polarons [41]. The Hamiltonian includes the infinite-range Coulomb, V_c and electron-phonon interactions. The implicitly present infinite on-site repulsion (Hubbard U) prohibits double occupancy and removes the need to distinguish the fermionic spin. Introducing spinless fermion operators $c_{\mathbf{n}}$ and phonon operators $d_{\mathbf{m}\nu}$, the Hamiltonian is written as

$$H = \sum_{\mathbf{n}\neq\mathbf{n}'} T(\mathbf{n}-\mathbf{n}')c_{\mathbf{n}}^{\dagger}c_{\mathbf{n}'} + \sum_{\mathbf{n}\neq\mathbf{n}'} V_c(\mathbf{n}-\mathbf{n}')c_{\mathbf{n}}^{\dagger}c_{\mathbf{n}}c_{\mathbf{n}'}^{\dagger}c_{\mathbf{n}'} + \omega_0 \sum_{\mathbf{n}\neq\mathbf{m},\nu} g_{\nu}(\mathbf{m}-\mathbf{n})(\mathbf{e}_{\nu}\cdot\mathbf{e}_{\mathbf{m}-\mathbf{n}})c_{\mathbf{n}}^{\dagger}c_{\mathbf{n}}(d_{\mathbf{m}\nu}^{\dagger}+d_{\mathbf{m}\nu}) + \omega_0 \sum_{\mathbf{m},\nu}\left(d_{\mathbf{m}\nu}^{\dagger}d_{\mathbf{m}\nu}+\frac{1}{2}\right). \tag{1}$$

The e-ph term is written in real space, which is more convenient in working with complex lattices.

In general, the many-body model Eq.(1) is of considerable complexity. However, we are interested in the limit of the strong e-ph interaction. In this case, the kinetic energy is a perturbation and the model can be grossly simplified using the canonical transformation [12] in the Wannier representation for electrons and phonons,

$$S = \sum_{\mathbf{m}\neq\mathbf{n},\nu} g_{\nu}(\mathbf{m}-\mathbf{n})(\mathbf{e}_{\nu}\cdot\mathbf{e}_{\mathbf{m}-\mathbf{n}})c_{\mathbf{n}}^{\dagger}c_{\mathbf{n}}(d_{\mathbf{m}\nu}^{\dagger}-d_{\mathbf{m}\nu}).$$

The transformed Hamiltonian is

$$\tilde{H} = e^{-S}He^{S} = \sum_{\mathbf{n}\neq\mathbf{n}'} \hat{\sigma}_{\mathbf{n}\mathbf{n}'}c_{\mathbf{n}}^{\dagger}c_{\mathbf{n}'} + \omega_0\sum_{\mathbf{m}\alpha}\left(d_{\mathbf{m}\nu}^{\dagger}d_{\mathbf{m}\nu}+\frac{1}{2}\right) + \sum_{\mathbf{n}\neq\mathbf{n}'} v(\mathbf{n}-\mathbf{n}')c_{\mathbf{n}}^{\dagger}c_{\mathbf{n}}c_{\mathbf{n}'}^{\dagger}c_{\mathbf{n}'} - E_p\sum_{\mathbf{n}} c_{\mathbf{n}}^{\dagger}c_{\mathbf{n}}. \tag{2}$$

The last term describes the energy gained by polarons due to e-ph interaction. E_p is the familiar polaron level shift

$$E_p = \omega_0 \sum_{\mathbf{m}\nu} g_{\nu}^2(\mathbf{m}-\mathbf{n})(\mathbf{e}_{\nu}\cdot\mathbf{e}_{\mathbf{m}-\mathbf{n}})^2, \tag{3}$$

which is independent of $\mathbf{n}$. The third term on the right-hand side in Eq.(2) is the polaron-polaron interaction:

$$v(\mathbf{n}-\mathbf{n}') = V_c(\mathbf{n}-\mathbf{n}') - V_{ph}(\mathbf{n}-\mathbf{n}'), \tag{4}$$

where

$$V_{ph}(\mathbf{n}-\mathbf{n}') = 2\omega_0 \sum_{\mathbf{m},\nu} g_{\nu}(\mathbf{m}-\mathbf{n})g_{\nu}(\mathbf{m}-\mathbf{n}') \times (\mathbf{e}_{\nu}\cdot\mathbf{e}_{\mathbf{m}-\mathbf{n}})(\mathbf{e}_{\nu}\cdot\mathbf{e}_{\mathbf{m}-\mathbf{n}'}).$$

The phonon-induced interaction V_{ph} is due to displacements of common ions by two electrons. Finally, the transformed hopping operator $\hat{\sigma}_{\mathbf{nn}'}$ in the first term in Eq.(2) is given by

$$\begin{aligned} \hat{\sigma}_{\mathbf{nn}'} &= T(\mathbf{n}-\mathbf{n}')\exp\Bigg[\sum_{\mathbf{m},\nu}[g_\nu(\mathbf{m}-\mathbf{n})(\mathbf{e}_\nu\cdot\mathbf{e}_{\mathbf{m}-\mathbf{n}}) \\ &\quad - g_\nu(\mathbf{m}-\mathbf{n}')(\mathbf{e}_\nu\cdot\mathbf{e}_{\mathbf{m}-\mathbf{n}'})]\,(d^\dagger_{\mathbf{m}\alpha}-d_{\mathbf{m}\alpha})\Bigg]. \end{aligned} \tag{5}$$

This term is a perturbation at large λ. It is absent in an isolated MQD, modeled as a single degenerate atomic level, so that the canonical transformation solves the problem exactly for any number of electrons (see below). In a crystal the term allows for a bipolaron tunnelling and high temperature superconductivity [22]. In particular crystal structures like perovskites, the tunnneling appears already in the first order in $T(\mathbf{n})$, so that $\hat{\sigma}_{\mathbf{nn}'}$ can be averaged over phonons. The result is

$$t(\mathbf{n}-\mathbf{n}') \equiv \langle\langle \hat{\sigma}_{\mathbf{nn}'}\rangle\rangle_{ph} = T(\mathbf{n}-\mathbf{n}')\exp[-g^2(\mathbf{n}-\mathbf{n}')], \tag{6}$$

$$\begin{aligned} g^2(\mathbf{n}-\mathbf{n}') &= \sum_{\mathbf{m},\nu} g_\nu(\mathbf{m}-\mathbf{n})(\mathbf{e}_\nu\cdot\mathbf{e}_{\mathbf{m}-\mathbf{n}})\times \\ &\quad [g_\nu(\mathbf{m}-\mathbf{n})(\mathbf{e}_\nu\cdot\mathbf{e}_{\mathbf{m}-\mathbf{n}}) - g_\nu(\mathbf{m}-\mathbf{n}')(\mathbf{e}_\nu\cdot\mathbf{e}_{\mathbf{m}-\mathbf{n}'})]. \end{aligned}$$

By comparing Eqs.(6) and Eqs.(3,4), the mass renormalization exponent can be expressed via E_p and V_{ph} as follows

$$g^2(\mathbf{n}-\mathbf{n}') = \frac{1}{\omega_0}\left[E_p - \frac{1}{2}V_{ph}(\mathbf{n}-\mathbf{n}')\right]. \tag{7}$$

When V_{ph} is larger than V_c the full interaction becomes negative and polarons form pairs. The real space representation allows us to elaborate more physics behind the lattice sums in Eq.(3) and Eq.(4). If a carrier (electron or hole) acts on an ion with a force $\mathbf{f}$, it displaces the ion by some vector $\mathbf{x} = \mathbf{f}/s$. Here s is the ion's force constant. The total energy of the carrier-ion pair is $-\mathbf{f}^2/(2s)$. This is precisely the summand in Eq.(3) expressed via dimensionless coupling constants. Now consider two carriers interacting with the *same* ion, see Fig.1a. The ion displacement is $\mathbf{x} = (\mathbf{f}_1+\mathbf{f}_2)/s$ and the energy is $-\mathbf{f}_1^2/(2s) - \mathbf{f}_2^2/(2s) - (\mathbf{f}_1\cdot\mathbf{f}_2)/s$. Here the last term should be interpreted as an ion-mediated interaction between the two carriers. It depends on the scalar product of $\mathbf{f}_1$ and $\mathbf{f}_2$ and consequently on the relative positions of the carriers with respect to the ion. If the ion is an isotropic harmonic oscillator, as we assume here, then the following simple rule applies. If the angle ϕ between $\mathbf{f}_1$ and $\mathbf{f}_2$ is less than $\pi/2$ the polaron-polaron interaction will be attractive, if otherwise it will be

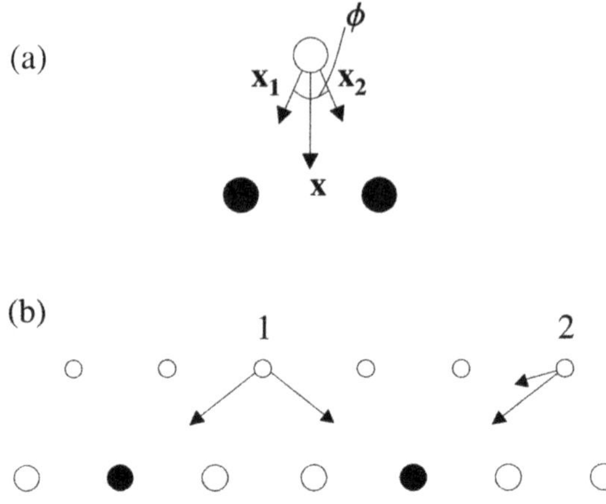

Figure 1. The mechanism of polaron-polaron interaction. (a) Together, the two polarons (solid circles) deform the lattice more effectively than separately. An effective attraction occurs when the angle ϕ is less than $\pi/2$. (b) A mixed situation. Ion 1 results in repulsion between two polarons while ion 2 results in attraction.

repulsive. In general, some ions will generate attraction, and some repulsion between polarons, Fig. 1b.

The overall sign and magnitude of the interaction is given by the lattice sum in Eq.(4), the evaluation of which is elementary. One should also note that according to Eq.(7) an attractive interaction reduces the polaron mass (and consequently the bipolaron mass), while repulsive interaction enhances the mass.

3. Steady current through MQD

Let us now apply the polaron formalism to MQD [40]. Here bipolarons might not exist because of a finite lifetime of electrons on a molecule connected with the leads, but the attractive correlations could strongly modify the current-voltage characteristics. We employ the Landauer-type expression for a steady current through a region of interacting electrons, derived by Meir and Wingreen [42] as (in units $\hbar = k_B = 1$)

$$I(V) = -\frac{e}{\pi} \int_{-\infty}^{\infty} d\omega \left[f_1(\omega) - f_2(\omega) \right] \mathrm{ImTr} \left[\hat{\Gamma}(\omega) \hat{G}^R(\omega) \right], \qquad (8)$$

where $f_{1(2)}(\omega) = \{\exp[(\omega + \Delta \mp eV/2)/T] + 1\}^{-1}$, T is the temperature, Δ is the position of the lowest unoccupied molecular level with respect to the

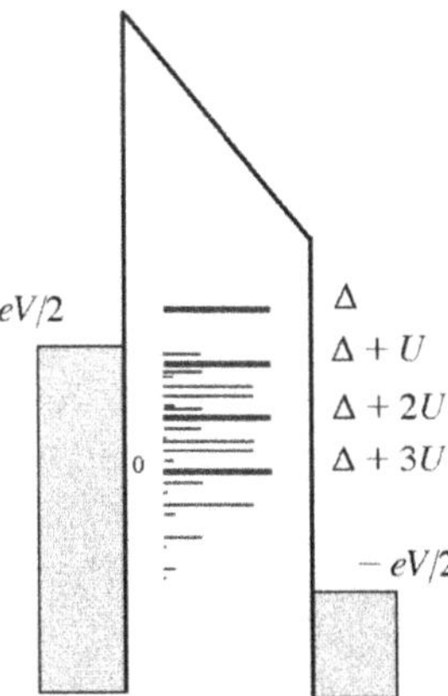

Figure 2. Schematic of the energy levels and phonon side-bands for molecular quantum dot under bias voltage V ($eV/2\Delta = 0.75$) with the coupling constant $\gamma^2 = 11/13$. The level is assumed to be 4-fold degenerate ($d = 4$) with energies $\Delta + rU, r = 0, \ldots, (d-1)$ (thick bars). Thin bars show the vibron side-bands with the size of the bar proportional to the weight of the particular contribution in the density of states (see text) in the case of one vibron with frequency $\omega_0/\Delta = 0.2$ at $T = 0$. Only the bands in the energy window $(eV/2, -eV/2)$ (shown) will contribute to current at zero temperature.

chemical potential. $\hat{\Gamma}(\omega)$ depends on the density of states (DOS) in the leads and on the hopping integrals connecting one-particle states in the left (1) and the right (2) leads with the states in MQD, Fig. 2.

This formula includes, by means of the Fourier transform of the full molecular retarded Green's function (GF), $\hat{G}^R(\omega)$, the e-ph and Coulomb interactions inside the MQD and coupling to the leads. Since the leads are metallic, electron-electron and e-ph interactions in the leads, and interactions of electrons in the leads with electrons and phonons in the MQD can be neglected. We are interested in the tunneling near the conventional threshold, $eV = 2\Delta$, Fig.2, within a voltage range about an effective attractive potential $|U|$ caused by phonons/vibrons (see below).

The attractive energy is the difference of two large interactions, the Coulomb repulsion and the phonon mediated attraction, of the order of 1eV each. Hence, $|U|$ is of the order of a few tens of one eV. We neglect the energy dependence of $\hat{\Gamma}(\omega) \approx \Gamma$ on this scale, and assume that the coupling to the leads is weak, $\Gamma \ll |U|$. In this case $\hat{G}^R(\omega)$ does not depend on the leads. Moreover we assume that there is a complete set of one-particle molecular states $|\mu\rangle$, where $\hat{G}^R(\omega)$ is diagonal. With these assumptions we can reduce Eq.(8) to

$$I(V) = I_0 \int_{-\infty}^{\infty} d\omega \left[f_1(\omega) - f_2(\omega)\right] \rho(\omega), \tag{9}$$

allowing for a transparent analysis of essential physics of the switching phenomenon. Here $I_0 = e\Gamma$ and the molecular DOS, $\rho(\omega)$, is given by

$$\rho(\omega) = -\frac{1}{\pi} \sum_{\mu} \mathrm{Im}\, \hat{G}_{\mu}^{R}(\omega), \tag{10}$$

where $\hat{G}_{\mu}^{R}(\omega)$ is the Fourier transform of $\hat{G}_{\mu}^{R}(t) = -i\theta(t) \left\langle \left\{ c_{\mu}(t), c_{\mu}^{\dagger} \right\} \right\rangle$, $\{\cdots,\cdots\}$ is the anticommutator, $c_{\mu}(t) = e^{iHt} c_{\mu} e^{-iHt}$, $\theta(t) = 1$ for $t \geqslant 0$ and zero otherwise.

We calculate $\rho(\omega)$ exactly(see below) in the framework of the Hamiltonian, which includes both the Coulomb U^C and e-ph interactions as

$$\begin{aligned} H &= \sum_{\mu} \varepsilon_{\mu} \hat{n}_{\mu} + \frac{1}{2} \sum_{\mu \neq \mu'} U_{\mu\mu'}^{C} \hat{n}_{\mu} \hat{n}_{\mu'} \\ &+ \sum_{\mu,q} \hat{n}_{\mu} \omega_q (\gamma_{\mu q} d_q + H.c.) + \sum_{q} \omega_q (d_q^{\dagger} d_q + 1/2). \end{aligned} \tag{11}$$

Here ε_{μ} are one-particle molecular energy levels, $\hat{n}_{\mu} = c_{\mu}^{\dagger} c_{\mu}$ the occupation number operators, c_{μ} and d_q annihilates electrons and phonons, respectively, ω_q are the phonon (vibron) frequencies, and $\gamma_{\mu q}$ are e-ph coupling constants (q enumerates the vibron modes). This Hamiltonian conserves the occupation numbers of molecular states $\hat{n}_{\mu}$. Hence it is compatible with Eq.(9).

4. MQD density of states

We apply the canonical polaron unitary transformation e^S, as in Section 1, integrating phonons out. The electron and phonon operators are transformed as

$$\tilde{c}_{\mu} = c_{\mu} X_{\mu}, \tag{12}$$

and

$$\tilde{d}_q = d_q - \sum_{\mu} \hat{n}_{\mu} \gamma_{\mu q}^{*}, \tag{13}$$

respectively. Here

$$X_{\mu} = \exp\left[\sum_{q} \gamma_{\mu q} d_q - H.c.\right].$$

The Lang-Firsov canonical transformation shifts ions to new equilibrium positions with no effect on the phonon frequencies. The diagonalization is exact in

MQD:

$$\tilde{H} = \sum_i \tilde{\varepsilon}_\mu \hat{n}_\mu + \sum_q \omega_q (d_q^\dagger d_q + 1/2) + \frac{1}{2} \sum_{\mu \neq \mu'} U_{\mu\mu'} \hat{n}_\mu \hat{n}_{\mu'}, \tag{14}$$

where

$$U_{\mu\mu'} \equiv U^C_{\mu\mu'} - 2 \sum_q \gamma^*_{\mu q} \gamma_{\mu' q} \omega_q \tag{15}$$

is the interaction of polarons comprising their interaction via molecular deformations (vibrons) and non-vibron (e.g. Coulomb repulsion) $U^C_{\mu\mu'}$. To simplify the discussion, we shall assume, that the Coulomb integrals do not depend on the orbital index, i.e. $U_{\mu\mu'} = U$.

The molecular energy levels are shifted by the polaron level-shift due to a deformation well created by polaron,

$$\tilde{\varepsilon}_\mu = \varepsilon_\mu - \sum_q |\gamma_{\mu q}|^2 \omega_q. \tag{16}$$

Applying the same transformation in the retarded GF we obtain

$$\begin{aligned} G^R_\mu(t) &= -i\theta(t) \left\langle \left\{ c_\mu(t) X_\mu(t),\, c^\dagger_\mu X^\dagger_\mu \right\} \right\rangle \\ &= -i\theta(t) [\left\langle c_\mu(t) c^\dagger_\mu \right\rangle \left\langle X_\mu(t) X^\dagger_\mu \right\rangle \\ &\quad + \left\langle c^\dagger_\mu c_\mu(t) \right\rangle \left\langle X^\dagger_\mu X_\mu(t) \right\rangle], \end{aligned} \tag{17}$$

where now electron and phonon operators are averaged over the quantum state of the transformed Hamiltonian $\tilde{H}$. There is no coupling between polarons and vibrons in the transformed Hamiltonian, so that

$$\begin{aligned} &\left\langle X_\mu(t) X^\dagger_\mu \right\rangle \\ &= \exp\left[\sum_q \frac{|\gamma_{\mu q}|^2}{\sinh \frac{\beta \omega_q}{2}} \left[\cos\left(\omega t + i \frac{\beta \omega_q}{2} \right) - \cosh \frac{\beta \omega_q}{2} \right] \right], \end{aligned} \tag{18}$$

where $\beta = 1/T$, and $\left\langle X^\dagger_\mu X_\mu(t) \right\rangle = \left\langle X_\mu(t) X^\dagger_\mu \right\rangle^*$.

Next, we introduce the N-particle GFs, which will necessarily appear in the equations of motion for $\left\langle c_\mu(t) c^\dagger_\mu \right\rangle$, as

$$G^{(N,+)}_\mu(t) \equiv -i\theta(t) \sum_{\mu_1 \neq \mu_2 \neq \ldots \mu} \left\langle c_\mu(t) c^\dagger_\mu \prod_{i=1}^{N-1} \hat{n}_{\mu_i} \right\rangle, \tag{19}$$

and

$$G_{\mu}^{(N,-)}(t) = -i\theta(t) \sum_{\mu_1 \neq \mu_2 \neq \ldots \mu} \left\langle c_{\mu}^{\dagger} c_{\mu}(t) \prod_{i=1}^{N-1} \hat{n}_{\mu_i} \right\rangle . \tag{20}$$

Then, using the equation of motion for the Heisenberg polaron operators, we derive the following equations for the N-particle GFs,

$$\begin{aligned} i\frac{dG_{\mu}^{(N,+)}(t)}{dt} &= \delta(t)(1-n_{\mu}) \sum_{\mu_1 \neq \mu_2 \neq \ldots \mu} \prod_{i=1}^{N-1} n_{\mu_i} \\ &+ [\tilde{\tilde{\varepsilon}}_{\mu} + (N-1)U] G_{\mu}^{(N,+)}(t) + U G_{\mu}^{(N+1,+)}(t), \end{aligned} \tag{21}$$

and

$$\begin{aligned} i\frac{dG_{\mu}^{(N,-)}(t)}{dt} &= \delta(t) n_{\mu} \sum_{\mu_1 \neq \mu_2 \neq \ldots \mu} \prod_{i=1}^{N-1} n_{\mu_i} \\ &+ [\tilde{\tilde{\varepsilon}}_{\mu} + (N-1)U] G_{\mu}^{(N,-)}(t) + U G_{\mu}^{(N+1,-)}(t), \end{aligned} \tag{22}$$

where $n_{\mu} = \left\langle c_{\mu}^{\dagger} c_{\mu} \right\rangle$ is the expectation number of electrons on the molecular level μ.

We readily solve this set of coupled equations for MQD with one d-fold degenerate energy level and with the e-ph coupling $\gamma_{\mu q} = \gamma_q$, which does not break the degeneracy. Assuming that $n_{\mu} = n$, Fourier transformation of the set yields for $N = 1$

$$G_{\mu}^{(1,+)}(\omega) = (1-n) \sum_{r=0}^{d-1} \frac{Z_r(n)}{\omega - rU + i\delta}, \tag{23}$$

$$G_{\mu}^{(1,-)}(\omega) = n \sum_{r=0}^{d-1} \frac{Z_r(n)}{\omega - rU + i\delta} \tag{24}$$

where $\delta = +0$, and

$$Z_r(n) = \frac{(d-1)!}{r!(d-1-r)!} n^r (1-n)^{d-1-r}. \tag{25}$$

In approximation, where we retain a coupling to a single mode with the characteristic frequency ω_0 and $\gamma_q = \gamma$, the *molecular DOS* is readily found as an imaginary part of the Fourier transform of Eq.(17) using Eqs.(23,24) and Eq.(18):

$$\rho(\omega) = \mathcal{Z} d \sum_{r=0}^{d-1} Z_r(n) \sum_{l=0}^{\infty} I_l(\xi)$$

$$\times\left[e^{\frac{\beta\omega_0 l}{2}}\left[(1-n)\delta(\omega-rU-l\omega_0)+n\delta(\omega-rU+l\omega_0)\right]\right.$$
$$+(1-\delta_{l0})e^{-\frac{\beta\omega_0 l}{2}}\left[n\delta(\omega-rU-l\omega_0)\right.$$
$$\left.\left.+(1-n)\delta(\omega-rU+l\omega_0)\right]\right], \quad (26)$$

where

$$Z=\exp\left[-\sum_{\mathbf{q}}|\gamma_q|^2\coth\frac{\beta\omega_q}{2}\right], \quad (27)$$

$\xi=|\gamma|^2/\sinh\frac{\beta\omega_0}{2}$, $I_l(\xi)$ is the modified Bessel function, and δ_{lk} is the Kroneker symbol. The important feature of the DOS, Eq.(26), is its nonlinear dependence on the occupation number n, which leads to the switching effect and hysteresis in the I-V characteristics for $d>2$, as will be shown below. It contains full information about all possible correlation and inelastic effects in transport, in particular, all the vibron-assisted tunneling processes and phonon sidebands, and describes the renormalization of hopping to the leads.

5. Nonlinear rate equation and switching

Generally, the electron density n_μ obeys an infinite set of rate equations for many-particle GFs which can be derived in the framework of a tunneling Hamiltonian including correlations [38]. In the case of MQD only weakly coupled with leads one can apply the Fermi-Dirac golden rule to obtain an equation for n. Equating incoming and outgoing numbers of electrons in MQD per unit time we obtain the self-consistent equation for the level occupation n as

$$(1-n)\int_{-\infty}^{\infty}d\omega\left\{\Gamma_1 f_1(\omega)+\Gamma_2 f_2(\omega)\right\}\rho(\omega)$$
$$-n\int_{-\infty}^{\infty}d\omega\left\{\Gamma_1[1-f_1(\omega)]+\Gamma_2[1-f_2(\omega)]\right\}\rho(\omega)=0 \quad (28)$$

where $\Gamma_{1(2)}$ are the transition rates from left (right) leads to MQD. Taking into account that $\int_{-\infty}^{\infty}\rho(\omega)=d$, Eq.(28) for the symmetric leads, $\Gamma_1=\Gamma_2$, reduces to

$$2nd=\int d\omega\rho(\omega)(f_1+f_2), \quad (29)$$

which automatically satisfies $0\leq n\leq 1$. Explicitly, the self-consistent equation for the occupation number is

$$n=\frac{1}{2}\sum_{r=0}^{d-1}Z_r(n)[na_r+(1-n)b_r], \quad (30)$$

where

$$a_r = \mathcal{Z}\sum_{l=0}^{\infty} I_l(\xi)\Big(e^{\frac{\beta\omega_0 l}{2}}[f_1(rU - l\omega_0) + f_2(rU - l\omega_0)] + (1-\delta_{l0})e^{-\frac{\beta\omega_0 l}{2}}[f_1(rU + l\omega_0) + f_2(rU + l\omega_0)]\Big), \quad (31)$$

$$b_r = \mathcal{Z}\sum_{l=0}^{\infty} I_l(\xi)\Big(e^{\frac{\beta\omega_0 l}{2}}[f_1(rU + l\omega_0) + f_2(rU + l\omega_0)] + (1-\delta_{l0})e^{-\frac{\beta\omega_0 l}{2}}[f_1(rU - l\omega_0) + f_2(rU - l\omega_0)]\Big). \quad (32)$$

The current is expressed as

$$j \equiv \frac{I(V)}{dI_0} = \sum_{r=0}^{d-1} Z_r(n)[na'_r + (1-n)b'_r], \quad (33)$$

where

$$a'_r = \mathcal{Z}\sum_{l=0}^{\infty} I_l(\xi)\Big(e^{\frac{\beta\omega_0 l}{2}}[f_1(rU - l\omega_0) - f_2(rU - l\omega_0)] + (1-\delta_{l0})e^{-\frac{\beta\omega_0 l}{2}}[f_1(rU + l\omega_0) - f_2(rU + l\omega_0)]\Big), \quad (34)$$

$$b'_r = \mathcal{Z}\sum_{l=0}^{\infty} I_l(\xi)\Big(e^{\frac{\beta\omega_0 l}{2}}[f_1(rU + l\omega_0) - f_2(rU + l\omega_0)] + (1-\delta_{l0})e^{-\frac{\beta\omega_0 l}{2}}[f_1(rU - l\omega_0) - f_2(rU - l\omega_0)]\Big). \quad (35)$$

There is only one physical ($0 < n < 0.5$) solution of the rate equation (30) and no switching for a nondegenerate, $d = 1$, and double-degenerate, $d = 2$, MQDs. However, the switching appears for $d > 2$. For example, for $d = 4$ the rate equation is of the fourth power in n,

$$\begin{aligned} 2n = {} & (1-n)^3[na_0 + (1-n)b_0] \\ & +3n(1-n)^2[na_1 + (1-n)b_1] \\ & +3n^2(1-n)[na_2 + (1-n)b_2] \\ & +n^3[na_3 + (1-n)b_3]. \end{aligned} \quad (36)$$

Differently from the non-degenerate or double-degenerate MQD, the rate equation (36) for $d = 4$ has two stable physical roots in a certain voltage range and the current-voltage characteristics show a hysteretic behavior. We

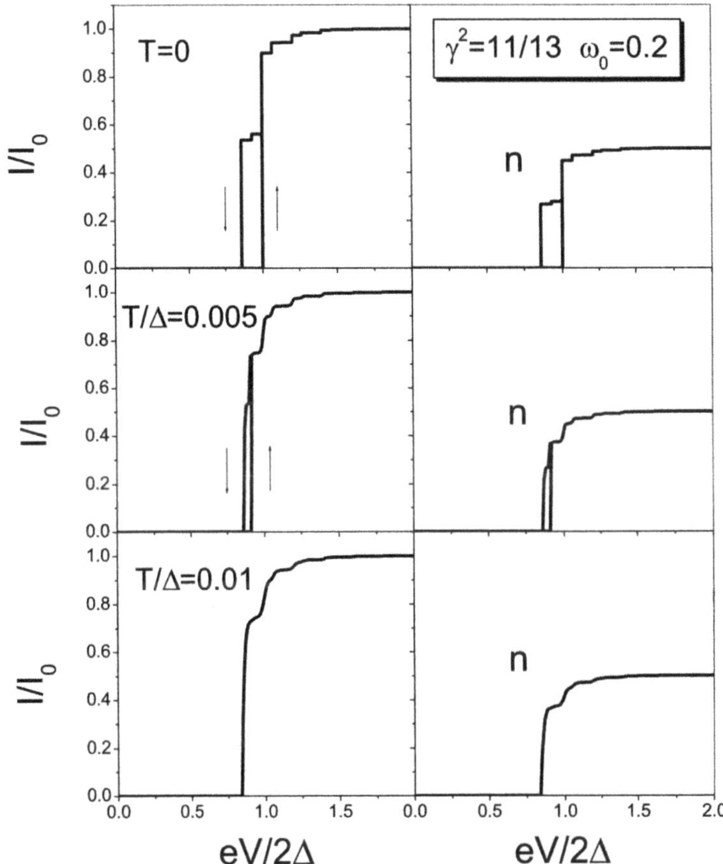

Figure 3. The bistable I-V curves for tunneling through molecular quantum dot (Fig. 2) with the electron-vibron coupling constant $\gamma^2 = 11/13$ and $\omega_0/\Delta = 0.2$. The up arrows show that the current picks up at some voltage when it is biased, and then drops at lower voltage when the bias is being reduced. The bias dependence of current basically repeats the shape of the level occupation n (right column). Steps on the curve correspond to the changing population of the phonon side-bands, which are shown in Fig. 2.

show the numerical results for $\omega_0 = 0.2$ (in units of Δ, as all the energies in the problem) and $U^C = 0$ for the coupling constant, $\gamma^2 = 11/13$ in Fig. 3. This case formally corresponds to a negative Hubbard $U = -2\gamma^2\omega_0 \approx -0.4$ (we selected those values of γ^2 to avoid accidental commensurability of the correlated levels separated by U and the phonon side-bands). The threshold for the onset of bistability appears at a voltage bias $eV/2\Delta = 0.86$ for $\gamma^2 =$

11/13 and $\omega_0 = 0.2$). The inelastic tunneling processes through the level, accompanied by emission/absorption of the vibrons, manifest themselves as steps on the I-V curve, Figs. 3. Those steps are generated by the phonon side-bands originating from correlated levels in the dot with the energies Δ, $\Delta + U$, ..., $\Delta + (d-1)U$. Since ω_0 is not generally commensurate with U, we obtain quite irregular picture of the steps in I-V curves. The bistability region shrinks down with temperature.

In conclusion, we have reviewed the multi-polaron theory of tunneling through a molecular quantum dot (MQD) taking phonon side-bands and attractive polaron correlations into account. The degenerate MQD with strong electron-vibron coupling shows a hysteretic volatile memory if the degeneracy of the molecular level is larger than two, $d > 2$. The hysteretic behavior strongly depends on the electron-vibron coupling and characteristic vibron frequencies. The current bistability vanishes above some critical temperature. It would be very interesting to look for an experimental realization of the model, possibly in a system containing a certain conjugated central part, which exhibits the attractive correlations of carriers with large degeneracy $d > 2$. Interesting candidate systems are C_{60} molecule ($d = 6$) where the electron-phonon interaction is strong [29], short nanotubes or other fullerenes ($d \gg 1$), and mixed-valence molecular complexes. Switching should be fast, 10^{-13} s or faster.

Acknowledgments

This work has been supported by DARPA, and by the Leverhulme Trust (UK).

References

[1] Landau L D 1933 *J. Physics (USSR)* **3** 664

[2] Pekar S I 1946 *Zh. Eksp. Teor. Fiz.* **16** 335

[3] Fröhlich H 1954 *Adv. Phys.* **3** 325

[4] Feynman R P 1955 1955 *Phys. Rev.* **97** 660

[5] Devreese J T 1996 in *Encyclopedia of Applied Physics*, vol. **14**, p. 383 (VCH Publishers) and references therein

[6] Tjablikov S V 1952 *Zh.Eksp.Teor.Fiz.* **23** 381

[7] Yamashita J and T. Kurosawa 1958 *J. Phys. Chem. Solids* **5** 34

[8] Sewell G L 1958 *Phil. Mag.* **3** 1361

[9] Holstein T 1959 *Ann. Phys.* **8** 325; *ibid* 343

[10] Friedman L and Holstein T 1963 *Ann. Phys* **21** 494

[11] Emin D and Holstein T 1969 *Ann. Phys* **53** 439

[12] Lang I G and Firsov Yu A 1962 *Zh. Eksp. Teor. Fiz.* **43** 1843; 1963 *Sov. Phys. JETP* **16** 1301

[13] Eagles D M 1963 *Phys. Rev.* 130 1381; 1969 *Phys. Rev.* **181** 1278; 1969 *Phys. Rev.* **186** 456

[14] Appel J 1968 in *Solid State Physics* **21** (eds. Seitz F, Turnbull D and Ehrenreich H, Academic Press)

[15] Firsov Yu A (ed) 1975 *Polarons* (Moscow: Nauka)

[16] Böttger H and Bryksin V V 1985 *Hopping Conduction in Solids (*Berlin: Academie-Verlag)

[17] Mahan G D 1990 *Many Particle Physics (*New York: Plenum Press)

[18] Alexandrov A S and Mott N F 1995 *Polarons and Bipolarons* (Singapore: World Scientific)

[19] Müller K A 2002 *Physica Scripta* T**102** 39, and references therein

[20] Alexandrov A S 1996 *Phys. Rev.* B**53** 2863

[21] Alexandrov A S 2003 *Theory of Superconductivity: From Weak to Strong Coupling* (Bristol and Philadelphia: IoP Publishing)

[22] Alexandrov A S 1983 *Zh. Fiz. Khim.* **57** 273 ; 1983 *Russ. J. Phys. Chem.* **57** 167; 1998 *Models and Phenomenology for Conventional and High-temperature Superconductivity* (Course CXXXVI of the Intenational School of Physics 'Enrico Fermi', eds. G. Iadonisi, J.R. Schrieffer and M.L. Chiofalo, Amsterdam: IOS Press), p. 309

[23] see contributions in 1995 *Polarons and Bipolarons in High-T_c Superconductors and Related Materials* (eds. Salje E K H, Alexandrov A S and Liang W Y, Cambridge: Cambridge University Press), and in 1995 *Anharmonic properties of High Tc cuprates* (eds. Mihailovic D, Ruani G, Kaldis E and Muller K A, Singapore: World Scientific)

[24] Lehn J-M 1990 *Angew. Chem. Int. Ed. Engl.* **29** 1304

[25] Tour J M 2000 *Acc. Chem. Res.* **33** 791; Tour J M *et al.* 1995 *J. Am. Chem. Soc.* **117** 9529

[26] Aviram A and Ratner M 1998 Eds. *Molecular Electronics: Science and Technology* (Ann. N.Y. Acad. Sci., New York)

[27] Collier C P *et al.* 1999 *Science* **285** 391 ; Chen J *et al.* 1999 *Science* **286** 1550; D.I. Gittins D I *et al.* 2000 *Nature (London)* **408** 677 ; He H X , Tao T J, Nagahara L A, Amlani I and Tsui R (unpublished).

[28] Zhitenev N B, Meng H, and Bao Z 2002 *Phys. Rev. Lett.* **88** 226801

[29] Park J, Pasupathy A N, Goldsmith J I, Chang C, Yaish Y, Retta J R, Rinkoski M, Sethna J P, Abruña H D, McEuen P L, and Ralph D C 2000 *Nature (London)* **417** 722

[30] Glazman L I and Shekhter R I 1987 *Zh. Eksp. Teor. Fiz.* **94**292 [Sov. Phys. JETP **67**, 163 (1988)]; Wingreen N S, Jacobsen K W, and Wilkins J W 1989 *Phys. Rev. B* **17** 11834

[31] Xi Li, Chen H, and Zhou S 1995 *Phys. Rev. B***52** 12202

[32] Kang K 1998 *Phys. Rev. B***57**, 11891

[33] Ermakov V N 2000 *Physica E***8** 99

[34] Di Ventra M, Kim S-G, Pantelides S T, and Lang N D 2001 *Phys. Rev. Lett.* **86** 288

[35] Ness N, Shevlin S A, and Fisher A J 2001 *Phys. Rev. B* **63** 125422

[36] Lundin U and McKenzie R H 2002 *Phys. Rev. B* **66**, 075303

[37] see this volume

[38] Alexandrov A S, Bratkovsky A M, and Williams R S 2003 Phys. Rev. B **67** 075301

[39] Stewart D *et al.* (unpublished).

[40] Alexandrov A S and Bratkovsky A M 2003 *Phys. Rev. B* **67** 235312

[41] Alexandrov A S and Kornilovitch P E 2002 *J. Phys.: Condens. Matter* **14** 5337

[42] Meir Y and Wingreen N S 1992 *Phys. Rev. Lett.* **68** 2512

THE DYNAMICS OF INELASTIC QUANTUM TUNNELING

S. A. Trugman[1], Li-Chung Ku[1,2,*], and J. Bonča[3]

[1] *Theoretical Division, Los Alamos National Laboratory, Los Alamos, New Mexico 87545, U.S.A.*
[2] *Department of Physics, University of California, Los Angeles, California 90024, U.S.A.*
[3] *FMF, University of Ljubljana and J. Stefan Institute, 1000, Ljubljana, Slovenia*
sat@lanl.gov

Abstract We describe how to obtain a numerically exact solution for an electron tunneling through a molecule or nanodot coupled to arbitrary phonon and spin degrees of freedom. Fully quantum and semiclassical tunneling are compared at zero and finite temperature. Frames from movies are presented for elastic and inelastic tunneling through a molecule, and for polaron-phonon scattering in a translation invariant system.

Keywords: polaron; tunneling; inelastic quantum tunneling; polaron phonon scattering

1. Introduction

There has been recent experimental and theoretical interest in electrons tunneling through molecular or nano devices while coupled to vibrational degrees of freedom. The interaction leads to phonon-assisted resonant tunneling, which is important in device applications [1, 2, 3]. It can also be considered a simple method to explore electronic transport in bulk in the presence of electron-phonon scattering. The vibrational degrees of freedom are quantum mechanical, although for simplicity they are often treated classically, relying on the Born-Oppenheimer approximation. A mixed treatment in which electrons are treated quantum mechanically and phonons are treated classically may or may not yield accurate results, as we will see below. For example, a very wide initial electron wavepacket does not excite any phonons in this approximation. (The force driving the phonon coordinates is proportional to the local electron

*Present address: Department of Physics, University of California, Irvine, California 92697, U.S.A.

A.S. Alexandrov et al. (eds.), Molecular Nanowires and Other Quantum Objects, 167–176.

density, which vanishes as the electron wavepacket becomes wide.) In contract, the fully quantum result has a substantial amplitude for phonon excitation even when the width of the electron wavepacket is arbitrarily large.

Fortunately, it is often possible to treat both electrons and phonons fully quantum mechanically in a tractable way. This can be done for polarons and bipolarons in extended systems [4, 5, 6, 7, 8, 9, 10], including if desired extra orbital and vibrational degrees of freedom as in Jahn-Teller systems [11, 12]. The problem of an electron tunneling through a molecule or quantum dot with essentially arbitrary interactions with local phonons, captive charge and spin degrees of freedom, potential barriers, etc., can also be solved numerically essentially exactly [13, 14].

2. Tunneling at Zero Temperature

A simple example of a Hamiltonian describing tunneling through a molecule is

$$H = -\sum_{i,j} t_{ij}(c_i^\dagger c_j + h.c.) + \sum_j \epsilon_j c_j^\dagger c_j - \lambda c_0^\dagger c_0 (a + a^\dagger) + \omega_0 a^\dagger a, \quad (1)$$

where $c_j^\dagger$ creates an electron on site j. The molecule is at the origin, which for simplicity has a single active orbital and a single vibrational mode of frequency ω_0, with electron-phonon coupling λ. The molecule may have a different diagonal energy than the leads as given by ϵ_j, the connection to the molecule may be different than within the leads (t_{ij}), and there may be a voltage drop between leads.

This Hamiltonian is straightforward to solve for a single incoming electron [13]. One makes a copy of each tight-binding orbital for $m = 0, \ldots, M$, where m is the number of phonon quanta, and M is a cutoff chosen large enough that it does not affect the results (see Figure (1)). At zero temperature, an electron with momentum k is incident from the left on the lowest lead, $\psi_j = e^{ikj}$. The Schrodinger equation using the Hamiltonian Eq. (1) provides enough equations to uniquely solve for the complex amplitude of the transmitted and reflected waves on each lead, where $m = 0$ corresponds to elastic and $m > 0$ to inelastic transmission or reflection. The transmission probability as a function of incoming electron energy shows a resonant peak. There is a second peak an energy ω_0 higher, which is designated the "one-phonon inelastic peak" by Tsui and collaborators who first measured it [1]. A surprise is that the "one-phonon inelastic peak" may sometimes consist primarily of electrons emerging in the *elastic* channel [13]. A phonon can be emitted and then reabsorbed before the electron exits. There are also strong interference effects between differ-

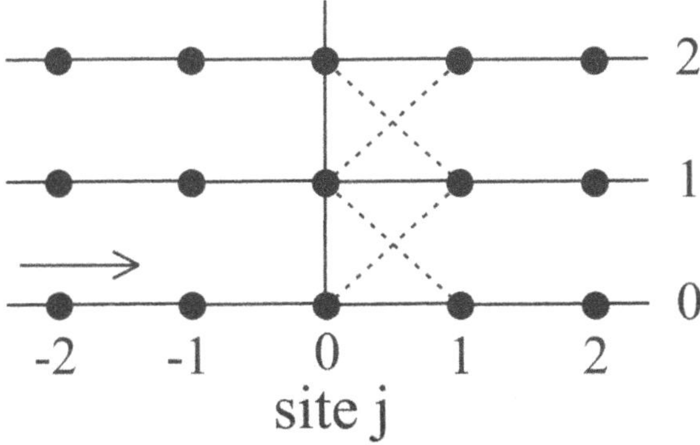

Figure 1. State space for a tunneling electron coupled to a phonon. Each dot represents a basis state in the many-body Hilbert space. The dots in the lowest row are tight-binding orbitals with no phonon excitation. The row above has one phonon excitation and diagonal energies ω_0 higher, etc. The molecule coupled to the phonon is on site 0. The vertical line from site 0 is an off-diagonal hopping matrix element of amplitude $-\lambda$, the electron-phonon coupling. The vertical line above that has amplitude $-\lambda\sqrt{2}$. The many-body (many phonon) scattering problem can be solved as if all of the dots represent Wannier orbitals in a one-body tight binding model. For the case of SSH type electron-phonon coupling between sites 0 and 1, in which the phonon modulates the hopping amplitude t rather than the on-site energy, the off-diagonal matrix elements proportional to the electron-phonon coupling are shown as dotted lines.

ent electron-phonon processes, which would be absent in simple rate equation treatments.

In addition to solving the time-independent Schrodinger equation for scattering eigenfunctions as described above, one can also solve the time-dependent Schrodinger equation $i\ d\psi/dt = H\psi$ in the full Hilbert space for arbitrary initial conditions. Figure (2) shows frames from a movie of an electron encountering a molecule coupled to a vibrational mode. The initial temperature is zero (the phonon is initially in the ground state). Frame (a) shows a bare electron wavepacket moving to the right. Frame (b) shows interference fringes as the elastic component of the backscattered wavepacket interferes with the incoming wavepacket. Frame (c) shows the electron amplitude slowly leaving the molecule (central spike). Frame (d) shows the transmitted and reflected electron probability in the elastic and inelastic channels. The group velocity is different in each channel, so the transmitted and reflected electron probability eventually separates into disjoint peaks, one for each propagating elastic or inelastic channel. This is a consequence of the quantization of the phonon energy levels, and is not present in a classical approximate treatment of the phonon. The incoming electron does not have sufficient energy to propagate to infinity in the 4-phonon channel, but it does penetrate a short distance into that and higher channels, which affects the probabilities in the allowed channels. For this Hamiltonian, there is left-right symmetry in the inelastic channels.

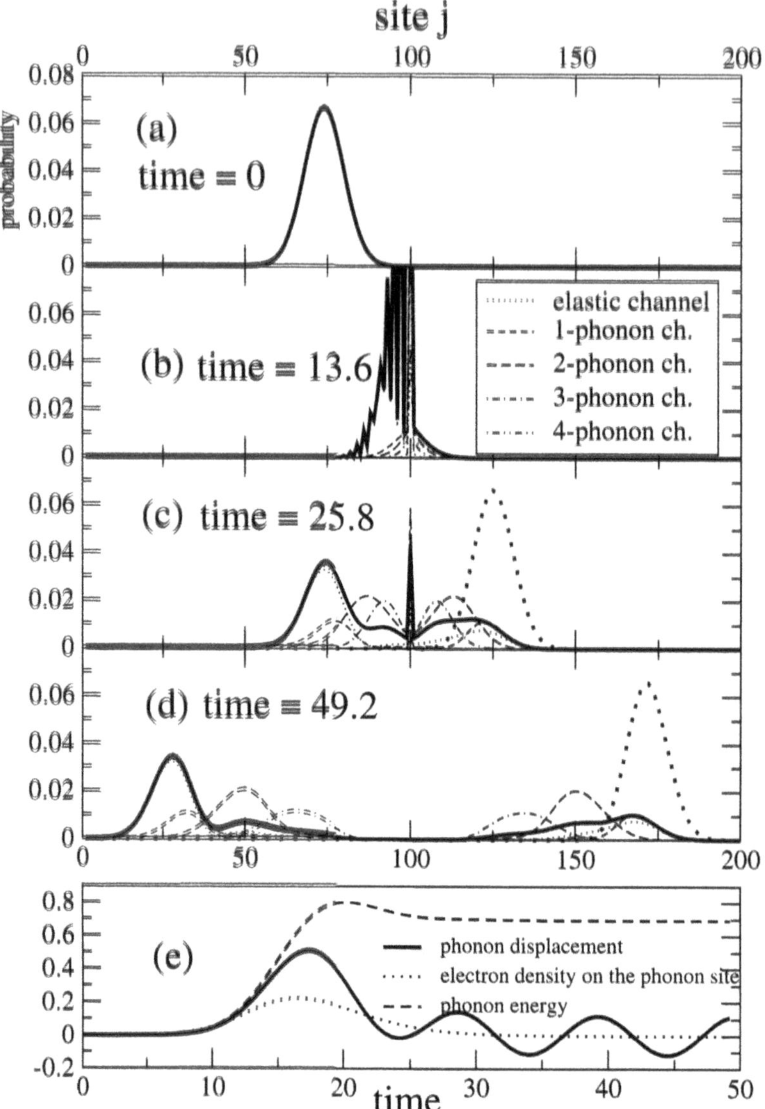

Figure 2. An electron encounters a molecule coupled to a phonon mode. The parameters are phonon frequency $\omega_0 = 0.6$, electron-phonon coupling $\lambda = 0.5$, hopping $t = 1$ except for the connections between the molecule in the center and the leads, which are $t_1 = 0.3$. The initial wavepacket has kinetic energy $\langle E \rangle = 0.165$, slightly above the center of the band, and corresponding to resonant tunneling. The total electron probability as a function of position site j is shown as a heavy line. (The probability in each inelastic channel has been multiplied by 5 for better visibility.) The small filled squares indicate where the unperturbed wavepacket would have been in the absence of scattering. Panel (e) shows the phonon displacement $\langle x \rangle = \langle a + a^\dagger \rangle$, the electron density on the molecule $\langle c_0^\dagger c_0 \rangle$, and the phonon energy $\langle a_0^\dagger a_0 \rangle$ as a function of time.

3. Finite Temperature

As discussed above, a simple approximation is to treat the phonon classically. If the phonon is slow, it acts as a static potential of strength $V = -\lambda X$, where the phonon displacement X is a gaussian random variable satisfying $\frac{1}{2}kX^2 = \frac{1}{2}k_BT$. Assuming the phonon is static, all scattering is elastic. At zero temperature (phonon displacement is zero), transmission is 100% when $E_{kin} = 0$ as shown in Fig. 3(a). The full width of the transmission spectrum is roughly t_1, where t_1 is the electron hopping from the molecule to the leads and t is the hopping within the leads. We focus on the adiabatic regime ($\omega_0 \ll t$ and $\omega_0 \ll t_1$), where the classical approximation is expected to be valid. Even in this regime, however, the classical result differs significantly from that of the quantum mechanical (QM) approach. The QM result (which is numerically exact) exhibits the following features: (1) It does not show the phonon sidebands of Ref. [13] because they are too dense to resolve when $\omega_0 \ll t_1$. (2) The maximum tunneling probability is much smaller than that of the static-phonon approach. It also occurs at E_{kin} near 0, which is different than the spectrum in Ref. [13] in the nonadiabatic regime. (3) The full width of the spectrum is broader than that of semiclassical approximation because the electron-phonon coupling energy (λ^2/ω_0) is greater than t_1.

Figure 3(b) shows the tunneling spectrum at finite temperature $k_BT = 2\omega_0$. There is good agreement on the total tunneling probability between the QM and semiclassical approaches. It seems that the static-phonon approach is a good approximation at finite temperature. However, the QM approach gives much more information about electron transport than the semiclassical method. For example, the QM result demonstrates that the probability of inelastic tunneling is greater than that of elastic tunneling (at the given parameters). In contrast, the semiclassical approach does not give any inelastic tunneling. Inelastic processes change the kinetic energy, and lead to dispersion of both the tunneling and backscattered electron. The group velocity of each inelastic process (the electron can exchange one or more phonon quanta while tunneling) is different. Weak localization properties are affected, because phase shifts differ in different channels and because electrons in distinct channels cannot interfere with each other. To understand how many phonon quanta are created (or annihilated) by the scattering, we compute $\Delta N_{ph} \equiv \langle (N_f - \langle N_i \rangle)^2 \rangle^{\frac{1}{2}}$, where N_i is initial (before scattering) phonon number and N_f is the final (after scattering) phonon number. We note that ΔN_{ph} (dot-dashed line in Fig. 3(b)) can be larger than one and that it depends on E_{kin}.

In a translation invariant system with electron-phonon coupling, a single charge carrier forms a polaron. At zero temperature without disorder, the polaron does not scatter, and the resistivity is zero. As temperature increases in a system with only optical phonons, thermally excited phonons appear with

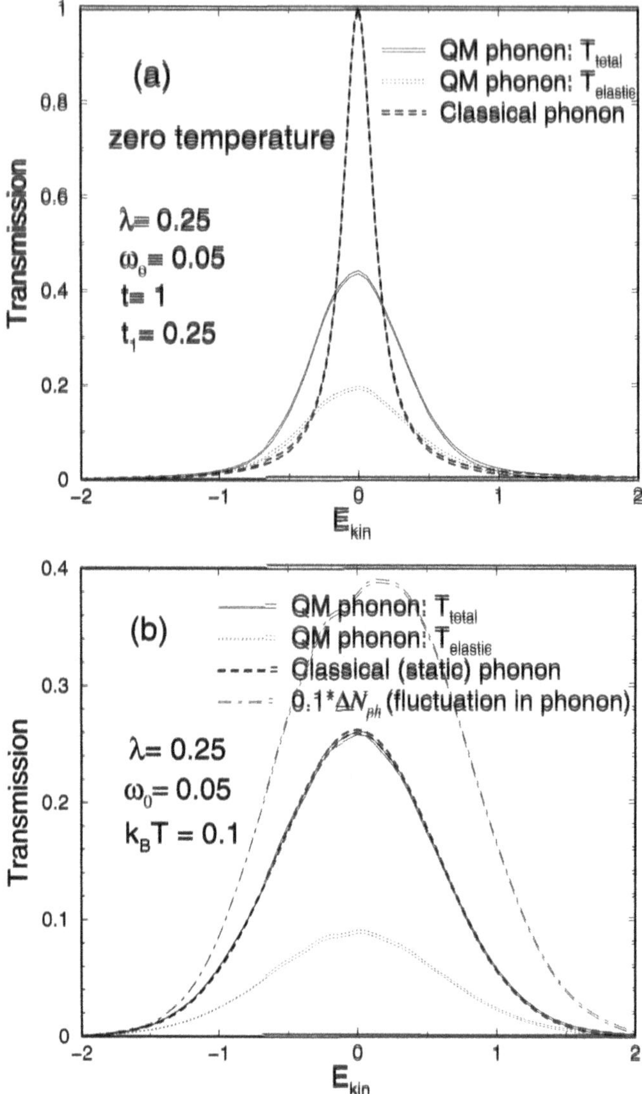

Figure 3. Transmission as a function of the kinetic energy of the incoming electron E_{kin}. Quantum-mechanical and semiclassical approaches are compared at (a) zero temperature, and (b) finite temperature $k_BT = 2\omega_0$.

a density $n \sim e^{-\hbar\omega_0/k_BT}$. Polarons scatter from these phonons, leading to finite resistivity. In Fig. (4), we investigate the polaron-phonon scattering process in a translation invariant system, by integrating the time-dependent

Schrodinger equation after performing a Lang-Firsov transformation. The incoming wavepacket is a polaron (electron plus dressed phonon cloud) of momentum $k = \pi/2$, which encounters a thermally excited optical phonon that is initially at the origin, shown in panel (a). There is a substantial probability for the polaron to be backscattered, the fundamental process that causes the electrical resistivity $\rho(T)$ of a dilute system of polarons to increase with temperature. As can be seen in panels (b) and (c), the thermally excited phonon does not recoil backward from the polaron, but rather moves toward it (actually swapping places as the polaron approaches from the left). This is seen more clearly in panel (d), which plots the center of mass of the thermally excited phonon as it moves left, $\sum_j \langle j a_j^\dagger a_j \rangle$. This process causes a negative phonon drag, or a heat current flowing in the opposite direction to the polaron charge current (for a positively charged polaron). This should be observable in a thermopower measurement. Note that for intermediate and strong electron-phonon coupling, a polaron and an additional thermally excited phonon can form a bound state (not shown), which can cause a very large phonon drag of the conventional sign.

4. Future Prospects

It is straightforward to generalize the molecular tunneling problem in a number of ways. One can include more orbitals on the molecule and more phonon modes coupled in arbitrary ways, although problems with greater than 10^8 inelastic channels may require more than a desktop computer. "Captive" local charge and spin degrees of freedom can be included [14]. Nonlinear phonons and nonlinear electron-phonon coupling present no problems. Strangely, ordinary harmonic phonons appear as a special case in which the energy level spacing happens to be uniform.

The conductivity of some molecules is affected by the rotation of π-orbitals with respect to their neighbors. The coupling is of the form

$$H_{el-ph} = -t(x)(c_i^\dagger c_j + h.c.), \tag{2}$$

where the hopping t is a function of the rotation angle x. In the usual SSH coupling, the hopping amplitude $t(x) = t_0 + t_1 x$, but for molecular rotations the leading expansion about the energy minimum is instead $t(x) = t_0 + t_2 x^2$, where $x^2 \sim (a + a^\dagger)^2$. The methods described above could be employed to solve this problem, including quantum zero-point motion, finite temperatures, and exact quantum dynamics, but results have not been obtained as yet.

On another issue, it has not proven straightforward to generalize the above essentially exact inelastic scattering results for one incoming electron to an entire fermi sea of electrons [15]. We believe that this generalization may be difficult, rather than merely awaiting a technical trick. A phonon mode on

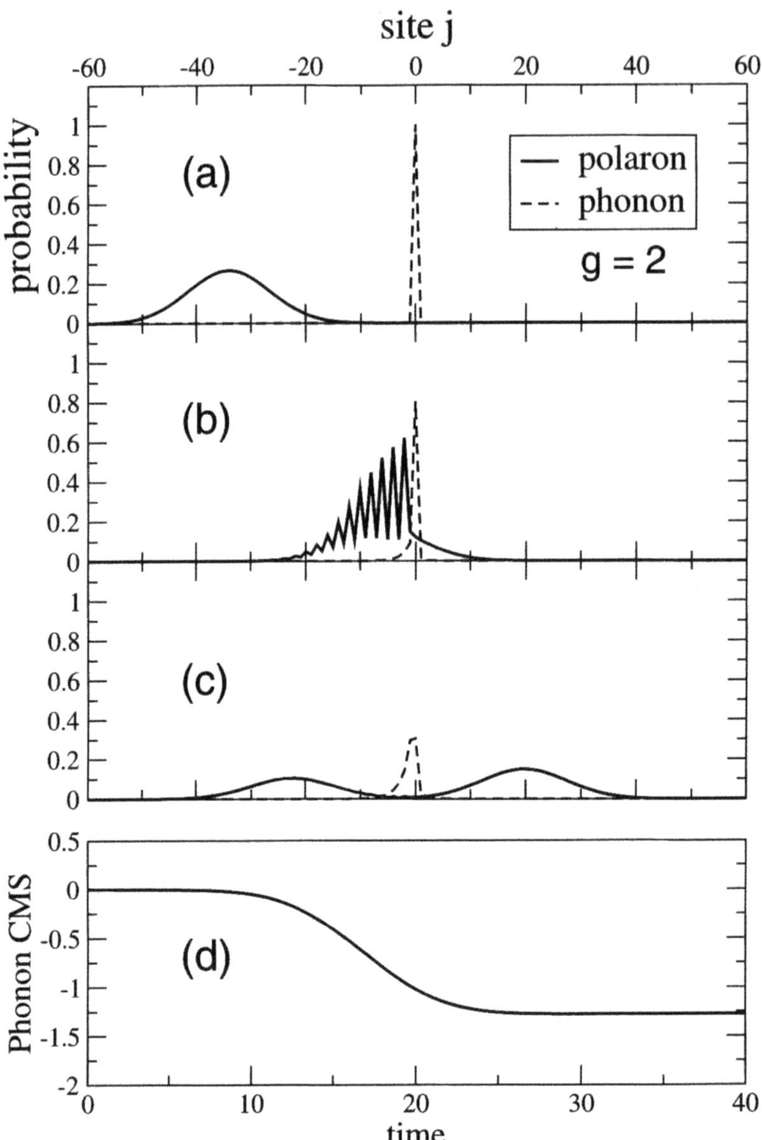

Figure 4. Scattering of a polaron from a thermally excited phonon. In contrast to Fig. (2), the system shown here is translation invariant with an optical phonon mode on every site. The parameters are $g = \lambda/\omega_0 = 2$, with time measured in units of e^{g^2} (equal to the polaron mass). The times are 0, 14.5, and 27.5 in panels (a), (b), and (c) respectively.

a single site induces local superconductivity, which spills out into the leads via the proximity effect. This may cause a gap or pseudogap in the tunneling density of states. Similarly, a locally repulsive interaction can result in a Kondo screening cloud that extends into the leads. It may be that having many-body interactions on a single site with a fermi sea is not much simpler than many-body interactions throughout. One approach is to first consider the two electron scattering problem, which is solvable by techniques similar to the one electron problem, albeit with more effort. Consider the case in which the incoming electrons are initially uncorrelated. Preliminary results indicate that when both electrons are transmitted, they become anticorrelated in the elastic channel (tend to avoid each other), and positively correlated in inelastic channels (tend to overlap). An accurate treatment of tunneling in the presence of a fermi sea should include such correlations. There may be situations with a voltage drop in which repeated interactions with electrons cause the phonon energy to increase without bound, if an additional damping mechanism is not included.

Acknowledgments

The authors would like to acknowledge valuable discussions with A. Alexandrov, I. Bezel, A. Bishop, D. Bowler, F. Bronold, H. Fehske, G. Kalosakas, C. Lambert, I. Martin, and D. Mozyrsky. This work was supported by the US DOE.

References

[1] V. J. Goldman, D. C. Tsui, and J. E. Cunningham, Phys. Rev. B **36**, 7635 (1987).

[2] J. Chen, M. A. Reed, A. M. Rawlett, and J. M. Tour, Science **286**, 1550 (1999).

[3] I. Martin and A. Schnirman, cond-mat/0310229.

[4] A. S. Alexandrov and Sir Nevill Mott, *Polarons and Bipolarons*, (World Scientific, London, 1995).

[5] F. Marsiglio, Physica C **244**, 21 (1995).

[6] G. Wellein and H. Fehske, Phys. Rev. B **56**, 4513 (1997).

[7] A. W. Romero, D. W. Brown, and K. Lindenberg, J. Chem. Phys. **109**, 6540 (1998).

[8] J. Bonča, S. A. Trugman, and I. Batistic, Phys. Rev. B **60** 1633 (1999).

[9] L.-C. Ku, S. A. Trugman, and J. Bonča, Phys. Rev. B **65**, 174306 (2002).

[10] Li-Chung Ku, Ph.D. Thesis, UCLA (2003).

[11] P. E. Kornilovitch, Phys. Rev. Lett. **84**, 1551 (2000).

[12] S. El Shawish, J. Bonča, L.-C. Ku, and S. A. Trugman, Phys. Rev. B **67**, 14301 (2003).

[13] J. Bonča and S. A. Trugman, Phys. Rev. Lett. **75**, 2566 (1995).

[14] R. Žitko and J. Bonča, Phys. Rev. B **68**, 085313 (2003).

[15] See however, A. P. Horsfield, D. R. Bowler, and A. J. Fisher (unpublished).

EXPLICIT AND HIDDEN SYMMETRIES IN QUANTUM DOTS AND QUANTUM LADDERS

K. Kikoin, Y. Avishai,
Ben-Gurion University of the Negev, Beer-Sheva 84105, Israel
kikoin@bgumail.bgu.ac.il

M.N. Kiselev
Institut für Theoretische Physik, Universität Würzburg, D-97074 Würzburg, Germany
kiselev@physik.uni-wuerzburg.de

Abstract The concept of dynamical and hidden symmetries in quantum dots and quantum ladders is introduced and developed. These symmetries are manifested in tunneling processes. If one studies the excitations in a given charge sector of nanoobject, then only the spin variables and/or electron-hole pairs are involved in the excitation spectrum. The spins in individual rung of a quantum ladder (QL) or in isolated complex quantum dot (CQD) form certain multiplets characterized by usual $SU(2)$ symmetry. This symmetry is broken due to spin transfer through QL or due to electron cotunneling through CQD. We show that dynamical symmetries of spin multiplets are unveiled in these processes. These symmetries are described by $SO(n)$ or $SU(n)$ groups in various conditions. We develop mathematical tools (fermionization procedure) for description of dynamical symmetries. The families of effective spin Hamiltonians of CQD and QL are derived in terms of generators of dynamical groups, and specific properties like Kondo tunneling through CQD and Haldane gap formation in QL are discussed.

Keywords: Spin ladders, quantum dots, Kondo tunneling, spin gap, excitons

1. Introductory notes

The symmetry of low-dimensional systems is a key to their peculiar properties [1]. It predetermines their thermodynamics, response to external fields, transport properties, phase diagrams, etc. As a rule, description of strongly interacting electrons in these systems should be constructed on non-commutative algebras, and specific structure of these algebras have direct consequences for observable physical properties of nanoobjects. In many physically interesting

A.S. Alexandrov et al. (eds.), Molecular Nanowires and Other Quantum Objects, 177–189.

cases not only the symmetry of a given Hamiltonian but also the *dynamical symmetry* of low-energy excitations is relevant. Let us consider a system with Hamiltonian $\mathcal{H}_0$ whose eigenstates $|\Lambda\rangle = |M\mu\rangle$ form a basis to an irredicible representation of some Lie group G (μ numerates the lines of this representation). It is convenient to express the generators of Lie algebras via Hubbard operators $X^{\Lambda\Lambda'} = |\Lambda\rangle\langle\Lambda'|$. Then the Hamiltonian under consideration is expressed in terms of diagonal Hubbard operators

$$\mathcal{H}_0 = \sum_{\Lambda=M\mu} E_\Lambda |\Lambda\rangle\langle\Lambda| = \sum_\Lambda E_M X^{\Lambda\Lambda}, \tag{1}$$

so that

$$[X^{\Lambda\Lambda'}, \mathcal{H}_0] = -(E_M - E_{M'})X^{\Lambda\Lambda'}. \tag{2}$$

Then the symmetry group of the Hamiltonian is generated by the operators $X^{M\mu,M\mu'}$, which commute with $\mathcal{H}_0$, whereas the dynamical symmetry of $\mathcal{H}_0$ is generated by the whole set of operators $\{X\}$. This dynamical symmetry may be exposed, when $\mathcal{H}_0$ describes a quantum object, which is a part of larger system with the Hamiltonian $\mathcal{H}$, and its symmetry is violated by interaction with this environment. If the interaction scale is characterized by some energy $\mathcal{E}$, than the dynamical symmetry is determined by transitions between those states from the manifold E_Λ, which fall into the interval $\mathcal{E}$. We divide the Hubbard operators acting within this low-energy interval into subsets $\{S\}$ and $\{R\}$. Here S-operators generate the symmetry group G, whereas S- and R-operators together generate the dynamical group D. In this paper we study spin properties of quantum dots and quantum ladders, so the group G is in fact $SU(2)$ group of a spin momentum. It will be shown that the dynamical symmetry of this object is that of $SO(n)$ group. We will construct the corresponding algebras by means of Hubbard operators, rewrite Hamiltonians $\mathcal{H}$ in terms of group generators, discuss the possible ways of fermionization of these Hamiltonians and consider some specific properties of quantum dots and quantum ladders.

2. From spin rotator to Kondo tunneling

The symmetry of spin rotator is an intrinsic property of many low-dimensional spin systems. As was shown in [2], this symmetry predetermines the low-energy dynamics of zero-D quantum dots with even occupation in a tunneling contact with metallic Fermi resevoirs. Let us consider a double quantum dot (DQD)occupied by two electrons in a neutral state in a T-shaped parallel geometry (Fig.1) as a representative example.

In this geometry two valleys of DQD are coupled by tunneling V. In the limit of strong Coulomb blockade Q, such that $V \ll Q$, the energy spectrum of isolated DQD consists of ground state singlet with energy E_S, spin triplet with the energy E_T separated by the exchange gap $\delta = 2V^2/Q$ from E_S and

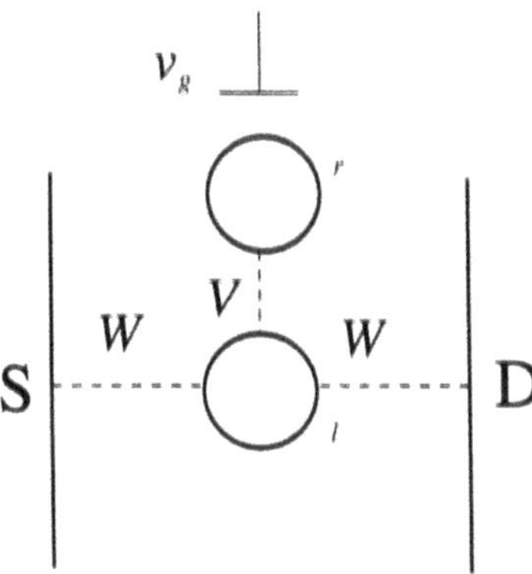

Figure 1. Parallel Double Quantum Dots in contact with source (S) and drain (D) metallic leads. V and W are tunneling coupling constants, v_g is a gate voltage.

two charge transfer excitons with large excitation energies $\sim Q$, the charging energy for a given well of DQD. Thus the indices Λ in the Hamiltonian $\mathcal{H}_0$ (1) acquire the values $\Lambda = S, T\mu$ with $\mu = 1, 0, \bar{1}$ standing for three projections of spin one.

The dynamical symmetry of the $\{S, T\}$ manifold is that of $SO(4)$ group. Two vectors generating this group are constructed by means of Hubbard operators (2) in the following way:

$$\begin{aligned} S^+ &= \sqrt{2}\left(X^{10} + X^{0-1}\right), \; \S_z = X^{11} - X^{-1-1}. \\ R^+ &= \sqrt{2}\left(X^{1S} - X^{S-1}\right), \; R_z = -\left(X^{0S} + X^{S0}\right). \end{aligned} \quad (3)$$

Here $\S$ is the spin 1 operator, while $\mathbf{R}$ is the R-operator describing S/T transitions. The spin algebra is o_4, which is characterized by the commutation relations

$$[S_\alpha, S_\beta] = ie_{\alpha\beta\gamma}S_\gamma, \; [R_\alpha, R_\beta] = ie_{\alpha\beta\gamma}S_\gamma, \; [R_\alpha, S_\beta] = ie_{\alpha\beta\gamma}R_\gamma \quad (4)$$

(α, β, γ are Cartesian coordinates, $e_{\alpha\beta\gamma}$ is a Levi-Civita tensor). These vectors are orthogonal, $\mathbf{S} \cdot \mathbf{R} = 0$, the Casimir operator is $\mathbf{S}^2 + \mathbf{R}^2 = 3$.

A gate voltage v_g applied to DQD turns the level positions essentially asymmetric, and the charging energy Q may be nearly compensated for at least one charge transfer singlet exciton (say, right, $\Lambda = E_r$). In this case we encounter a "Coulomb resonance" excitation, where the spin singlet and charge transfer exciton (also singlet!) are strongly intermixed, but the spin triplet is untouched by this resonance tunneling. This means that we deal with a manifold $\{S, T, E_r\}$. Then one more R-vector $\mathbf{R}_1$ and a scalar A should be included in the set of group generators. These generators are expressed in terms of Hubbard operators as follows:

$$R_1^+ = \sqrt{2}\left(X^{1E_r} - X^{E_r1}\right), \quad R_{1z} = -\left(X^{0E_r} + X^{E_r0}\right),$$
$$A = i(X^{SE_r} - X^{E_rS}). \tag{5}$$

To close the alebra the commutation relations (4) which are valid also for $R_{1\alpha}$ should be completed by

$$[R_{l\alpha}, R_{1\beta}] = i\delta_{\alpha\beta}A, [R_{1\alpha}, A] = iR_{l\alpha}, \tag{6}$$
$$[A, R_{l\alpha}] = iR_{1\alpha}, \quad [A, S_{l\alpha}] = 0.$$

The system of commutation relations (4), (6) is that of o_5 algebra, and the manifold $\{S, T, E_r\}$ obeys $SO(5)$ dynamical symmetry, provided all three levels are involved in interaction in a framework of the Hamiltonian H. The Casimir operator for $SO(5)$ group is $\mathbf{S}^2 + \mathbf{R}^2 + \mathbf{R_1}^2 + A^2 = 4$.

In terms of these operators $\mathcal{H}_0$ acquires the form

$$\mathcal{H}_0 = \frac{1}{2}\left(E_T\mathbf{S}^2 + E_S\mathbf{R}^2\right) + Q(\hat{N} - 2)^2, \tag{7}$$

and

$$\mathcal{H}_0 = \frac{1}{2}\left(E_T\mathbf{S}^2 + E_S\mathbf{R}^2 + E_{E_r}\mathbf{R}_1^2\right) + Q(\hat{N} - 2)^2. \tag{8}$$

for $SO(4)$ and $SO(5)$ group, respectively. The last terms in (7) and (8) control the number of electrons given by the operator $\hat{N}$ in DQD.

As is seen from this equation, spin is still conserved in isolated DQD. However, a tunnel contact with metallic leads breaks the spin conservation and reveals the dynamical symmetry of DQD. The mechanism of this non-conservation is *electron cotunneling* with spin flips, when an electron with spin σ enters DQD, whereas another electron with spin σ' leaves it. This process is known to be a source of Kondo effect in tunnel barriers and quantum dots [3]. Eliminating charge degrees of freedom by means of the Schrieffer-Wolff transformation, one usually comes to an exchange-like cotunneling Hamiltonian of the type $J_{cot}\mathbf{S} \cdot \mathbf{s}$, where $J_{cot} \sim W^2$, and W is a lead-dot tunneling amplitude.

Since $E_T - E_S = \delta > 0$, the Kondo effect seems to be irrelevant in DQD with even occupation. However, one should remember that the tunneling W induces additional contribution into indirect exchange between two wells in DQD. As is shown in Refs. [2] this contribution may change the sign of δ provided the excitation E_r is soft enough, but the condition $V/(E_T - E_S) \ll 1$ is still valid. Then the exciton E_r is eliminated from the manifold, the symmetry of DQD is reduced from $SO(5)$ to $SO(4)$ and the Schrieffer-Wolff transformation yields the effective spin Hamiltonian

$$\mathcal{H} = \mathcal{H}_0 + J_{cot}^T\mathbf{S} \cdot \mathbf{s} + J_{cot}^{ST}\mathbf{R} \cdot \mathbf{s}, \tag{9}$$

where J_{cot}^{ST} and J_{cot}^{ST} are two indirect exchange coupling parameters which are renormalized by Kondo screening. This screening is given by both vectors **S** and **R**.

The problem of Kondo tunneling within the framework of the Hamiltonian (9) is solved already, at least in the weak coupling limit (see [2] and references therein), so we do not enter the details here. For our further purposes it is important, that this example demonstrates how the dynamical symmetry of S/T pair is revealed in interaction with continuum, which breaks the rotational symmetry of isolated spin system. This interaction inserts its own energy scale $\mathcal{E}$ in the problem (the Kondo temperature T_K in example considered above), and the dynamical symmetry of spin rotator becomes relevant when the T/S energy splitting is comparable with T_K. In more complicated quantum dots the spin manifolds consist of several S/T pairs, and the dynamical symmetry of such dots is described by $SO(n)$ groups (see Ref. [4] where the cases of $n = 5, 7$ are described).

3. From spin rotator to spin ladder

Being armed by the above mathematical tools, we see that any rung of a two-leg spin ladder (Fig. 2a) possesses the same $SO(4)$ symmetry, because two spins 1/2 form a S/T manifold. Therefore the dynamical symmetry is an intrinsic property of spin ladders and decorated spin chains shown in Figs 2b,d. Here we derive the family of Hamiltonians for these systems and discuss various manifestations of this symmetry in their energy spectrum.

Generic Hamiltonian for spin systems under consideration is the Heisenberg-type spin 1/2 ladder Hamiltonian

$$H^{(SL)} = J_t \sum_{\langle i1,i2\rangle} \mathbf{s}_{i1} \cdot \mathbf{s}_{i2} + J_l \sum_{\alpha} \sum_{\langle i\alpha,j\alpha\rangle} \mathbf{s}_{i\alpha} \cdot \mathbf{s}_{j\alpha} \tag{10}$$

Here index $\alpha = 1, 2$ enumerates the legs of the ladder, and the sites $\langle i1, i2\rangle$ belong to the same rung (Fig.2a).

A chain of dimers of localized spins illustrated by Fig. 2b is described by the simplified version of this Hamiltonian

$$H^{SRC} = J_t \sum_{\langle i1,i2\rangle} \mathbf{s}_{i1} \cdot \mathbf{s}_{i2} + J_l \sum_{\langle ij\rangle} \mathbf{s}_{i1} \cdot \mathbf{s}_{j1} \tag{11}$$

The geometry of alternate rungs is chosen in the system (11) to avoid exchange interaction between spins $\mathbf{s}_{i2}$ and $\mathbf{s}_{j2}$.

The transverse coupling may emerge either from direct exchange (in case of localized spins) or from indirect Anderson-type exchange induced by tunneling (similarly to the case of quantum dots). In the latter case the sign of J_t is antiferromagnetic (AFM), in the former case it may be ferromagnetic (FM) as well. The same is valid for J_l.

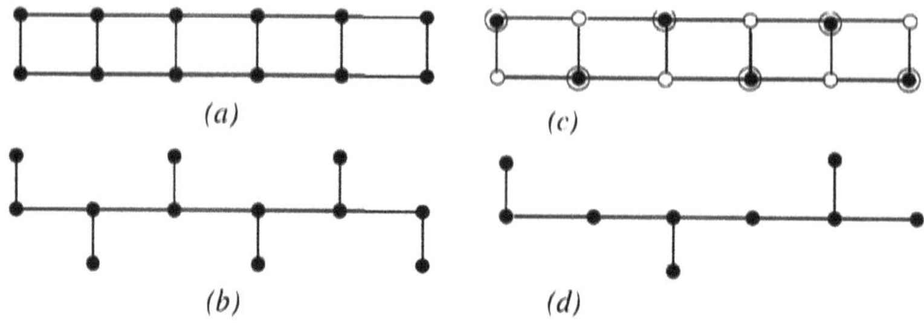

Figure 2. Spin Ladder (a), Spin Rotator Chain (b), Spin ladder in the CDW phase (c) and Alternate Spin Rotator Chain (d).

We start with diagonalization of the Hamiltonian of perpendicularly aligned dimer (cf. Ref. [5]). The $SO(4)$ symmetry stems from the obvious fact that the spin spectrum of a dimer $\{i1, i2\}$ is formed by the same S/T pair as the spin spectrum of DQD studied in the previous section. This analogy prompts us a canonical transformation connecting two pairs of spin vectors, $\{\mathbf{s}_{i1}, \mathbf{s}_{i2}\}$ and $\{\mathbf{S}_i, \mathbf{R}_i\}$: Two sets of spin operators are connected by a simple rotation

$$\mathbf{s}_{i1} = \frac{\mathbf{S}_i + \mathbf{R}_i}{2}, \quad \mathbf{s}_{i2} = \frac{\mathbf{S}_i - \mathbf{R}_i}{2}, \tag{12}$$

Then the Hamiltonian $\mathcal{H}_i$ of a single dimer i is the same as the Hamiltonian (7) of DQD. The total spin of a dimer is not conserved in a spin chain, so the dynamical symmetry of an individual rung is revealed by the modes propagating along the chain [5]. Applying the rotation operation (12) to the Hamiltonians (10) and (11), we transform them to a form

$$\mathcal{H} = \mathcal{H}_0 + \mathcal{H}_{int} \tag{13}$$

Here $\mathcal{H}_0 = \sum_i \mathcal{H}_i$ is common for both models. It is useful to include the Zeeman term in $\mathcal{H}_i$,

$$\mathcal{H}_i = \frac{1}{2}\left(E_S\mathbf{R}_i^2 + E_T\mathbf{S}_i^2\right) + hS_{iz}. \tag{14}$$

We confine ourselves by a charge sector $N_i = 2$ and omit the Coulomb blockade term for the sake of brevity. The interaction part of SL Hamiltonian transforms under the rotation (12) to the following expression

$$\mathcal{H}_{int}^{SL} = \frac{1}{4}J_l \sum_{\langle ij\rangle}(\mathbf{S}_i\mathbf{S}_j + \mathbf{R}_i\mathbf{R}_j) \tag{15}$$

The interaction part of the SRC Hamiltonian is

$$\mathcal{H}_{int}^{SRC} = \frac{1}{4}J_l \sum_{\langle ij\rangle}(\mathbf{S}_i\mathbf{S}_j + 2\mathbf{R}_i\mathbf{S}_j + \mathbf{R}_i\mathbf{R}_j) \tag{16}$$

One may also consider the *alternate SRC model* (ASRC, see Fig. 2(c)). Its interaction Hamiltonian acquires after rotation (12) the form

$$\mathcal{H}_{int}^{ASRC} = \frac{1}{4} J_l \sum_{\langle ij \rangle} (\mathbf{S}_i \mathbf{S}_j + \mathbf{S}_i \mathbf{R}_j) \tag{17}$$

Now we see that all three effective Hamiltonians belong to the same family. In all cases initial ladder or "semi-ladder" Hamiltonian is transformed into really one-dimensional spin-chain Hamiltonian, which, however, takes into account the hidden symmetry of a dimer. The effective Hamiltonians (15), (16), (17) contain operators $\mathbf{R}$ describing the dynamical symmetry of dimers. The dynamical symmetry turns the spectrum of this Hamiltonians to be richer than that of a standard Heisenberg chain. Like in many other cases, rotation transformation eliminates the antisymmetric combination of two generators.

The transformation (12) reveals the hidden symmetry of spin 1/2 ladder (15). It maps the ladder Hamiltonian onto a pair of coupled chain Hamiltonians: one is the conventional spin 1 chain, the other is a pseudospin chain. Spin $\mathbf{S}_i$ and pseudospin $\mathbf{R}_i$ are coupled kinematically by the commutation relations and by the local Casimir constraint

$$\mathbf{S}_i^2 + \mathbf{R}_i^2 = 3. \tag{18}$$

It is instructive to compare the Hamiltonian (15) with the effective Hamiltonian of spin 1 chain, which arises after decomposition of spin-one operators into a pair of spin 1/2 operators, $\mathbf{S}_i = \mathbf{s}_i + \mathbf{r}_i$ [6]. This decomposition operation transforms initial Hamiltonian into a form similar to H^{SRC} but for spin-one-half operators $\mathbf{s}_i$, $\mathbf{r}_i$. The difference between two cases is that these effective spins commute (unlike the operators $\mathbf{S}_i$, $\mathbf{R}_i$). In other terms, the difference is that the local symmetry of spin-one chain is $SO(3)$ whereas the local symmetry of SRC is $SO(4)$. The spin rotator chains (16), (17) are in some sense intermediate between spin chains and spin ladders. In this cases the spin-pseudospin symmetry is obviously broken by the cross terms $2\mathbf{S}_i\mathbf{R}_j$.

The excitation spectrum of spin ladders may be calculated in terms of operators $\mathbf{S}_i$ and $\mathbf{R}_i$. For example, the well known expression for a gap ΔE in the excitation spectrum in the limit of strong transverse exchange $J_t \gg J_l$ for AFM interaction [5] is

$$\Delta E = J_t + \frac{(J_l/4)^2 \sum_{ij,\alpha\beta} \left(\langle T_{ij}^{\alpha\beta} | \mathbf{R}_i \mathbf{R}_j | S_i S_j \rangle \right)^2}{(E_T - E_S)} = J_t + \frac{3J_l^2}{8J_t} \tag{19}$$

(here $T_{ij}^{\alpha\beta}$ and $S_i S_j$ stand for possible triplet projections and spin states at the sites i, j, respectively). The singlet-triplet excitations above this gap are given by the dispersion law $\omega(k) = \Delta E + J_l \cos k$.

In all cases the simplified versions of Heisenberg Hamiltonians may be considered. The simplified SL models are well known [7]. The anisotropic versions of the Hamiltonian (16) are: *Ising-like SRC model:*

$$H = \frac{1}{4} J_l \sum_{\langle ij \rangle} (S_i^z S_j^z + 2 S_i^z R_j^z + R_i^z R_j^z); \tag{20}$$

Anisotropic SRC model:

$$H = \frac{1}{4} J_l \sum_{\langle ij \rangle} \Big[(S_i^+ S_j^- + S_i^+ R_j^- + S_i^- R_j^+ + R_i^+ R_j^-) \tag{21}$$

$$+ \Delta (S_i^z S_j^z + 2 S_i^z R_j^z + R_i^z R_j^z) \Big] ;$$

SRC in strong magnetic field: the $SO(4)$ group reduces to the $SU(2)$ group in a magnetic field, when the Zeeman splitting *exactly* compensates the exchange gap in a single dimer, $h_0 = |E_T - E_S|$. Then at low T, the states $|i0\rangle$ and $|i-1\rangle$ are quenched, and only two components, $R^{\pm}$ survive in the manifold (3). As a result, the Hamiltonian (16) is mapped onto an XY-model for spin 1/2:

$$H_{XY}^{(R)} = \frac{1}{4} J_l \sum_{\langle ij \rangle} (R_i^+ R_j^- + H.c.). \tag{22}$$

This means that starting from a singlet ground state for $J_t \equiv E_T - E_S > 0$, one may induce development of spin liquid-like excitations by applying strong magnetic field. In a near vicinity of this point of degeneracy, H_{int} acquires the features of an *XY model in transverse magnetic field.*

4. Fermionization

To describe the elementary excitations in SRC, one should generalize the $SU(2)$-like semi-fermionic representation for S operators [8]

$$S^+ = \sqrt{2}(f_0^\dagger f_{\bar{1}} + f_1^\dagger f_0), \; S_z = f_1^\dagger f_1 - f_{\bar{1}}^\dagger f_{\bar{1}}, \tag{23}$$

where $f_1^\dagger$, $f_{\bar{1}}^\dagger$ denote creation operators for fermions with spin "up" and "down" respectively whereas f_0 stands for spinless fermion. Fermionization of $SO(4)$ group is completed by introducing one more spinless fermion f_s which represents the singlet state. As a result, R-operators are given by the following equations:

$$R^+ = \sqrt{2}(f_1^\dagger f_s - f_s^\dagger f_{\bar{1}}), \; R^z = -(f_0^\dagger f_s + f_s^\dagger f_0). \tag{24}$$

Then the single-site Hamiltonians may be represented in a form

$$H_i = -\delta f_{is}^\dagger f_{is} + h(f_{i1}^\dagger f_{i1} - f_{i\bar{1}}^\dagger f_{i\bar{1}}) \tag{25}$$

The Casimir operator (18) transforms to the local constraint $\sum_{\Lambda=\pm,0,s} f^\dagger_\Lambda f_\Lambda = 1$. To fermionize the generators of $SO(5)$ group, one should add two more spin fermions and one more spinless fermion describing transitions to the excitonic state $|E\rangle$.

We start the studies of elementary excitations in SRC with the aniso-tropic XXZ version. of general effective Hamiltonian. The simplest one is the case (17). The problem is reduced to a standard XY-model for spin one half, and the spinon spectrum may be easily obtained either by bosonization or by spinon-type fermionization. In former case one deals with hard-core bosons, and in the latter one the problem is mapped onto the non-interacting incompressible fermions at half-filling.

Next is a more complicated case of XXZ-SRC model (21) specifically on its simplified alternate version, which is obtained from the Hamiltonian (17). The Hamiltonian of this model is

$$H = \frac{1}{4} J_l \sum_{\langle ij \rangle} (S_i^+ S_j^- + S_i^+ P_j^- + S_i^- P_j^+ + \Delta(S_i^z S_j^z + 2S_i^z P_j^z). \qquad (26)$$

The S-S part of this Hamiltonian describes the S=1 chain, with the Haldane gap in the excitation spectrum (see, e.g.,[9, 10]). The question is, how do the S-P interaction modifies the gap. We consider the case of FM dimers, when the triplet is the ground state. In this case one has one more gap mode, where the gap equals J_t. This mode is coupled to Haldane branch only via S-P exchange terms in (26).

The spin liquid fermionization approach adopted here is a convenient tool for description of Haldane spectrum. Unlike the S=1/2 model, where the spin-liquid state is easily described by global U(1) invariant modes $T_{ij}T_{ji} = \sum_\sigma f^\dagger_{i\sigma} f_{j\sigma}|^2$, in case of S=1, one deals with variables which *effectively break this symmetry.* One can rewrite the Hamiltonian of SRC model with $\Delta = 0$ in a form

$$H = \frac{1}{4} J_l \sum_{ij} \left[\left(f^\dagger_{i1} f_{j1} + f^\dagger_{i\bar{1}} f_{j\bar{1}} \right) \bar{B}_j^{0S} B_i^{0S} + (f^\dagger_{i\bar{1}} f^\dagger_{j1} C_j^{0S} B_i^{0S} + H.c.) \right] \quad (27)$$

where $B_j^{0S} = f_{0j} + f_{Sj}$, $C_j^{0S} = f_{0j} - f_{Sj}$. The terms in the first line of Eq. (27) describe coherent propagation of spin fermions accompanied by a backflow on neutral fermions, whereas the terms in the second line are "anomalous" (they do not conserve spin fermion number). For example the propagator $\langle S_i^+ S_j^- \rangle$ contains anomalous components $f^\dagger_{i1} f^\dagger_{j\bar{1}} f_{j0} f_{i0} \to F^*_{ij,1\bar{1}} F_{ji,00}$ along with normal ones $f^\dagger_{i1} f_{j1} f^\dagger_{j0} f_{i0}$. Here $F_{ij,\Lambda\Lambda'} = f_{j\Lambda} f_{i\Lambda'}$. The first term in (27) describes the kinetic energy spinon excitations, and two last anomalous term breaking U(1) symmetry are responsible for the Haldane gap.

To reveal the effect of dynamical symmetry on the Haldane gap, one has to note that the terms B^{0S} and B^{0S} appear both as a counterflow in the first term and as gauge symmetry breaking terms in the second line. In spin 1 ladder the counterflow term $\sim f_{i0}^{\dagger} f_{j0}$ predetermines the width of spinon band described by the first line of Eq. (27). Apparently, the one extra channel (tripet/singlet transitions in B^{0S}) enhances this effect, because in this case the local constraint imposes further restrictions of phase fluctuations.

The gap itself is due to anomalous correlations described by the second line of Eq. (27). Here the appearance of second channel of spinless excitations results in formation of even and odd operators B_j^{0S} and C_j^{0S}. The Haldane gap closes when the $|0\rangle$ and $|S\rangle$ states are degenerate (the odd operator C_j^{0S} nullifies the anomalous terms responsible for its formation). This means that appearance of $0S$ channel *favors* closing of the Haldane gap.

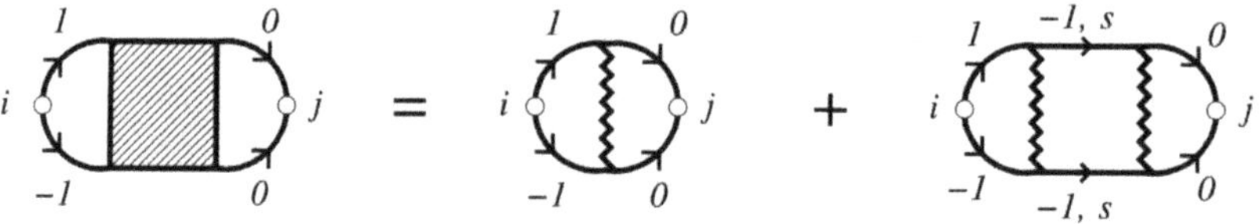

Figure 3. Lowest order contributions to anomalous propagator (27).

In a strong coupling case of $J_t \gg J_l$ both trends above may be considered at least in the lowest order of perturbation theory. In case of spin ladders [5] the 1-st and 2-nd-order in $g = J_l/J_t$ anomalous diagrams are represented in Fig.3.

5. SO(n) dynamical symmetries for a two-leg quantum ladder

It was mentioned in Section II that the dynamical symmetry of DQD becomes $SO(5)$, if charge transfer excitonic state is involved (see Eq. 8). In this section we discuss the origin of this symmetry in spin ladders. This problem arose in a context of $SO(5)$ symmetric $t - J$ model of 2D cuprate superconductors [11]. Later on the version of this theory was formulated for cuprate two-leg ladders [12]. Here we show that the *dynamical* $SO(5)$ group arises in description of Heisenberg ladder, but excitonic states are involved in this symmetry instead of Cooper states.

Let us consider a two-leg quantum ladder depicted in Fig.2a under condition of strong Coulomb blockade imposed on each rung i. We allow electron tunneling t_{ij}^{α} along both legs. This tunneling is described by the Hamiltonian

$$\mathcal{H}_{tun} = \sum_{ij} \sum_{\alpha\sigma} t^{\alpha} d_{i\alpha\sigma}^{\dagger} d_{j\alpha\sigma}. \tag{28}$$

(only nearest-neighbor hopping along the leg is allowed). This hopping results in the appearance of *charged* rungs because each hopping act creates a hole on a rung j and an electron on a rung i. To treat this charging properly the Coulomb blockade term in the Hamiltonian $\mathcal{H}_i$ (14) should be restored (see Eq. 8), and the terms with excess electron and excess hole should be added. It is more convenient to represent the Hamiltonian $\mathcal{H}_i$ of an individual rung i in terms of diagonal Hubbard operators [see (1)]

$$\mathcal{H}_i = \sum_i \left[\sum_\Lambda E_\Lambda X_i^{\Lambda\Lambda} + \sum_\gamma E_\gamma X_i^{\gamma\gamma} + \sum_\Gamma E_\Gamma X_i^{\Gamma\Gamma} \right] \tag{29}$$

Here the index $\gamma = \alpha\sigma$ stands for the states with one electron with spin σ on a site $i\alpha$ of the rung i, the index $\Gamma = \alpha\sigma$ stands for three-electron states of a rung, where two electrons occupy site $i\alpha$ and one electron with spin σ is located in a site $i\bar{\alpha}$ ($\bar{\alpha} = 2$ if $\alpha = 1$ and v.v). The energy levels E_γ and E_Γ are separated by a Coulomb blockade gap $\sim Q$ from the two-electron states E_Λ. The Hamiltonian (28) in these terms is

$$\mathcal{H}_{tun} = \sum_{ij,\alpha} \sum_{\gamma\Gamma\Lambda} t^\alpha X_i^{\Gamma\Lambda} X_j^{\gamma\Lambda} + H.c. \tag{30}$$

It is seen from (30) that the intersite hopping "charges" two neighboring rungs in a ladder, which was initially neutral, and one should pay the energy $\sim Q$ for each hopping act, like in the generic Hubbard model at half-filling. This energy loss is reduced if an electron-hole pair is created at a given rung i. In this case the electron-hole attraction $V < 0$ partially compensates charging energy Q. Let us assume the hierarchy $Q \gg Q - |V| \gg t$. Then the states $|\Gamma\rangle$ may be excluded from the manifold in favor of excitonic states $|iE_\alpha\rangle$ similar to the states $|E_r\rangle$ introduced in (5). Here $\alpha = 1(2)$ for the electron occupying site $i1(i2)$. If the ground state of a rung is singlet, $|iS\rangle$, then electron and hole have antiparallel spins and the excitation energy is $Q' = Q - |V|$. Even combination of two states $|iE_{(1,2)}\rangle$ form a singlet exciton $|iE\rangle$. Such exciton can propagate coherently along the ladder unlike single electron, whose tunneling leaves a trace of charged states according to (30). Indeed, translation of e-h pair from a rung i to a neighboring rung $i+1$ can be presented as coherent tunneling of electron from a site $i\alpha$ to a site $i+1, \alpha$ and another electron in the opposite direction (from $i+1, \bar{\alpha}$ to $i, \bar{\alpha}$. The exciton propagation is described by the following term in effective Hamiltonian:

$$H_{ex} = \sum_i K^S X_i^{SE} X_{i\pm1}^{ES} \tag{31}$$

with effective exchange coupling constant $K^S = |t_1 t_2|/Q'$, and the dispersion law describing coherent exciton propagation is $\epsilon_S(k) = 2K^S \cos k$. As was shown in Section 2, the manifold $\{iS, iT, iE\}$ possesses the local dynamical symmetry $SO(5)$ [see Eqs. (5), (6)], and this symmetry allows existence of coherent collective singlet exciton mode. The Hamiltonian (31) acquires a form $H_{ex} = (K^S/4)\sum_{ij} \widetilde{A}_i \widetilde{A}_j$ in terms of generators of SO(5) group (5), where $\widetilde{A}_i = (\mathbf{R}_i^2 - 1)A_i$. There is one more collective mode, namely triplet exciton $|E_\mu\rangle$ ($\mu = \pm 1, 0$) separated by the gap $\sim J_t$ from the singlet exciton. In case of triplet ground state ($J_t < 0$), this mode becomes the lowest one, and the Hamiltonian similar to (31) may be derived for triplet exciton propagation with operators $X_i^{TE_\mu}$ replacing X_i^{SE}. In this case the manifold $\{iS, iT, iE_\mu\}$ consists of one singlet and two triplets, and the corresponding dynamical group is $SO(7)$ [4].

If exchange and excitonic gaps are comparable in magnitude, then the interplay between exciton and magnon modes is possible, and dynamical symmetry will result in observable physical effects. Like in cuprate ladder, [13], the excitonic instability can develop for certain values of model parameters, which results in phase separation and, in particular in formation of CDW phase illustrated by Fig. 2c (where double and empty circles stand for doubly occupied and empty sites respectively).

6. Concluding remarks

We rederived a family of Hamiltonians for quantum dots and quantum ladders in terms of $SO(4)$ group, which describes the dynamical symmetry of a spin rotator [2]. We exploited the fact that in case, when the Hamiltonian $\mathcal{H}$ contains blocks $\mathcal{H}_i$ formed by two sites occupied by spins 1/2, one may use its eigenstates (singlet-triplet manifolds) as a basis for representing the spin invariants entering $\mathcal{H}$. These invariants contain the Runge-Lenz-like vectors $\mathbf{R}_i$ along with the usual spin vectors $\mathbf{S}_i$. If the electron-hole pairs are also included in the set of eigenstates, then the local dynamical symmetry of $\mathcal{H}_i$ is characterized by the $SO(n)$ group with $n = 5$ or 7 for a singlet and triplet ground state of $\mathcal{H}_i$, respectively. The elementary excitations in quantum dots and quantum ladders are described by means of generators of $SO(n)$ groups and the interplay between different branches of excitation spectra is a direct manifestation of local dynamical symmetry violated by non-local interactions.

References

[1] A.O. Gogolin, A.A. Nersesyan and A.M. Tsvelik, *Bosonization in Strongly Correlated Systems*, Cambridge University press, Cambridge, 1998.

[2] K. Kikoin and Y. Avishai, Phys. Rev. Lett. **86**, 2090 (2001); Phys. Rev. **B65**, 115329 (2002).

[3] J. Appelbaum, Phys. Rev. Lett. **17**, 91 (1966); L.I. Glazman and M.E. Raikh, JETP Lett. **47**, 452 (1988); T.K. Ng and P.A. Lee, Phys. Rev. Lett. **61**, 1768 (1988).

[4] T. Kuzmenko, K. Kikoin and Y. Avishai, Phys. Rev. Lett. **89**, 156602 (2002).

[5] T. Barnes, E. Dagotto, J. Riera, and E.S. Swanson, Phys. Rev. B **47**, 3196 (1993).

[6] A. Luther and D. Scalapino, Phys. Rev. B **16**, 1153 (1977); J. Timonen and A. Luther, J. Phys. C: Solid State Phys. **18**, 1439 (1985).

[7] E. Dagotto, Rep. Progr. Phys. **62**, 1525 (1999).

[8] V.N. Popov and S.A. Fedotov, Sov. Phys. JETP **67**, 535 (1988); M. Kiselev and R. Oppermann, Phys. Rev. Lett **85** (2000), M. Kiselev et al, Eur. Phys. J **B22**, 53 (2001).

[9] C.D. Batista and G. Ortiz, cond-mat/0207106, and references therein.

[10] F.D.M. Haldane, Phys. Rev. Lett. **50**, 1153 (1983); H.J. Schulz, Phys. Rev. B **34**, 6372 (1986); V.A. Kashurnikov *et al*, Phys. Rev. B **59**, 1162 (1999).

[11] E. Demler and S.-C. Zhang, Phys. Rev. Lett. **75**, 4126 (1995).

[12] D. Scalapino, S.-C. Zhang and W. Hanke, Phys. Rev. B**58**, 443 (1998)

[13] M. Tsuchiizu and A. Furusaki, Phys. Rev. B **66**, 245106 (2002).

HOLE BAND ENGINEERING IN SELF-ASSEMBLED QUANTUM DOTS AND MOLECULES

F.M. Peeters, M. Tadić,* K.L. Janssens, and B. Partoens

Departement Natuurkunde Universiteit Antwerpen (Campus Drie Eiken) Universiteitsplein 1 B-2610 Antwerpen Belgium

peeters@uia.ua.ac.be

Abstract The electronic structure in type-II self-assembled quantum dots and molecules are discussed. As an example we consider disk-shaped InP/GaInP quantum dots. Depending on the thickness of the quantum dot holes are located inside (pillar case) or outside (flat-dot case), which implies that by varying the dot size one can engineer the position of the holes to establish either type-I or type-II confinement. In quantum-dot molecules we find that the strain leads to an upward shift of the lowest energies in all explored electron shells in both the double and triple dot molecules in the case of thick spacers, while for thin spacers the quantum mechanical coupling prevails, and a downward shift is observed. The exciton states are found to be strongly influenced by holes, which are able to turn bonding behavior of exciton levels into antibonding in a certain range of spacer thickness. The diamagnetic shift is computed for the coupled quantum dots, and the theoretical results are compared with available magneto-photoluminescence data.

Keywords: quantum dot; quantum-dot molecule; strain; valence band; exciton

1. Introduction

Quantum dots are small man-made structures in which electrons are confined in all three spatial directions, therefore a discrete spectrum of charge carriers occurs, like in natural atoms, therefore they are nicknamed *artifical atoms*. Two quantum dots in the vicinity of each other form artificial or *quantum-dot molecule*. Various techniques have been utilized to produce quantum dots, but the Stranski-Krastanow growth mode, which enables quantum dots to self-assemble [1], seems to be the most promising contender for technologies of the future, as lasers and optical amplifiers [2], photonic detectors [3], resonant tunneling devices [4], or memory elements [5]. The growth of self-assembled

*On leave of absence from: Faculty of Electrical Engineering, University of Belgrade, Serbia.

A.S. Alexandrov et al. (eds.), Molecular Nanowires and Other Quantum Objects, 191–202.

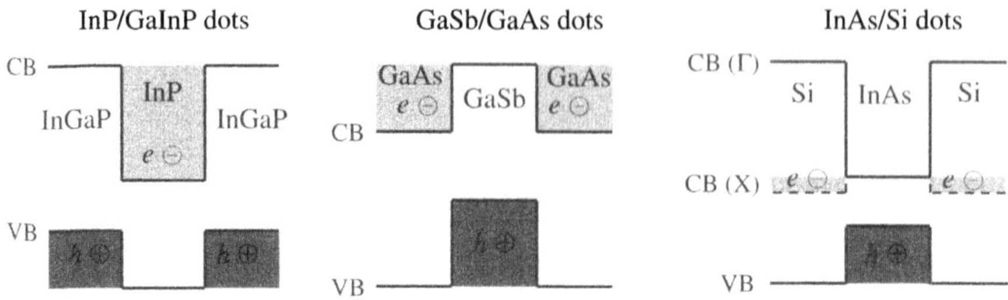

Figure 1. Different types of conduction and valence band alignements in type II quantum dots.

quantum dots (SAQD's) is promoted by the strain fields, which allow one dot to form directly above another [6, 7], resulting in stacked quantum dots.

With respect to the band alignment, quantum dots are classified as type-I or type-II. Type-I quantum dots, in which both the electrons and holes are localized inside the quantum dots, attracted most attention, while relatively few works considered type-II quantum dots, where the electrons and holes are spatially separated. Related to the alignment of the conduction and valence band, two classes of type-II quantum dots can be distinguished, as illustrated in Fig. 1. In the first class, such as e.g. the InP/InGaP dots in the strain-free case, the electrons are confined inside InP, while the quantum dot expells the holes to the surrounding matrix of InGaP. In the second class, the holes are confined within the dot and the electrons are located outside. The notable examples of such systems are GaSb/GaAs [8] and InAs/Si [9]. In the latter, however, the lowest energy conduction-band states are localized in the X point of the Brillouin zone, which makes theoretical treatment of this system more involved. Experimental studies on such type-II quantum dots have been performed in Refs. [10, 11, 12] for the InP/InGaP system and in Refs. [8, 13] for GaSb/GaAs quantum dots. A heuristic explanation of the hole localization based on magnetophotoluminescence measurements was given in Refs. [11]. The experimental data indicate that the conduction-band electrons are localized inside the dots and the holes in the spacers.

Various models have been employed to compute the electronic structure of quantum dots, with inherent limitations and advantages. Yet, even the simplest single-band effective-mass approach is able to offer a qualitatively good insight into the localization of the single particle states in self-assembled quantum dots. For both the single-band effective-mass and the multiband $\mathbf{k} \cdot \mathbf{p}$ approach, a small set of principally well defined input parameters is used and the strain may straightforwardly be taken into account. In the multiband $\mathbf{k} \cdot \mathbf{p}$ theory, however, the mixing between Bloch states is explicitly included, hence it was successfully adopted to the calculation of electron and hole states in SAQD's of various shapes and composed of various materials [14, 15, 16]. Tadić *et al.*

recently employed the 6×6 $\mathbf{k} \cdot \mathbf{p}$ model to compute the valence band structure in single cylindrical InP/InGaP quantum dots and estimated the influence of the elastic anisotropy on the hole states [17].

Here we study the single particle and exciton spectra in single InP/InGaP quantum dots and quantum-dot molecules. The exact determination of the shape is spoiled by the acting strain fields. For the purpose of simplicity, we assume that quantum dots have cylindrical shape, which is indeed found in the fabricated quantum dots (see Ref. [11]). Our objective is to explore the electron, hole, and exciton energies as function of the dimensions of the quantum dots and the quantum-dot molecules. InP/InGaP SAQD's exhibit a very rich behavior which is not found in other SAQD's, e.g. depending on the size of the system, they can be either type-I or type-II, as recently found by Janssens *et al.* [18]. Therefore one can effectively engineer the position of the holes by varying the dimensions of the dots. Variation of the electron and hole localization with the size of quantum dots is explored by both the single-band effective mass model and the multiband $\mathbf{k} \cdot \mathbf{p}$ theory. For the quantum-dot molecules, we go beyond the heuristic predictions of Ref. [11], with the aim to provide a consistent explanation of the spatial localization of electrons, holes, and excitons in stacks composed of flat circular quantum dots. We consider quantum-dot molecules consisting of two and three cylindrical InP quantum dots embedded into an InGaP matrix. The magnetic field dependence of the exciton energy is compared with the magnetophotoluminescence experiments of Ref. [11].

2. Theoretical formalism

Strain modelling

In two-dimensional quantum wells, biaxial strain removes the zone center degeneracy between the heavy-hole and the light-hole bands, and elevates the bottom of the conduction band within the well, but, providing an infinite lateral extension of the system, no strain propagates beyond the well in these systems, and also it is uniformly distributed within the slab. In quantum dots, however, the strain is nonuniform, and extends into the matrix, therefore the splitting between the tops of the heavy-hole and light-hole bands varies in space, and barriers in the conduction-band are erected in the matrix.[17]

For the description of the strain at the level of the crystal unit cell, a computa- ...intensive atomistic approach is required. A handful of other approaches ... at one's disposal [19]. One of them being the contin- ...l [19], which perfectly fits a continuum model of ...$\mathbf{k} \cdot \mathbf{p}$ theory, and also may take account of the crys- ... the continuum mechanical model for the isotropic ... of inclusion. This approach was adopted by Davies,

who showed that the distribution of strain can be extracted from a scalar potential which obeys the Poisson equation with the lattice mismatch replacing the charge density. Tadić *et al.* recently compared atomistic models which rely on Keating's and Stillinger-Weber potentials, the CM model, and the approach of Davies, and found that certain discrepancies exist between the four models for both InAs/InGaAs and InP/InGaP SAQD's [19]. However, the deviations in the InP/InGaP SAQD's, which have a smaller lattice mismatch than InAs/InGaAs SAQD's, turned out to be rather small.

Without strain, the InP/InGaP quantum dot is a type-II structure, but the strain reverts the band alignment to type-I. Because of the small value of the unstrained valence band offset, experimental verification of the type of the band alignment is not an easy task, but as confirmed by the recent single-band calculations for the InP/InGaP SAQD's by Janssens *et al.*, the experimental results are not considerably affected by the assumed type of unstrained band alignment [18].

Electronic structure modelling

For the description of the conduction band we employ the single band approach, while the hole localization is analyzed by both single-band and multiband approaches [17, 18]. The explicit forms of the single-band and multiband Hamiltonians are given in Refs. [17, 18]. Based on the values of Luttinger parameters in InP and InGaP, small shear strains, low in-plane anisotropy of the relevant combinations of the diagonal strain tensor components, and the assumed cylindrical shape of the quantum dot, the multiband Hamiltonian is transformed into axially symmetric form, which commutes with the z component of the total angular momentum $F_z = f\hbar$. The parity σ, is a good quantum number in our system, therefore the eigenstates of the multiband Hamiltonian are labelled as nX_f^σ, where n is the number of the state for given f and σ, and X denotes the minimum value of the orbital momentum in the chosen basis set for the valence band states [17].

For the calculation of the exciton properties, we used two approaches. One is a mean-field type approach, which is equivalent, for this problem, to the Hartree-Fock approximation [20, 21]. The other approach we use in the framework of the multiband $\mathbf{k} \cdot \mathbf{p}$ theory is an exact-diagonalization-type calculation.

3. Single quantum dot

If band mixing is neglected, the separate equations are solved for the heavy holes and light holes. We found that when the thickness h of the disk increases, the heavy holes move towards the radial boundary of the disk. This eff
is purely due to strain, which increases the potential maximum at the r
boundary with respect to the potential inside the disk. For thicker di
height of the potential in the disk systematically decreases, making it

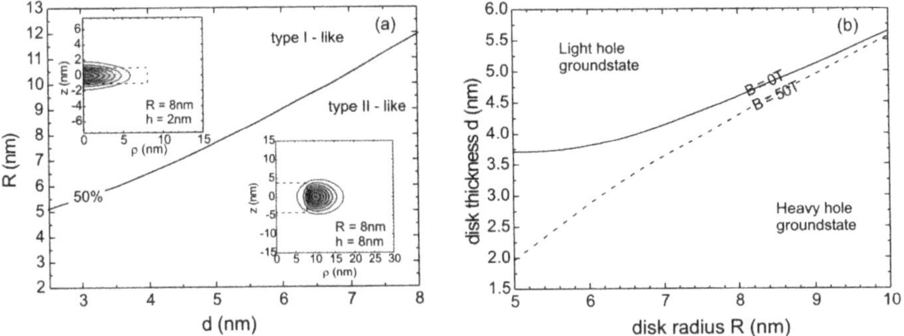

Figure 2. (a) Phase diagram for the probability for the heavy hole to be located at the radial boundary of the disk as a function of the disk radius and thickness. The curve denotes a probability of 50%. The insets show the heavy-hole wavefunctions for, respectively, a type-I-like system and a type-II-like system (dashed lines are a sideview of half of the disk). (b) Phase diagram indicating whether the heavy- or light-hole exciton is the ground state, as a function of the disk radius and thickness. The solid curve denotes the result for $B = 0$, whereas the dashed curve gives the result for $B = 50$ T.

preferable for the heavy holes to move out of the system and resulting in a type-II system. We summarize our results in the phase diagram of Fig. 2(a). In order to construct this figure, we computed the probability to find the heavy hole at the radial boundary of the disk,

$$P_{side} = 2\pi \int_{-\infty}^{+\infty} dz_h \int_{R}^{+\infty} d\rho_h \rho_h |\Psi(\rho_h, z_h)|^2. \tag{1}$$

The line in Fig. 2(a) indicates where the probability is $P_{side} = 50\%$. At the left side of this line, the heavy holes are mainly localized inside the disk (left inset in Fig. 2(a)), and the system is type-I-like. At the right side of the 50% line, the heavy hole is predominantly located at the radial boundary outside the disk (right inset in Fig. 2(a)), and the system is type-II-like. The results for the exciton confinement are summarized in Fig. 2(b), which shows a phase diagram of the heavy-to-light hole transition as a function of both R and d. We find that the light-hole exciton becomes the ground state for increasing thickness of the disk. The full curve in Fig. 2(b) shows the separation between the heavy and light-hole exciton ground states in the absence of a magnetic field. Notice that the heavy-hole exciton is in the heavy-hole type-I region for $R > 6.4$ nm. When the heavy hole is at the radial boundary outside the dot, the exciton energy is higher that that for the (type-II) light-hole exciton, except in a very narrow range of d values for $R < 6.2$ nm. The dashed curve shows the result for $B = 50$ T. Thus a magnetic field lowers the light-hole ground state with respect to the heavy hole.

4. Quantum-dot molecules

Double-dot molecules

The influence of the strain on the double-dot molecule (DDM) consisting of cylindrical 3 nm high InP/InGaP quantum dots is shown in Figs. 3(a-c) where the effective potentials at $(\rho = 0, z)$ are shown for the conduction band, the heavy hole band, and the light hole band. Here, strain is extracted from the anisotropic continuum mechanical model.

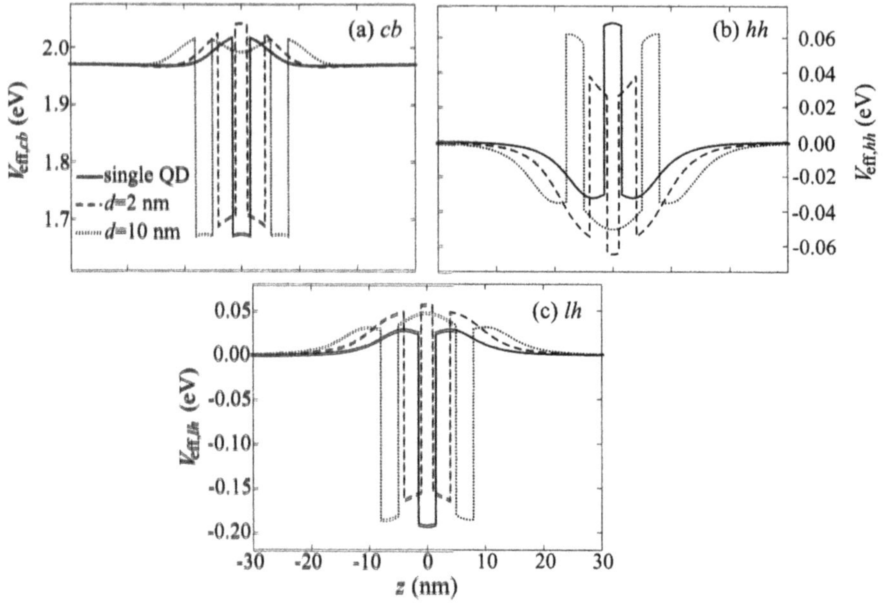

Figure 3. The effective potentials for the electrons (a), heavy holes (b), and light holes (c) in the single quantum dot (SQD) (solid lines) are compared with those in the double-dot molecule (DDM) for 2 nm thick spacer (dashed line) and for 10 nm thick spacer (dotted line) along the z axis. The dot radius equals 8 nm and the dot height is 3 nm.

In Figs. 3(a-c) the effective potentials for the SQD are compared with the DDM's for 10 and 2 nm thick spacers. Even for the spacer as thick as $d = 10$ nm, the strain fields around the two dots interact, but they do not appreciably affect the bottom of the conduction band within the two dots, which appears almost flat in Fig. 3(a). However, when the distance between the dots decreases to $d = 2$ nm, the strain lifts and modifies the shape of the effective potential to a triangular form inside the dots, and it considerably increases the bottom of the *cb* in the spacer. By reversing the energy axis, one may notice that the confinement of the heavy holes is modified by the strain in a similar way as for the *cb* electrons (see Fig. 3(b)). For both values of the spacer thickness, and also in the SQD case, the strain confines the heavy holes inside the dot, like in a type-I system. Type-II confinement is found for the light holes in the SQD, as shown in Fig. 3(c), where the light holes are expelled by the strain to the matrix, near the dot-matrix interface, i.e. towards the top and bottom of the quantum dots. When the two dots join together, the strain creates an effective quantum well for the light holes in the spacer, and deepens the two effective potential wells in the matrix.

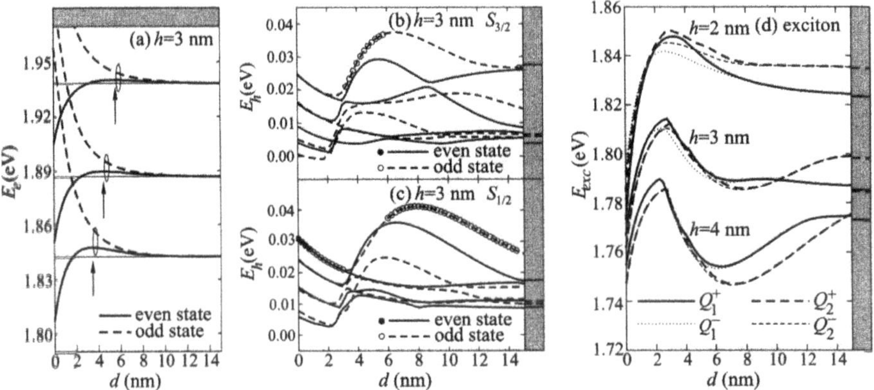

Figure 4. The spacer thickness dependence of the energy levels in the double-dot molecule. (a) Electron energy levels in the s, p and d shells for $h = 3$ nm thick quantum dots. The electron energies in the SQD are shown by the thin horizontal lines, and the values of the coupling length by arrows. (b) $S_{3/2}$ and (b) $S_{1/2}$ valence-band energy levels in the DDM consisting of 3 nm high quantum dots as function of the spacer thickness. The solid lines in (a), (b), and (c) indicate even states, while the dashed lines are for the odd states. The ground state of holes is denoted by solid dots (•) if it is even, and by open circles (○) if it is odd. (d) The lowest $Q_2^{\pm}$ and $Q_2^{\pm}$ exciton quartets. The exciton energies are given for $h = 2$ nm, $h = 3$ nm, and $h = 4$ nm.

In the axially symmetric model the electron energies depend on the orbital momentum, as shown in Figs. 4(a) for $l = 0$, $l = 1$, and $l = 2$ doublets in the DDM composed of $h = 3$ nm thick quantum dots. The ground electron doublets are displayed in Fig. 4(a) and the convergence of the states in the doublets is demonstrated by adding thin horizontal lines, which represent the electron energies in the SQD. Note that in the simpler case, without inclusion of strain, the states should split symmetrically around the SQD energies. This is clearly not so when strain is included. The strain lifts the energies of the even states above the level in the SQD, which reverts bonding to antibonding behavior for thick spacers. The increase of the ground state energies for intermediate values of the spacer thickness indicate that the quantum mechanical coupling is ineffective for thick spacers, but opposes the strain, leading to a local maximum for a certain value of d, and finally reverting the behavior to bonding-like for thin spacers. The distance $d = d_0$, where the overshoot is largest is indicated by arrows in Fig. 4(a).

Because of the mixing and different effective potentials in the three valence bands, the electronic structure of the holes is more complex than for the conduction-band electrons. In most of the explored range of d-values, the highest energy valence-band has either the $S^{\pm}_{\pm 3/2}$ or $S^{\pm}_{\pm 1/2}$ symmetry, which are shown in Figs. 4(b) and (c), respectively. The coupling in the $1S^{\pm}_{3/2}$ states is established by the light holes, which are less confined as the spacer thickness decreases, yet, as Fig. 3(b) shows, even for $d = 15$ nm there is still a sufficiently deep well to localize the light holes. Due to the strain and the higher orbital momentum of the heavy holes in the $1S^{-}_{3/2}$ state, the coupling in it is less efficient than in the $1S^{-}_{1/2}$ state. Decreasing the thickness of the spacer, the maxima in the $S_{1/2}$ and $S_{3/2}$ ground states of both parities are first reached, and then the energies decrease. For all energy levels shown in Figs. 4(b) and (c) the valence band energies increase again for $d <\approx 3$ nm, at which d-value most anticrossings between the hole states take place. For $d < 3$ nm, there is a near degeneracy of states in the doublets, which arises from the localization in the two large "effective quantum dots" in the matrix.

Both electron and hole zone center states exhibit double Kramers degeneracy, therefore *each exciton energy is fourfold degenerate* in the exact diagonalization approach. Examples of these quartets are:

$$Q_1^+ = \left[\boxed{S^+_{-1\uparrow}}_{(x,y)} \quad S^-_{+2\uparrow} \quad \boxed{S^-_{+1\downarrow}}_{(x,y)} \quad S^+_{-2\downarrow} \right], \tag{2a}$$

$$Q_1^- = \left[S^-_{-1\uparrow} \quad S^+_{+2\uparrow} \quad S^+_{+1\downarrow} \quad S^-_{-2\downarrow} \right], \tag{2b}$$

$$Q_2^+ = \left[S^+_{0\uparrow} \quad S^-_{+1\uparrow} \quad S^+_{-1\downarrow} \quad S^-_{0\downarrow} \right], \tag{2c}$$

and

$$Q_2^- = \left[\boxed{S^-_{0\uparrow}}_{(z)} \quad \boxed{S^+_{+1\uparrow}}_{(x,y)} \quad \boxed{S^-_{-1\downarrow}}_{(x,y)} \quad \boxed{S^+_{0\downarrow}}_{(z)} \right], \tag{2d}$$

and the optically active states in the quartets are indicated by boxes. For the quartet given in Eq. (2a), for example, the $S^+_{-1\uparrow}$ and $S^-_{+1\downarrow}$ are bright exciton states, while the $S^-_{2\uparrow}$ and $S^+_{-\downarrow}$ appear as dark. The superscripts here are chosen according to the ordering of the states in the single quantum dot, i.e. $1Q_1^+$ and $1Q_2^+$ states have the lowest energies among the four quartets. Quite generally and similar to electrons, exciton levels overshoot the SQD energies, but larger overshoots which are shifted towards lower d are found here than in the electron spectra, and consequently cannot be solely attributed to the electrons. As a result of the strong anticrossings with the higher exciton states, sharper peaks of exciton energies versus d are found for the 3 nm and 4 nm thick quantum dots than in the quantum-dot molecules consisting of the 2 nm thick ones.

Triple-dot molecules

The electron energy levels in triple-dot molecules (TDM's) are ordered in triplet shells. The lowest energy triplets in the s, p, and d shells, are shown in Fig. 5(a) for $h = 3$ nm. The shaded areas represent the energy continuum. The thin horizontal lines in Fig. 5(a) denote the SQD energies in the different shells. The energy overshoots and coupling lengths are similar to those computed for the case of DDM. Surprisingly, no real decrease of the ground state energies in the TDM's with respect to those in the DDM's is found, which led us to conclude that the effect of the competition between strain and quantum mechanical coupling is larger in TDM's.

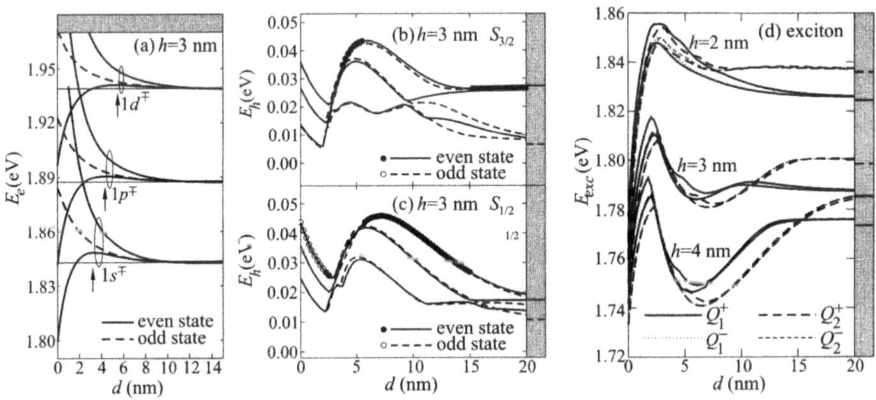

Figure 5. The same as Fig. 4, but now for the triple-dot molecule.

According to the number of dots present in the system and the explored spectra of the *cb* electrons, one may anticipate splitting of the SQD hole levels into triplets. However, the scheme of splitting of the ground triplet of hole states in the TDM is qualitatively similar to the one found in the case of DDM, with two levels in each triplet arranged in a nearly degenerate doublet. Such pairing arises from the different *hh* effective potentials in the central and satellite quantum dots, and its presence is found in both the $S_{3/2}$ and $S_{1/2}$ shells, as shown by Figs. 5(b) and (c), respectively.

The exciton energies in DDM's are not considerably altered when an extra dot is added to the stack, as is demonstrated by Fig. 5(d) for TDM's consisting of 2, 3, and 4 nm high quantum dots. By inspecting Figs. 4(d) and 5(d), one finds similar variations of the exciton energies in DDM's and TDM's with d. *First*, all exciton energies exhibit triplet ordering and overshoots with respect to the SQD exciton energy levels. *Second*, the exciton energies oscillate for $h = 3$ and 4 nm, while exciton energies decrease for $h = 2$ nm and $d > 2$ nm, due to the different behavior of hole levels for large d. However, splittings between the exciton states are smaller in TDM's than in DDM's, which we ascribe to the doublet structure of the hole states in TDM's.

Comparison with experiment

Here we show the results of our single-band Hartree calculation of the exciton states in the coupled dots, when d varies. The inset of Fig. 6(a) displays the result for the transition energy for a system of three coupled dots with interdot distance $d = 1.5$ nm. In the single-band model, the light-hole state is the groundstate of the system, which agrees with our $\mathbf{k} \cdot \mathbf{p}$ approach, where a dominant light-hole contribution to the mixed hole ground state is found for a wide range of d-values. A comparison with the experimental data of Ref. [11] (sample D) shows that our theoretical result underestimates the experimental result by ≈ 15 meV. We attribute this to uncertainties regarding the disk/and or material parameters and if we increase the disk thickness by 0.5 nm our theoretical results shifts almost on top of the experimental data.

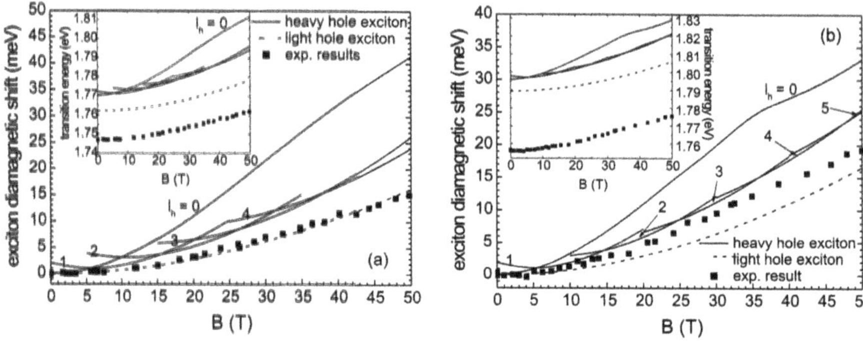

Figure 6. (a) Exciton diamagnetic shift as function of the magnetic field for three vertically stacked identical quantum dots with interdot spacing $d = 1.5$ nm, dot radius $R = 8$ nm, and dot thickness $h = 2.5$ nm. The solid and dashed curves denote respectively the heavy and light hole exciton, whereas the squares indicate the experimental result of Ref. [11]. The different solid curves show the results for different heavy-hole angular momentum, as indicated. The inset shows the result for the transition energy. (b) Same as (a), but now for an interdot spacing of $d = 3$ nm.

The theoretical and experimental results for the diamagnetic shift, i.e. $\Delta E = E(B) - E(B = 0)$ of the exciton energy are shown in Figs. 6(a) and (b) as a function of the magnetic field. The solid and dashed curves denote respectively our results for the heavy and the light hole exciton, and the squares are the experimental result by Hayne *et al.* [11]. The agreement between the light hole exciton curve and the experimental result is very good, in particular in view of the fact that no fitting parameters are introduced in the theory. As we know from the result for the transition energy, the light hole exciton is also the groundstate of the system. Therefore, we conclude that experimentally the light hole exciton is observed, which is located above and below the stack of the dots. The different solid curves in Fig. 6(a) show the result for different hole angular momenta l_h. These magnetic-field-induced angular momentum transitions are a consequence of the fact that the heavy hole for this case is situated at the outer radial edge of the system (see inset of Fig. 2). We find that the final heavy hole groundstate (consisting of different angular momentum states) approaches now more closely the light-hole result and the experimental result, although it is not the groundstate of the system.

We further explored a quantum-dot molecule consisting of three quantum dots separated by $d = 3$ nm, and the results are shown in Fig. 6(b). Again we find that the light hole exciton is the groundstate of our system. Comparison with the experimentally observed transition energy (sample C in Ref. [11]) indicates that the theoretical result underestimates the experiment by ≈ 30 meV at $B = 0T$, which can be removed by increasing h in our calculation. The theoretical result for the diamagnetic shift agrees with the experimental result up to $B \approx 10$ T, but beyond this field, our theoretical results are lower than the experimental results.

5. Conclusion

We studied single-particle and exciton states in quantum dots and molecules. Investigation of the effect of strain on the confinement potentials tells us that, starting from a negative (type-II) unstrained valence-band offset, the confinement potential for the heavy holes is reversed (i.e. becomes type-I), while for the light holes, the system becomes even stronger type-II. For a certain disk thickness, it becomes thus preferable for the heavy holes to move towards the radial boundary, into the barrier material, making the system also type-II. The strain induces an upward shift of the ground electron energy levels in symmetric InP/InGaP quantum-dot molecules. The hole energies exhibit both local maxima and minima as function of d, which is a result of the mixing between the valence band and strain variation with the spacer thickness. Such variations of the hole energy levels may give rise to oscillations of the exciton energy levels, while, depending on the quantum dot height, the exciton energy levels may also be flattened as d varies. This manifestly shows that the InP/InGaP quantum dots and quantum-dot molecules are much richer systems than e.g. the well studied InAs/GaAs system.

Results for the diamagnetic shift show for the systems of three stacked disks a fairly to very good agreement between the theoretical light hole exciton curve and the experimental results. Together with the finding that the light hole exciton forms the groundstate, this allows us to conclude that experimentally the light hole exciton was observed.

Acknowledgments

This work was supported by the European Commission GROWTH programme NANOMAT project, Contract No. G5RD-CT-2001-00545, the University of Antwerp (GOA and VIS) and the Belgian Interuniversity Attraction Poles (IUAP). One of us (B.P.) is a postdoctoral fellow with the FWO-Vl.

References

[1] D. Bimberg, M. Grundmann, and N.N. Ledentsov, *Quantum Dot Heterostructures* (Wiley, London, 1999).

[2] D. Bimberg and N. Ledentsov, J. Phys.: Condens. Matter **15**, R1063 (2003).

[3] A.M. Adawi, E.A. Zibik, L.R. Wilson, A. Lemaitre, J.W. Cockburn, M.S. Skolnick, M. Hopkinson, G. Hill, S.L. Liew, and A.G. Cullis, Appl. Phys. Lett. **82**, 3415 (2003).

[4] M. Borgstrom, T. Bryllert, T. Sass, B. Gustafson, L.-E. Wernersson, W. Seifert, and L. Samuelson, Appl. Phys. Lett. **78**, 3232 (2001).

[5] H. Pettersson, L. Baath, N. Carlsson, W. Seifert, and L. Samuelson, Appl. Phys. Lett. **79**, 78 (2001).

[6] Q. Xie, A. Madhukar, P. Chen, and N.P. Kobayashi, Phys. Rev. Lett. **75**, 2542 (1995).

[7] R. Heitz, A. Kalburge, Q. Xie, M. Grundmann, P. Chen, A. Hoffmann, A. Madhukar, and D. Bimberg, Phys. Rev. B **57** 9050 (1998).

[8] L. Müller-Kirsch, R. Heitz, A. Schliwa, O. Stier, D. Bimberg, H, Kirmse, and W. Neumann, Appl. Phys. Lett. **78**, 1418 (2001).

[9] R. Heitz, N.N. Ledentsov, D. Bimberg, M.V. Maximov, A.Yu. Egorov, V.M. Ustinov, A.E. Zhukov, Zh.I. Alferov, G.E. Cirlin, I.P. Shoshnikov, N.D. Zakharov, P. Werner, and U. Gösele, Appl. Phys. Lett. **74**, 1701 (1999).

[10] S. Nomura, L. Samuelson, C. Pryor, M.-E. Pistol, K. Uchida, N. Miura, T. Sugano, and Y. Aoyagi, Appl. Phys. Lett. **71**, 2316 (1997).

[11] M. Hayne, R. Provoost, M.K. Zundel, Y. M. Manz, K. Eberl, and V.V. Moshchalkov, Phys. Rev. B **62** 10324 (2000).

[12] M. Sugisaki, H.-W. Ren, K. Nishi, S. Sugou, T. Okuno, and Y. Masumoto, Physica B **256-258**, 169 (1998).

[13] F. Hatami, M. Grundmann, N.N. Ledentsov, F. Heinrichsdorf, R. Heitz, J. Böhrer, D. Bimberg, S. S. Ruvimov, P. Werner, V.M. Ustinov, P.S. Kop'ev, and Zh. I. Alferov, Phys. Rev. B **57**, 4635 (1998).

[14] M. Grundmann, O. Stier, and D. Bimberg, Phys. Rev. B **52**, 11969 (1995).

[15] O. Stier. M. Grundmann, and D. Bimberg, Phys. Rev. B **59**, 5688 (1999).

[16] C. Pryor, M.-E. Pistol, and L. Samuelson, Phys. Rev. B **56**, 10404 (1997).

[17] M. Tadić, F.M. Peeters, and K.L. Janssens, Phys. Rev. B **65**, 165333 (2002).

[18] K.L. Janssens, B. Partoens, and F.M. Peeters, Phys. Rev. B **67**, 235325 (2003).

[19] M. Tadić, F.M. Peeters, K.L. Janssens, M. Korkusiński, and P. Hawrylak, J. Appl. Phys. **92**, 5819 (2002).

[20] K.L. Janssens, B. Partoens, and F.M. Peeters, Phys. Rev. B **64**, 155324 (2001).

[21] K.L. Janssens, B. Partoens, and F.M. Peeters, Phys. Rev. B **66**, 075314 (2002).

QUANTUM DOT IN THE KONDO REGIME COUPLED TO UNCONVENTIONAL SUPERCONDUCTING ELECTRODES

Tomosuke Aono[1], Anatoly Golub[1] and Yshai Avishai[1,2]
[1]*Department of Physics, Ben-Gurion University, Beer-Sheva, Israel*
[2]*Ilse Katz Center for Nanotechnology, Ben-Gurion University, Beer-Sheva, Israel*
yshai@bgumail.bgu.ac.il

Abstract Quantum dots connected on both sides to *normal* metallic leads turn out to be a central research topic in contemporary condensed matter physics. What happens if one or both leads are *superconducting*? In a series of papers the authors developed a theoretical basis for the relevant physical situation. As it turn out, the resulting physical observables strongly depend on the symmetry of the superconducting electrodes order parameter. In our previous publications we have studied the case of s-wave superconducting electrodes. Here, in this work, the physics of junctions containing p-wave superconducting and normal leads weakly coupled to an Anderson impurity in the Kondo regime is elucidated. For p-wave (unlike s wave) superconducting leads, mid-gap surface states play an important role in the tunneling process and help the formation of the Kondo resonance. The current, shot-noise power and Fano factor are calculated and displayed as functions of the applied voltage V in the sub-gap region $eV < \Delta$ (the superconducting gap). In addition, the Josephson current for a quantum dot in the Kondo regime weakly coupled on both sides to p-wave superconductors is computed as function of temperature and phase. The peculiar differences between the cases of s-wave and p-wave superconducting leads are pointed out.

Keywords: Qunatum dot, strongly correlated electrons, Kondo effect, superconductivity, unconventional superconductors.

1. Introduction

In this research we focus on transport through an interacting quantum dot in the Kondo regime weakly coupled to p-wave superconducting leads[1]. It is quite conceivable that the role of the quantum dot can be played also by a single molecule or by a Carbon nano-tube. Recently, it has been unambiguously established that the Kondo physics [2] plays an important role in electron transport through quantum dots, where instead of a magnetic impurity one encounters

A.S. Alexandrov et al. (eds.), Molecular Nanowires and Other Quantum Objects, 203–217.

localized electrons [3, 4, 5]. Observations of the Kondo effect in transport through quantum dots[6, 7], in carbon nano-tubes (CNT)[8], in vertical dots [9], and in single molecules[10] demonstrate the feasibility of exploiting tunable physical parameters in these systems in order to yield important information on the Kondo physics and other many-body related phenomena.

The Kondo physics in a quantum dot attached on both its sides to *normal* metallic leads received much recent attention. Many experiments and theoretical investigations are devoted to its study ever since the first experiments were reported in 1998. Recently, it has been realized that novel physical effects emerge if (one or both) electrodes is a superconductor[11]. Hereafter we abbreviate by N a normal metallic lead, by S a superconducting lead and by K a quantum dot in the Kondo regime weakly attached to N and/or S leads. SKS and SKN junctions can now be fabricated in laboratories. Moreover, there are also natural candidates: Fabrication of superconducting junctions with a weak link formed by CNT have already been reported [12, 13].

In a series of recent works [14, 15] we have developed a formalism for the study of transport in SKN and SKS junctions. It is then useful to state clearly at this point the peculiar aspects of the present work compared with the previous ones. In Ref. [14] the underlying physics of SKS and SKN junctions was analyzed out of the Kondo regime, that is, at $\Delta \gg T_K$ (the Kondo temperature), while in Ref.[15] attention is focused on SKS junctions for which the leads are composed of s-wave superconductors. Here we are particularly interested in electron transport in the Kondo regime in SKN junctions for which the S lead is composed of a p-wave superconductor. The physics of junctions with p-wave superconducting leads is essentially distinct from that pertaining to junctions with s-wave superconducting leads (especially in the Kondo regime) as is carefully explained and underlined below. Thus, while we heavily rely on the computational machinery developed earlier, the results obtained here are novel.

One of the crucial differences between the case of normal and superconducting leads touches upon the question of how electrons are transport from one lead to the other. The central electron-transport mechanism in SKS and SKN junctions is that of Andreev reflections when two particles tunnel together coherently to form a Cooper pair in the superconductor. Of course, Andreev reflections play an important role in bulk SIS and SN junctions (here I denotes an insulating layer). In the former case they are responsible for occurrence of direct Josephson current while in the latter case they enhance the conductance by a factor 2.

What is the role of Andreev reflections in quantum dots such as SKS and SKN junctions? The answer to this question is rather interesting and exposes a subtle distinction between the cases of s-wave and p-wave superconductors. The pertinent physics is governed by the interplay of Andreev reflections and the formation of the Kondo-resonance in the spectral density of states of the dot electron [11]. In the case of an SKN junction when one electrode is an s-wave superconductor and the Kondo impurity is weakly coupled to the S and N electrodes, the physics is determined by competition between two phenomena. The first one is the formation of a Kondo singlet which screens the bare impurity spin and drives the system toward the unitary limit at very low temperatures. The second one is the existence of the superconducting gap which implies a vanishingly small density of low energy electron states [16, 17, 18]. These are precisely the electron states which are needed in order to screen the Kondo impurity. A relevant parameter in this context is the ratio between the Kondo temperature and the superconducting gap,

$$t_K \equiv \frac{T_K}{\Delta}. \tag{1}$$

When $t_K < 1$, the Kondo effect is suppressed by the superconducting gap while for $t_K > 1$, the Kondo effect (close to the unitary limit) and superconductivity coexist. Indeed, for $t_K > 1$, electron states outside the gap can participate in the screening interaction. Several works have studied these aspects both for $t_K < 1$ [18, 16], and for $t_K > 1$ [17, 15, 19, 20]. In a recent experiment[13], a crossover around $t_K = 1$ has been realized in SKS junctions.

Consider, on the other hand, an SKN junction in which the S electrode consists of an *unconventional* superconductor. To be more specific, we mean superconductors with triplet pairing, such that the order parameter has a p-wave orbital symmetry. Such p-wave superconductors have recently been discovered by Maeno *et al* [1] in Sr_2RuO_4. The fact that in these superconductor the order parameter is not rotationally symmetric implies a special importance for its orientation when it is integrated into a junction. In particular, let us assume that the superconductor is oriented relative to the interface in such a way that the pair potential reverses its sign on the Fermi surface. In this case, zero-energy states (ZES) are formed (that is, *inside the gap*), which are localized near the surface of the unconventional superconductor. These states can now participate in screening the impurity spin through the Kondo effect and emergence of sub-gap current is expected. For example, in the experimentally feasible setup of $S-CNT-N$ where just a few tunneling channels are present, charge is carried mainly by quasiparticles moving perpendicular to the interface. This restricts the possible values of the angle θ between the superconducting surface and the direction of the injected quasiparticles [21]. Formation of ZES is possible when $\Delta(\theta) = -\Delta(\pi - \theta)$. If the impurity is almost point-like, the relevant injection angle is of course $\theta = 0$. For impurities of finite extent, one may also consider formation of ZES in d-wave superconductors. Thus, the physics of SKN junctions with an S electrode whose order parameter has a non-trivial symmetry is affected by the formation of ZES in the Kondo regime. In short, the physics of SKN junction will be prominently different between s-wave on the one hand and p or d-wave superconductors on the other hand.

The question now arises is whether the above mentioned distinction between the cases of s-wave and p-wave superconducting electrodes can be elucidated. We answer it positively by calculating a number of transport observables in SKN junctions for several values of t_K. Beyond investigating the conductance dependence on the applied bias we also explore the shot-noise and the Fano factor. Moreover, at zero bias we also consider SKS junctions and analyze the Josephson (direct) current dependence on the phase difference between the two superconductors as well as on the temperature. In section II the model Hamiltonian is written in terms of the slave boson formalism. The Green functions pertaining to p wave superconducting leads are introduced and the mean-field slave boson approximation (MFSBA) is briefly discussed. Calculations and presentations of conductance, shot-noise power and Josephson current are respectively detailed in sections III, IV and V.

2. Model Hamiltonian and p wave Green functions

The formalism employed below is the slave boson mean-field approximation, explained and justified in our earlier works[14, 15]. Therefore we will skip most of it except the definitions of Green functions peculiar for the case of p-wave superconducting lead. The model Hamiltonian of SKN or SKS junctions is represented by the Anderson model with the superconducting lead:

$$H = H_L + H_R + H_d + H_t + H_c, \tag{2}$$

in which H_j ($j = L, R$) are the Hamiltonians of the electrodes which depend on the electron field operators $\psi_{a\sigma}(\mathbf{r}, t)$ at $\mathbf{r} = (x, y)$ with the spin $\sigma = \pm$. For a superconducting lead we have,

$$H_j = \sum_{j=L,R} \int_j dr \quad \left(\psi^\dagger_{j,+}(r)\psi_{j,-}(r)\right) \times \begin{pmatrix} \xi(\nabla) & \Delta_j \\ \Delta^*_j & -\xi(\nabla) \end{pmatrix} \begin{pmatrix} \psi_{j,+}(r) \\ \psi^\dagger_{j,-}(r) \end{pmatrix}, \tag{3}$$

where $\xi(\nabla) = -\nabla^2/2m - \mu$ with the chemical potential μ depends on the bias voltage V, and Δ is the Cooper pairing potential. For s-wave superconductor, Δ is isotropic (in fact it is a constant) while for p-wave superconductor, Δ is anisotropic and depends on the two dimensional momentum vector: $\Delta(\alpha) = |\Delta|(k_x + ik_y)/|k| = |\Delta| \exp(i\alpha)$ with with the azimuthal angle $\alpha = \arctan(k_y/k_x)$. This pairing potential changes the sign: $\Delta(\alpha) = -\Delta(\pi - \alpha)$. For a normal lead, $\Delta = 0$ of course.

The quantum dot consists of a single energy level $\epsilon_0 < 0$ with Coulomb interaction U. We assume that $U \to \infty$ to exclude double occupancy of electrons in the dot. In this scheme, the annihilation operator d_σ of electron in the dot is written as $d_\sigma = b^\dagger c_\sigma$ with the slave boson operator b and the pseudo fermion operator c_σ and the constraint term of H_c:

$$H_c = \lambda(\sum_\sigma c_\sigma^\dagger c_\sigma + b^\dagger b - 1), \tag{4}$$

where λ is a Lagrange multiplier [22]. The corresponding dot and tunneling Hamiltonians, H_d and H_t are expressed as:

$$H_d = \epsilon_0 \sum_\sigma c_\sigma^\dagger c_\sigma,$$
$$H_t = \sum_{j\sigma} \mathcal{T}_j c_\sigma^\dagger b \psi_{j\sigma}(\mathbf{0}, t) + h.c., \tag{5}$$

where $\mathcal{T}_j$ is the tunneling amplitude. If the quantum dot is almost point-like, the relevant value of α for the paring potential is zero.

Since electron field in the dot couples only with $\psi_{j\sigma}(\mathbf{0}, t)$, (referred to as *surface states*), we can integrate out electron fields inside the lead [14]. Following Ref. [15] let us consider the dynamical "partition function"

$$Z \sim \int \mathcal{D}[F] \exp(i\mathcal{S}), \tag{6}$$

where the path integral is carried out over all fields $[F]$ and the action $\mathcal{S}$ is obtained by integrating the Lagrangian pertaining to the Hamiltonian (2) along the Keldysh contour. In performing the functional integrations the boson field operators are treated as c-numbers. As a result one arrives at an effective action expressed in terms of the Green functions of the leads.

$$S_{\text{eff}} = -i\text{Tr}\ln \hat{G}^{-1} - \int dt[\hat{\lambda}\sigma_z(\hat{b}\hat{b} - 1)]. \tag{7}$$

Here $\hat{\lambda} = (\lambda_1, \lambda_2)$, $\hat{b} = (b_1, b_2)$ and σ_z are diagonal matrices acting in Keldysh space. The inverse propagator $\hat{G}^{-1}$ depends on the Green functions of the electrodes.[14]. Performing the standard basis rotation in Keldysh space one finds,

$$\hat{G}^{-1}(\epsilon, \epsilon') = \delta(\epsilon - \epsilon')(\epsilon - \tau_z\tilde{\epsilon} - \frac{\Gamma b^2}{2}\tau_z\hat{g}_+(\epsilon)\tau_z), \tag{8}$$

where $\tilde{\epsilon} = \epsilon_0 + \lambda$ is the renormalized level position (in the Kondo limit one has $\tilde{\epsilon} \simeq 0$) and $\Gamma = (\Gamma_L + \Gamma_R)/2 \propto \mathcal{T}_{\mathcal{L},\mathcal{R}}{}^2$ is the usual transparency parameter. The 2×2 matrix representation (in Keldysh space) for g is composed of diagonal elements $\hat{g}^{R/A}(\epsilon)$ and an upper off-diagonal element $\hat{g}^K(\epsilon) = (\hat{g}^R - \hat{g}^A)th(\epsilon/2T)$. Here and below we define

$$\hat{g}_\pm = \gamma_L \hat{g}_L \pm \gamma_R \hat{g}_R, \tag{9}$$

with asymmetry parameters $\gamma_j = \Gamma_j/\Gamma$. The matrix $\hat{g}_R$ has the standard structure with retarded and advanced superconductor Green functions which in the $s - wave$ case reads,

$$\hat{g}^{R/A}(\epsilon) = i\frac{(\epsilon \pm i0) + |\Delta|\tau_x}{\sqrt{(\epsilon \pm i0)^2 - |\Delta|^2}}, \tag{10}$$

The retarded and advanced Green functions $g^{R/A}(\epsilon)$ for the surface states of the p-wave superconductor with incident angle α is represented by 2×2 matrices:

$$\hat{g}^{R/A}(\epsilon) = \hat{g}_1^{R/A}1 + \hat{g}_2^{R/A}\tau_x, \quad \text{with} \tag{11}$$

$$\hat{g}_1^{R/A} = \frac{i\sqrt{(\epsilon \pm 0)^2 - |\Delta|^2}\cos\alpha - \epsilon\sin\alpha}{\epsilon\cos\alpha + i\sqrt{(\epsilon \pm 0)^2 - |\Delta|^2}\sin\alpha},$$

$$\hat{g}_2^{R/A} = \frac{i|\Delta|}{\epsilon\cos\alpha + i\sqrt{(\epsilon \pm 0)^2 - |\Delta|^2}\sin\alpha}, \tag{12}$$

employing the unit matrix 1 and the x component of the Pauli matrix τ_x. The density of states $\rho(\epsilon)$ of surface states is then given by $\rho(\epsilon) = \Im\hat{g}_{11}^R(\epsilon)$. For p-wave superconductor with $\alpha = 0$, $\rho(\epsilon)$ includes mid-gap (ZES):

$$\rho(\epsilon) = \frac{\sqrt{\epsilon^2 - \Delta^2}}{|\epsilon|}\theta(|\epsilon| - \Delta) + \pi\Delta\delta(\epsilon), \tag{13}$$

while for the s-wave case, it includes no ZES, having the well-known BCS form. This difference becomes important when the Kondo effect takes place, because ZES strongly interact with the Kondo resonant state which appears also at $E_F = 0$.

Performing the variation of the effective action with respect to the fields b and λ a couple of self-consistency equations are obtained that determine these fields. In order to explicitly write down these self-consistency equations let us introduce the bare Kondo temperature $T_K^0 = Dexp[-\pi||\epsilon_0|/(2\Gamma)]$ and define a parameter X by $\Gamma b^2 = T_K^0 X$, where D is the energy bandwidth. Then the MFSBA equations take the form

$$X = -\frac{i\Gamma}{2T_K^0}\text{Tr}\hat{G}^K\tau_z, \tag{14}$$

$$\lambda = \frac{i\Gamma}{8}\text{Tr}[\hat{G}^K\tau_z(\hat{g}_+^R + \hat{g}_+^A) + (\hat{G}^R + \hat{G}^A)\tau_z\hat{g}_+^K]\tau_z, \tag{15}$$

where the trace includes energy integration as well. Eq. (14) effectively determines the Kondo temperature (through the parameter X), and reflects the constraint which prevents double occupancy in the limit $U \to \infty$. The second self-consistency equation (15) defines the renormalized energy level position $\tilde{\epsilon}$. Let us briefly discuss the validity range of the present analysis.

A comment on the validity of the mean field slave boson approximation scheme and the relation to other approximation schemes is in order. It is known that it gives an adequate description in the Fermi liquid regime of the Kondo effect. (the strong coupling limit.) The Kondo effect is suppressed by Δ as well as by the applied bias voltage eV. If $T_K < \max(\Delta, eV)$, the mean field approximation looses its validity. Fortunately, interesting features appear in the sub-gap voltage regime $eV \leq \Delta$, as we will see below. Hence the validity of the approximation for the sub-gap region is intimately related with the value of t_K. Quantitatively , our approximation is reliable for sufficiently large t_K. On the other hand, when $t_K < 1$, the Kondo effect is strongly suppressed by superconductivity and consequently we need another approximation scheme such as the non-crossing approximation (NCA) developed in Ref.[18]. This approximation, however, fails to describe the Fermi liquid picture, specified by the region of $t_K \gg 1$. Thus, the mean field approximation constitutes a bridge in calculation methods between the low temperature regime (when the system is a Fermi liquid) and the regime where the NCA is valid.

3. Conductance

The tunneling current I through the quantum dot is given by $I = \langle \hat{I}(t) \rangle$ with

$$\hat{I}(t) = \frac{-ie}{\hbar} \sum_{j\sigma} [T_j c_\sigma^\dagger b \psi_{j\sigma}(\mathbf{0}, t) - h.c.]. \tag{16}$$

and has has a simple representation [15] in terms of the dot Green function,

$$I = i\frac{eXt_K}{8\hbar} \mathrm{Tr}[(\hat{G}^R \tau_z - \tau_z \hat{G}^A)\hat{g}_-^K - \hat{G}^K \tilde{g}], \tag{17}$$

where for SKN junctions we denote

$$\tilde{g} = -\gamma_R(\hat{g}^R \tau_z - \tau_z \hat{g}^A) - 2i\gamma_L \tau_z \tag{18}$$

Being combined with eqs. (14) and (15) the result (17) can be conveniently used for computing the transport current and the differential conductance of an SKN junction in the Kondo regime for different values of t_K and eV/Δ. Here we present the result of our calculations putting special emphasis on the distinction between s-wave and p-wave superconducting leads.

Figure. 1 shows the conductance $G = dI/dV$ and its dependence on V in SKN junctions for both s-wave and p-wave superconducting leads for $t_K = 100, 5, 3$, and 2 with $\Gamma/T_K^0 = 200$. When $t_K = 100$ (the upper curves in Fig.1), the $G - V$ curve shows no difference between s-wave and p-wave superconductors; $G = 4e^2/h$ when $eV < \Delta$ and G decreases gradually when

$eV > \Delta$. In this limit, the Kondo resonance reaches the unitary limit, and consequently, this SKN junction reduces to an SN junction with pure ballistic contact, which has already been analyzed [11]. Indeed, in the limit of large $t_K \gg 1$, the expression for the current (16) reduces to that derived in Ref. [11]. For lower values of t_K the distinction between s-wave and p-wave leads becomes prominent. For s-wave superconductor, as t_K decreases, the Kondo

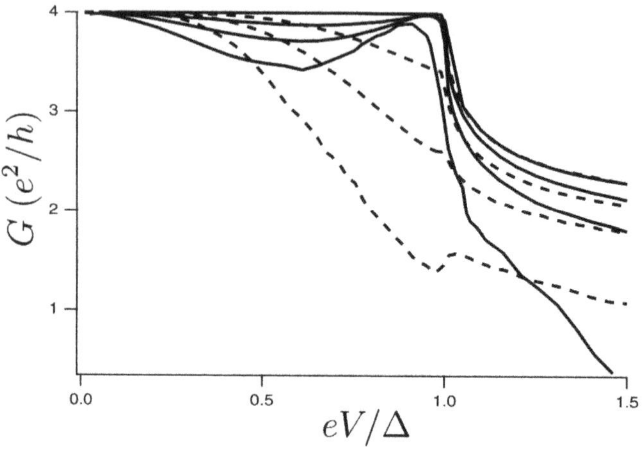

Figure 1. The conductance G (in units of e^2/h) versus the bias V (in units of Δ/e) for an s-wave (dash curves) and p-wave (solid lines) SKN junctions at sub-gap voltages with $\Gamma/T_K^0 = 200$. The parameter t_K takes values 2,3,5,100 (from down up). The upper line corresponding to t_K =100 coincides for s and p wave superconducting leads.

state is driven away from the unitary limit. For $t_K = 5$ the $G - V$ curve noticeably deviates from the one of $t_K = 100$, reflecting the suppression of the Kondo effect due to both Δ and V. For $t_K = 2$ the competition between gap-related suppression of the Kondo effect and the effective transparency of the junction becomes essential, leading to further decrease of the conductance. However, it is interesting to note that for an S(s-wave)KN junction at $t_K = 2$ the conductance displays a small peak at the gap edge. Its interpretation is that the Kondo correlations strongly compete with superconductivity and influence the quasiparticle correlations when the energy exceeds the gap. Such an effect takes place even when $t_K < 1$ (see [18]).

For p-wave superconductor, the $G-V$ curves are less influenced by variations of t_K in contrast to the situation found for s-wave superconductor, indicating superconductivity plays a minor role in the suppression of the Kondo resonance and the result reported in Ref. [11] persists for smaller values of t_K. The upshot is that ZES support the formation of a Kondo singlet for lower values of t_K, and effectively turn the junction to be more transparent, approaching the unitary limit [11].

4. Shot-noise

The shot-noise power is defined as the symmetrized current-current correlation function

$$K(t_1, t_2) \quad = \quad \hbar[\langle \hat{I}(t_1)\hat{I}(t_2)\rangle - \langle \hat{I}\rangle^2], \tag{19}$$

with the current operator $\hat{I}$ defied in equation (16). The Fourier transform of $K(t_1, t_2)$ gives the shot noise power spectrum $K(\omega)$. The general expression for the zero frequency shot-noise $K(0)$ has been obtained within the mean field slave boson approximation [15]. It is convenient to write it as $K = (K_1 + K_2)e^2\Delta/(8\hbar)$ for which the expressions derived are,

$$K_1 = \frac{Xt_K}{2}\mathrm{Tr}\{(\hat{g}_+^R - \hat{g}_+^A)(\hat{G}^R - \hat{G}^A) - \hat{g}_+^K\hat{G}^K\}, \tag{20}$$

$$K_2 = -\frac{(Xt_K)^2}{8}\mathrm{Tr}\{(\hat{G}^K\tilde{g})^2 - 2\tau_z\tilde{g}\tau_z\hat{G}^A\tilde{g}\hat{G}^R - \\ [2\tilde{g}\hat{G}^R\tau_z\hat{g}_-^K\hat{G}^K - (\hat{G}^A\hat{g}_-^K\tau_z)^2 + \mathrm{h.c.}]\}. \tag{21}$$

Expressions (20) and (21) (supplemented by the self-consistency eqs. (14) and (15)) are then solved numerically for a set of parameters Γ/T_K^0, t_K.

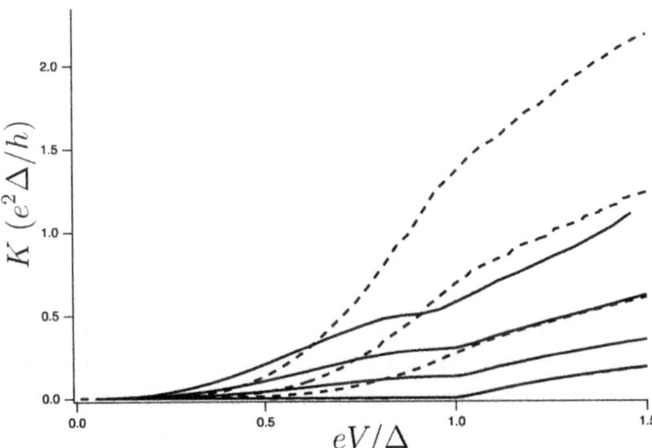

Figure 2. The shot-noise power $K(0)$ (in units of $e^2\Delta/h$) as a function of V (in units of Δ/e) for an SKN junction and for several values of t_K (t_K grows from top to bottom). Dashed (solid) curves correspond to s-wave (p-wave) superconducting lead. The parameters and notations are the same as in Fig. 1. The curves with t_K=100 for s and p wave superconductors coincide.

In Figs. 2 and 3, the zero-energy shot-noise power K and the Fano factor $K/(2eI)$ are displayed versus the applied voltage V for t_K=100, 5, 3 and 2 and

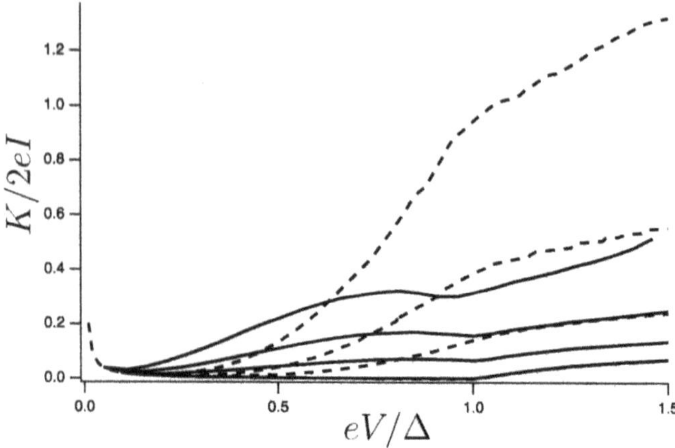

Figure 3. The Fano factor $K/2eI$ as a function of V (in units of Δ/e) for an SKN junction and for several values of t_K (t_K grows from top to bottom). Dashed (solid) curves correspond to s-wave (p-wave) superconducting lead. The parameters and notations are the same as in Fig. 1.

for Γ/T_K^0=200. These results are clearly correlated with those for the $G-V$ curve and can be summarized as follows: In the limit $t_K \gg 1$ the characteristics of shot-noise power spectrum for both s-wave and p-wave superconductors are consistent with those obtained for purely ballistic junctions which exhibit strong suppression of the shot-noise power in the sub-gap region. At lower t_K the physics is distinct. For $t_K = 5$ the noise spectrum for s-wave superconducting lead still shows features typical for a junction with relatively high transparency, while the results for t_K=2 are somewhat resemble those for a low transparency junction. Such dependence is explicitly exposed in the plot of the Fano factor versus the applied voltage (see Fig. 3). Though the Fano factor does not reach the maximum value of 2, it is strongly enhanced for the smaller value of $t_K = 2$. For p-wave superconductor the shot-noise power (as function of voltage) reflects the same physics as in the conductance: ZES turn the Kondo resonance to be less vulnerable to the impact of superconductivity and the junction remains close to the unitary limit almost in the whole range of values of t_K considered here (see the solid curves on Figs. 2 and Fig. 3).

5. Josephson current

In this section we study an equilibrium property (Josephson effect) of an SKS junction in which both electrodes are p-wave superconductors (for comparison we also represent the results for s-wave superconducting leads).

The Free energy F for the Hamiltonian (2) is given by

$$F = -T\sum_{\omega} \ln\left[\omega^2(1+\alpha(\omega))^2 + \tilde{\epsilon}^2 + \beta(\omega)^2\Delta^2\cos^2(\delta/2)\right] + \lambda b^2, \tag{22}$$

with $\alpha(\omega) = \tilde{\Gamma}\sqrt{\omega^2+\Delta^2}/\omega^2$, $\beta(\omega) = \tilde{\Gamma}/\omega$, $\tilde{\epsilon} = \epsilon + \lambda$, $\tilde{\Gamma} = b^2\Gamma$ and the phase difference δ between two superconductors. Then the self-consistent equations, $\delta F/\delta\lambda = 0$ and $\delta F/\delta b^2 = 0$ read, respectively

$$\tilde{\epsilon} + \frac{2\Gamma}{\pi}\log\frac{\tilde{\epsilon}}{T_K} = \sum_{\omega}\Big[2\Gamma\frac{(1+\alpha(\omega))\sqrt{\omega^2+\Delta^2} + \beta(\omega)\Delta^2/\omega\cos^2\frac{\delta}{2}}{(1+\alpha(\omega))^2\omega^2 + \tilde{\epsilon}^2 + \beta(\omega)^2\Delta^2\cos^2\frac{\delta}{2}} - \frac{2\Gamma|\omega|}{\omega^2+\tilde{\epsilon}^2}\Big], \tag{23}$$

$$\tilde{\Gamma} = \Gamma\sum\left[\frac{2\tilde{\epsilon}}{(1+\alpha(\omega))^2\omega^2 + \tilde{\epsilon}^2 + \beta(\omega)^2\Delta^2\cos^2\frac{\delta}{2}}\right] \tag{24}$$

Self consistent equations for the case of s-wave superconducting leads were derived in Ref. [19].

The Josephson current is given by, $I = (2e/\hbar)\partial F/\partial\delta$, that is,

$$I = \frac{e}{\hbar}\sum_{\omega}\frac{(\beta(\omega)\Delta)^2\sin\delta}{\omega^2(1+\alpha(\omega))^2 + \tilde{\epsilon}^2 + (\beta(\omega)\Delta)^2\cos^2\frac{\delta}{2}} \tag{25}$$

The self-consistency equation and the expression for the Josephson current can easily be extended to the case of an anisotropic coupling :$\Gamma_{L/R} = \Gamma(1\pm p)$ (the anisotropy parameter $0 < p < 1$). For this, one should replace $\cos^2\frac{\delta}{2} \to (\cos^2\frac{\delta}{2} + p^2\sin^2\frac{\delta}{2})$ and $I \propto \sin\delta \to I \propto (1-p^2)\sin\delta$.

In the unitary limit $T_K \gg \Delta, \tilde{\epsilon} = 0$ we can approximate $\alpha(\omega) = \tilde{\Gamma}\sqrt{\omega^2+\Delta^2}/\omega^2 \gg 1$ since the relevant energy scale for the superconductor is $\omega \leq \Delta$. As a result, the Josephson current (25) for p-wave superconductor is

$$J = \frac{2e\Delta}{\hbar}\frac{(1-p^2)}{4}\frac{\tanh(\beta\Delta f(\delta)/2)}{f(\delta)}\sin\delta \tag{26}$$

with $f(\delta) = \sqrt{1+\cos^2(\delta/2) + p^2\sin^2(\delta/2)}$, which is different from the one for s-wave [23].

In Fig. 4, the Josephson current is calculated as a function of the phase difference δ for a small anisotropy parameter $p = 0.1$ and for values of t_K =100, 5, 3, and 2 at temperatures close to $T = 0$.

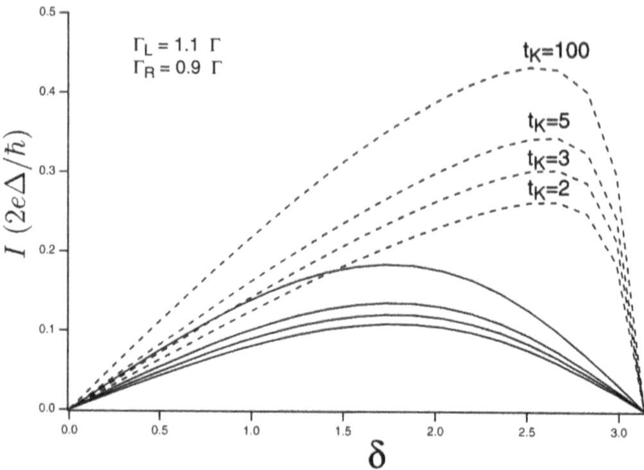

Figure 4. Josephson current versus phase difference δ at $T \rightarrow 0$. Dash and solid curves correspond to s and p-wave superconductors, respectively. The parameters are the same as in Fig.1. The value of t_K decreases from top downward.

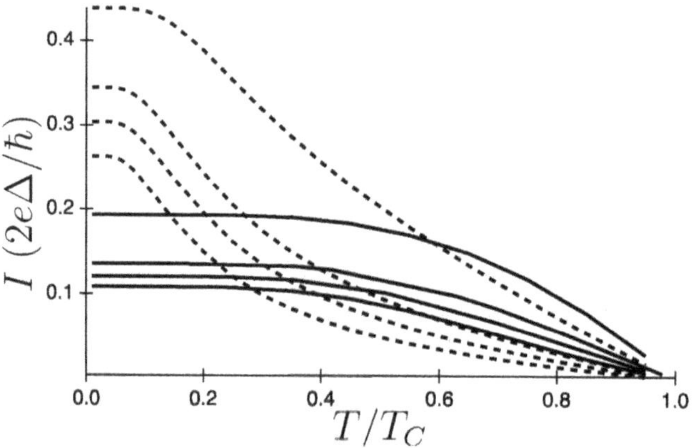

Figure 5. Temperature dependence of the maximum Josephson current for s-wave (dashed curves) and p-wave (solid curves) superconducting leads for t_K =100, 5, 3 and 2 (from top to bottom). The phase δ of each curve is chosen so as to give the maximum value of Josephson current at $T = 0$ in Fig. 4. Other parameters are the same as in Fig. 1.

For an s-wave superconductor (dashed lines in figure 4) the current at t_K =100 is nearly the same as the one at the unitary limit [23]; It deviates from the sinusoidal form and reaches its maximum around $\delta = \pi$. At lower values of t_K, the current characteristic is similar but its amplitude is suppressed. Nevertheless, in this region of parameters, the Kondo effect overcomes superconductivity and the impurity spin is screened. Hence there are no traces of a π-junction.

For p-wave superconductor (solid lines in figure 4), the current at t_K =100 is approximately given by Eq. (26), which is qualitatively different from that obtained for the s-wave superconductor; The current has its maximal value around $\pi/2$ and the curve is very close to being sinusoidal. This result is very similar to the one encountered in SIS junctions. The mid-gap states act as if they increase the effective normal region of the junction and therefore, the junction becomes more resistive, which makes the contact of the junction less transparent, again in marked distinction from the s-wave case. Indeed, the amplitude of the Josephson current is reduced compared with its value for the s-wave junction.

Similar analysis holds for the temperature dependence of the Josephson current. Figure 5 displays the maximum Josephson current $I(T)$ as function of temperature for s-wave (dashed lines) and p-wave (solid lines) superconducting leads. For p-wave junctions, $I(T)$ assumes relatively large values near $T = T_C$ while for s-wave junctions, it decreases more rapidly as T increases. The difference is more prominent when t_K becomes smaller. These results are explained by the same reasoning above; the temperature dependence is more similar to the usual SIS junction for the p-wave case, while for s-wave superconductors it is closer to the one in SNS junctions. This difference stems from the existence of ZES. The situation is similar to the one encountered in Josephson current through an SIS system [24].

In conclusion, we have analyzed an important physical problem involving strong correlations, the Kondo effect and unconventional superconductivity in the regime $T < \Delta < T_K$. These aspects can be realized in an SKN junction consisting of an Anderson impurity in the Kondo regime. The S electrode consists of a p-wave superconductor specially oriented with respect to the contact surface allowing the formation of zero energy bound-states in the middle of the gap. First, we have investigated non-equilibrium aspects and elucidated the nonlinear $G - V$ characteristics. Moreover, we calculated the shot-noise power spectrum of SKN junctions at voltages $eV \leq \Delta$. It is found that at sufficiently large t_K the Kondo resonance effectively turns the junction behavior to be similar to that of highly transparent non-interacting weak links for both s and p-wave superconductors. However, when the ratio of the Kondo temperature to the superconducting gap becomes smaller the behavior of these two types of junctions is quite different: the Kondo resonance persists much more effectively for junctions with p-wave leads than it does for s-wave leads. Similar effects emerge also in the (equilibrium) Josephson current which has also been calculated above.

Acknowledgments

We would like to thank Y. Tanaka for very helpful and stimulating discussions. This research is supported by DIP German Israel Cooperation project **Quantum electronics in low dimensions** by the Israeli Science Foundation grant **Many-Body effects in non-linear tunneling** and by the US-Israel BSF grant **Dynamical instabilities in quantum dots**. One of us (TA) is supported by JSPS Research Fellowships for Young Scientists.

References

[1] Y. Maeno, H. Hashimoto, K. Yoshida, S. Nishizaki, T. Fujita, J. G. Bednorz, and F. Lichtenberg, Nature **372**, 532 (1994).

[2] A. C. Hewson, *The Kondo Problem to Heavy Fermions* (Cambridge University Press, Cambridge, 1993).

[3] L. I. Glazman and M. E. Raikh, Pis'ma Zh. Eksp. Teor. Fiz. **47**, 378 (1988) [JETP Lett. **47** 452 (1988)]; T. K. Ng and P. A. Lee, Phys. Rev. Lett. **61**, 1768 (1988).

[4] S. Hershfield, J. H. Davies, and J. W. Wilkins, Phys. Rev. Lett. **67**, 3720 (1991).

[5] Y. Meir, N. S. Wingreen, and P. A. Lee, Phys. Rev. Lett. **70**, 2601 (1993).

[6] D. Goldhaber-Gordon, Hadas Shtrikman, D. Mahalu, David Abusch-Magder, U. Meirav and M. A. Kastner, Nature **391**, 156 (1998); S. M. Cronenwett, T. H. Oosterkamp and L. P. Kouwenhoven, Science **281**, 540 (1998); J. Schmid, J. Weis, K. Eberl and K. v. Klitzing, Physica B, **256**, 182 (1998); F. Simmel, R. H. Blick, J. P. Kotthaus, W. Wegscheider, and M. Bichler, Phys. Rev. Lett. **83**, 804 (1999).

[7] M. A. Kastner, Comm. Cond. Matt. **17**, 349 (1996).

[8] J. Nygard, D. H. Cobben, and D. E. Lindelof, Nature **408**, 342 (2000).

[9] W. G. van der Wiel, S. De Franceschi, T. Fujisawa, J. M. Elzerman, S. Tarucha, and L. P. Kouwenhoven, Science **289**, 2105 (2000).

[10] G. A. Fiete, J. S. Hersch, E. J. Heller, H. C. Manoharan, C. P. Lutz, and D. M. Eigler, Phys. Rev. Lett. **86**, 2392 (2001); M. A. Schneider, L. Vitali, N. Knorr, and K. Kern, Phys. Rev. B **65**, 121406 (2002); J. Park, A. N. Pasupathy, J. I. Goldsmith, C. Chang, Y. Yaish, J. R. Petta, M. Rinkoski, J. P. Sethna, H. D. Abruña, P. L. McEuen, D. C. Ralph, Nature **417**, 722 (2002); W. Liang, M. P. Shores, M. Bockrath, J. R. Long, and H. Park, Nature **417**, 725 (2002).

[11] T. M. Klapwijk, G. E. Blonder, and M. Tinkham, Physica **109+110B**, 1157 (1982).

[12] A. Yu. Kasumov, R. Deblock, M. Kociak, B. Reulet, H. Bouchiat, I. I. Khodos, Yu. B. Gorbatov, V. T. Volkov, C. Journet, and M. Burghard, Science **284**, 1508 (1999).

[13] M. R. Buitelaar, T. Nussbaumer, and C. Schönenberger, Phys. Rev. Lett. **89**, 256801 (2002).

[14] Y. Avishai, A. Golub, and A. D. Zaikin, Phys. Rev. B **63**, 134515 (2001); Europhys. Lett. **54**, 640 (2001).

[15] Y. Avishai, A. Golub, and A. D. Zaikin, Phys. Rev. **B** 67, 041301 (2003).

[16] R. Fazio and R. Raimondi, Phys. Rev. Lett. **80**, 2913 (1998); **82**, 4950 (E)(1999).

[17] P. Schwab and R. Raimondi, Phys. Rev. B **59**, 1637 (1999).

[18] A. A. Clerk, V. Ambegaokar, and S. Hershfield, Phys. Rev. B **61**, 3555 (2000).

[19] A. V. Rozhkov and D. P. Arovas, Phys. Rev. B **62**, 6687 (2000).

[20] G. Campagnano, D. Giuliano, and A. Tagliacozzo, cond-mat/0106532.

[21] Y. Tanaka, T. Hirai, K. Kusakabe, and S. Kashiwaya, Phys. Rev. B **60**, 6308 (1999).

[22] P. Coleman, Phys. Rev. B **29**, 3035 (1984).

[23] L. Glazman and K. A. Matveev, Pis'ma Zh. Eksp. Teor. Fiz. **49**, 570 (1989) [JETP Lett. **49**, 659 (1989)].

[24] Y. Tanaka, and S. Kashiwaya, Phys. Rev. B **56**, 892 (1997); S. Kashiwaya, and Y. Tanaka, Rep. Prog. Phys. **63**, 1641 (2000).

QUANTUM CROSSBARS: SPECTRA AND SPECTROSCOPY

S. Gredeskul[1], I. Kuzmenko[1], K. Kikoin[1], Y. Avishai[1,2]
[1] *Department of Physics,*
Ben Gurion University of the Negev, Beer-Sheva 84105, Israel
[2] *Ilse Katz Center for Meso and Nanoscale Science and Technology,*
Ben Gurion University of the Negev, Beer-Sheva 84105, Israel
sergeyg@bgumail.bgu.ac.il

Abstract Quantum crossbars (QCB) is a system of two crossing arrays of parallel metallic nanotubes or quantum wires coupled by electrostatic interaction in the crosses. In the charge sector, QCB possesses a rich Bose type spectrum of plasmon excitations in the corresponding 2D Brillouin zone. The spectrum conserves a Luttinger liquid (LL) zero energy fixed point, and demonstrates both 1D and 2D behavior depending on the direction of the plasmon wave vector. Plasmon excitations in QCB may be involved in resonance diffraction of incident electromagnetic waves and in optical absorption in the infrared (IR) part of the spectrum. Capacitive interaction of QCB with semiconductor substrate does not destroy the LL character of the long wave QCB excitations. However, the dielectric losses on a substrate surface are significantly modifed due to appearance of additional Landau damping regions.

Keywords: Luttinger liquid, plasmon spectrum, dimensional crossover, infrared absorption, Landau damping

1. Introduction

Low dimensional electronic systems form a wide class of natural and artificial objects in contemporary condensed matter physics (quantum dots, quantum wires, nanotubes, polymer threads, inversion layers, superthin films, surface phase layers are well known examples). Each of these systems possesses specific nontrivial features essentially depending on the system dimension. Recent developments of experimental techniques including self-assembling, etching, lithography, imprinting and even sewing techniques [1, 2, 3] made it possible to fabricate novel systems that have in a sense mixed dimension with unnusual electronic correlation properties. These systems are $2D$ networks composed of $1D$ constituents like single wall carbon nanotubes (SWCNT) [3, 4], molecular

A.S. Alexandrov et al. (eds.), Molecular Nanowires and Other Quantum Objects, 219–232.

chains [5] or quantum wires. Similar networks arise also naturally as, e.g., crossed striped phases of doped transition metal oxides [6].

In the simplest case, such a network can be modelled as double $2D$ grid, i.e., two superimposed crossed arrays of parallel conducting quantum wires or nanotubes (square grids consisting of 2 arrays were considered in various physical contexts in recent papers [7]-[10]). This grid represents a novel unique nano-object – quantum crossbars (QCB). A constituent element of QCB possesses the Luttinger liquid (LL) spectrum [11, 12]. A single array of parallel quantum wires is still a LL-like system qualified as a sliding phase [13] provided only the electrostatic interaction between adjacent wires is taken into account. If an inter-wire tunneling is possible, the electronic spectrum of an array is that of 2D Fermi liquid (FL) [14, 15]. Similar low-energy long-wave properties are characteristic of QCB as well. Its phase diagram inherits some properties of sliding phases in case when the wires and arrays are coupled only by capacitive interaction [13, 16]. When inter-array electron tunneling is possible, say, in crosses, dimensional crossover (DC) from LL to 2D FL occurs [10, 13]. Some possibilities of observation of the crossover in transport measurements were discussed in [17].

In the present paper we review the results of studying $2D$ square QCB with capacitive inter-array interaction in the crosses. Such QCB possesses the LL zero energy fixed point, and a rich Bose-type excitation spectrum (plasmon modes) arises at finite energies in 2D Brillouin zone (BZ). The spectral and correlation properties of QCB strongly depend on the direction of the wave vector in reciprocal space, demonstrating $1D \rightarrow 2D$ dimensional crossover (DC). DC manifests itself in appearance of additional lines of optical absorption, non-zero space correlators and specific Rabi oscillations [18]. Characteristic frequencies of QCB plasmons are in IR region. Direct absorption of infrared radiation by QCB is possible only at the center of the BZ . However, each of the QCB arrays serves as a diffraction lattice for its partner. This results in the possibility of observing a series of single absorption lines at the BZ center. In order to scan all the QCB spectrum one can use an external diffraction lattice (DL) with a period exceeding that of the QCB. In the general case, absorption spectrum consists of a set of single lines but in special directions they are replaced (completely or partially) by a set of doublets due to DC. Capacitive interaction of QCB with a semiconductor substrate does not destroy the LL character of the long wave QCB excitations. However, dielectric losses on a substrate surface are significantly altered due to appearance of additional Landau damping regions. The latter is initiated by diffraction processes on QCB superlattice and manifests itself as strong narrow absorption peaks lying below the damping region of an isolated substrate.

2. Quantum crossbars

A square QCB is formed by two perpendicular arrays of $1D$ quantum wires or carbon nanotubes periodically crossed with a period a. In fact, QCB is a cross-structure of suspended SWCNT lying in two parallel planes separated by an inter-plane distance d and placed on an insulator or a semiconductor substrate (see Fig. 1). Nevertheless, some generic properties of QCB may be described under the assumption that a QCB is a genuine $2D$ system. A single wire is characterized by its radius r_0, length L, Fermi velocity v of electrons in the wire, and LL interaction parameter g. The minimal nanotube radius is $r_0 \approx 0.35$ nm [19], maximal nanotube length is $L \approx 1$ mm, and the LL parameter are estimated as $g \approx 0.3$, $v \approx 8 \cdot 10^7$ cm/sec [12]. In a typical setup [4] the characteristic lengths mentioned above have the values $d \approx 2$ nm, $a \approx 20$ nm, so that the inequalities $r_0 \ll d \ll a \ll L$ are satisfied. To

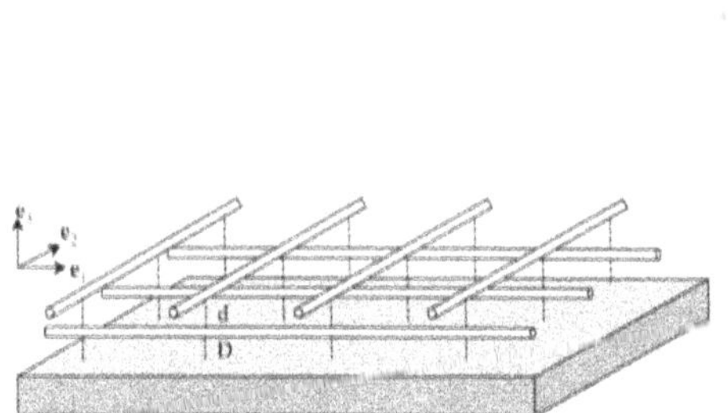

Figure 1. QCB on a substrate.

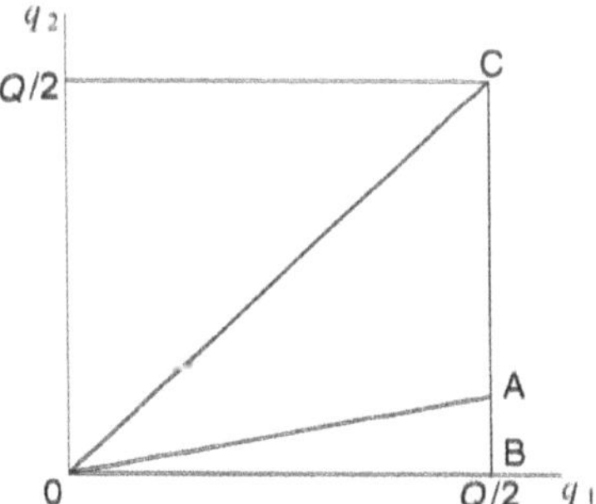

Figure 2. The first quarter of the BZ.

describe the QCB geometry we choose a coordinate system so that the axes x_j and the corresponding basic unit vectors $\mathbf{e}_j$ are oriented along the j-th array ($j = 1, 2$), the x_3 axis is perpendicular to the QCB plane) the x_3 coordinate is zero for the second array, $-d$ for the first one , and $-(d + D)$ for the substrate. In what follows the distance D between the QCB and the substrate is chosen as $D = 1$ nm. The basic vectors of the reciprocal superlattice for a square QCB are $Q\mathbf{e}_{1,2}$, $Q = 2\pi/a$ so that any reciprocal superlattice vector $\mathbf{m}$ is a sum $\mathbf{m} = \mathbf{m}_1 + \mathbf{m}_2$, where $\mathbf{m}_j = m_j Q \mathbf{e}_j$ (m_j integer). Similarly, any vector $\mathbf{k}$ of the reciprocal space has the form $\mathbf{k} = \mathbf{q} + \mathbf{m}$ where $\mathbf{q} = q_1\mathbf{e}_1 + q_2\mathbf{e}_2$ belongs to the first BZ $|q_{1,2}| \leq Q/2$ (see Fig. 2). A characteristic QCB plasmon wave vector is of order $Q \sim 10^6$ cm^{-1} and its characteristic frequency $\omega = vQ \sim 10^{14}$ sec^{-1} lies in the far IR region.

The QCB Hamiltonian,

$$H_{QCB} = H_1 + H_2 + H_{12} \tag{1}$$

is the sum of two LL Hamiltonians of arrays $j = 1, 2$

$$H_j = \frac{\hbar}{2} \sum_{\mathbf{qm_j}} \left\{ vg\pi^\dagger_{j\mathbf{q}+\mathbf{m}_j} \pi_{j\mathbf{q}+\mathbf{m}_j} + \frac{\omega^2_{q_j+m_jQ}}{vg} \theta^\dagger_{j,\mathbf{q}+\mathbf{m}_j} \theta_{j,\mathbf{q}+\mathbf{m}_j} \right\}, \quad (2)$$

and the Hamiltonian of the inter-array interaction H_{12}. Here $\{\theta_j, \pi_j\}$ are the conventional canonically conjugate boson fields and $\omega_k = v|k|$. The form of H_{12} strongly depends on the distance d between arrays and the electron screening within a single wire. In our case $d \approx 2$ nm, electron tunneling between arrays is completely suppressed, the charge sector of the excitation spectrum is separated from the spin sector. Then, electrons in a nanotube are screened at a distance of order nanotube radius r_0 [20]. As a result the inter-array interaction is a short-range contact capacitive coupling in the crosses of the bars which can be written as [18]

$$H_{12} = \frac{\hbar\phi}{vg} \sum_{\mathbf{q},\mathbf{m}} \xi_{q_1+m_1Q} \xi_{q_2+m_2Q} \theta^\dagger_{1,\mathbf{q}+\mathbf{m}_1} \theta_{2,\mathbf{q}+\mathbf{m}_2}, \quad (3)$$

where $\phi = 2ge^2r_0^2/(\hbar vad)$ and $\xi_k = \omega_k \mathrm{sign}k$. For QCB formed by carbon nanotubes, this interaction is small, $\phi \sim 0.007$.

3. Spectrum and correlations

The excitations in a single wire Hamiltonian (2) are 1D LL plasmons of the j-th array with wave numbers $q_j + m_jQ$ and frequencies $v|q_j + m_jQ|$. These plasmons can be treated in terms of "empty array" excitations described by quasimomenta q_j and dispersion law $\omega_{s_j}(q_j) = v|q_j + m_jQ|$. The band number is $s_j = 1+[2\omega_{s_j}(q_j)/vQ]$ (square brackets denote integer part of a number). The direct products of such eigenstates of two arrays form the eigenstates of a $2D$ "empty lattice". They are characterized by quasimomenta $\mathbf{q} + \mathbf{m}$ and band number $s = 1 + [2(\omega_{s_1}(q_1) + \omega_{s_2}(q_2))/vQ]$. The QCB Hamiltonian (1) describes a system of coupled harmonic oscillators. It can be diagonalized exactly. The diagonalization procedure is, nevertheless, rather cumbersome due to the mixing of plasmons belonging to different bands and arrays. However, separability of the interaction (3) facilitates the procedure and allows one to describe the QCB spectrum analytically [18]. Moreover, being small, this interaction grossly conserves the unperturbed $1D$ systematics of levels and states, at least in the low energy region corresponding to the first few energy bands. This means that perturbed eigenstates can be described in terms of the same quantum numbers (array number, band number and quasimomentum) as the plasmons of an "empty lattice". Such a description fails in two specific regions of the reciprocal space $\mathbf{k} = k_1\mathbf{e}_1 + k_2\mathbf{e}_2$. The first one is the vicinity of the high symmetry lines $k_j = nQ/2$ with n integer (the lines with $n = \pm 1$ include

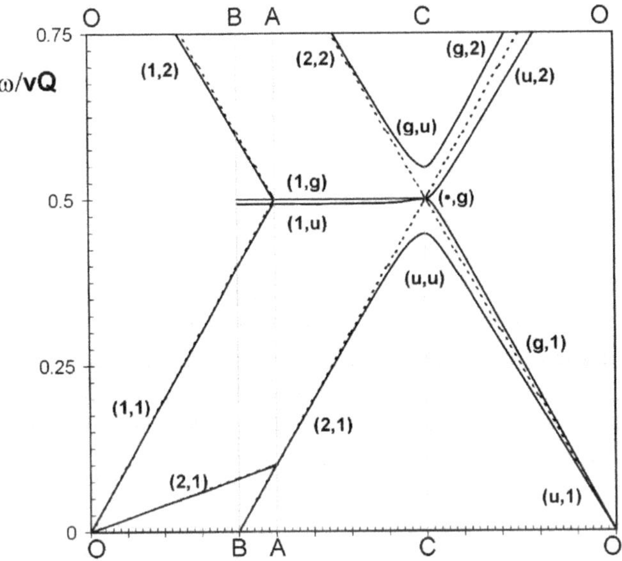

Figure 3. The energy spectrum of QCB (solid lines) and noninteracting arrays (dashed lines) along the lines OA, FC, and OC in Fig. 2.

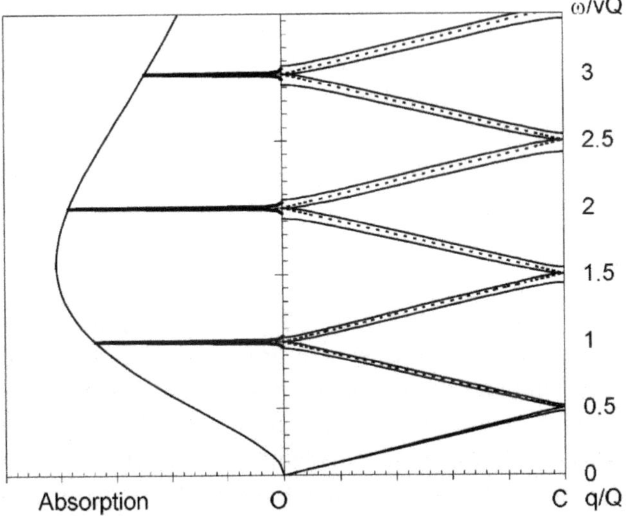

Figure 4. The lowest three bands of the QCB spectrum along the BZ diagonal OC and the corresponding single lines of direct IR absorption (see section 4 below).

the BZ boundaries). Around these lines, the *inter-band* mixing is significant and two modes from adjacent bands are degenerate. The second region is the vicinity of the resonant lines $k_1 \pm k_2 = nQ$ where the eigenfrequencies of the unperturbed plasmons

$$\omega_j(\mathbf{k}) = v|k_j|, \quad j = 1, 2,$$

from the same band propagating along two arrays coincide $\omega_1(\mathbf{k}) = \omega_2(\mathbf{k} + \mathbf{m})$. The *inter-array* mixing is significant around the resonant lines where two modes corresponding to different arrays are degenerate. The inter-array interaction introduces real two dimensionality into the problem, lifts the degeneracy of the bare modes and splits degenerate frequencies.

This picture of QCB spectrum is illustrated in Figs. 3, 4 where the dispersion laws corresponding to a few lowest bands are displayed. One can see well pronounced inter-array splitting around the BZ diagonal OC and inter-band splitting in the vicinity of the BZ corner C. A salient manifestation of the 2D character of QCB system is related to the possibility of periodic energy transfer between the two arrays. Consider an initial perturbation which excites a plane wave with wave vector $\mathbf{q}$ within the first array in the system of *non-interacting* arrays. If $\mathbf{q}$ is not too close to the resonance line of the BZ, weak inter-array interaction slightly changes the bose-field θ_1 of the first array and leads to the appearance of a small $\theta_2 \sim \phi$ component. However, for $\mathbf{q}$ lying on the resonance line (diagonal OC in Fig. 2), both components within the main approximation have the same order of magnitude. This corresponds to 2D propagation of a plane wave with wave vector $\mathbf{q}$, *modulated* by a "slow" frequency $\sim \phi\omega$. As a result, beating arises due to periodic energy transfer from one array to another during a long period $T \sim (\phi\omega)^{-1}$ (see Fig. 5). These peculiar "Rabi oscillations" may be considered as one of the fingerprints of the physics exposed in QCB systems.

4. Long wave absorption

QCB plasmons can be treated as a set of dipoles distributed within the QCB constituents. Therefore a natural way to probe the QCB spectrum is to study QCB interaction with an external IR radiation. In the case of an insulator substrate transparent in the IR region, one can treat the QCB as an isolated grid (without substrate) interacting directly with the incident radiation. Consider the simplest geometry where an external wave falls normally onto the QCB plane, and its electrical field $\mathbf{E} = E_0\mathbf{e}_1 \cos(\mathbf{kr} - \omega t)$ is parallel to the lower (first) array. In this geometry the field $\mathbf{E}$ is *longitudinal* for array 1 and *transverse* for array 2. The eigenfrequencies of the transverse modes in array 2 substantially exceed the IR frequency of the incident wave and even the standard LL ultraviolet cutoff frequency. Thus, the incident wave can be treated as a static polarization field for this array, and the factor $\cos\omega t$ can be omitted.

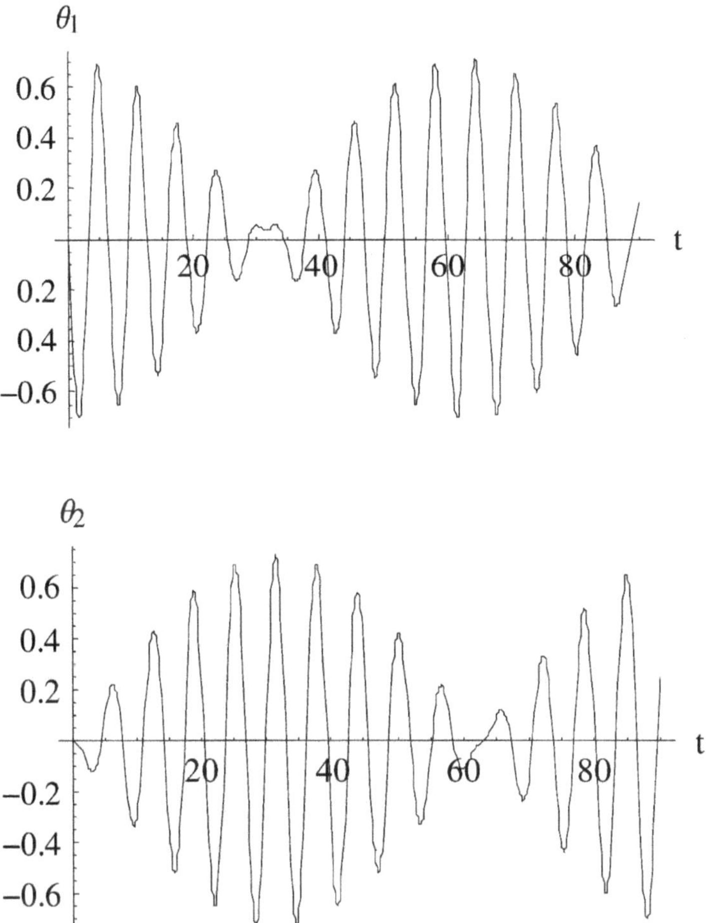

Figure 5. Rabi oscillations: time evolution of the bose-fields $\theta_{1,2}$ at a fixed QCB point.

The polarization waves in array 2 form a longitudinal diffraction field for array 1 with quasi wave vectors nQ (n is the diffraction order). The characteristic order of magnitude Q of a QCB plasmon wave vector is much larger than the wave vector **k** of the incident light, and we put the latter equal to zero from the onset. The light wavelength is much longer than a nanotube diameter and the geometrical shadow effect can be neglected. As a result the total field which affects array 1 consists of an external field and a diffraction field produced by a static charge induced in array 2.

An incident light excites plasmons only at the BZ center $\mathbf{k} = \mathbf{0}$, but due to Umklapp processes they have replicas at frequencies $\omega_n = nvQ$. Standard procedure in the vicinity of the resonance $|\omega - \omega_n| \ll \omega_n$ immediately yields the relative absorption of Lorentz type. Due to the exponential decay of diffraction field with the diffraction order, only the first few diffraction harmonics contribute to absorption. Nevertheless, one can probe at least the first five spectral lines. Three of them are displayed in Fig. 4.

5. Scanning of the spectrum

In order to probe plasmons with nonzero wave vectors, an external DL may be imposed parallel to the QCB with DL period $A \gg a$. The wave vector of M-th harmonic of the diffraction field $\mathbf{K}(M) = 2\pi M/A(\sin\varphi, \cos\varphi)$ can be represented as $\mathbf{K}(M) = \mathbf{q}(M)+\mathbf{m}_1(M)+\mathbf{m}_2(M)$. Here the angle φ describes orientation of the DL stripe, $\mathbf{q}$ belongs to the BZ and $\mathbf{m}_j$, $j = 1, 2$ are QCB reciprocal lattice vectors. Thus an incident light excites the plasmons of the j-th array with wave vectors $\mathbf{q}(M)+\mathbf{m}_j(M)$. The nature of excited plasmons, their frequencies and absorption structure strongly depend on the direction of the vector $\mathbf{K}(M)$, i.e. on the DL orientation. In a generic case, when $\mathbf{K}(M)$ does not reach neither a resonance direction nor the BZ boundary for any M, each of the points $\mathbf{K}(M)$ corresponds to a couple of plasmons mostly propagating along the j-th array, $j = 1, 2$, with unperturbed frequencies $\omega_{K_j(M)} = vK_j(M)$. Therefore, increasing the frequency of an incident light one observes a set of single absorption lines that consists of two almost equidistant subsets with frequencies corresponding to excitation of plasmons of the first or second arrays.

Varying the angle φ one can reveal the dimensional crossover. It manifests itself as appearance of doublets in the absorption spectrum. The angle $\varphi = \pi/4$ corresponds to the main resonance where $\omega(K_1) = \omega(\mp K_2)$ for all integer M and modes propagating along the two arrays are always degenerate. Inter-array interaction lifts the degeneracy. As a result, increasing the frequency of an incident light one observes an equidistant set of absorption doublets (Fig. 6).

For some specific directions, the points $\mathbf{K}(M)$ satisfy the higher resonance condition $\omega(K_1(M)) = \omega(nQ \mp K_2(M))$ at some values of M. Here, in a zero order approximation with respect to the inter-array interaction we expect appearance of two sets of absorption lines with frequencies $p\omega_j = vK_j(pM_0)$, $j = 1, 2$, and p natural, corresponding to excitation of plasmons within the $pm_j(M_0)$-th band of the j-th array. However, due to the resonance condition, plasmon in the first array with wave vector $K_1(pM_0)$ and frequency $\omega_1 = vK_1(pM_0)$ is coupled with plasmon in the second array with the same frequency and wave vector $K_2' = \mp(npQ - K_1(pM_0))$ (inter-array degeneracy). Similarly, plasmon in the second array with wave vector $K_2(pM_0)$ and frequency $\omega_2 = vK_2(pM_0)$ is coupled with plasmon in the first array with the same frequency and wave vector $K_1' = npQ \mp K_2(pM_0)$. This degeneracy of two modes corresponding to the same band but to different arrays is lifted by the inter-array interaction. As a result, increasing the frequency of an inci-

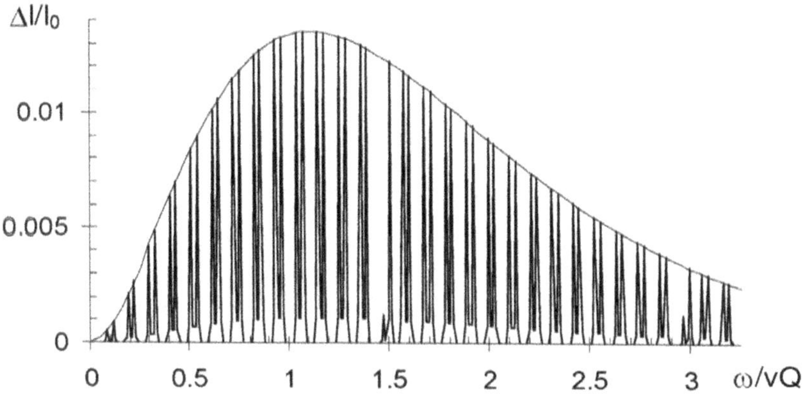

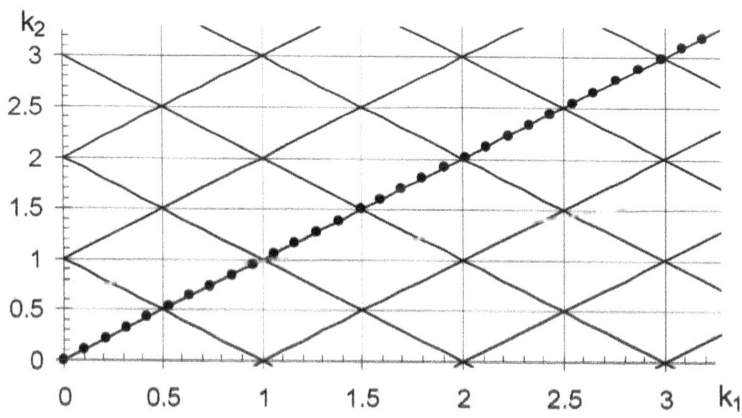

Figure 6. Absorption doublets in the resonant direction. One component is supressed when the wave vector is close to the corners of the BZs.

dent wave, one should observe two equidistant sets of single absorption lines with two sets of equidistant doublets built in these series. The corresponding absorption spectrum is displayed in Fig. 7.

Thus studying absorption of light by QCB one can expose not only dimensional crossover with respect to an angle (direction) [18], but also a new type of crossover with *an external frequency* as a control parameter.

6. QCB bands of Landau damping

One more approach to study QCB spectrum is to excite an appropriate surface wave (plasmon) in semiconductor substrate interacting with QCB. Plasmons of

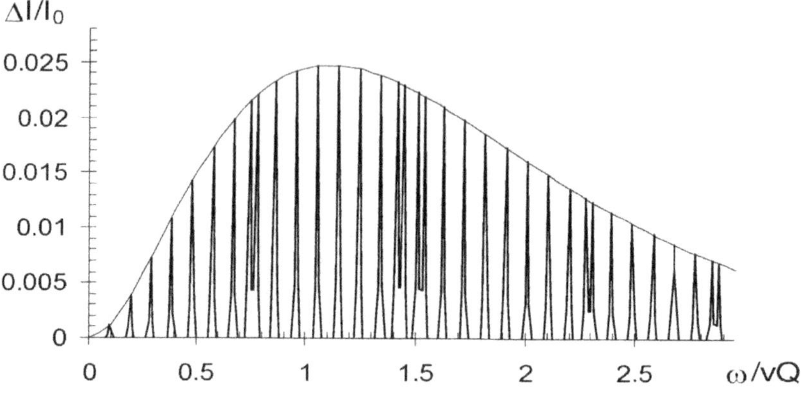

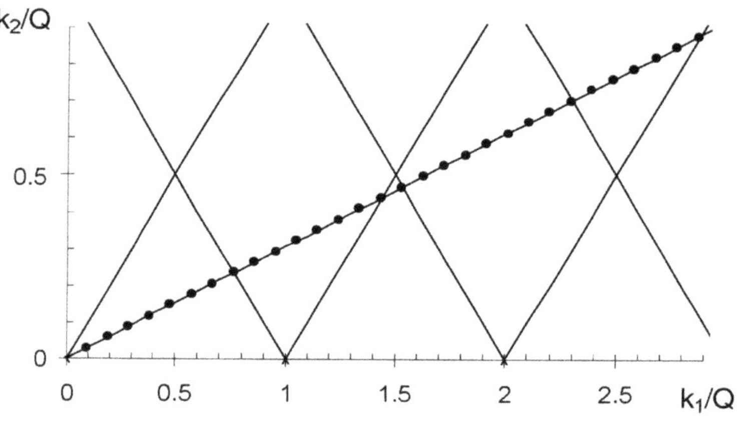

Figure 7. Absorption spectrum in a higher resonance direction. Two sets of doublets with periods 8 and 15 are well pronounced.

an isolated substrate with wave vector **k** are stable within a circle $k < k^*$ (for GaAs $k^* \approx 0.1 r_B^{-1}$, where $r_B = \hbar^2/me^2$ is the Bohr radius [21]). Outside the circle, a collective surface wave decays into electron and hole (Landau damping). Due to coupling of this wave with QCB plasmons, conventional picture of substrate dielectric losses modes is substantially altered. A $2D$ periodic structure of QCB thrusted on a substrate, provokes Umklapp processes which result in appearance of new Landau damping regions inside the initial region, i.e. at $k < k^*$. The structure of new damping regions and their existence strongly depend on the QCB period a. Varying this parameter, one can realize a rich variety of possible damping scenarios.

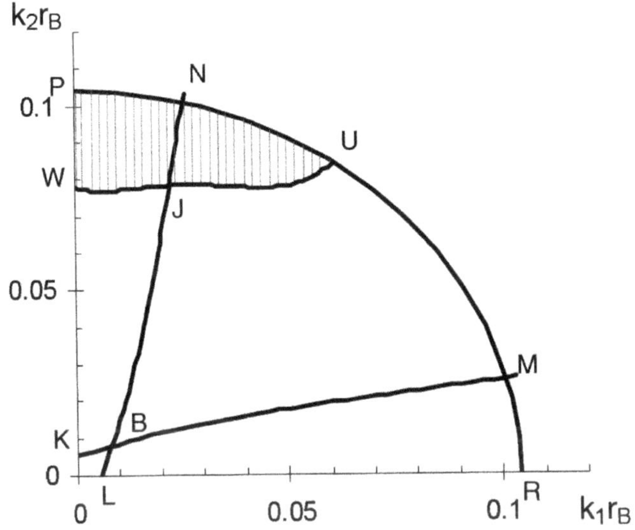

Figure 8. New tails of Landau damping for $a = 20$ nm.

In figures 8, 9, Landau damping region of an isolated substrate lies outside of the arc $PNMR$ while new damping regions appear within the quarter circle $OPNMRO$. Along the line LBN (KBM) the substrate plasmon interacts resonantly with the first (second) array QCB plasmon. We assumed the distance between the substrate and the QCB to be small, $D = 1.0$nm. The values of all othes parameters are the same as in Section 2. This enables one to distribute the roles between two QCB arrays: the nearest (first) array interacts with the substrate while the second one is responsible for Umklapp processes.

Let us trace the evolution of the new damping area. For small enough QCB period $a < a_1 = 17.3$ nm this area does not exist. For $a_1 < a < a_2 = 23.6$ nm the damping tails touching the initial Landau damping region appear in certain directions of the $\mathbf{k}$ plane (Fig. 8). They correspond to the Umklapp vector $-Q\mathbf{e}_2$. For $a_2 < a < a_3 = 34.6$ nm the new damping region is related to the same Umklapp vector $-Q\mathbf{e}_2$, but it now has a strip-like structure (Fig. 9). For larger QCB period, $a_3 < a < a_4 = 47.1$ nm the new damping area consists of three parts which correspond to two Umklapp vectors $-2Q\mathbf{e}_2$ and $-Q\mathbf{e}_2$. Further increase of QCB period leads to further extension of new damping regions and for $a > a_5 = 65.2$ nm Landau damping emerges in the whole quarter circle OPR .

Consider in details the new Landau damping region for $a = 30$ nm (Fig. 9). The points I, J, T, Q, M, U of intersection between the resonant lines LBN and KBM, the curves SIT and $WJQU$, and the arc $PNTMUR$ define six rays OI, OJ, OT, OQ, OM, OU and corresponding six angles $\varphi_1 \approx 16°$, $\varphi_2 \approx 33°$, $\varphi_3 \approx 38°$, $\varphi_4 \approx 69°$, $\varphi_5 \approx 76°$, $\varphi_6 \approx 81°$. The new damping region has a specific structure between each pair of adjacent rays (together with coordinate semiaxes).

Fixed direction of the excited surface wave predetermines the structure of the new Landau damping region. For $0 < \varphi < \varphi_3$ it is well separated from the initial region. Between φ_3 and φ_6 it turns to a tail which touches the initial damping region. Finally, for $\varphi < \varphi_6 < \pi/2$ additional regions of Landau damping do not appear at all. There are two special sectors of resonance interaction between the surface wave and the QCB plasmon of the first array $\varphi_1 < \varphi < \varphi_2$ or of the second one, $\varphi_4 < \varphi < \varphi_6$. Here the damping profile contains a sharp resonant peak on a wide pedestal while outside of these sectors the damping is small and described by a smooth curve. The existence

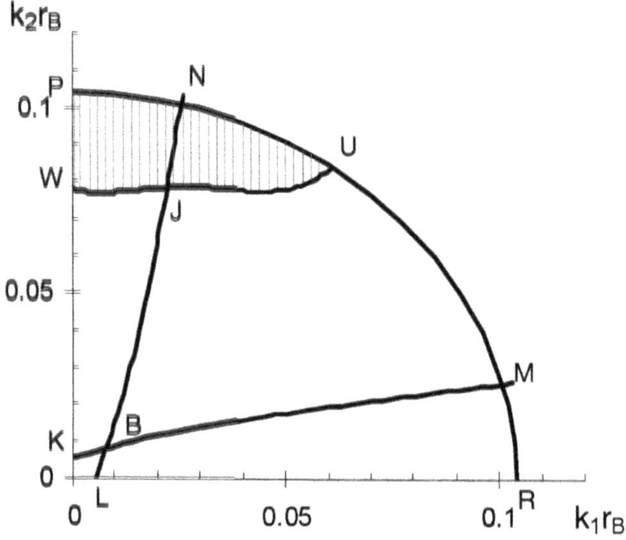

Figure 9. New regions of Landau damping for $a = 30$ nm.

of additional QCB bands (tails) of Landau damping and appearance of the resonant peak within the band (tail) is a clear manifestation of an interplay between genuine $2D$ surface plasmons and quasi-$2D$ QCB plasmons.

7. Summary

QCB is a novel nano-object which has in a sense an intermediate dimension that yields its peculiar spectral and correlations properties. The charge sector of QCB possesses a rich Bose type spectrum of plasmon excitations. In the long wave (low energy) region they conserve LL properties and demonstrate both

1D and 2D behavior in most part of the BZ. Dimensional 1D to 2D crossover in QCB results in the appearance of non-zero transverse space correlators and specific Rabi oscillations. Infrared absorption is a good tool for studying the QCB spectra. Together with employing an external DL, it enables one to scan the QCB spectrum in all the BZ. One more possibility to probe QCB spectrum is to measure dielectric losses in a semiconductor substrate interacting with QCB. This interaction leads to appearance of new regions of Landau damping. Their structure is very sensitive to the QCB period and direction of the substrate plasmon. The new damping regions are initiated by diffraction processes on QCB superlattice and manifest themselves as strong but narrow absorption peaks lying below the damping region of an isolated substrate.

Acknowledgments

This research is partially supported by grants from Israeli Science Foundation and US-Israel Binational Science Foundation.

References

[1] M.R. Diehl, S.N. Yaliraki, R.A. Beckman, M. Barahona, and J.R. Heath, Angew. Chem. Int. Ed. **41**, 353 (2002).

[2] B. Q. Wei, R. Vajtai, Y. Jung, J. Ward, R. Zhang, G.Ramanath, and P. M. Ajayan, Nature **416**, 495 (2002).

[3] A.B. Dalton, S. Collins, E. Muñoz, J.M. Razal, V.H.Ebron, J.P. Ferraris, J.N. Coleman, B.G. Kim, and R.H. Baughman,Nature **423**, 703 (2003).

[4] T. Rueckes, K. Kim, E. Joselevich, G. Y. Tseng, C. L. Cheung, and C. M. Lieber, Science **289**, 94 (2000).

[5] Y. Luo, C. P. Collier, J.O. Jeppesen, K.A. Nielsen, E. Delonno, G. Ho, J. Perkins, H-R. Tseng, T. Yamamoto, J.F. Stoddardt, J.R. Heath, ChemPhysChem **3**, 519 (2002).

[6] J.M. Tranquada, J. Physique IV, **12** 239 (2002).

[7] J.E. Avron, A. Raveh, and B. Zur, Rev. Mod. Phys. **60**, 873 (1988).

[8] Y. Avishai, J.M. Luck, Phys. Rev. B **45**, 1074 (1992).

[9] F. Guinea, and G. Zimanyi, Phys. Rev. B **47**, 501 (1993).

[10] A.H. Castro Neto, and F. Guinea, Phys. Rev. Lett.**80**, 4040 (1998).

[11] M. Bockrath, D.H. Cobden, J. Lu, A.G. Rinzler, R.E. Smalley,L. Balents, P.L. McEuen, Nature **397**, 598 (1999).

[12] R. Egger, A. Bachtold, M.S. Fuhrer, M. Bockrath, D.H. Cobden, and P.L. McEuen, in *Interacting Electrons in Nanostructures*, p. 125, R. Haug, and H. Schoeller (Eds.), Springer (2001).

[13] R. Mukhopadhyay, C.L. Kane, and T. C.Lubensky, Phys. Rev. B **63**, 081103(R) (2001).

[14] X. G. Wen, Phys. Rev. B **42**, 6623 (1990).

[15] H. J. Schultz, Int. J. Mod. Phys. **1/2**, 57 (1991).

[16] I. Kuzmenko, S. Gredeskul, K. Kikoin, and Y. Avishai, Low Temp. Phys. **28**, 539 (2002) [Fiz. Nizk. Temp. **28**,752 (2002)].

[17] R. Mukhopadhyay, C.L. Kane, and T. C. Lubensky, Phys. Rev. B **64**, 045120 (2001).

[18] I. Kuzmenko, S. Gredeskul, K. Kikoin, and Y. Avishai, Phys. Rev. B **67**, 115331 (2003).

[19] S.G. Louie in *Carbon Nanotubes*, M.S. Dresselhaus, G. Dresselhaus, Ph. Avouris (Eds.), Topics Appl. Phys. **80**, 113 (2001), Springer, Berlin 2001.

[20] K. Sasaki, Phys. Rev. B **65**, 195412 (2002).

[21] F. Stern, Phys.Rev. Lett. **18** 546 (1967).

QUANTIZED CONDUCTANCE IN ATOMIC-SCALE POINT CONTACTS FORMED BY LOCAL ELECTROCHEMICAL DEPOSITION OF SILVER

Christian Obermair[1], Robert Kniese[1], Fang-Qing Xie[1], Thomas Schimmel[1,2]
[1] *Institute for Applied Physics, University of Karlsruhe, D-76128 Karlsruhe, Germany*
[2] *Institute of Nanotechnology, Forschungszentrum Karlsruhe, D-76021 Karlsruhe, Germany*
Thomas.Schimmel@physik.uni-karlsruhe.de

Abstract We report on conductance quantization at room temperature in atomic-scale silver point contacts electrochemically fabricated within a nanoscale gap between two gold electrodes on a glass substrate. The formation of stable contacts exhibiting quantized conductance at integer multiples of the conductance quantum $G_0 = 2e^2/h$ ($\approx 1/12.9$ kΩ) was observed. While transient contacts with other conductance values were also found, a clear preference for values close to integer multiples of G_0 was observed for contacts stable for up to several hours. Stable conductance levels were observed not only at $1G_0$, but also at higher integer multiples of G_0 (up to $8G_0$). When applying electrochemical deposition or dissolution potentials, sharp transitions were induced between different quantized conductance levels while between these transitions, horizontal plateaus of constant conductance were found.

Keywords: Quantized Conductance, Atomic-Scale Point Contact, Electrochemical Deposition, Silver

1. Introduction

The continuing miniaturization of electronics has reached a stage where the discussion of devices on the molecular and even on the atomic scale becomes increasingly relevant [1]. Contacts consisting of individual atoms and molecules have become an object of intensive investigation by numerous groups [1-16]. Considerable progress was achieved in recent years in the fabrication and study of metallic point contacts on the atomic scale [6-16]. As the size of these atomic-scale metallic constrictions is smaller than the scattering length of the conduction electrons, transport through such contacts is ballistic and as the width of the contacts is on the length scale of the electron wavelength, the quantum nature of the electron is directly observable. As a result, quantization

A.S. Alexandrov et al. (eds.), Molecular Nanowires and Other Quantum Objects, 233–242.

of the conductance in multiples of the conductance quantum $G_0 = 2e^2/h$ is predicted, where e is the charge of an electron and h is Planck's constant.

Experimentally, two different approaches are available for the fabrication of these metallic quantum point contacts: mechanically controlled deformation of thin metallic junctions [6-10] and electrochemical fabrication techniques [11-16].

For the mechanical fabrication of atomic-scale junctions and wires, a Scanning Tunneling Microscope (STM) is used in many experiments. A metallic STM tip (e.g. gold) is pushed into a sample of the same metal and subsequently withdrawn [6,7]. Immediately before the contact between the metallic tip and the metallic sample breaks, an atomic-scale neck or wire is forming, and quantized conductance is observed in many cases. Recently, such necks were even investigated directly with transmission electron microscopy [17,18], revealing the atomic structure of thin atomic wires forming before the contact breaks. Another frequently used method for the fabrication of atomic-scale contacts is the Mechanically Controlled Break Junction (MCB) technique, which allows the controlled opening and closing of a free-standing metal bridge on a substrate. For these experiments, the sample containing the metallic bridge is built into an external setup which induces the mechanical deformation thus allowing the controlled opening and closing of the bridge leading to the opening and closing of the contact. Again, immediately before breakage of the contact, for different metals, an atomic-scale neck is forming, and a stepwise variation of the conductance is found [8-10]. The STM-based techniques and the break junction technique have in common that a mechanical deformation of the metal junction is required for the fabrication of the contact and that an external mechanical setup is required to induce and to control this deformation.

An alternative route for the fabrication of metallic quantum point contacts is based on electrochemical deposition and etching. By etching of thin metallic wires [12] or by deposition of metals within metallic junctions [12-16] or between the tip and the sample of an electrochemical STM [11], atomic-scale contacts can be formed without the need for a mechanical deformation of the contact. This absence of externally induced mechanical deformations is an important advantage both for the fabrication and investigation of such contacts and for their basic understanding. As it has been shown [17], these deformations lead to significant deviations of the atomic periodicity as compared to the value determined in bulk samples without mechanical stress and to the formation of defects within the contact area. Defects in the contact area, in turn, are expected to severely influence the conductance of the atomic-scale contact, as theoretical calculations indicate [21]. According to these calculations, for observing horizontal quantized conductance plateaus at integer multiples of G_0, it is necessary to avoid surface roughness as well as surface and volume defects within the contact area [21].

While the electrochemical deposition especially of atomic-scale copper point contacts was comprehensively investigated [11,12,14], little is known about quantum point contacts of silver. In mechanically formed silver contacts [18-20], typically a probability maximum in the conductance distribution is found only for $1G_0$, but not at higher multiples of G_0. Electrochemical fabrication of silver point contacts was only reported once [15], using a previously etched gap between two gold wires, also exhibiting only one integer plateau near $1G_0$ and conductance at non-integer levels (i.e. not at integer multiples of G_0) above this value.

However, silver is of special interest for the electrochemical fabrication of well-ordered quantum point contacts due to its high electrochemical exchange current densities, as electrochemical exchange currents may provide a means for structural reconfigurations and for the healing of atomic-scale defects within the contact area (see below).

Here, we report first results on the electrochemical deposition of silver quantum point contacts between two nanoscale electrodes directly deposited on a substrate. Due to the small size of the device setup with overall dimensions of the order of 100 nm (see below), a high thermal and mechanical stability is achieved. Quantized conductance of silver contacts with plateaus at higher integer multiples of G_0 is reported for the first time.

2. Experimental

The experimental setup is illustrated in Fig. 1. By applying an electrochemical potential, silver was deposited within the gap between two microscopic gold electrodes on a glass substrate. The gold electrodes (thickness approx. 10 – 50 nm) served as working electrodes and were covered with an insulating polymer except for the immediate contact area. Typical gap widths were of the order of 100 nm. The potentials of the working electrodes with respect to the (quasi-)reference and counter electrodes inserted into the electrolyte were set by a computer-controlled bipotentiostat.

The variation of the electrochemical potential difference between the reference electrode and the gold working electrodes was performed by varying the potential of the reference electrode relative to the ground potential. Silver wire of 0.25 mm in diameter with 99.9985% purity was used for the counter electrode and the quasi-reference electrode. Deposition was performed from an aqueous solution of silver nitride. The electrolyte consisted of 1mM $AgNO_3$ + 0.1 M HNO_3 in bi-distilled water. When applying an electrochemical potential of typically 2 - 40 mV between the electrochemical reference electrode and the two gold electrodes (gold electrodes with negative bias relative to the reference electrode), silver islands formed on the two gold electrodes, two islands finally meeting each other by forming an atomic-scale contact. During deposition, the

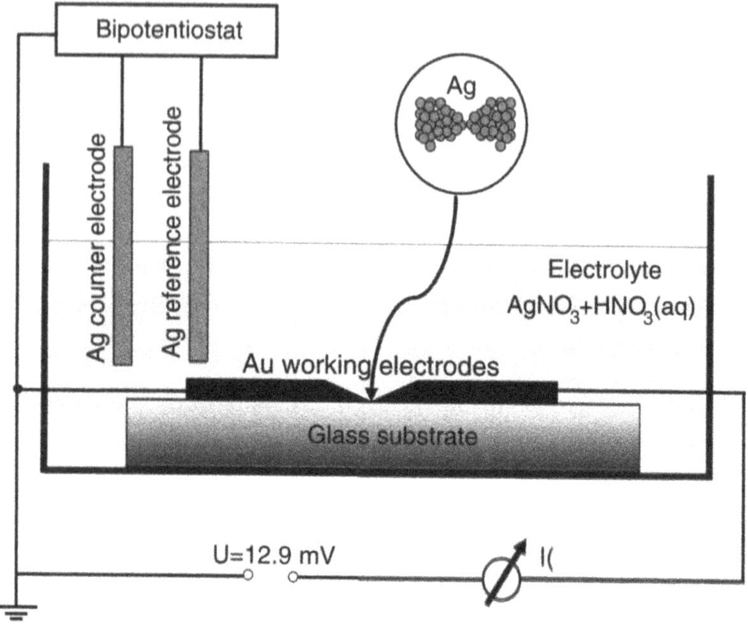

Figure 1. Illustration of the experimental setup. Silver quantum point contacts were electrochemically grown within a nanoscale gap between two electrodes deposited on a substrate.

conductance between the two gold electrodes was continuously monitored, thus allowing to stop deposition automatically via a computer-controlled feedback as soon as a predefined conductance value was reached.

3. Aspects of thermal, mechanical and chemical stability

The concept of electrochemically growing silver contacts within the gap between two nanoscale gold electrodes deposited on a substrate has several advantages concerning the stability of the contacts:

The electronic transport properties of the quantum point contact are strongly influenced by mechanical deformation. Therefore, it is important to keep the distance between the two electrodes meeting each other to form the atomic-scale junction stable with maximum tolerances of the order of 0.1 nm. Consequently, aspects of thermal and mechanical stability are key issues for contact performance.

As the size of the free-standing silver bridge is of the order of 100 nm and as the gold nanocontacts and leads in our setup are firmly connected with the substrate, the free-standing bridge which has to be taken into account for

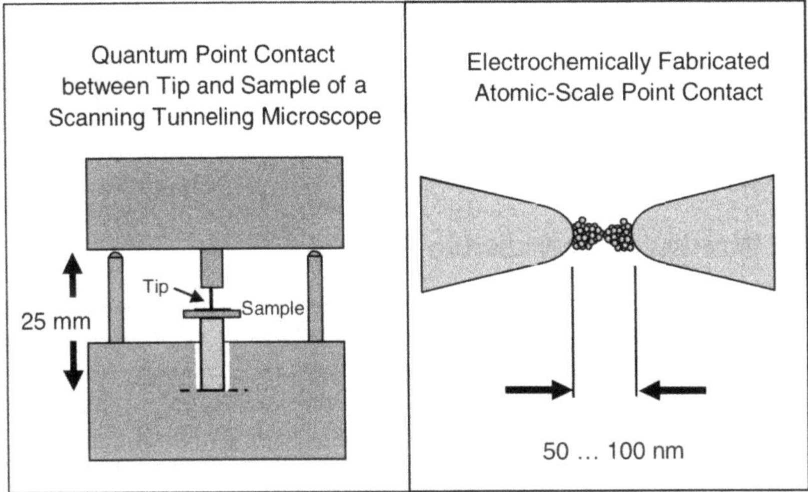

Figure 2. Comparison of the size of the mechanically free-standing loop of a scanning tunneling microscope setup (left) and the setup of Fig. 1 (right). The small size leads to an increase of the stability against thermal expansion and mechanical vibrations by a factor of typically 250 000.

thermal expansion and for mechanical vibrations of one side of the contact against the other is only of the order of 100 nm. This is different from the case of a scanning tunneling microscope (STM) setup and from the case of a mechanically controlled break junction setup. In both cases, the mechanical loop controlling the distance between the two contacting electrodes involves additional external mechanical components such as scanner and tip holder in the case of the STM and a mechanical adjustment setup in the case of the mechanically controlled break junctions. This is illustrated in Fig. 2. While in the case of the STM the mechanical loop determining the distance between the two electrodes forming the contact is of the order of 25 mm, in the case of our electrochemical setup with nanoscale electrodes grown on a substrate, it is of the order of 100 nm. As a consequence, both the thermal and the mechanical stability are increased as compared to the STM setup by a factor of typically 250 000 (!).

In addition, no external feedback and no adjustment mechanisms or control mechanism are required to stabilize the distance between the two contacting electrodes. The entire device (and not only the contact itself, as in the case of an STM) has an overall size on the nanometer scale, thus meeting a basic requirement for potential future applications in miniaturized setups and circuits.

Another advantage is connected with the use of silver as the contact material. In contrast to mechanical deformation of contacts, defects in atomic-scale contacts which are in equilibrium with an electrolyte can be more easily healed due to electrochemical exchange currents, leading to an strongly increased effective surface atom mobility. In this way, the rate of reconfiguration of atoms for the formation of stable and defect-free contacts on the atomic scale is much higher in the case of the electrochemical formation of point contacts as compared to the mechanical formation. Silver, which is well known for its high electrochemical exchange current densities [22], is especially well suited for this purpose.

4. Results and Discussion

Fig. 3 gives an example of an electrochemically deposited silver contact exhibiting a conductance of $1G_0$. The conductance is plotted as a function of time. After deposition the potential of the quasi-reference electrode was decreased to a potential of -29 mV relative to the working electrode, leading to the subsequent dissolution of the contact. As a result, a discontinuous jump of the conductivity to zero is observed. Subsequently, the potential of the quasi-reference electrode was again increased relative to the working electrode to a value of 2 mV, leading a deposition of silver on the working electrodes and finally to the formation of a new contact, indicated by a rise of the conductance to approx. $1G_0$. After this value was reached, deposition was stopped. The deviation from the exact value of $1G_0$ in this case was less than 1%.

Fig. 4 gives an example of conductance quantization at higher integer multiples of G_0, starting at $8G_0$. The conductance is plotted as a function of time. During slow electrochemical dissolution of the contact, discontinuous jumps of the conductance from $8G_0$ to approx. $4G_0$ and from $4G_0$ to $1G_0$ were observed before the contact finally completely dissolved. During the jump from $8G_0$ to $4G_0$, an instability occurred which lead to a transitional opening of the contact, indicating atomic rearrangements within the immediate contact area before a new, stable contact formed. The contact at approx. $4G_0$ was stable for 36 minutes before the jump to $1G_0$ occurred. It is remarkable that within the accuracy of the measurements, no slope of the conductance plateaus is observed. Horizontal conductance plateaus are followed by sharp transitions.

The fact that conductance plateaus of silver in previous experiments [18-20] have only been observed at $1G_0$, but never at higher integer multiples of G_0 may be explained by the calculations by Brandbyge and coworkers [21] who conclude that conductance is strongly affected by defects in the contact area. Slow electrochemical deposition and dissolution – in contrast especially to mechanical deformation – is expected to allow for a healing of defects, thus explaining conductance quantization even at higher integer multiples of G_0.

The time stability of another contact is shown in Fig. 5. Followed by initial fluctuations of the conductance value (see left side of the diagram, time scale: 0 – 200 s), a stable contact was formed with a conductance of $5G_0$. The contact remained stable for more than one hour. In this case, the electrochemical

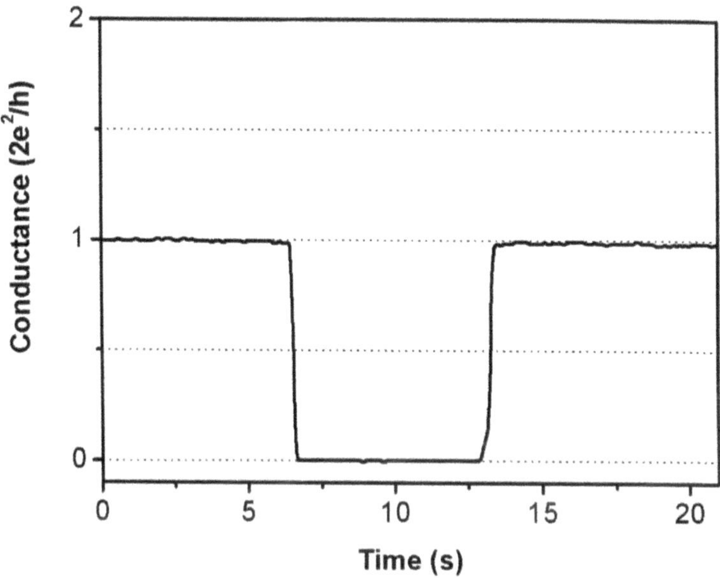

Figure 3. Quantized conductance of a silver quantum point contact at $G_0 = 2e^2/h$ ($\approx 1/12.9$ kΩ). By variation of the electrochemical potential, the contact is dissolved and subsequently reproducibly formed again.

potential was actively controlled by a feedback loop to avoid both further silver deposition and contact dissolution.

The data of Fig. 5 support the idea that due to electrochemical deposition a contact is formed which in many cases reaches a stable configuration only after several reconfigurations on the atomic scale. These reconfigurations can be observed in the experiment in the form of the resulting changes in conductance. While stable configurations preferably exhibited conductance values at integer multiples of G_0, the transitional stages did not show such a preference.

5. Conclusions

To conclude, stable silver quantum point contacts were grown electrochemically within the gap between two nanoscale gold electrodes deposited on a substrate. Due to the small overall size of our devices of the order of 100 nm, high mechanical and thermal stability was achieved. Quantized conductance both at $1G_0$ and at integer multiples of this value (up to $8G_0$) was observed at room temperature. Frequently, stable quantized levels were reached only after multiple atomic rearrangements of the contact, indicated by short peri-

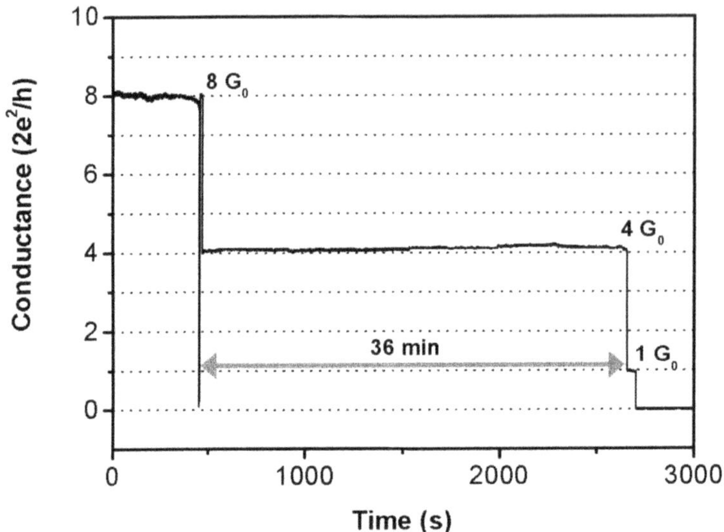

Figure 4. Example of conductance quantization at higher integer multiples of G_0, starting at $8G_0$. The conductance is plotted as a function of time. During slow electrochemical dissolution of the contact, discontinuous jumps of the conductance between integer, quantized plateaus from $8G_0$ to approx. $4G_0$ and from $4G_0$ to $1G_0$ are observed before the contact finally completely dissolves.

ods of changes and fluctuations of the conductance prior to the formation of a stable contact. The experiments thus also give insight into the processes of atomic-scale contact formation and the effect of atomic-scale rearrangements on electronic transport and conductance quantization.

Acknowledgments

The authors would like to thank H. Kuhn, L. Nittler, P. Kerber, P. Pfundstein and W. Send for experimental support. This work was supported by the Deutsche Forschungsgemeinschaft within the Center for Functional Nanostructures and by the Research Award of the State of Baden-Württemberg (Landesforschungspreis).

References

[1] Joachim, C., Gimzewski, J. K. & Aviram, A. Electronics using hybrid-molecular and mono-molecular devices. *Nature* **408**, 541-548 (2000).

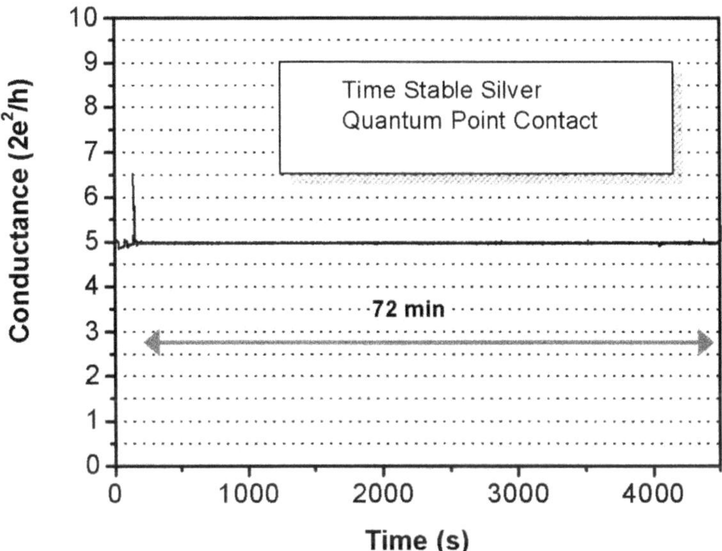

Figure 5. Time stability of a contact at $5G_0$. Followed by initial fluctuations of the conductance value (see left side), a contact is formed with a conductance of $5G_0$ which is stable for more than one hour.

[2] Tans, S. J. *et al.* Individual single-wall nanotubes as quantum wires. *Nature* **386**, 474-476 (1997).

[3] Bachtold, A., Hadley, P., Nakanishi, T. & Dekker, C. Logic Circuits with carbon nanotube transistors. *Science* **294**, 1317-1320 (2001).

[4] Chen, J., Reed, M. A., Rawlett, A. M. & Tour, J. M. Large on-off ratios and negative differential resistance in a molecular electronic device. *Science* **286**, 1550-1552 (1999).

[5] Gao, H. J. *et al.* Reversible, nanometer-scale conductance transitions in an organic complex. *Phys. Rev. Lett.* **84**, 1780-1783 (2000).

[6] Agraït, N., Rodrigo, J. G. & Vieira, S. Conductance steps and quantization in atomic-size contacts. *Phys. Rev. B* **47**, 12345-12348 (1993).

[7] Pascual, J. I. *et al.* Quantum contact in gold nanostructures by scanning tunneling microscopy. *Phys. Rev. Lett.* **71**, 1852-1855 (1993).

[8] Krans, J. M. *et al.* One-atom point contacts. *Phys. Rev. B* **48**, 14721-14724 (1993).

[9] Krans, J. M., van Ruitenbeek, J. M., Fisun, V. V., Yanson, I. K. & de Jongh, L. J. The signature of conductance quantization in metallic point contacts. *Nature* **375**, 767-769 (1995).

[10] Scheer, E. *et al.* The signature of chemical valence in the electrical conduction through a single atom contact. *Nature* **394**, 154-157 (1998).

[11] Li, C. Z. & Tao, N. J. Quantum transport in metallic nanowires fabricated by electrochemical deposition / dissolution. *Appl. Phys. Lett.* **72**, 894-896 (1998).

[12] Li, C. Z., Bogozi, A., Huang, W. & Tao, N. J. Fabrication of stable metallic nanowires with quantized conductance. *Nanotechnology* **10**, 221-223 (1999).

[13] Morpurgo, A. F, Marcus, C. M. & Robinson, D. B. Controlled fabrication of metallic electrodes with atomic separation. *Appl. Phys. Lett.* **74**, 2084-2086 (1999).

[14] Li, C. Z., He, H. X. & Tao, N. J. Quantized tunneling current in the metallic nanogaps formed by electrodeposition and etching. *Appl. Phys. Lett.* **77**, 3995-3997 (2000).

[15] Li, J., Kanzaki, T. Murakoshi K. & Nakato, Y. Metal-dependent conductance quantization of nanocontacts in solution. *Appl. Phys. Lett.* **81**, 123-125 (2002).

[16] Elhoussine, F., Mátéfi-Tempfli, S., Encinas, A. & Piraux, L. Conductance quantization in magnetic nanowires electrodeposited in nanopores. *Appl. Phys. Lett.* **81**, 1681-1683 (2002).

[17] Ohnishi, H., Kondo, Y. & Takayanagi, K. Quantized conductance through individual rows of suspended gold atoms. *Nature* **395**, 780-783 (1998).

[18] Rodrigues, V., Bettini, J., Rocha, A. R., Rego, L. G. C. & Ugarte, D. Quantum conductance in silver nanowires: Correlation between atomic structure and transport properties. *Phys. Rev. B* **65**, 153402-1-4 (2002).

[19] Enomoto, A., Kurokawa, S. & Sakai, A. Quantized conductance in Au-Pd and Au-Ag alloy nanocontacts. *Phys. Rev. B* **65**, 125410 (2002).

[20] García, N., Costa-Krämer J. L., Gil A., Marqués M. I., Correia A. Conductance quantization in metallic nanowires. in L. L. Sohn *et al.* (eds.), *Mesoscopic Electron Transport* (pp.581-616, Kluwer Academic Publishers, The Netherlands, 1997).

[21] Brandbyge, M., Jacobsen, K. W. & Norskov, J. K. Scattering and conductance quantization in three-dimensional metal nanocontacts. *Phys. Rev. B* **55**, 2637-2650 (1997).

[22] Hamann, C. H., Hamnett, A., Vielstich, W. *Electrochemistry* (Wiley-VCH, Weinhein, 1998).

SHELL-EFFECTS IN HEAVY ALKALI-METAL NANOWIRES

A.I. Yanson[⋆,‡], I.K. Yanson[†] & J.M. van Ruitenbeek[⋆]
[⋆]*Kamerlingh Onnes Laboratorium, Leiden University,*
PO Box 9504, NL-2300 RA Leiden, The Netherlands
[†]*B. Verkin Institute for Low Temperature Physics and Engineering,*
National Academy of Sciences, 310164, Kharkiv, Ukraine
[‡]*Present address: Dept. of Physics, 510 Clark Hall, Cornell University, Ithaca, NY 14853*
yanson@ilt.kharkov.ua

Abstract We supplement our previous observations of the shell effect in alkali-metal nanowires (Li, Na, K) [4, 6, 7, 8] with data extended to the heavy alkalis Rb and Cs. Our observations include: i) a non-monotonous dependence of conductance-histogram peak heights on atomic weight, ii) a rapid transition to an atomic shell structure at elevated temperatures, and iii) a "reverse" atomic-electronic shell transition, caused by the closeness to the liquid state.

Keywords: schell-effect, metallic nanowire, alkaline

1. Introduction

Conductance quantization was first observed in a 2-dimensional electron gas in a semiconductor heterostructure [1]. Later the effect was extended to a 3-dimensional electron gas in metals using nanowires produced by the break-junction technique [2]. The alkali metals used in the latter study have a single valence electron very weakly bound to the nuclei. They are well suited to satisfy the free-electron gas model. It has been long known that for clusters of alkali metals the conduction electrons are spherically distributed around the ions, smoothing the corrugation of ionic roughness. In a nanowire this leads to a cylindrical shape of the binding potential for finite electron movement in transverse dimensions. For thicker nanowires, the side surface becomes faceted. This transformation is stipulated by elevated temperatures, where the ions occupy the positions corresponding to a minimum of the lattice free energy.

At helium temperatures nanowires obtained by the break-junction technique assume all possible diameters, down to a single atom in cross-section, on many times stretching and renewing the contact between massive electrodes. The

A.S. Alexandrov et al. (eds.), Molecular Nanowires and Other Quantum Objects, 243–254.

summation of conductance versus elongation traces (we call these "scans") yields a conductance histogram whose peaks indicate the enhanced stability of nanowires with given diameters (conductance). These diameters correspond to a "magic" number of atoms in the narrowest cross-section whose conductance electrons fully occupy the so-called electron-energy shells in the transverse dimensions [3, 4]. By shells we imply the bunches of electron energy levels in the 2-dimensional cross-section whose positions on the energy scale are separated from each other by much wider energy gaps.

At elevated temperatures the probability of observing the electronic "magic" diameters becomes greatly enhanced, since the atoms have enough mobility to occupy the "magic" cross-section during subsequent rearrangements, which relax the strength created by pulling. Accordingly, the shell structure is greatly enhanced with rising temperature.

As in metallic clusters, the amplitude of electronic shell oscillations in thicker nanowires decreases with radius R, and the geometric (atomic) oscillations take over. The side surface becomes faceted and a complete layer of atoms (or a completed facet) corresponds to the enhanced stability of the nanostructure (cluster or wire) [5, 6]. There is an important difference between clusters and nanowires. While in the first case atoms on the surface are exchanged with the gas phase in the mass spectroscopy device through evaporation or condensation, in the second one the surface atoms are exchanged with the banks by surface diffusion and mechanical deformation during stretching of the wire on pulling. This imposes additional geometrical constraints to which the geometric shell structure should be compatible [7]. This is observed when recording return-histograms, which differ from the usual ones that are built from scans in the forward (pulling) direction, in being collected while applying a pushing force to compress the wire (in reverse movement).

In the present paper we extend the previous studies of conductance quantization, electronic shell/supershell and atomic (geometric) shell effects in Li, Na, and K [4, 6, 7, 8] to the heavy alkali metals Rb and Cs. We supplement our data with a number of new observations, among them are: i) a non-monotonous succession of peak heights in the conductance histogram against the atomic weight in the series Li→Na→K→Rb→Cs; ii) the evolution of conductance histograms and their Fourier spectra as a function of temperature; iii) the reverse transition from atomic to electronic shell oscillations at the highest attainable temperature for the heaviest alkali Cs, which we connect with the liquid state of the nanowire.

2. Low-temperature conductance histograms

Conductance histograms for the five alkali metals measured by us, and their atomic weights and melting points are shown in Fig.1. All of them exhibit con-

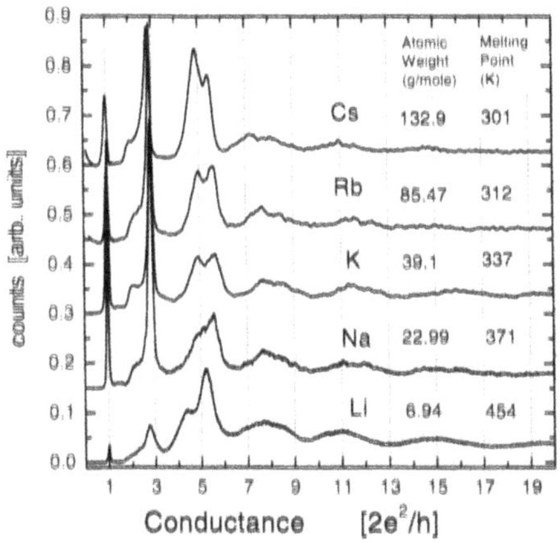

Figure 1 Conductance histograms for the alkali metals at helium temperatures (4.5 K). The curves are shifted vertically for clarity and normalized to the area up to $G = 20$ conductance quanta.

ductance quantization for electron wave functions propagating along a metallic waveguide with circular cross-section of variable radius [2]. The peaks near conductances G equal to 1, 3, 5 and 6 conductance quanta $G_0 = 2e^2/h = (12.9 \times 10^3\ \Omega)^{-1}$ are clearly seen. Note that peaks at $g = G/G_0$=2, and 4 are (almost) absent. For Na and K the peak heights at g=1, 3 are approximately equal (Fig.2) while for Li the peak at 1 is always lower than that at 3. For the heaviest metals Rb and Cs this behavior could be understood in terms of the atomic mobility in a one-atom contact, taking into account their noticeably lower melting point, but for Li it seems strange. Another anomaly for Li is the much higher shift of the peak positions at g =5 and 6 to lower conductances. Presumably, the mobility of the Li atoms is higher than that of Na and K even at helium temperatures, in spite of its higher melting point. This could be due to tunnelling of the light Li atoms through shallow barriers in the case where the number of interatomic bonds is greatly reduced. This could be tested by measurements at still lower temperatures where tunnelling of atoms would be seen as a temperature independent height of the first peak.

The intense peaks at g=1, 3, 5÷6 should be distinguished from the shallow maxima at 8, 11, 15 ... (Fig,1), which are due to the low-temperature electronic shell structure. Below we shall show how such a structure evolves with rising temperature, but here we note that the shell structure extends down even for the conductance peaks at g=3 and 5÷6. Especially this is evident for $g \approx 5$, which according to calculation for Na [9] should be much smaller than 6, and according to the experiment given in Ref.[10] does not carry a fully open conductance channel as that of g=6. From the data presented in Fig.1, the intensity of the

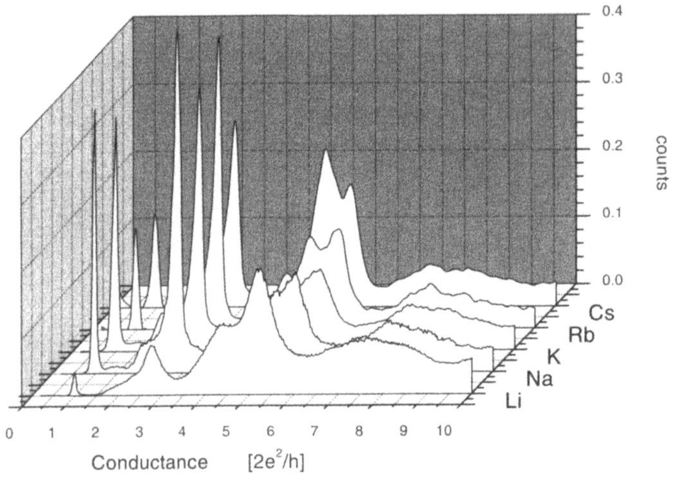

Figure 2. The same data as in Fig.1 are plotted in 3 dimensions to show how the heights of several initial peaks change with the atomic number.

peak near g=5 is of the same order as that near 6, and increases (compared to the peak at 6) in the direction Na→Cs, according to the lowering of the melting point. In other words, the increased mobility of the atoms (due to the melting point lowering) stimulates the shell-effect contribution to the peak at 5 in the same way as a rise in the temperature affects other shell-structure peaks (8, 11, 15, etc.).

3. Rubidium

Shell and supershell structure

Conductance histograms for Rb both for pulling (i.e. g decreasing) and pushing (g increasing) force are shown in Fig.3 (a). The histograms are obtained by cycling a Rb contact many times between g=100 and 0.2. Along the abscissa we use the square root of g since for this coordinate the shell oscillations are expected to be periodic, and we find a period $\Delta\left(g^{1/2}\right) \approx 0.59$ [4] (Fig.3 (b)). This period is close to the average periodicity predicted by quasi classical calculation for triangular and rectangular orbits in circular cross section [7]. There is a noticeable difference between the regular (pulling) and return (pushing) histograms. The observed hysteresis signifies that not only the equilibrium temperature influences the atomic movement, but the dynamics of atomic rearrangement itself strongly determines the observed histograms. In particular, the visible number of shell oscillations while pushing is substantially greater

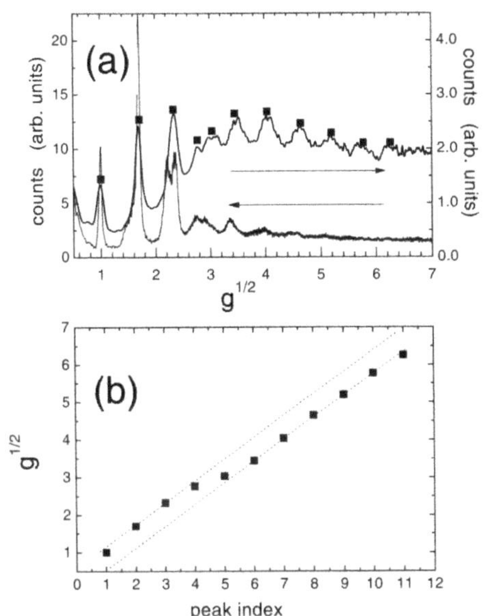

Figure 3 (a) Low temperature (4.5 K) histograms for Rb nanowires recorded while stretching (arrow pointing left) and compressing (arrow pointing right) the contacts. The peak positions for compressing force are marked by filled squares. The quantity along the abscissa is the square root of the reduced conductance, $g = G/G_0$. The units on the ordinate scale are given arbitrary, but the same relative to each other. (b) Positions of maxima in the return histogram versus peak number (index). The dotted lines are the linear fit to the upper branch of the data which is displaced vertically to fit the lower branch.

than that for pulling force (see the symbol positions which mark the oscillation maxima in Fig.3 (a)). The peak heights in the return histogram follow the shell oscillation envelope more smoothly, than those in the regular one.

Around peak indices $4 \div 6$ a shift of the straight line fit is observed. This may be a π phase-shift due to a the node of shell oscillation amplitude (the so-called, supershell effect [8]).

For completeness we show the Fourier spectrum for the return histogram of Fig.3 (a). A smooth background is subtracted (see Fig.4 (a)) for clarity.

The spectrum in Fig.4 (b) displays two principal maxima at frequencies $g^{-1/2} = 1.2$ and 1.7 which are characteristic of the electronic shell effect, where the triangular and rectangular trajectories are seen as a single maximum at $g^{-1/2} = 1.7$. The structure at higher frequencies may be considered as harmonics of the principal spectrum.

Evolution of the shell structure with temperature

Up to a temperature of <40 K the electronic shell effect prevails in the conductance histogram. An example of this is shown in Fig.5. The peaks at $g^{-1/2} = 1.2$ and 1.8, characteristic of the electronic shell oscillations, are the strongest ones in the forward (pulling, Fig.5 (b)) and return (pushing, Fig.5 (c)) Fourier spectra. The enhanced mobility of the Rb atoms stimulated by increasing temperature leads to a maximum intensity of the peak at $g^{1/2} = 2.3$

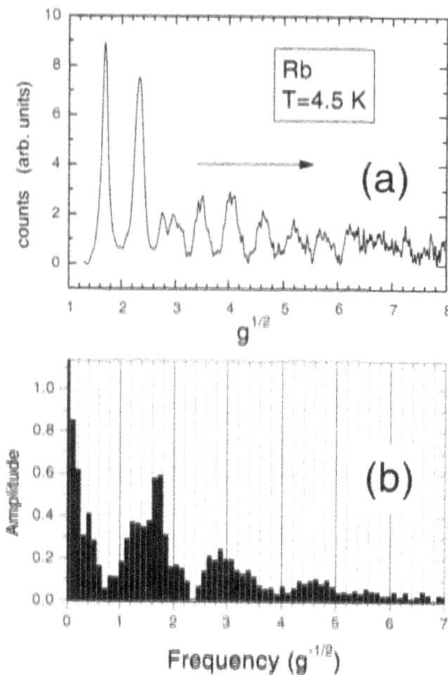

Figure 4 (a) The same return histogram as in Fig.3 with a smooth background subtracted, and its Fourier spectrum (b). Note the maxima at $g^{-1/2} = 1.2$ and 1.7, which are characteristic of the electronic shell effect.

relative to the peaks at $g^{1/2} = 1$ and 1.7 in the pulling histogram of Fig.5 (a), unlike the pulling histogram at $T = 4.5$ K (Fig.3 (a)).

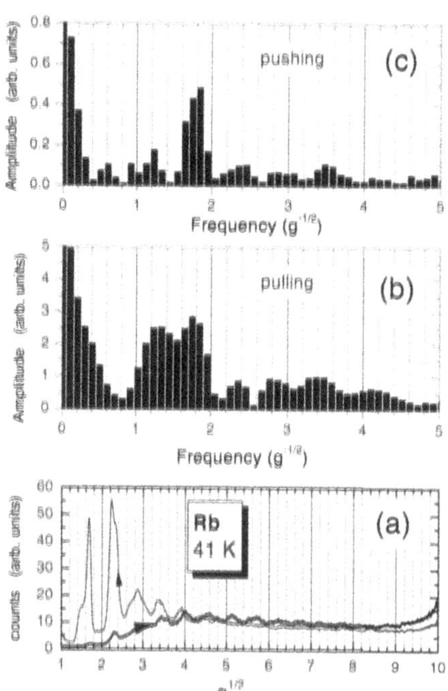

Figure 5 Rb histograms for pulling and pushing force (see the arrows superimposed on the data curves in panel (a)) at 41 K and their Fourier spectra (b) and (c), respectively.

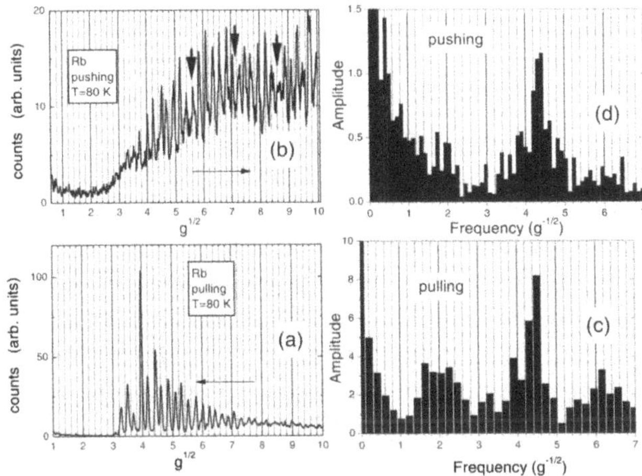

Figure 6. Rb histograms for pulling (a) and pushing (b) force (see the horizontal arrows in panels (a) and (b)) at 80 K. In the latter case the vertical arrows mark the positions of minima in the atomic shell amplitude. (c) and (d) present the corresponding Fourier spectra for pulling and pushing tension.

With rising temperature (T =60 K, not shown) a new peak at $g^{-1/2} = 4.4$ emerges with the same intensity as the electronic shell peaks. This implies that atomic shell structure [6] occurs with approximately the same probability as the electronic one. Moreover, the center of the maximum at $g^{-1/2} = 1.7 \div 1.8$ is shifted to a higher frequency having a full width at the half height in the range of $g^{-1/2} = 1.7 \div 2.3$. A new spectral peak at $g^{-1/2} = 2.2$ is located at half the principal atomic shell frequency ($g^{-1/2} = 4.4$). Its origin will be discussed below.

Finally, at a still higher temperatures (80 K), only the atomic shell frequencies remain in the forward and return Fourier spectra (Fig.6 (c) and (d)) which are shown with original conductance histograms in panels (a) and (b). The principal frequency is $g^{-1/2}$ =4.4.

Turning to the conductance histogram (Fig.6 (a)), one notices that at the given temperature the wires with conductances up to $g^{1/2} = 3$ are not stable, and for the return histogram (Fig.6 (b)) a clear modulation of the peak amplitudes is visible. The envelope period embraces approximately 6 atomic shell oscillations. As we mentioned in Refs.[6, 7], the 6-fold period approximately corresponds to a full atomic layer coverage of the hexagonal wire, and each oscillation corresponds to the coverage of a single facet in the hexagonal symmetry of the crystal structure of the neck. Here, again we can see an example that the forward and reverse histograms differ from each other.

It is interesting that further measurements at $T = 80$ K the nanowires revert to the electronic shell effect in the regular (pulling) histogram with some retardation in time of about 15÷20 minutes, during which the metal might be purified by continuous cycling [11]. This reappearance concerns only the forward (stretching) histograms, while for the backward (compressing) movement no stable nanowire appears up to conductances $g \sim 100$.

It is important to realize how the nanowire behaves during recording of the scan cycles (forward-return cycles). In reality, most of the stretching scans do not survive down to conductances $g = 0.2$. Instead, they show a break at larger conductances. Atoms migrate from the protrusions left on the banks after breaking to the bulk of the electrodes until the reverse movement of break-junction electrodes recovers the electrical contact. A gap (expressed as Volts applied to the piezo driver of the controllable break-junction) appears between the end of the pulling and the onset of the pushing movement (see, for example, Figs.7, 8, and 9). This gap becomes larger the higher the mobility of the atoms at given experimental conditions (temperature, adsorbates on the surface). In the case where the electronic shell structure reappears the gap becomes much larger than that for the scans of atomic shell structure recorded at lower temperatures. This implies that the reversed reappearance of the electronic shell structure corresponds to faster atomic mobility, possibly close to the liquid state for the surface atoms. Finally, if we decrease the temperature once again (say, from 80 to 60 K), the atomic shell structure is recovered.

4. Cesium: scans and histograms

Rb and Cs are very similar in their properties. All the features that we ascribe to Rb can also be found in Cs nanowires and vice versa. In this section we would like to show how scans of the Cs break junctions evolve with temperature along with the histograms. The characteristics of the Cs nanowire at $T = 50$ K are shown in Fig.7. The directions of recording are shown with arrows superimposed on the data curves. The stretching wire corresponds to an increase in the piezo voltage. One can see a hysteresis in the forward and backward directions, and a gap of about 0.5 V, due to retractions of atoms to the electrodes while the contact breaks. Although an accurate calibration has not been attempted and varied in different experiments, for the same series the change in V_{piezo} scale induced by the temperature is quite certain. The histogram is constructed from several thousands scans, which vary widely and one of them is shown by the dotted curve in Fig.7. As a whole, they produce a reproducible histogram shown in the same graph. Some of the strongest peaks in the forward direction (at $g^{1/2} = 2.2, 3$) are seen in the scan as a steeper part, but most of the steps do not correspond to any feature in the histograms. This means that these

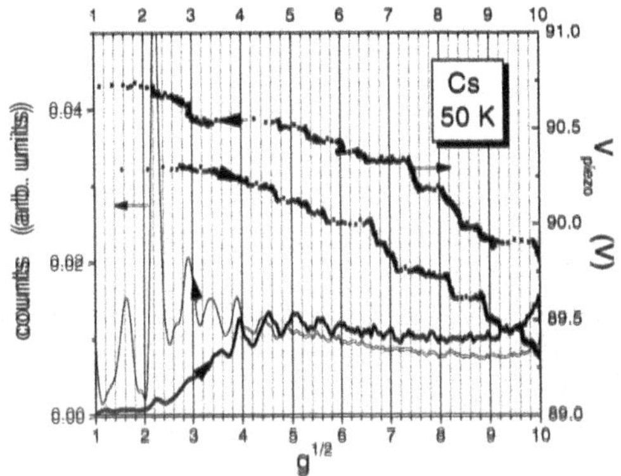

Figure 7 Cs histograms at 50 K displaying electronic shell oscillations, with representative scans in both directions. The ordinate scale (right axis) for the scans is given in Volts applied to the piezo driver, which is proportional to wire length. An increase of V corresponds to an elongation of the nanowire. The scan data are shown by dots, while the histograms are represented by continuous curves.

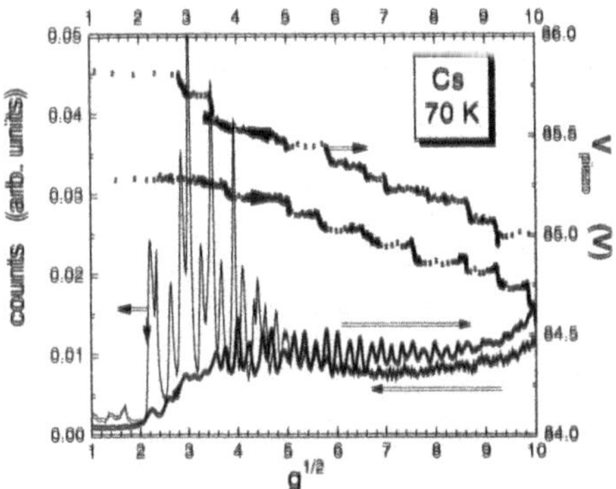

Figure 8 The same as in Fig.7, at $T = 70$ K. The histogram exhibits atomic shell oscillations for a similar experimental series as in Figs.7, and 9.

steps are completely random and give a monotonous background seen at high ($g^{1/2} = 7 \div 10$) conductances.

For this particular junction, an increase in temperature up to 70 K transforms the shell structure to the atomic one (Fig.8). There is no immediate agreement between the steps in the scans to any of the extrema in the atomic-structure part of the histogram. Electronic shell structure is observed at $g^{1/2} = 3$, 3.4, 3.8 and corresponding steps are seen in the stretching scan. At a still higher temperatures (80 K, Fig.9), the electronic shell oscillations reappear (as in the case with Rb) with a noticeable increase of the gap in the piezo voltage between forward and return scans. We recall that the scans shown are representative of many different scans used to built the histogram.

The most spectacular series of temperature dependent histograms for Cs is shown in Fig.10. It demonstrates an example of how the low temperature

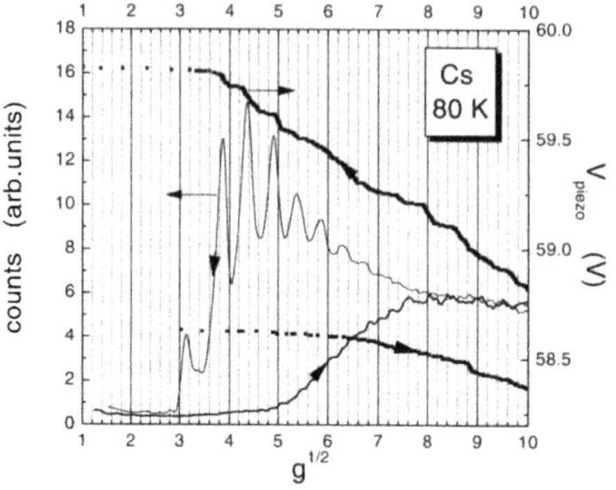

Figure 9 The restoring of electronic shell oscillations in Cs at $T = 80$ K. Note the large voltage gap (especially at low conductances) between the stretching and compressing scans.

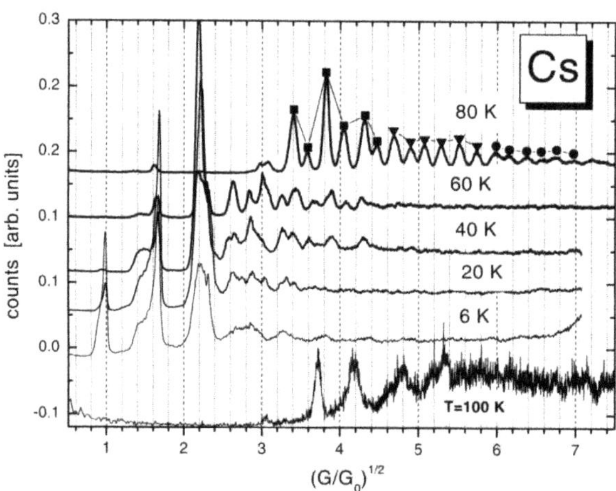

Figure 10. A series of Cs histograms at different temperatures. Note the signatures of atomic oscillations at relatively low temperatures. For $T = 80$ K the clear grouping of oscillations by sixes can be seen. To emphasize this the maxima are marked at the top by different symbols (squares, triangles and circles). At $T = 100$ K presumably a phase transition to the liquid state occurs and the electronic shell period takes over.

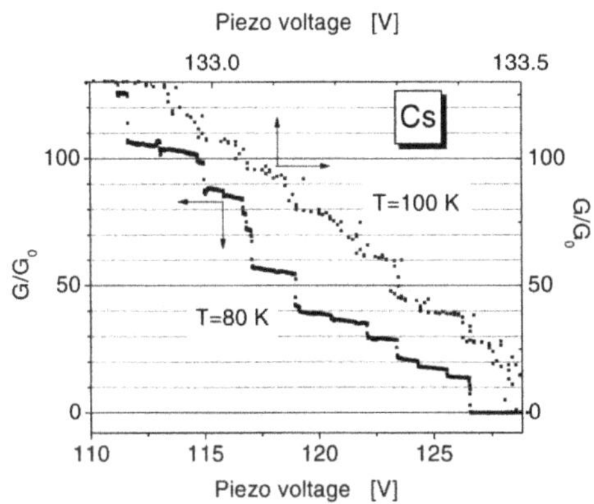

Figure 11 Representative elongation scans for $T = 80$ and 100 K. The first is characteristic of the solid state of the nanowire, while the latter seems to be in the liquid state. Note the huge increase of noise and, more important, the different range of abscissa scale for the scans.

electronic oscillations gradually transform to the atomic ones, starting already at rather low temperatures. At $T = 80$ K the atomic oscillations prevail and are clearly grouped by sixes, shown with different symbols on top of each of the maxima. At a still higher temperatures (100 K) the histogram shows the above mentioned reversed transformation to the electronic shell structure. We suggest that the wire becomes fully liquid. More precisely, the speed of surface diffusion of the atoms becomes much faster than the experimental timescale. This suggestion is justified by the corresponding scans (Fig.11), which confirm that at $T = 80$ K the neck is solid because of the well defined step structure, while at $T = 100$ K such a dependence becomes noisy and destroyed. Note, also, the dramatic shrinking of the length of stretching of the wire, which in the solid state amounts about 12 V on the piezovolt scale, while in the liquid state (100 K) this is only <0.5 V.

Fig.10 presents an example of how at low temperatures (20 K) the atomic shell oscillations for heavy alkali are superimposed on the electronic ones. In part, this is due to the heavy atomic weight since we never observed such a transition in the lightest metal, Li. In cluster physics, the transition between electronic and atomic shells is connected to the cooling of the cluster beam. It is believed that electronic shell structure is observed in the liquid state of a cluster, while the atomic structure corresponds to the solidification of the metal cluster. In nanowires in most cases we deal with solid necks since the step-like scans conclusively persuade us of the succession of elastic elongation and yielding stages. The latter is impossible in the liquid state. The higher temperatures greatly enhance the thermal fluctuations of the atoms forcing them to explore many locations, some of which may have a lower total free energy due to minima in the electronic contribution. Energy which is released during yielding may cause further heating. Upon rearrangement the atoms on average fall into the deepest free energy minimum, where the shell structure plays a noticeable role.

5. Summary

We supplement our previous study of light alkali metals Li, Na, and K with heavy alkalis Rb and Cs. All the features discovered in that study were observed here. Besides, some additional observations confirming our previous results were obtained.

The low-temperature conductance quantization features for the smallest cross-sections are the same for all five alkalis studied. There exist some differences in relative intensities of the peaks near $g = 3, 5$, and 6 between the conductance histograms of different metals, in particular for the peak near 5. We explain the enhancement of the latter peak for heavy alkalis by the electronic shell effect extending to lower radii due to a lower melting point.

The evolution of conductance histograms and their Fourier spectra with rising temperature, showing the transition to atomic (geometric) shell oscillations, exhibits the fundamental frequency 4.5 $g^{-1/2}$. For heavy alkalis the atomic shells are observed at lower temperatures than for the lighter ones, due to their lower melting point.

One of the unexpected observation is the reverse transitions from the atomic shell structure to the electronic one with rising temperature. We explain this by relative heating of heavy alkali metal nanowire having lower melting point.

References

[1] B.J. van Wees, H. van Houten H., C.W.J. Beenakker, J.G. Williamson, L.P. Kouwenhoven, D. van der Marel, and C.T. Foxon. Phys. Rev. Lett. **60**, 848 (1988).

[2] J.M. Krans, C.J. Muller, I.K. Yanson, Th.C.M. Govaert, R. Hesper, and J.M. van Ruitenbeek. Phys. Rev. B **48**, 14721 (1993).

[3] W.D. Knight, K. Clemenger, W.A. de Heer, W.A. Saunders, M.Y. Chou, and M.L. Cohen. Phys. Rev. Lett. **52**, 2141 (1984); W.A. de Heer, Rev. Mod. Phys. **65**, 677 (1993)

[4] A.I. Yanson, I.K. Yanson, and J.M. van Ruitenbeek, Nature (London) **400**, 1440 (1999).

[5] T.P. Martin. Phys. Rep. **273**, 199 (1996).

[6] A.I. Yanson, I.K. Yanson and J.M. van Ruitenbeek, Phys. Rev. Letters **87**, 216805 (2001).

[7] A.I. Yanson, I.K. Yanson, and J.M. van Ruitenbeek, Fiz. Nizk. Temp. **27**, 1092 (2001); Low Temp. Phys. **27**, 807 (2001).

[8] A.I. Yanson, I.K. Yanson, and J.M. van Ruitenbeek, Phys. Rev. Lett. **84**, 5832 (2000).

[9] J.A. Torres and J.J. Saenz, Phys. Rev. Lett. **77**, 2245 (1996).

[10] B. Ludoph, M.H. Devoret, D. Esteve, C. Urbina, and J.M. van Ruitenbeek. Phys. Rev. Lett. **82**, 1530 (1999).

[11] C. Untiedt, G. Rubio, S. Vieira, and N. Agraït, Phys. Rev. B **56**, 2154 (1997).

CONDUCTANCE OF NANOSYSTEMS WITH INTERACTION

A. Ramšak[1,2] and T. Rejec[1]
[1] *Jožef Stefan Institute, Ljubljana, Slovenia*
[2] *Faculty of Mathematics and Physics, University of Ljubljana, Ljubljana, Slovenia*
anton.ramsak@fmf.uni-lj.si

Abstract The zero-temperature linear response conductance through an interacting mesoscopic region attached to noninteracting leads is investigated. We present a set of formulas expressing the conductance in terms of the ground-state energy of an auxiliary system, namely a ring threaded by a magnetic flux and containing the correlated electron region. We prove that the formalism is exact if the ground state of the system is a Fermi liquid. We show that in such systems the ground-state energy is a universal function of the magnetic flux, where the conductance is the relevant parameter. The method is illustrated with results for the transport through an interacting quantum dot and a simple Aharonov-Bohm ring with Kondo-Fano resonance physics.

Keywords: Conductance; Nanosystems; quantum dot

1. Introduction

In the last decade technological advances enabled controlled fabrication of small regions connected to leads and the conductance, relating the current through such a system to the voltage applied between the leads, proved to be the most important property of such systems. There is a number of such examples, e.g. metallic islands prepared by e-beam lithography or small metallic grains,[1] semiconductor quantum dots,[2] or a single large molecule such as a carbon nanotube or DNA. It is possible to break a metallic contact and measure the transport properties of an atomic-size bridge that forms in the break,[3] or even measure the conductance of a single hydrogen molecule. [4] Recent measurements of conductance through single molecules proved that strong electron correlations can play an important role in such systems. [5]

The transport in noninteracting mesoscopic systems is theoretically well described in the framework of the Landauer-Büttiker formalism. The conductance G is at zero temperature determined with the Landauer-Büttiker formula [6]

A.S. Alexandrov et al. (eds.), Molecular Nanowires and Other Quantum Objects, 255–268.

$$G = G_0 \left| t\left(\epsilon_F\right)\right|^2, \qquad G_0 = \frac{2e^2}{h}. \tag{1}$$

The key quantity here is the single particle transmission amplitude $t(\epsilon_F)$ for electrons at the Fermi energy. The formula proved to be very useful and reliable, as long as electron-electron interaction in a sample is negligible. However, the Landauer-Büttiker formalism cannot be directly applied to systems where the interaction between electrons plays an important role. Several approaches have been developed to allow one to treat such systems. The Kubo formalism provides us with a conductance formula which is applicable in the linear response regime and was intensively studied by Oguri.[7, 8] A more general approach applicable also to non-equilibrium cases was developed by Meir and Wingreen. [9] Recently, ab initio methods to study the transport through small molecular junctions were also applied.[10]

2. Conductance formulas for Fermi liquid systems

The relevant system is schematically presented in Fig. 1(a). A mesoscopic interacting region, which could be a molecule, a quantum dot, a quantum dot array or a similar 'artificial molecule' system, is attached to noninteracting leads. As shown in Ref. [11] (hereafter referred to as RR), the conductance of such a system can be determined solely from the ground-state energy of an auxiliary system, formed by connecting the leads of the original system into a ring and threaded by a magnetic flux, Fig. 1(b). The main advantage of this method is the fact that it is often much easier to calculate the ground-state energy (for example, using variational or quantum Monte Carlo methods) than the Green's function, which is needed in the Kubo and Keldysh approaches. The method is applicable only to a certain class of systems, namely to those exhibiting Fermi liquid properties, at zero temperature and in the linear response regime. However, in this quite restrictive domain of validity, the method promises to be easier to use than the methods mentioned above.

The basic property that characterizes Fermi liquid systems is that the states of a noninteracting system of electrons are continuously transformed into states of the interacting system as the interaction strength increases from zero to its actual value.[12] One can then study the properties of such a system by means of the perturbation theory, regarding the interaction strength as the perturbation parameter. Dynamics of Fermi liquid systems at low temperature and in the linear response regime is governed by quasiparticles. However, the question how quasiparticles propagate in a correlated system is a non-trivial one. The answer can be extracted from the Green's function for a particular problem if it is known. An alternative way, which we advocate in this paper, is to analyze the excitation spectrum of a system directly. If E_M and E_{M+1} are the ground-state energies of an interacting ring system containing M and $M+1$ electrons,

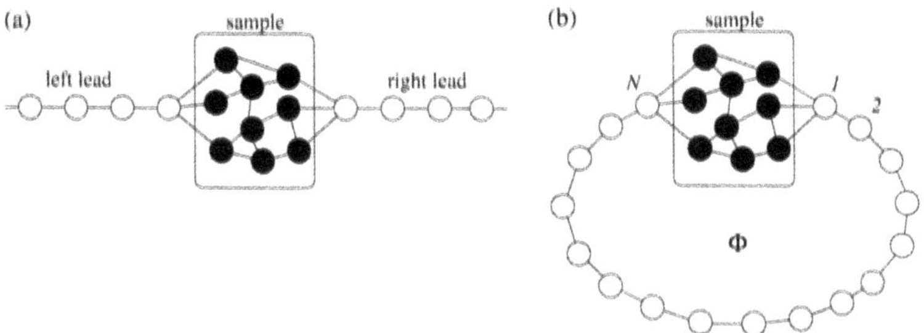

Figure 1. (a) Schematic picture of a sample with interaction connected to noninteracting leads. (b) The sample embedded in a ring formed by joining the left and right leads of the system (a). Auxiliary magnetic flux $\Phi = \frac{\hbar}{e}\phi$ penetrates the ring.

respectively, the energy difference can be attributed to the first quasiparticle energy level $\tilde{\epsilon}$ above the Fermi energy,

$$\tilde{\epsilon} = E_{M+1} - E_M. \tag{2}$$

The variation of the quasiparticle energy with flux threading the ring allows us to determine the conductance of the system. The complete proof of the formalism is given in RR and a brief overview is presented in the next Section. Here we show how the method can be implemented in practice.

The key property of ring systems presented in Fig. 1(b) is the universality expressed in the variation of the ground-state energy with auxiliary magnetic flux through the ring. Here we assume a system obeys the time reversal symmetry. The more general case is presented in the last Section. For an even number of electrons in the system and a large number of sites in the ring $N \to \infty$ the ground-state energy takes a universal form

$$E(\phi) - E\left(\frac{\pi}{2}\right) = \frac{\Delta}{\pi^2}\left(\arccos^2\left(\mp\sqrt{g}\cos\phi\right) - \frac{\pi^2}{4}\right), \tag{3}$$

where the average level spacing at the Fermi energy $\Delta = [N\rho(\epsilon_F)]^{-1}$ is determined by the density of states at the Fermi energy in an infinite noninteracting lead $\rho(\epsilon_F)$ and $g = G/G_0$ is the dimensionless conductance. For systems with an odd number of electrons, the ground-state energy is given with

$$E(\phi) - E\left(\frac{\pi}{2}\right) = \frac{\Delta}{\pi^2}\arcsin^2\left(\sqrt{g}\cos\phi\right). \tag{4}$$

It should be mentioned that the ground-state energy of an interacting ring system exactly corresponds to the expression for persistent currents in noninteracting rings, as derived by Gogolin. [14] The only parameter determining the ground-state energy is the conductance g of the original system, Fig. 1(a). In Fig. 2 the ground-state energy as a function of the flux ϕ for even and odd numbers of electrons is presented. It should be pointed out that as successive quasiparticle levels are being occupied, the points $\phi = 0$ and $\phi = \pi$ interchange their roles (as is also the case for noninteracting systems), and that the periodicity in even and odd cases are π and $\pi/2$, respectively.

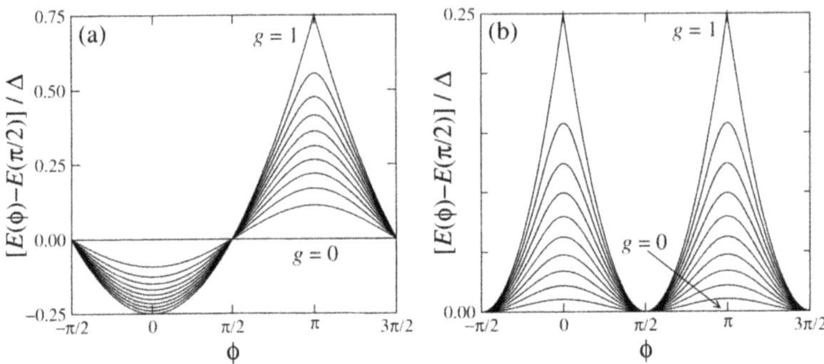

Figure 2. The ground-state energy of an interacting system as a function of flux ϕ for an even (a) and an odd (b) number of electrons for g going from 0 to 1 in steps of 0.1 and $N \to \infty$.

If the ground state of the system in Fig. 1(b) is known, the conductance of the original open system can be extracted from Eq. (3) [or Eq. (4)]. There are several ways how to determine g from Eq. (3). The simplest seems to be the use of the relation

$$g = \sin^2 \left(\frac{\pi}{2} \frac{E(\pi) - E(0)}{\Delta} \right), \tag{5}$$

where $E(0)$ and $E(\pi)$ are the ground-state energies of the ring system for $\phi = 0$ and $\phi = \pi$, respectively. The first advantage of this formula is the fact that the energies can be calculated using periodic and antiperiodic boundary conditions, respectively, and thus the wave functions of the system can be taken real. Additional advantage is fast convergence with N, as briefly discussed in the last Section. This formula was derived in the $g \to 0$ limit as $g = \left(\frac{\pi}{2\Delta}[E(\pi) - E(0)]\right)^2$ by Favand and Mila for noninteracting systems and applied to interacting Hubbard chains.[15] More recently, a similar approach was performed in Ref. [16]. In Fig. 3(a) the use of formula Eq. (5) is schematically presented.

The derivative of the ground-state energy with respect to flux gives the persistent current in the ring $j(\phi) = \frac{e}{\hbar}\frac{\partial E}{\partial \phi}$. [17] The second formula relates the conductance to the persistent current at $\phi = \frac{\pi}{2}$,

$$g = \left(\frac{\pi}{\Delta} \frac{\hbar}{e} j\left(\frac{\pi}{2}\right) \right)^2 . \tag{6}$$

This relation was recently derived for noninteracting systems [18] and successfully applied to systems with interaction.[18, 19]

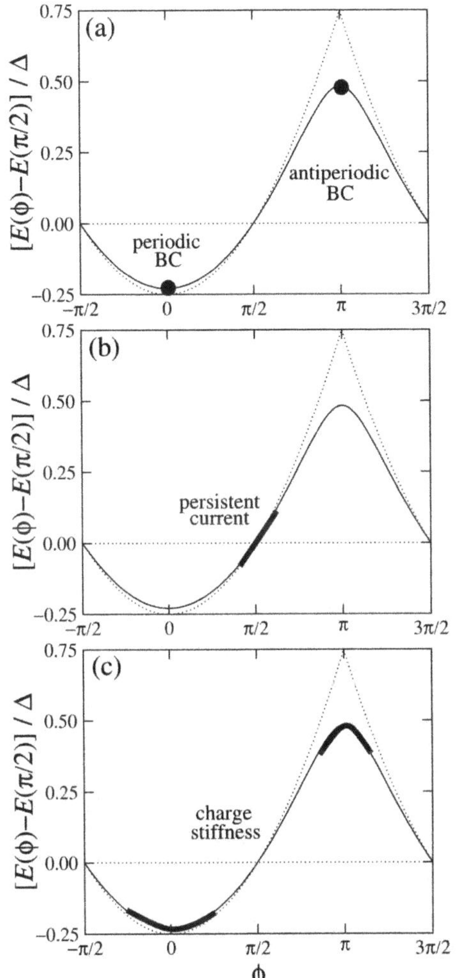

Figure 3 The conductance can be extracted from: (a) the two-point formula, Eq. (5), (b) the persistent current formula, Eq. (6), or (c) the charge stiffness formula, Eq. (7).

At $T = 0$ the charge stiffness is an important quantity describing the charge transport in correlated systems.[20] It is defined as the second derivative of the ground-state energy of the system with respect to the flux in the minimum of the energy vs. flux curve, $D = \frac{N}{2}\,\partial^2 E/\partial\phi^2|_{E=\min}$.[21] The sensitivity of the ground state energy to flux has been applied also in the context of electron localisation.[22] One can also define the corresponding quantity for the energy maximum as $\tilde{D} = -\frac{N}{2}\,\partial^2 E/\partial\phi^2|_{E=\max}$. From Eq. (3) the conductance can be related to the charge stiffness with an implicit relation,

$$\frac{1}{\Delta}\,\frac{\partial^2 E}{\partial\phi^2}\bigg|_{E=\min,\,\max} = \pm\frac{2}{\pi^2}\sqrt{\frac{g}{1-g}}\,\arccos\left(\pm\sqrt{g}\right). \tag{7}$$

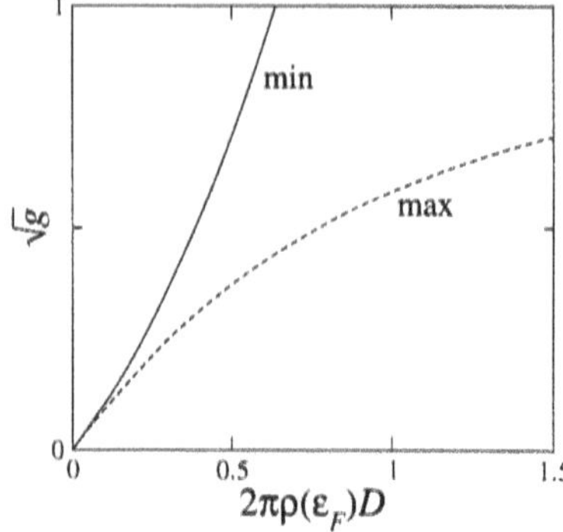

Figure 4 Conductance vs. charge stiffness using D at the energy minimum (full line) and $\tilde{D}$ at the maximum (dashed line). Note the quadratic dependence $g \propto D^2$ for $g \to 0$.

Here the upper and the lower signs correspond to the second derivative at a minimum and at a maximum of the energy vs. flux curve, respectively. In general, this equation has to be solved numerically and the solutions are presented in Fig. 4. In the limit of a very small conductance and in the vicinity of the unitary limit, analytic formulas are available

$$g = \begin{cases} \left[2\pi\rho\left(\varepsilon_F\right)D\right]^2, & g \to 0, \\ \left(\frac{1}{2} + \frac{3\pi}{4}\left[2\pi\rho\left(\varepsilon_F\right)D\right]\right)^2, & g \to 1. \end{cases} \tag{8}$$

Note that there is a quadratic relation between the conductance and the charge stiffness in the low conductance limit. The corresponding formulas for the maximum of the energy vs. flux curve are

$$g = \begin{cases} \left[2\pi\rho\left(\varepsilon_F\right)\tilde{D}\right]^2, & g \to 0, \\ \left(1 - \frac{2}{\left[2\pi\rho(\varepsilon_F)\tilde{D}\right]^2}\right)^2, & g \to 1. \end{cases} \tag{9}$$

It should be stressed that the validity of all formulas presented in this Section is based on an assumption that the number of sites in the ring is sufficiently large according to the condition [11]

$$N \gg \frac{1}{\rho\left(\varepsilon_F\right)} \frac{\partial\sqrt{g\left(\varepsilon_F\right)}}{\partial\varepsilon_F}. \tag{10}$$

This means that if $g\left(\varepsilon_F\right)$ exhibits sharp resonances, as is the case, e.g., in chaotic systems,[23] the calculation has to be performed on a large auxiliary ring system and in such cases the method might be impractical compared to other methods. On the other hand, for systems with strong interaction the method promises to be extremely efficient already for ring systems of a moderate size. [24, 16, 19]

3. Proof of the formalism

The complete proof of the formalism is presented in RR, here we briefly describe the main steps. The proof strongly relies on an assumption that the ground state of the system under investigation is a Fermi liquid.[25]

We start with the linear response conductance of a general interacting system of the type shown in Fig. 1(a). The conductance can be calculated from the Kubo formula [26]

$$g = \lim_{\omega \to 0} \frac{i\pi}{\omega + i\delta} \Pi_{II} (\omega + i\delta), \tag{11}$$

where $\Pi_{II}(\omega + i\delta)$ is the retarded current-current correlation function. For Fermi liquid systems, the current-current correlation function can be calculated within the perturbation theory. At $T = 0$, only the bubble diagram gives a non-vanishing contribution [8] and the conductance can be expressed in terms of the Green's function $G_{n'n}(z)$ of the system,

$$g = \left| \frac{1}{-i\pi\rho(\epsilon_F)} e^{-ik_F(n'-n)} G_{n'n}(\epsilon_F + i\delta) \right|^2, \tag{12}$$

where n and n' are sites in the left and the right lead, respectively.

In Fermi liquid systems obeying the time-reversal symmetry,[13] the imaginary part of the retarded self-energy at $T = 0$ vanishes at the Fermi energy and is quadratic for frequencies close to the Fermi energy.[27] Using the Fermi energy as the origin of the energy scale, i.e. $\omega - \epsilon_F \to \omega$, we can express this as

$$\mathrm{Im}\Sigma_{ij}(\omega + i\delta) \propto \omega^2. \tag{13}$$

Close to the Fermi energy, the self-energy can be expanded in powers of ω resulting in an approximation to the Green's function,

$$\mathbf{G}^{-1}(\omega + i\delta) = \omega\mathbf{1} - \mathbf{H}^{(0)} - \mathbf{\Sigma}(0 + i\delta) - \\ -\omega \left. \frac{\partial \mathbf{\Sigma}(\omega + i\delta)}{\partial \omega} \right|_{\omega=0} + \mathcal{O}\left(\omega^2\right). \tag{14}$$

Here $\mathbf{H}^{(0)}$ contains matrix elements of the noninteracting part of the Hamiltonian. The Green's function for ω close to the Fermi energy can then be expressed as

$$\mathbf{G}^{-1}(\omega + i\delta) = \mathbf{Z}^{-1/2} \tilde{\mathbf{G}}^{-1}(\omega + i\delta) \mathbf{Z}^{-1/2} + \mathcal{O}\left(\omega^2\right), \tag{15}$$

where we defined the quasiparticle Green's function

$$\tilde{\mathbf{G}}^{-1}(\omega + i\delta) = \omega\mathbf{1} - \tilde{\mathbf{H}} \tag{16}$$

as the Green's function of a *noninteracting quasiparticle* Hamiltonian

$$\tilde{\mathbf{H}} = \mathbf{Z}^{1/2} \left[\mathbf{H}^{(0)} + \mathbf{\Sigma} \left(0 + i\delta \right) \right] \mathbf{Z}^{1/2}, \tag{17}$$

and introduced the renormalization factor matrix $\mathbf{Z}$. Matrix elements of $\mathbf{Z}$ differ from those of an identity matrix only if they correspond to sites of the central region.

The reason for introducing the quasiparticle Hamiltonian is to obtain an alternative expression for the conductance in terms of the quasiparticle Green's function. Eq. (15) relates the values of the true and the quasiparticle Green's function at the Fermi energy,

$$\mathbf{G} \left(0 + i\delta \right) = \mathbf{Z}^{1/2} \tilde{\mathbf{G}} \left(0 + i\delta \right) \mathbf{Z}^{1/2}. \tag{18}$$

Specifically, $G_{n'n} \left(0 + i\delta \right) = \tilde{G}_{n'n} \left(0 + i\delta \right)$ if both n and n' are sites in the leads, as a consequence of the properties of the renormalization factor matrix $\mathbf{Z}$ discussed above. Eq. (12) then tells us that the zero-temperature conductance of a Fermi liquid system is identical to the zero-temperature conductance of a noninteracting system defined with the quasiparticle Hamiltonian for a given value of the Fermi energy.

These conclusions are valid if the central region is coupled to semi-infinite leads. Here we generalize the concept of quasiparticles to a finite ring system with N sites and M electrons, threaded by a magnetic flux ϕ. One can define the quasiparticle Hamiltonian for such a system,

$$\tilde{\mathbf{H}} \left(N, \phi; M \right) = \mathbf{Z}^{1/2} \left[\mathbf{H}^{(0)} \left(N, \phi \right) + \mathbf{\Sigma} \left(0 + i\delta \right) \right] \mathbf{Z}^{1/2}. \tag{19}$$

Here the self-energy and the renormalization factor matrix are determined in the thermodynamic limit where, as we prove in RR, they are independent of ϕ and correspond to those of an infinite two-lead system.

Suppose now that we knew the exact values of the renormalized matrix elements in the quasiparticle Hamiltonian (19). As this is a noninteracting Hamiltonian, we could then apply the conductance formulas presented in the previous Section (the proof of validity of energy formulas for noninteracting systems is given in RR and the corresponding result for persistent currents in Ref. [14]). Such a procedure would provide us with the exact conductance of the original interacting system. However, to obtain the values of the renormalized matrix elements, one needs to calculate the self-energy of the system, which is a difficult many-body problem. In RR we study the excitation spectrum of a finite ring system with interaction and threaded with a magnetic flux. We show that

$$\begin{aligned} E \left(N, \phi; M + 1 \right) - E \left(N, \phi; M \right) = \\ = \tilde{\varepsilon} \left(N, \phi; M; 1 \right) + \mathcal{O} \left(N^{-\frac{3}{2}} \right), \end{aligned} \tag{20}$$

where $E(N,\phi;M)$ and $E(N,\phi;M+1)$ are the ground-state energies of the interacting Hamiltonian for a ring system with N sites and flux ϕ, containing M and $M+1$ electrons, respectively, and $\tilde{\epsilon}(N,\phi;M;1)$ is the energy of the first single-electron level above the Fermi energy of the finite ring quasiparticle Hamiltonian (19). The error in Eq. (20) is small enough that the proof of the ground-state energy formulas for noninteracting systems, which involves only the properties of a set of neighboring single-electron energy levels, remains valid also for interacting Fermi liquid systems, provided a system is a Fermi liquid for all values of the Fermi energy below its actual value.

4. Examples

Noninteracting system

In this Section we discuss the convergence properties of the conductance formulas. As the first example we take a double-barrier potential scattering problem presented in Fig. 5. Results of various formulas for different number of sites in the ring are presented in Fig. 6.

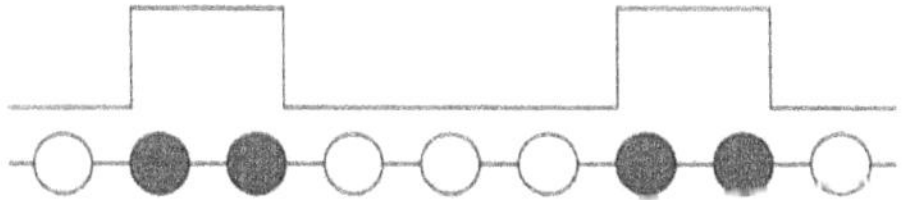

Figure 5. A double barrier noninteracting system. The height of the barriers is $0.5t$, where t is the hopping matrix element between neighboring sites.

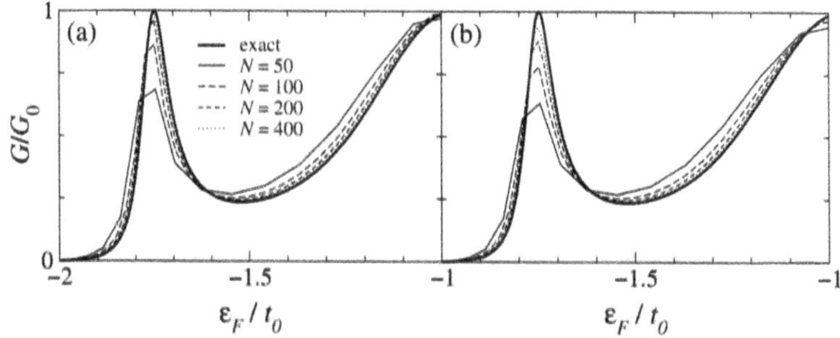

Figure 6. (a) The conductance through the system in Fig. 5 calculated from the two-point formula, Eq. (5), and (b) from the persistent current formula, Eq. (6). Note the different convergence behavior of the two formulas.

The exact zero-temperature conductance for this system exhibits a sharp resonance peak superimposed on a smooth background conductance. We notice immediately that as the number of sites in the ring increases, the convergence is generally faster in the region where the conductance is smooth than in the resonance region, which is consistent with the condition Eq. (10). Comparing the results obtained employing the two-point formula Eq. (5) and the persistent current formula Eq. (6) we observe that the convergence is better for the two-point formula expressing the conductance in terms of the difference of the energies at $\phi = 0$ and $\phi = \pi$. From the computational point of view there is an additional advantage of the two-point formula. In this case, all the matrix elements can be made real if one chooses such a vector potential that only one hopping matrix element if modified by the flux as then the additional phase factor is $e^{\pm i\pi} = -1$.

Anderson impurity model

As a nontrivial example of the use of the formalism we calculate the zero-temperature conductance of a single impurity Anderson model realized as a quantum dot attached to leads as shown in Fig. 7.

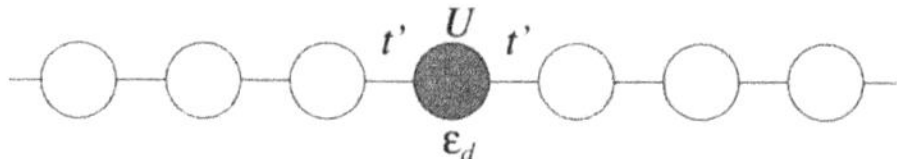

Figure 7. The Anderson impurity model realized as a quantum dot coupled to two leads. The dot is described with the energy level ϵ_d and the Coulomb energy of a doubly occupied level U. t' is the hopping between the dot and leads.

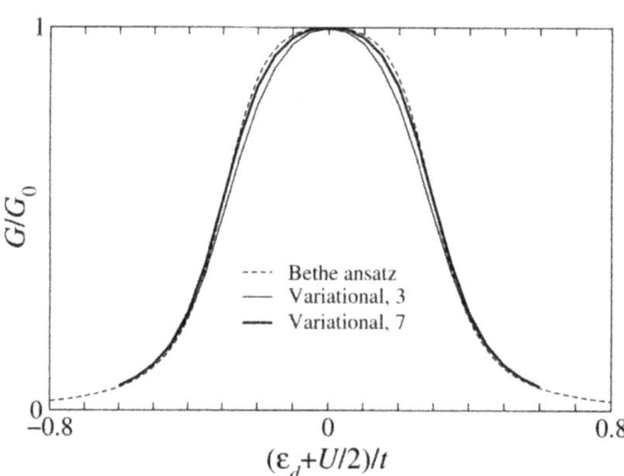

Figure 8 The zero-temperature conductance calculated from ground-state energy vs. magnetic flux in a finite ring system using the variational method described in RR with 3 and 7 basis functions. For comparison, the exact Bethe ansatz result is presented with a dashed line. The system is shown in Fig. 7, with $U = 0.64t$ and $t' = 0.2t$.

In Fig. 8 the results are compared to exact conductance of the Bethe ansatz approach.[29] To calculate the conductance, Eq. (5) was used, with the ground-state energies at $\phi = 0$ and $\phi = \pi$ obtained using a variational method described in RR. For each position of the ϵ_d level relative to the Fermi energy, we increased the number of sites in the ring until the conductance converged. The number of sites needed to achieve the convergence was the lowest in the empty orbital regime and the highest (about 1000 for the system shown in Fig. 8) in the Kondo regime. This is a consequence of Eq. (10) as a narrow resonance related to the Kondo resonance appears in the transmission probability of the quasiparticle Hamiltonian (17) in the Kondo regime.

Interacting Aharonov Bohm rings

One can generalize the conductance formulas to systems which exhibit time reversal asymmetry, such as is e.g. an Aharonov-Bohm (AB) type of system presented in Fig. 9(a).[13]

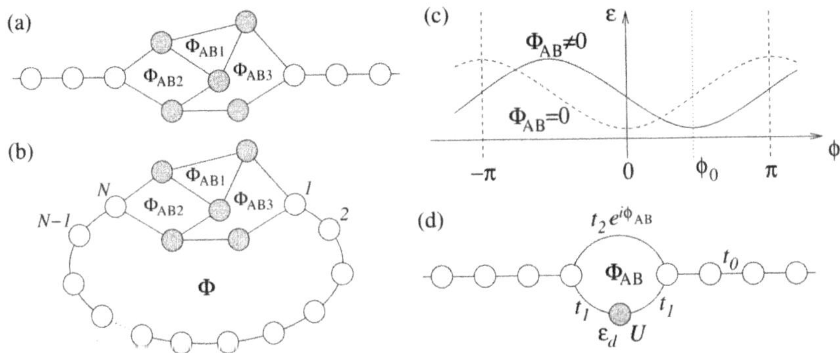

Figure 9. (a) Interacting mesoscopic region (gray-shaded sites), threaded by magnetic flux and coupled to noninteracting leads. (b) Auxiliary ring system. (c) Behavior of energy levels as the flux threading the ring is varied. (d) Example system with a quantum dot embedded in an AB ring.

If there is no AB flux threading the mesoscopic region, the time reversal symmetry is restored and the energy is an even function of ϕ. In the general case, the energy extremum is shifted to a non-trivial point $\phi_0\left(\varepsilon_F\right)$, as illustrated in Fig. 9(c). The ground-state energy is then generalized to $E(\phi) = \pi^{-2}\Delta \arccos^2\left(\mp\sqrt{g}\cos\left[\phi-\phi_0\left(\varepsilon_F\right)\right]\right) + \text{const.}$ for an even number of electrons in a system and to $E(\phi) = \pi^{-2}\Delta \arcsin^2\left(\sqrt{g}\cos\left[\phi-\phi_0\left(\varepsilon_F\right)\right]\right)+\text{const.}$ for an odd number of electrons in a system. From the former expression, the transmission probability can be extracted, and the conductance is given by [13]

$$g = \sin^2\left(\frac{\pi}{2}\frac{E\left(\phi_0+\pi\right)-E\left(\phi_0\right)}{\Delta}\right), \tag{21}$$

where $\phi_0 \equiv \phi_0\left(\epsilon_F\right)$ is determined by the position of the minimum (or maximum) in the energy vs. flux curve. The conductance can also be calculated from the more convenient four-point formula [32]

$$g = \sin^2\left(\frac{\pi}{2}\frac{E(\pi)-E(0)}{\Delta}\right)+\sin^2\left(\frac{\pi}{2}\frac{E(\pi/2)-E(-\pi/2)}{\Delta}\right), \tag{22}$$

and ϕ_0 is determined with the expression

$$\phi_0 = -\arctan \frac{\sin\left(\frac{\pi}{2}\frac{E(\pi/2)-E(-\pi/2)}{\Delta}\right)}{\sin\left(\frac{\pi}{2}\frac{E(\pi)-E(0)}{\Delta}\right)}. \tag{23}$$

If no AB flux is present in the mesoscopic region, we recover the two-point formula Eq. (5) since $E(\pi/2) = E(-\pi/2)$ in this case.

If the time reversal symmetry is broken due to AB flux, Eq. (13) is not valid and the proof of the previous Section has to be reconsidered. Repeating the steps as presented in detail in RR, the proof is restored and basically unchanged if the self energy obeys the relation

$$\frac{1}{2i}\left[\Sigma_{ij}(\omega + i\delta) - \Sigma_{ij}(\omega - i\delta)\right] \propto \omega^2. \tag{24}$$

It follows that the linear response conductance of an interacting AB system at zero temperature is given by the four-point formula Eq. (22). The condition Eq. (24) is fulfilled if the system is a Fermi liquid.

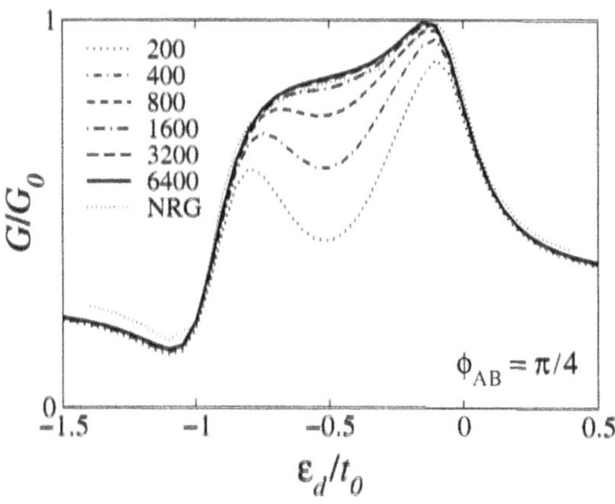

Figure 10 Zero-temperature linear response conductance of the system in Fig. 9(d) as a function of level position ε_d for various number of ring sites N and $\phi_{AB} = \pi/4$. The dotted line is the NRG result from Ref. [31]. Parameters: $t_1 = 0.177t_0$ ($\Gamma = 0.125t_0$), $t_2 = 0.298t_0$, $U = t_0 = 8\Gamma$.

In order to demonstrate the practical value of the method, we quantitatively analyze the conductance through an Aharonov-Bohm ring with a quantum dot placed in one of the arms[30, 31] as presented in Fig. 9(d). In Fig. 10 a convergence test of the method is shown. The convergence with N is fast in the empty orbital regime and becomes progressively slower as ϵ_d shifts toward the Kondo regime, as was also the case for a quantum dot attached to leads. The converged conductance curve is in excellent agreement with the numerical renormalization group result of Ref. [31].

5. Summary

We have demonstrated how the zero-temperature conductance of a sample with electron-electron interaction, attached to noninteracting leads can be determined. The method is extremely simple and is based on several formulas relating the conductance to the ground-state energy of an auxiliary ring system. The conductance is determined from the ground-state energy of an interacting system, while in more traditional approaches, one needs to know the Green's function of the system. The advantage of the present method comes from the fact that the ground-state energy is often relatively simple to obtain by various numerical approaches, including variational methods. Let us summarize the key points of the method:

(1) The "open" problem of the conductance through a sample coupled to semi-infinite leads is mapped on to a "closed" problem, namely a ring threaded by a magnetic flux and containing the same correlated electron region.

(2) For the case of a Fermi liquid interacting system, even with broken time-reversal symmetry, it is shown that the zero-temperature conductance can be deduced from the variation of the ground state energy with the flux in a large, but finite ring system.

(3) In order to prove this, the concept of Fermi liquid quasiparticles is extended to finite, but large systems. The conductance formulas give the conductance of a system of noninteracting quasiparticles, which is equal to the conductance of the original interacting system.

(4) The results of our method are compared to results of other approaches for problems such as the transport through a quantum dot containing interacting electrons. The comparison shows an excellent quantitative agreement with exact Bethe ansatz results. We have demonstrated the usefulness of the formula also by applying it to a prototype system exhibiting Kondo-Fano behavior. Results based on the four-point formula confirm the results of the numerical renormalization group method.

References

[1] J. von Delft and D. C. Ralph, Phys. Rep. **345**, 61 (2001).

[2] L. P. Kouwenhoven *et al.*, in *Mesoscopic Electron Transport*, edited by L. L. Sohn, L. P. Kouwenhoven, and G. Schön (Kluwer Academic, New York, 1997).

[3] N. Agraït, A. L. Yeyati, and J. M. van Ruitenbeek, Phys. Rep. **377**, 81 (2003).

[4] R. H. M. Smit *et al.*, Nature **419**, 906 (2002).

[5] W. Liang *et al.*, Nature **417**, 725 (2002).

[6] R. Landauer, IBM J. Res. Dev. **1**, 233 (1957); Philos. Mag. **21**, 863 (1970); M. Büttiker, Phys. Rev. Lett. **57**, 1761 (1986).

[7] A. Oguri, Phys. Rev. B **56**, 13422 (1997).

[8] A. Oguri, J. Phys. Soc. Jpn. **70**, 2666 (2001).

[9] Y. Meir and N. S. Wingreen, Phys. Rev. Lett. **68**, 2512 (1992); H. M. Pastawski, Phys. Rev. B **46**, 4053 (1992).

[10] N. D. Lang and Ph. Avouris, Phys. Rev. Lett. **84**, 358 (2000); J. Heurich *et al., ibid.* **88**, 256803 (2002).

[11] T. Rejec and A. Ramšak, Phys. Rev. B **68**, 035342 (2003).

[12] P. Nozières, *Theory of interacting Fermi systems* (W. A. Bejnamin, Inc., New York, 1964).

[13] T. Rejec and A. Ramšak, Phys. Rev. B **68**, 033306 (2003).

[14] A. O. Gogolin and N. V. Prokof'ev, Phys. Rev. B **50**, 4921 (1994).

[15] J. Favand and F. Mila, Eur. Phys. J. B **2**, 293 (1998).

[16] R. A. Molina *et al.*, Phys. Rev. B **67**, 235306 (2003).

[17] F. Bloch, Phys. Rev. **137**, A787 (1965); M. Büttiker, Y. Imry, and R. Landauer, Phys. Lett. **96A**, 365 (1983).

[18] O. P. Sushkov, Phys. Rev. B **64**, 155319 (2001).

[19] V. Meden and U. Schollwöck, Phys. Rev. B **67**, 193303 (2003).

[20] P. Prelovšek and X. Zotos, in *Lectures on the Physics of Highly Correlated Electron Systems VI*, edited by F. Mancini (American Institute of Physics, New York, 2002).

[21] W. Kohn, Phys. Rev. **133**, A171 (1964).

[22] J.T. Edwards and D.J. Thouless, J. Phys. C**5**, 807 (1972).

[23] R. A. Jalabert, A. D. Stone, and Y. Alhassid, Phys. Rev. Lett. **68**, 3468 (1992); J. A. Verges, E. Cuevas, M. Ortuño, and E. Louis, Phys. Rev. B **58**, R10143 (1998).

[24] J. Bonča, A. Ramšak, and T. Rejec, unpublished.

[25] L. D. Landau, Pis'ma Zh. Eksp. Teor. Fiz. **3**, 920 (1956); **5**, 101 (1957); P. Noziéres, J. Low Temp. Phys. **17**, 31 (1974).

[26] R. Kubo, J. Phys. Soc. Jpn. **12**, 570 (1957).

[27] K. Yamada and K. Yosida, Prog. Theor. Phys. **76**, 621 (1986); A. Oguri, J. Phys. Soc. Jpn. **66**, 1427 (1997).

[28] D. S. Fisher and P. A. Lee, Phys. Rev. B **23**, 6851 (1981).

[29] P. B. Wiegman and A. M. Tsvelick, Pis'ma Zh. Eksp. Teor. Fiz. **35**, 100 (1982); J. Phys. C **16**, 2281 (1983).

[30] B. R. Bułka and P. Stefanski, Phys. Rev. Lett. **86**, 5128 (2001).

[31] W. Hofstetter, J. König and H. Schoeller, Phys. Rev. Lett. **87**, 156803 (2001).

[32] It is also possible to calculate the conductance from the ground-state energies at three distinct values of flux ϕ by solving equations numerically.

STM IMAGING OF VORTEX STRUCTURES IN THIN FILMS

A. Troyanovsky[1], G. Van Baarle[2], T. Nishizaki[3], J. Aarts[2], P. Kes[2]
[1]*Kapitza Institute for Physical Problems, Russian Academy of Sciences, Kosygina 2, Moscow, 119334, Russia*
[2]*Kamerlingh Onnes Laboratory, Leiden University, PO Box 9504, 2300 RA Leiden, The Netherlands*
[3] *Institute for Materials Research, Tohoku University, Sendai 980-8577, Japan*
troyan@kapitza.ras.ru

Abstract We have developed the techniques of imaging vortices in thin superconducting films using surface passivation with Au thin layer. We applied our technique to study vortex configurations in thin film with weak vortex pinning (a-$Mo_{2.7}Ge$) and a strongly pinning material (NbN) at 4.2K in magnetic fields up to 1.4T.

Keywords: thin film, superconductor, vortex, STM, imaging

1. Introduction

Mesoscopic superconductors of the size comparable to the coherence length can have a small number of vortices and such systems reveal unusual vortex formations – multiquanta vortices and vortex molecules [1,2]. But the experimental studies of these systems, limited by a few experiments based on magnetization or transport measurements without resolving the real structure of vortex configurations. The magnetic distribution over the surface of mesoscopic superconductors is almost uniform due to the small size, thus it is almost impossible to use imaging techniques based on sensitivity to magnetic field such as magnetic decorations [3], scanning Hall microscopy [4], Lorentz microscopy [5], magneto-optics [6] and some others. The principal advantage of scanning tunneling microscopy (STM) in comparison with other imaging techniques is that STM measures the electronic density of states instead of the magnetic flux density, thus being sensitive on the scale of the coherence length, rather than the magnetic penetration depth. It gives possibility to study vortices in superconductors at large magnetic fields region almost up to H_{C2} with high spatial resolution. Because of this feature STM can be an excellent tool to

A.S. Alexandrov et al. (eds.), Molecular Nanowires and Other Quantum Objects, 269–274.

study vortex configuration in mesoscopic superconductors. However due to the requirements of having a clean, almost atomically flat surface, the STM was successfully used to study vortices in only several types of single-crystalline materials. Well-known examples are $NbSe_2$ [7,8], YBCO [9], BSCCO [10], (Lu/Y)Ni_2B_2C [11,12] and MgB_2 [13] crystals, while in thin films we know of only one report on observing vortex signatures [14].

In this work we describe our technique, which makes imaging vortex configurations in superconducting thin films possible by STM. The main idea is to use surface passivation with an ultrathin protection layer to overcome such major obstacles as surface roughness and oxidation, which make tunneling and scanning by STM impossible. We chose a-Mo_3Ge as superconducting material because of the extremely flat surface characterisic of amorphous films [15]. Moreover, the low intrinsic pinning properties ($J_C \sim 10^6$ A/m^2), the superconducting transition temperature T_C of 6.5K and the upper critical field ~6 T (at 4.2K) make a-Mo_3Ge a suitable candidate for vortex experiments. As a sample with polycrystalline structure we chose a NbN film. This strongly pinning material has T_C of 11K and H_{C2} of about 8.8T (at 4.2K) but it has J_C at least two orders of magnitude higher than a-Mo_3Ge [16].

2. Experimental technique

The 50-nm thick Mo_3Ge films were rf-sputtered from a composite Mo/Ge target in Ar atmosphere on Si substrate. Atomic force microscopy (AFM) investigation of a freshly sputtered Mo_3Ge film showed the surface with a roughness <0.4 nm over the scan range of ~0.5 μm. Next, we deposited a 5 nm Au capping layer immediately after depositing the a-$Mo_{2.7}Ge$. This yielded a rather flat, closed protection layer with a maximum height variation of the surface of around 3 nm on a 0.5x0.5 μm scan. The NbN films were sputtered from a Nb target in an Ar/N2 atmosphere. For a 50-nm NbN layer, passivation by a Au is possible, but better results are obtained by depositing intermediate 20 nm layer of a-Mo_3Ge followed by the Au capping layer. This results in a flatter surface than when Au is directly deposited on NbN, and it has a more uniform thickness, which is an advantage for the proximity effect.

After sputtering, the samples were mounted on the scanner of the STM at ambient condition, as a tip we used a PtIr wire. For our experiments we used the same STM setup as mentioned in [8]. The STM-head was designed as mechanically very stable compact unit with automatic coarse approach drive controlled by computer [17]. The STM was attached to the insert to a helium transport vessel and it was cooled down by slowly immersing it into liquid helium. A magnetic field was generated by a superconducting coil, which is permanently living in the helium transport vessel. The STM unit was directly immersed in the liquid helium, so a constant temperature 4.2K was ensured.

To carried out STM experiments we used commercial RHK control electronics with STM feedback loop. Tunneling contact was set with tunneling current 10-15 pA and a tip-sample bias of 1-2 mV.

3. Experimental results

To get vortex imaging we acquired full I-V tunneling spectra simultaneously with the topography measurement (constant current mode), typically we used a resolution of 128x128 points per image. From I-V curves we evaluated the ratio of zero-bias and high-bias (above superconducting gap Δ) slopes, providing information about the proximity-induced, local quasiparticale density of states at the position of the tip on the sample. This is rather slow way to get vortex image, the acquisition time was about 1 hour or more. In our previous experiments with $NbSe_2$ [8] we used constant average current mode, in that case the acquisition time was about 10 seconds. Now we could not use successively the same way for vortex imaging in thin films due to the high roughness of the surface of thin films. But quality of the surface was good enough for slow scanning and getting reliable surface topography. Experiments showed that the proximity-induced superconductivity is uniform on the Au surface of composed thin films in zero magnetic field.

In Fig.1 we present results obtained on a-$Mo_{2.7}$Ge at different magnetic fields – from 0.07 to 0.3T. The images demonstrate that we are able to clearly image vortex structure on thin film. These images were acquired at the same location of scanning area. At first we could observe the rotation of the vortex lattice (VL) on these images whereas at bulk crystal the orientation of VL is fixed [3]. At low field (up to $\sim$1.0T) we always obtained images of VL without defects, although the lattice as a whole may show some deformations (fig. 1, image at 0.2T). Since the scan direction is vertical, this deformation is a real property of the lattice rather than an artifact of the measurements. We propose that these deformations and rotations of the VL are due to the presence of large domains with different orientations that change in size and orientation after a field step.

The VL configuration in the case of NbN film was quite different. We always observed disordered configurations of vortices in this material as presented at Fig.2a. We did not found visible qualitative changes in VL structure of the NbN films up to 1.4T, the highest magnetic field measured. To observe the relaxation of VL after setting the field we acquired several consecutive images on the same location of the film, each having an acquisition time of around 90 minutes. Some of them had the same configuration of VL without visible changing of vortex position but on others we observed the changing position of vortices at restricted part of the scanning area. In Fig.2b we present a difference between two subsequent images acquired at the same place, this procedure makes rearrangement of vortices more visible The behavior of vortices described above

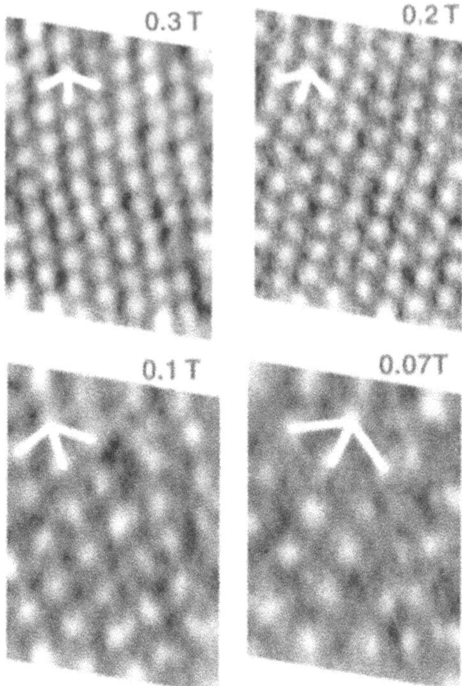

Figure 1. Images of vortex lattices in the a-Mo_3Ge film for various flux densities. Magnetic fields are indicated. Scanning area is about 700x860 nm^2. Lines on each image point the directions of the VL.

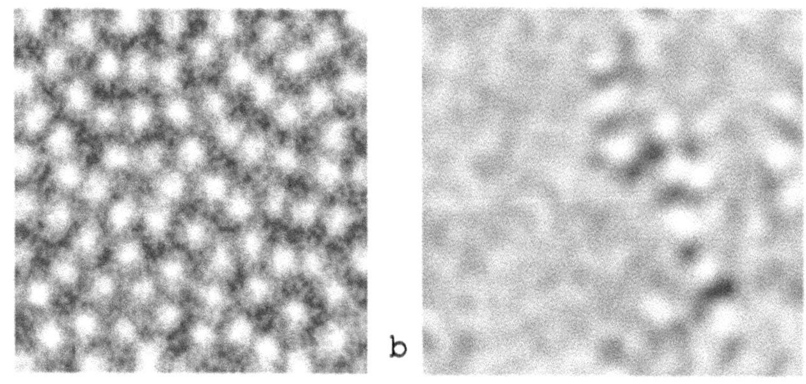

Figure 2. a).Vortex configuration in NbN thin film acquired at 0.5 T, 4.2 K, scan range is 700x500 nm^2. b). Difference between two consecutive images with time acquisition of about 90 min shows the changing of position vortices in time at 4.2 K and 0.5 T.

means that at measured fields (H/H_{C2}<0.15) grain boundary pinning is more important than the elastic interactions between vortices, leading to isolated vortex behavior [18].

4. Conclusions

We have developed a technique to image vortex configurations in superconducting thin films which were not accessible before. We used our technique to study two different types of superconducting films: amorphous $Mo_{2.7}Ge$ and polycrystalline NbN. We could get the clear images of vortex lattice on both samples. In $Mo_{2.7}Ge$ we observed well-ordered hexagonal lattice, whereas in NbN the vortex lattice was fully disordered. The technique can be used to observe vortex configurations in any type-II superconductor on which the protection conductance layer can be grown in a smooth layer. As a protection layer we used thin Au film but in principal any stable metal without thick oxide layer can be used. The imaging vortices in thin films without complicated surface cleaning procedures opens a way to study the microscopic behavior of the vortices in artificially structured samples.

Acknowledgments

These experiments were carried out in the Leiden University under FOM project (Stichting voor Fundamenteel Onder-zoek der Materie) and partially supported by JSPS (Japanese Society for Promotion of Science).

References

[1] A.S. Mel'nikov, V.M. Vinokur, Nature, **415** (2002), 60.

[2] L.F. Chibotaru, A. Ceulemans, V. Bruyndoncx, V.Moshchalkov, Nature, **408** (2000), 50.

[3] Essmann V., Trauble H., Phys. lett., A **24** (1967), 526.

[4] Chang A., Appl. Phys. Lett., **61** (1992), 1974.

[5] C. Sow, K.Harada, A.Tonomura, G.Crabtree and D.Grier, Phys. Rev. Lett., **80** (1998), 2693.

[6] P.Goa, H.Hauglin, M.Baziljevich, E.Il'yashenko, P.Gammel and T.Johansen, Supercond. Sci. Technol. **14** (2001), 729.

[7] H.F. Hess, R.B. Robinson, R.C. Dynes, J.M. Valles, J.V. Waszczak, Phys. Rev. Lett. **62** (1989), 214 .

[8] A. Troyanovski, J.Aarts, P.H.Kes, Nature, **399** (1999), 665.

[9] I. Maggio-Aprile, Ch. Renner, A. Erb, E. Walker, Ø. Fischer, Phys. Rev. Lett. **75** (1995), 2754.

[10] Ch. Renner, B. Revaz, K. Kadowaki, I. Maggio-Aprile, Ø. Fischer, Phys. Rev. Lett. **80** (1998), 3606.

[11] Y. de Wilde, M. Iavarone, U. Welp, V. Metlushko, A.E Koshelev, I. Aranson, G.W. Crabtree and P.C. Canfield, Phys. Rev. Lett., **78** (1977), 4273.

[12] H. Sakata, M.Oosawa, K. Matsuba, N.Nishida, H.Takeya and K. Hirata, Phys. Rev. Lett. **84** (2000), 1583.

[13] M.R. Eskilden, M.Kugler, G.Levy, S.Tanaka, J.Jun, S.M.Kazakov, J.Karpinski, Ø.Fischer, Physica C, **388-389** (2003), 143.

[14] S.Kashiwaya, M. Koyanagi and A. Shoji, Appl. Phys. Lett. **61** (1992), 1847.

[15] G.J.C. van Baarle, A.M. Troianovski, P.H.Kes, J.Aarts, Physica C, **369** (2002), 335.

[16] G.J.C. van Baarle, A.M. Troianovski, T.Nishizaki, P.H.Kes, and J.Aarts, Appl. Phys. Lett., **82** (2003), 7.

[17] Volodin A.P., Troyanovski A.M., Instr. Exp. Techn., **40** (1997), 724.

[18] A. Pruymboom, W. H.B. Hoondert, H.W. Zandbergen and P.H. Kes, Jpn. Jn. Appl. Phys. 26, Suppl. 26-3 (1987), 1531.

HYBRID SUPERCONDUCTOR/FERROMAGNET NANOSTRUCTURES

M. Lange, M. J. Van Bael, S. Raedts, V. V. Moshchalkov
Nanoscale Superconductivity and Magnetism Group, Laboratory for Solid State Physics and Magnetism, K.U.Leuven, Celestijnenlaan 200D, 3001 Leuven, Belgium
martin.lange@fys.kuleuven.ac.be

A. N. Grigorenko, S. J. Bending
Department of Physics, University of Bath, Claverton Down, Bath, BA2 7AY, United Kingdom

Abstract Vortex pinning is investigated in thin type-II superconducting Pb films covering arrays of Co/Pt dots with out-of-plane magnetization. After aligning all magnetic moments of the dots, the magnetization loop of the superconductor is strongly asymmetric with respect to the polarity of the applied field. Several matching effects at rational and integer multiples of the first matching field are observed as peaks or cusps in the magnetization loop. Local vortex imaging by scanning Hall probe microscopy reveals that the asymmetry of the magnetization loop can be ascribed to the attraction or repulsion of vortices by the dots, depending on the mutual orientation of the magnetic moments and the applied field.

Keywords: vortices, superconductors, ferromagnetic dots

1. Introduction

In the mixed state magnetic field penetrates type-II superconductors in the form of vortices (flux lines). The single-valueness of the superconducting order parameter ψ makes it necessary that the phase of ψ changes by integer multiples of 2π when following an arbitrary closed path in the superconductor. A non-zero value of such a change in phase means that the enclosed area contains a vortex. Thus, vortices in type-II superconductors are quantum objects. The motion of these vortices creates an electric field, causing dissipation and resulting in the appearance of a resistive state. Hence, it is of crucial importance for achieving high critical current densities j_c that all vortices in the superconductor are firmly pinned. Recent progress in nanostructuring has made it possible to create periodic artificial pinning centers in superconducting films such as arrays of microholes (*antidots*)[1, 2]. By tuning preparation parameters of the

A.S. Alexandrov et al. (eds.), Molecular Nanowires and Other Quantum Objects, 275–286.

artificial pinning sites (type, size, shape or distribution), one can optimize the pinning efficiency and enhance j_c up to its theoretical limit - depairing current. Matching effects between the vortex lattice and the array of pinning centers can be observed experimentally in electrical transport and magnetization measurements. These matching effects manifest themselves as anomalies in the field-dependence of resistivity, critical current, magnetization and susceptibility of the superconductor (for an overview, see [2] and references therein). Vortex pinning by artificial pinning arrays has also been investigated on a microscopic scale by local imaging techniques such as Lorentz microscopy [3] and scanning Hall probe microscopy (SHPM) [4, 5].
Investigations on magnetic pinning centers concentrated initially on dots with in-plane magnetization in contact with type-II superconducting Nb or Pb films [6, 7, 8, 9]. The main conclusion derived from these experiments was that the strong pinning is due to magnetic interactions between dots and vortices. Morgan and Ketterson reported on the use of Ni dots with an aspect ratio (height divided by diameter) of almost one, so that after out-of-plane saturation the major component of the magnetization can be expected to be perpendicular to the sample surface [10]. After this magnetization procedure, critical current measurements revealed that significantly stronger pinning is obtained when the applied field H and the magnetic moments m of the dots have the same polarity compared to the case of opposite polarity.
In this article we report on the vortex pinning properties of magnetic nanostructures that consist of Co/Pt multilayers with perpendicular magnetic anisotropy. The experiment of Morgan and Ketterson [10] already indicated that field polarity dependent effects can be expected because of the direct interaction of the magnetic moment of the dots with the field of the vortices. The dots used in this work have a clear perpendicular magnetic anisotropy and a small aspect ratio, so that the superconducting Pb film completely covers the dots and is non-perforated. By application of scanning Hall probe microscopy, we are able to investigate the vortex pinning on a microscopic scale, which will allow us to gain important information about the origin of the field polarity dependent pinning in these nanosystems [11].

2. Structural and magnetic properties of the magnetic dots

The investigated sample is a ferromagnetic/superconducting hybrid system consisting of an array of Co/Pt dots and a type-II superconducting Pb film. Details on sample preparation can be found elsewhere [11, 12, 13, 14]. An atomic force microscopy (AFM) image of a typical array of magnetic dots is shown in Fig. 1. The AFM image shows that the dot array has a very well defined structure. After structural and magnetic characterization of the dots, a

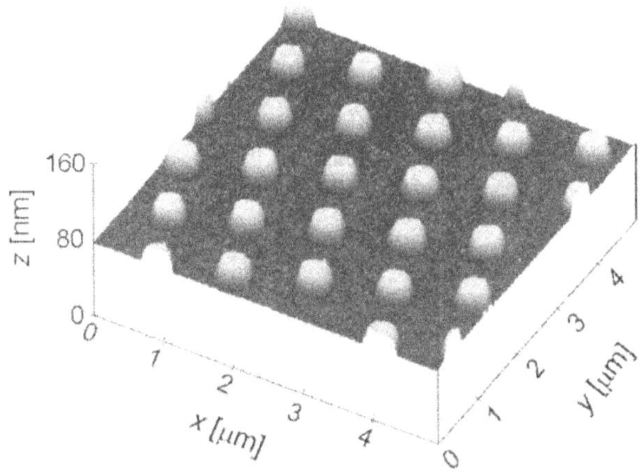

Figure 1. AFM micrographs of a square array of ferromagnetic Co/Pt dots . The dots consist of Pt(2.5 nm) / $[Co(0.4 nm)/Pt(1.0 nm)]_{10}$ layers and have a square shape with rounded corners with a side length of $\sim$0.4 μm.

10 nm Ge film, a type-II superconducting Pb film with thickness $d_{Pb} = 50$ nm, a protective 30 nm Ge layer and 50 nm Au were consecutively deposited on top of the dot array. The top Au layer facilitates the approach of the scanning Hall probe microscope (SHPM) in a tunnelling control mode for the vortex imaging experiments.
For the flux pinning experiments it is of crucial importance that the magnetic state of the dots remains unchanged when applying a small magnetic field perpendicular to the sample surface. Fig. 2 shows the hysteresis loop of the dot array that was used for the flux pinning experiments described in section 3 with the magnetic field H applied perpendicular to the sample surface measured by magneto-optical Kerr effect (MOKE) at room temperature.

The magnetic dots consist of a $[Co(0.3 nm)/Pt(1.1 nm)]_{10}$ multilayer on a 3 nm Pt buffer layer. They have a square shape with a side length of 0.4 μm and rounded corners and are arranged in a square lattice with a period of 1 μm. The dots have a coercive field of $\mu_0 H_{coe} = 0.23$ T and 100 % magnetic remanence, implying that very stable magnetic states of the dots are obtained after applying and removing a high magnetic field. In this particular sample, application of a magnetic field -100 mT $< \mu_0 H <$ 100 mT, which is much higher than the typical fields applied for measuring the flux pinning properties, is not able to change this stable state.

3. Vortex pinning properties

In all vortex pinning experiments, the magnetic moments $\vec{m}$ of the dots and the applied field H are both perpendicular to the sample surface (i.e., the z-

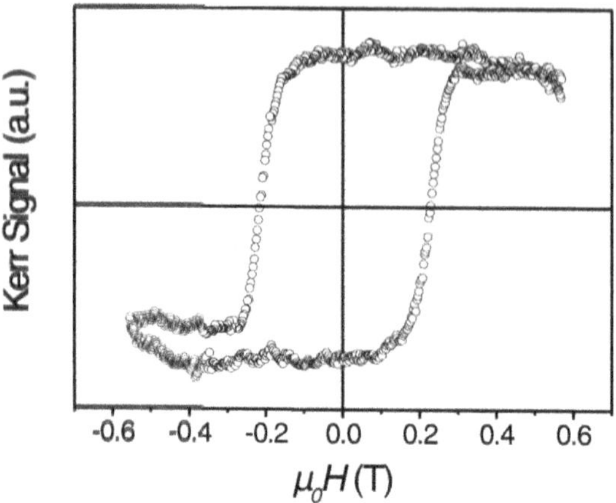

Figure 2. Magnetic hysteresis loop of a Co/Pt dot array measured by magneto-optical Kerr effect at room temperature before covering the dots with the Ge/Pb/Ge/Au layers. The different offset of the Kerr signal for increasing and decreasing sweep directions of the field at high negative field is due to an experimental artefact caused by a shift of the detected signal when the field sweep direction was reversed.

direction). This leads to magnetic interactions between the magnetic dots and the vortices in the superconducting film, depending on the mutual orientation of m_z and H. Additionally, the regular arrangement of the dots is expected to give rise to the appearance of matching effects in the magnetization curves. The pinning properties were studied by measuring the magnetization M of the sample as a function of H (which is always applied in the z-direction) in a Quantum Design SQUID magnetometer.

Fig. 3 shows the $M(H)$ curve of the hybrid sample at different temperatures after the sample has been saturated in a large negative field of $\mu_0 H = -2\,\mathrm{T}$. The field axis is normalized to the first matching field $\mu_0 H_1 = \phi_0/(1\mu\mathrm{m})^2 = 2.068\,\mathrm{mT}$, at which exactly one flux quantum ϕ_0 is generated by the external field per unit cell of the dot array. The most obvious feature of the curves shown in Fig. 3 is the asymmetry of the $M(H)$ curves with respect to the polarity of the applied field H, and the huge number of matching peaks and cusps at rational and integer multiples of the first matching field H_1. The asymmetry manifests itself in a large width of the magnetization loop ΔM (which is proportional to j_c in the Bean model) for negative H and a smaller ΔM for positive H. The asymmetry of the $M(H)$ curve is completely reversed with respect to the field polarity when the dots have been magnetized in very high positive field before measurement (not shown).

In the curves shown in Fig. 3, pronounced matching effects appear at integer

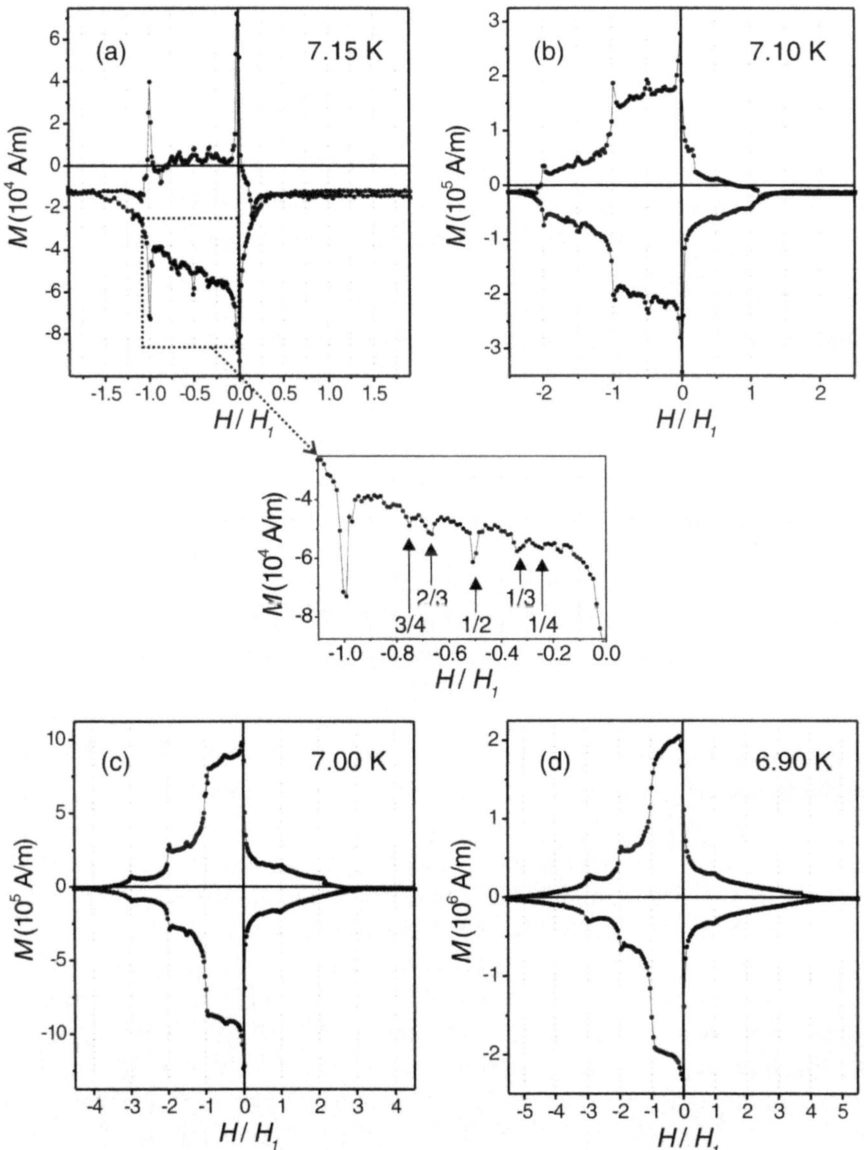

Figure 3. Magnetization curves of the Pb film on top of an array of Co/Pt dots after the sample was saturated in a high negative field before the measurement ($m_z < 0$) at different temperatures: (a) $T = 7.15$ K, (b) $T = 7.10$ K, (c) $T = 7.00$ K and (d) $T = 6.90$ K.

(-1, -2, and -3) and rational multiples (-1/2 and -3/2) of H_1. Moreover, small kinks in the $M(H)$ curve are present at H/H_1 = -1/4, -1/3, -2/3, -3/4, -5/4, -4/3, -5/3 and -7/4. For positive H, only weak matching features are observed at H/H_1 = 1/2 and 1. Note that the offset of the magnetization caused by the ferromagnetic contribution of the dots has not been subtracted.

These magnetization loops show some similarities with those of thin superconducting films with arrays of antidots. For example, for $-H_1 < H < 0$ the magnetization has a plateau, which is also seen for antidot samples close to the critical temperature when there are more pinning sites than vortices in the sample [15]. ΔM drops rapidly as interstitial vortices appear, as is observed for $H < -H_1$ in Fig. 3(a).

The appearance of matching effects in the $M(H)$ curves is due to the formation of stable vortex configurations at certain applied fields. Many of these matching features have already been reported for films containing antidot arrays [16], superconducting networks [17], or films covering lattices of magnetic Co and non-magnetic Ge dots [8, 18]. Based on these observations, we propose that the vortex configurations shown in Fig. 4 for some selected fields are realized when m_z and H have the same polarity. The matching effects at $H/H_1 = \pm 1/2$ are not shown, but will be discussed below in section 4. The configuration at $H/H_1 = 1/4$, where the vortices form a distorted triangular lattice, has been visualized in Nb films with antidot arrays [3]. At $H/H_1 = 1/3$, the vortices order in rows along the diagonals of the dot array, like in superconducting networks [17]. At $H/H_1 = 2/3$, the inverse of the vortex configuration at $H/H_1 = 1/3$ is realized, i.e., two adjacent diagonals of the dot array are filled with vortices, and the next diagonal is vacant. This configuration has been imaged by scanning Hall probe microscopy in a Pb film with a square antidot array [4]. The inverse pattern of the one at $H/H_1 = 1/4$ is probably realized at $H/H_1 = 3/4$. The vortex arrangement at the rational matching fields of $H/H_1 = 5/4, 4/3, 5/3$, and $7/4$ can be easily reconstructed by assuming that the saturation number of the dots, i.e., the maximum number of vortices that a dot can pin, is equal to one. If this is the case, then the excessive vortices in the field range $1 < H/H_1 < 2$ will be caged at the interstitial positions with the same symmetry as at the corresponding rational matching fields in the field range $H/H_1 < 1$. As an example, at $H/H_1 = 5/4$, every dot pins one vortex and the excessive vortices form the same distorted triangular lattice as for $H/H_1 = 1/4$, but at the interstitial positions of the dot array. The same considerations can be made for the matching fields at $H/H_1 = 4/3, 5/3$ and $7/4$. A further discussion of vortex configurations at other matching fields can be found in [19].

We have observed that the matching anomalies only appear close to T_c in the magnetization loops. This can be understood by the requirement of having long-range interactions between the vortices over several periods of the dot array.

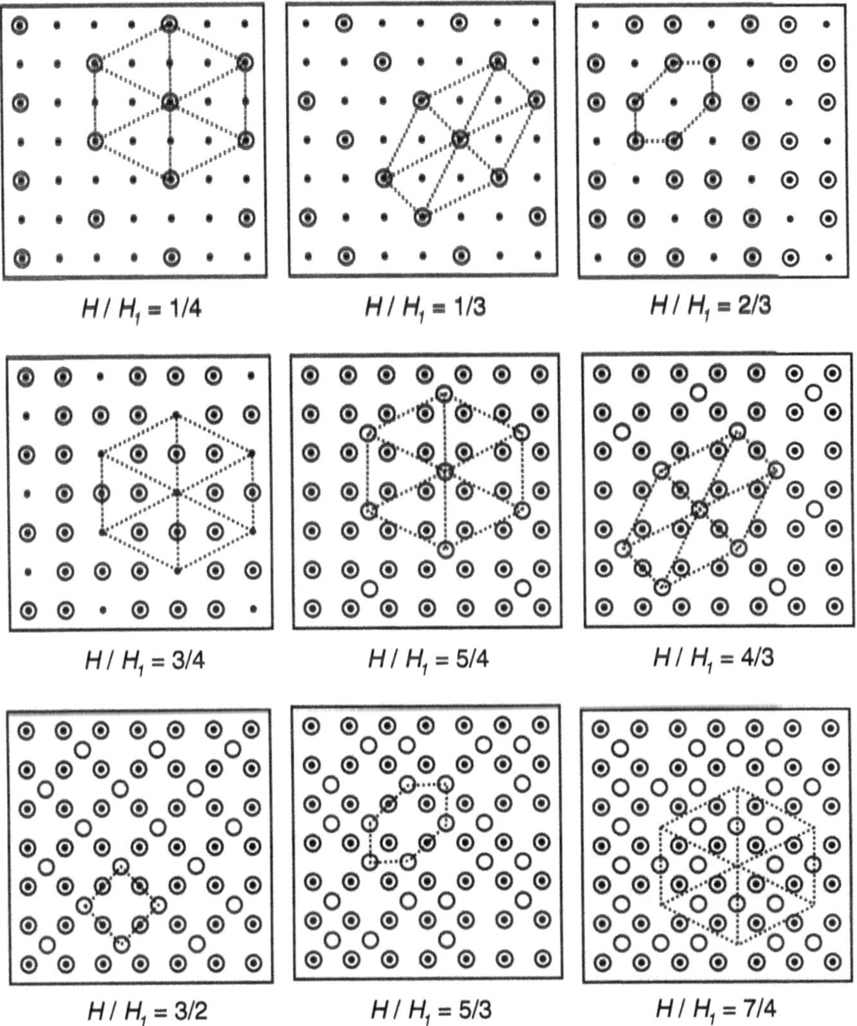

Figure 4. Schematic presentation of the suggested stable vortex lattices for a square periodic pinning potential at the experimentally observed rational matching fields when H and m_z are parallel. Open circles and dots represent vortices and magnetic dots, respectively. Dashed lines are guides to the eye emphasizing the symmetry of the vortex lattices.

The range over which vortices are interacting is determined by the temperature dependent penetration depth. Close to T_c, the penetration depth diverges, explaining why rational matching effects that correspond to vortex lattices with a large unit cell disappear first when decreasing the temperature, and at even lower temperatures also integer matching effects with smaller unit cell of the vortex lattice vanish.

Moreover, the flux gradient inside the sample plays an important role in these systems. The Bean model assumes a linear flux gradient in the sample, which is not realistic in a superconducting film with arrays of magnetic dots or antidots. It was suggested that in films with such periodic pinning sites, the flux gradient could be better described by a multi-terrace [20] or a single terrace flux profile [21]. However, matching effects will only appear in the sample when the flux gradient in the sample is small, because larger gradients prohibit the formation of larger regions with commensurate vortex arrangements. As one approaches T_c from below the vortex-vortex interaction becomes stronger since the overlap of the magnetic field of two vortices increases with increasing temperature. This results in a smaller flux gradient close to T_c, in agreement with the observed pronounced matching effects. At lower temperatures, the matching effects vanish due to the large flux gradient. This argument has also been given to explain the disappearance of matching effects at lower temperatures for superconducting films and multilayers with an antidot lattice [15].
Moreover, at very low temperatures the coherence length $\xi(0)$ becomes of the order of the grain size of the Pb films. As a consequence, core pinning at grain boundaries could be stronger than the magnetic pinning by the dot array, so that matching effects disappear.

4. Local vortex visualization of rational matching effects

SHPM was used to image specific configurations of the vortex lattice at matching fields with the magnetic dots in the $m_z < 0$ state. Here we will concentrate on rational matching effects at $|H| < H_1$, because the lateral resolution of the SHPM did not allow the imaging of single vortices at fields $H > H_1$.
Fig. 5 shows the SHPM images obtained after field cooling in $H/H_1 = -1/2$ (a) and $+1/2$ (b). For $H/H_1 = -1/2$ a clear matching effect is observed in the magnetization curve of the superconductor, see Fig. 3. This matching effect is due to a stable *checkerboard* configuration of the vortices, shown in Fig. 5(c), where every second dot pins one vortex. If the sample is field cooled in $H = +1/2\ H_1$, the vortices are visible as bright spots located at the interstitial positions of the dot array, see Fig. 5(b). At this field, the upper right area of the scanned image contains a vortex lattice in checkerboard configuration where every second interstitial position of the array is occupied as schematically depicted in Fig. 5(d). This observation is consistent with the presence of a small peak in the magnetization curve at this field in Figs. 3. However, the difference in mobility due to on-site and interstitial pinning causes the much more pronounced matching effect when m_z and H have the same polarity.
The field polarity dependent pinning can be explained by considering a balance of different energy terms. The asymmetry of the $M(H)$ curves proves that only an energy term that depends on the mutual orientation of H and m can deliver

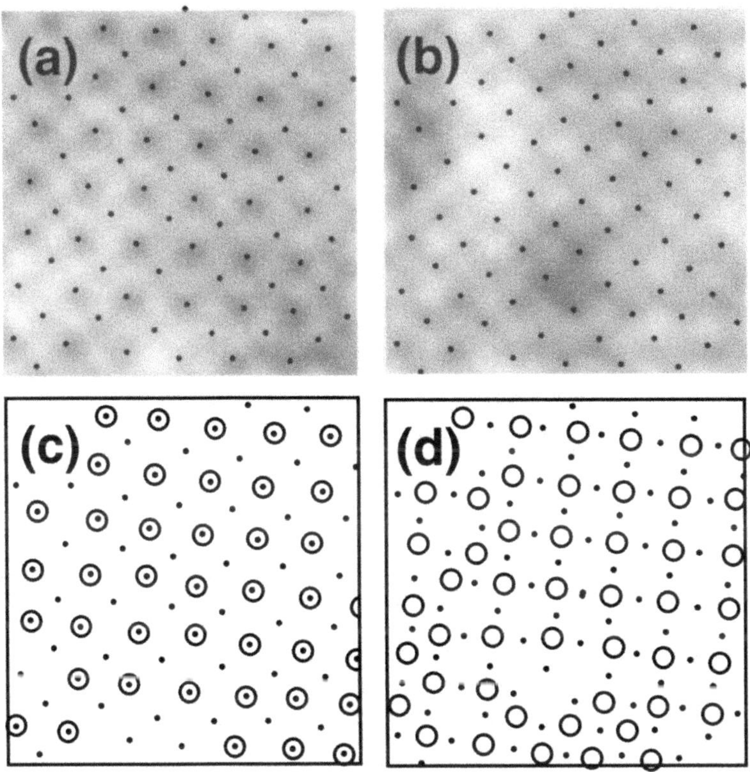

Figure 5. Images of the magnetic stray field above the sample surface after field cooling to $T = 6.80$ K (a) in $H/H_1 = -1/2$ and (b) in $H/H_1 = +1/2$. The small black points indicate the positions if the dots. The sample was negatively magnetized before measurement ($m_z < 0$). The two schematic drawings show the positions of the vortices (circles) with respect to the dot array in drawings in $H/H_1 = -1/2$ (c) and $H/H_1 = +1/2$ (d).

the main contribution to the pinning energy. Thus, the main term must have a magnetic origin. For a detailed qualitative discussion we refer to Ref. [11]. Quantitative calculations of the interaction between a single magnetic dipole with out-of-plane magnetization above an infinite superconducting plane and a vortex have been carried out within the London approximation [22].
Fig. 6(a) and (b) are obtained after field cooling in $H = -1/3\ H_1$ and $H = +1/3\ H_1$, respectively. Also in these images the field polarity dependent pinning is obvious. No ordered vortex state can be seen in Fig. 6(b) at $H = +1/3\ H_1$, which is in agreement with the absence of a matching effect at this field value in the magnetization curves. At $H = -1/3\ H_1$, the vortex configuration with minimum energy corresponds to a state where the vortices order in rows along the diagonal of the dot array with the vortices occupying every third row, depicted schematically in Fig. 4. However, this domain state is six-fold

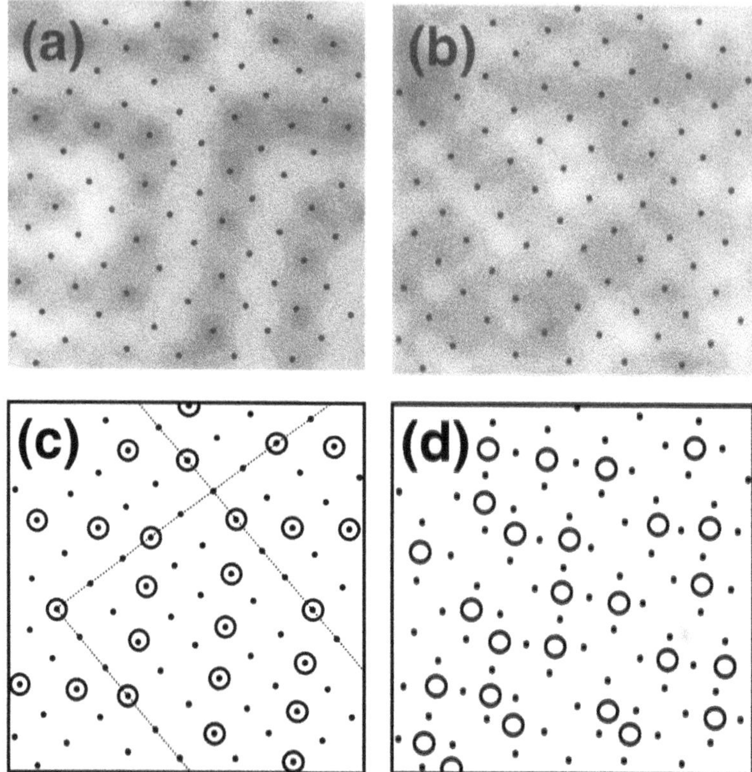

Figure 6. Images of the magnetic stray field above the sample surface after field cooling to $T = 6.80$ K (a) in $H/H_1 = -1/3$ and (b) in $H/H_1 = +1/3$. The small black points indicate the positions if the dots. The sample was negatively magnetized before measurement ($m_s < 0$). The two schematic drawings show the positions of the vortices (circles) with respect to the dot array in drawings in $H/H_1 = -1/3$ (c) and $H/H_1 = +1/3$ (d).

degenerate, because three different rows can be filled by the vortices along both diagonals of the dot array. In the image shown in Fig. 6 four of the degenerated configurations can be observed, see Fig. 6(c). Domain walls separating these different states are indicated by the dotted lines. The domain formation is probably promoted due to the efficient incorporation of mismatched excess vortices or vacancies at the corners of domain walls which outweighs the high energetic cost of creating them [23].

5. Summary

Vortex pinning was investigated in a superconducting Pb film covering an array of magnetic Co/Pt dots. When all magnetic moments of dots are aligned perpendicular to the sample surface, the magnetization curve of the hybrid system is clearly asymmetric with respect to the sign of the applied magnetic field.

This shows that the flux pinning strength is *field polarity dependent.* Moreover, a large number of integer and fractional matching effects is observed.
Direct local studies of the vortex patterns by scanning Hall probe microscopy indicate that the larger width of the magnetization loop for parallel m and H can be attributed to a stronger on-site pinning of flux lines at the dot positions. The much weaker pinning in the antiparallel case is related to the vortices caged at interstices due to the repulsive interaction with surrounding magnetic dots.

Acknowledgments

The authors thank Y. Bruynseraede for fruitful discussions. This work was supported by the Fund for Scientific Research (F.W.O.) - Flanders, by the Belgian IUAP and the European ESF "VORTEX" Programs, and by the Research Fund K.U.Leuven GOA. ML and MJVB are Postdoctoral Research Fellows of the F.W.O. - Flanders.

References

[1] A. F. Hebard, A. T. Fiory, and S. Somekh, IEEE Trans. Magn. **1**, 589 (1977).

[2] V. V. Moshchalkov, V. Bruyndoncx, L. Van Look, M. J. Van Bael, Y. Bruynseraede, and A. Tonomura, in *Handbook of Nanostructured Materials and Nanotechnology*, edited by H. S. Nalwa (Academic Press, San Diego, 2000), Vol. 3, Chap. 9, p. 451.

[3] K. Harada, O. Kamimura, H. Kasai, T. Matsuda, A. Tonomura, and V. V. Moshchalkov, Science **274**, 1167 (1996).

[4] A. N. Grigorenko, G. D. Howells, S. J. Bending, J. Bekaert, M. J. Van Bael, L. Van Look, V. V. Moshchalkov, Y. Bruynseraede, G. Borghs, I. I. Kaya, and R. A. Stradling, Phys. Rev. B **63**, 052504 (2001).

[5] S. B. Field, S. S. James, J. Barentine, V. Metlushko, G. Crabtree, H. Shtrikman, B. Ilic, and S. R. J. Brueck, Phys. Rev. Lett. **88**, 067003 (2002).

[6] O. Geoffroy, D. Givord, Y. Otani, B. Pannetier, and F. Ossart, J. Magn. Magn. Mater. **121**, 223 (1993).

[7] J. I. Martín, M. Vélez, J. Nogués, and I. K. Schuller, Phys. Rev. Lett. **79**, 1929 (1997).

[8] M. J. Van Bael, K. Temst, V. V. Moshchalkov, and Y. Bruynseraede, Phys. Rev. B **59**, 14674 (1999).

[9] M. J. Van Bael, J. Bekaert, K. Temst, L. Van Look, V.V. Moshchalkov, Y. Bruynseraede, G. D. Howells, A. N. Grigorenko, S. J. Bending, and G. Borghs, Phys. Rev. Lett. **86**, 155 (2001).

[10] D. J. Morgan and J. B. Ketterson, Phys. Rev. Lett. **80**, 3614 (1998).

[11] M. J. Van Bael, M. Lange, S. Raedts, V. V. Moshchalkov, A. N. Grigorenko, and S. J. Bending, Phys. Rev. B **68**, 014509 (2001).

[12] M. Lange, M. J. Van Bael, Y. Bruynseraede, and V. V. Moshchalkov, Phys. Rev. Lett. **90**, 197006 (2003).

[13] M. Lange, M. J. Van Bael, L. Van Look, K. Temst, J. Swerts, G. Güntherodt, V.V. Moshchalkov, and Y. Bruynseraede, Europhys. Lett. **53**, 646 (2001); **56**, 149 (2002).

[14] M. Lange, M. J. Van Bael, V. V. Moshchalkov, and Y. Bruynseraede, Appl. Phys. Lett. **81**, 322 (2002).

[15] M. Baert, V. V. Metlushko, R. Jonckheere, V. V. Moshchalkov, and Y. Bruynseraede, Phys. Rev. Lett. **74**, 3269 (1995).

[16] M. Baert, V. V. Metlushko, R. Jonckheere, V. V. Moshchalkov, and Y. Bruynseraede, Europhys. Lett. **29**, 157 (1995).

[17] B. Pannetier, in *Quantum Coherence in Mesoscopic Systems*, edited by B. Kramer (Plenum Press, New York, 1991), p. 457.

[18] M. J. Van Bael, *Regular arrays of magnetic dots and their flux pinning properties*, PhD thesis, Katholieke Universiteit Leuven, Leuven, 1998.

[19] T. Joseph and C. Dasgupta, Phys. Rev. B **67**, 214514 (2003).

[20] L. D. Cooley and A. M. Grishin, Phys. Rev. Lett. **74**, 2788 (1995).

[21] V. V. Moshchalkov, M. Baert, V. V. Metlushko, E. Rosseel, M. J. Van Bael, K. Temst, R. Jonckheere, and Y. Bruynseraede, Phys. Rev. B **54**, 7385 (1996).

[22] M. V. Milosevic, S. V. Yampolskii, and F. M. Peeters, Phys. Rev. B **66**, 174519 (2002).

[23] A. N. Grigorenko, S. J. Bending, M. J. Van Bael, M. Lange, V. V. Moshchalkov, H. Fangohr and P. A. J. de Groot, Phys. Rev. Lett. **90**, 237001 (2003).

PHASE TRANSITIONS IN MESOSCOPIC SUPERCONDUCTING FILMS

V. V. Kabanov
J. Stefan Institute, Jamova 39, 1000 Ljubljana, Slovenia
viktor.kabanov@ijs.si

T. Mertelj
Faculty of Mathematics and Physics, Jadranska 19, 1000 Ljubljana, Slovenia
and
J. Stefan Institute, Jamova 39, 1000 Ljubljana, Slovenia
tomaz.mertelj@ijs.si

Abstract We solve the Ginzburg-Landau equation (GLE) for the mesoscopic superconducting thin film of the square shape in the magnetic field for the wide range of the Ginzburg-Landau parameter ξ. We found that the phase with the antivortex exists in the broad range of parameters. When the coherence length decreases the topological phase transition to the phase with the same total vorticity and a reduced symmetry takes place. The giant vortex with the vorticity $m = 3$ is found to be unstable for any field, ξ and $\kappa_{eff} \geq 0.1$. Reduction of the Ginzburg-Landau parameter does not make the phase with the antivortex more stable contrary to the case of the cylindric sample of the type I superconductor.

Keywords: Mesoscopic Superconducting Films; vortices;

1. Introduction

Advances in nanotechnology and constantly shrinking semiconductor devices have motivated researches to study properties of mesoscopic superconducting samples. One line of research in this field has focused on the problem of the phase transitions in the mesoscopic superconducting sample under the influence of the external magnetic field. [1] We will focus on the case when the size of the sample $a \sim \xi, \lambda$, with ξ and λ being the superconducting coherence length and the London penetration depth respectively. In that case there are only a few vortices in the sample. The standard Abrikosov approach [2] must be modified because of the strong influence of sample boundaries. Thermodynamics of the

A.S. Alexandrov et al. (eds.), Molecular Nanowires and Other Quantum Objects, 287–296.

system is determined by the short-range repulsion of vortices and interaction of vortices with boundaries.

Recently it was shown that the influence of boundaries can lead to stabilization of the vortex-antivortex molecules in mesoscopic samples.[1] Analysis of the linearized Ginzburg-Landau equation (GLE) has shown that such molecules appear at particular values of the external magnetic field depending on the sample shape and size[3]. The solution of the GLE in the limit of the extreme type-II superconductor shows that such molecules have a very shallow minimum in the free energy[4, 5] and are very sensitive to the change of the sample shape[6].

In a square mesoscopic thin film with the total vorticity $m = 3$ the solution of the linearized GLE with the lowest free energy[1] is the symmetric solution with four vortices and one antivortex. According to ref. [4], away from the H_{c2} line the giant vortex with vorticity $m = 3$ is stable and has the lowest free energy. This implies that a topological phase transition *without change* of the vorticity and *without a reduction* of the symmetry should take place with the change of the external field or/and the coherence length away from the critical-field line.

Here we report the results of the extensive studies of different kinds of phase transitions for the thin film of the square shape. We focused to the region $4 < a/\xi < 8$ and $\kappa_{eff} > 0.05$ (see also[7]). Here $\kappa_{eff} = \lambda^2/d\xi$ where d is the film thickness. We found that the antivortex phase with $m = 3$ is stable in a broad range of parameters. The region of stability of the phase does not depend strongly on the value of the parameter κ_{eff}. The energy gain due to the antivortex formation is much smaller then the energy difference between two phases with different vorticities. The giant vortex with $m = 3$ is unstable for any field, ξ/a and $\kappa_{eff} \geq 0.1$. Phase transition to the phase with three separated vortices takes place when ξ/a is driven away from the critical field line. The reduction of κ_{eff} does not stabilize the antivortex phase for the thin film sample in the contrast to the case of the cylindric sample ref.[8].

2. Formalism and Solution

GLE for the normalized complex order parameter ψ has the following form:[4]

$$\xi^2(i\nabla + \frac{2\pi\mathbf{A}}{\Phi_0})^2\psi - \psi + \psi|\psi|^2 = 0 \tag{1}$$

here Φ_0 is the flux quantum, $\mathbf{A}$ is the vector potential and $\mathbf{H} = \nabla \times \mathbf{A}$ the magnetic field. We split the vector potential into the external part due to external currents, $\mathbf{A}_{ext}$, and the internal part due to the response of the superconducting film, $\mathbf{A}_{int}$. The second GLE equation for the total vector potential reads:

$$\nabla \times \nabla \times \mathbf{A} = -i\frac{\Phi_0}{4\pi\lambda^2}(\psi^*\nabla\psi - \psi\nabla\psi^*) - \frac{|\psi|^2\mathbf{A}}{\lambda^2}. \tag{2}$$

In addition to Eq.(1) we assume the boundary condition for the superconductor insulator junction on the sample edges:

$$(i\nabla + \frac{2\pi \mathbf{A}}{\Phi_0}) \cdot \mathbf{n}\psi = 0, \tag{3}$$

where $\mathbf{n}$ is the vector normal to the surface of the sample.

As it was described in ref.[4] we introduce $N \times N$ discrete points on the square and rewrite Eq. (1) in the form of the nonlinear discrete Schrödinger equation:

$$\sum_{\mathbf{l}} t_{\mathbf{i+l},\mathbf{i}}\psi_{\mathbf{i+l}} - \epsilon(\mathbf{i})t_{\mathbf{i},\mathbf{i}}\psi_{\mathbf{i}} - \psi_{\mathbf{i}} + \psi_{\mathbf{i}}|\psi_{\mathbf{i}}|^2 = 0, \tag{4}$$

where the summation index $\mathbf{l} = (\pm 1, 0), (0, \pm 1)$ points toward the nearest neighbors, where $t_{\mathbf{i}_1,\mathbf{i}} = (\xi N/a)^2 \exp(i\phi_{\mathbf{i}_1,\mathbf{i}})$ and $\phi_{\mathbf{i}_1,\mathbf{i}} = -\frac{2\pi}{\Phi_0}\int_{\mathbf{r}_\mathbf{i}}^{\mathbf{r}_{\mathbf{i}_1}} \mathbf{A}(\mathbf{r})d\mathbf{r}$. The boundary conditions are included in the discrete nonlinear Schrödinger equation as in ref.[4] where $\psi_{\mathbf{i}} = 0$ if $\mathbf{i}$ is outside of the sample and $\epsilon(\mathbf{i}) = 4 - \delta_{i_x,1} - \delta_{i_x,N} - \delta_{i_y,1} - \delta_{i_y,N}$ where $\mathbf{i} = (i_x = 1, \ldots, N, i_y = 1, \ldots, N)$.

After discretization of Eq. (2) we obtain the exact expression for the internal part of the vector potential[7]:

$$A^v_{int,\mathbf{i}} = \sum_{\mathbf{n}} K(\mathbf{i} \quad \mathbf{n}) J^v_{\mathbf{n}}, \tag{5}$$

where

$$J^v_{\mathbf{i}} = \frac{\Phi_0 a}{4\pi\lambda_{eff}N}\Im(\exp(-i\phi_{\mathbf{i+l_v},\mathbf{i}})\psi^*_{\mathbf{i}}\psi_{\mathbf{i+l_v}} - \exp(-i\phi_{\mathbf{i-l_v},\mathbf{i}})\psi^*_{\mathbf{i}}\psi_{\mathbf{i-l_v}}) \tag{6}$$

with $v \in \{x, y\}$ and $\mathbf{l_x} = (1, 0)$, $\mathbf{l_y} = (0, 1)$ and

$$K(\mathbf{n}) = \frac{N}{2\pi^2 a}\int_0^\pi dxdy \frac{\cos(n_x x)\cos(n_y y)}{\sqrt{4 - 2\cos(x) - 2\cos(y)}}. \tag{7}$$

Here we should point out that Eq.(5) contains 2D integration only[7]. All dependence on the thickness of the sample appears through the parameter $\lambda_{eff} = \lambda^2/d$ (d is the thickness of the sample)[9]. This is important difference from the case of the cylindric sample considered in the Ref.[8] where the function $K(\mathbf{n})$ is essentially different.

The numerical self consistent solution of the problem is obtained by iterating the solution of the nonlinear equation for the order parameter Eq.(4) and calculations of the current and the vector potential Eqs.(5,6). We used two ways of solving Eq.(4). The first is similar to that reported in ref.[4] and corresponds to the iterative solution of the linearized Eq.(4). The second relies on the fact that Eq.(4) represents the Euler equation for the free-energy functional with

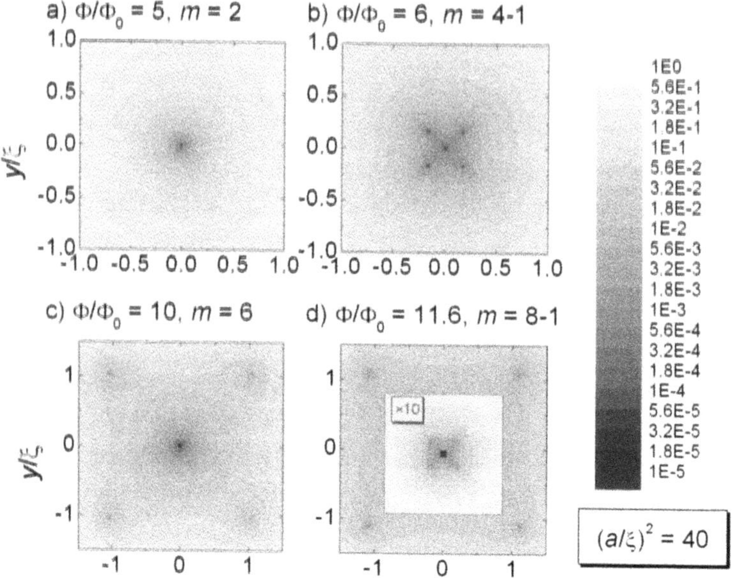

Figure 1. The modulus of the order parameter $|\psi|$ at different magnetic fields. The parameter Φ is the total external magnetic flux through the sample. The inset in d) shows the central region in an expanded scale. Note the logarithmic intensity scale.

included boundary conditions. Eq.(4) was therefore solved by the direct minimization of the corresponding functional using the conjugate-gradient method. Both techniques gave identical results.

3. Results

We investigate first the phase diagram in the regions where the solution with one antivortex and four vortices and one antivortex and eight vortices has been reported. In Fig.1 we present the change of a spatial pattern of the modulus of the order parameter $|\psi|$ when the vorticity changes from $m = 2$ (Fig.1a) to $m = 4 - 1$ (Fig.1b) and from $m = 6$ (Fig.1c) to $m = 8 - 1$ (Fig.1d).

When $m = 2$ and 6 (Fig.1a and Fig.1b) we observe a giant double vortex in the center. In the case when $m = 4l - 1$ with $l = 1, 2$ the symmetry induced square pattern of four vortices with the antivortex in the center is formed instead of the giant triple vortex. The results of the calculations show that the region of the phase diagram where the symmetry induced antivortex solution has the lowest energy is broader than expected from the solution of the linearized GLE. As it is shown in Fig. 2 for $\kappa_{eff} = \infty$ the antivortex phase is stable up to $(a/\xi)^2 \sim 55$, depending on Φ/Φ_0. For a finite κ_{eff} this region shifts to the higher field as $(a/\xi)^2$ increases (see Fig.2). In the Fig.3 we presented the

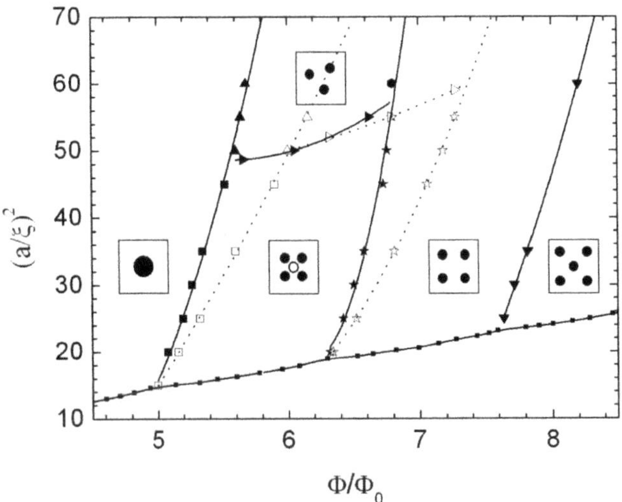

Figure 2. The calculated phase diagram. Different phases are marked with icons schematically indicating the vortex pattern where the full dot represents a vortex, the open dot represents an antivortex and the larger full dot represents a double vortex. The full symbols and continuos lines represent the phase boundaries for $\kappa_{eff} = \infty$ while the open symbols and dotted lines represent the phase boundaries for $\kappa_{eff} = 1$. In the later case only the phase boundaries of the region with the total vorticity 3 are shown.

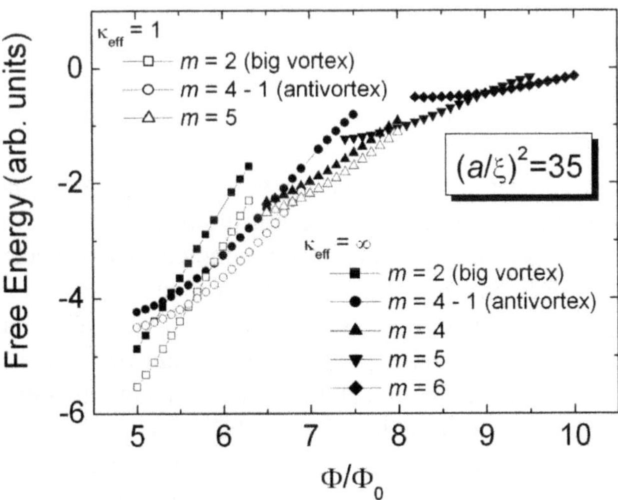

Figure 3. The free energy as a function of magnetic field for different vorticities m.

calculated value of the sample free energy as a function of external field for $(a/\xi)^2 = 35$ and for two values of $\kappa_{eff} = 1, \infty$. Topological phase transitions with the changes of the total vorticity $\Delta m = 1$ are clearly seen. Reduction of κ_{eff} leads to the shift of the transition point to the higher field.

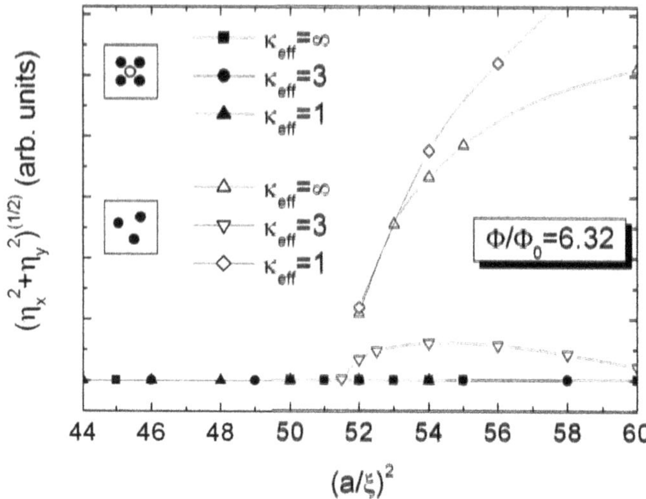

Figure 4. The magnitude of the order parameter around the transition where one vortex anihilates with the antivortex as a function of $(a/\xi)^2$ at the constant magnetic field.

The interesting behavior is observed when the external field is fixed and $(a/\xi)^2$ increases. Close to the H_{c2} the lowest minimum of the free energy corresponds to the solution with the vorticity $m = 4 - 1$ with the antivortex in the center of the square. Present calculations do not confirm the existence of the giant-vortex solution with $m = 3$ in this region of the phase diagram as reported previously [4]. The difference is due to increase of the number of discrete points N enabling detection of the antivortex. With increase of $(a/\xi)^2$ away from the H_{c2} line the phase transition to the multivortex state with the same vorticity ($m = 3$) and a lower symmetry takes place (see Fig. 2). In general, the free energy depends on the vorticity $m = n_+ - n_-$ and the total number of vortices in the system $n = n_+ + n_-$. The transition at $(a/\xi)^2 \sim 55$ and $\Phi/\Phi_0 \sim 5.5$ takes place at the constant vorticity $m = 3$ with the change of n from 5 to 3. The transition is therefore not only characterized by an order parameter, but also by the change of the number of vortices at the constant total vorticity m, suggesting that the transition is close to the first order. This statement is confirmed by the observation that above the transition point, $(a/\xi) > (a/\xi)_{crit}$, both solutions with $m = 3$ and $m = 4 - 1$ coexists. Since near the transition the free-energy difference between the phases with the same vorticity and different total number of vortices is small, it is difficult to determine the phase boundary between phases with $m = 4 - 1$ and $m = 3$ accurately. The transition could be easier observed by calculating the two component order parameter $\eta_x = \int x|\psi(x,y)|^2 dxdy$, $\eta_y = \int y|\psi(x,y)|^2 dxdy$ shown in Fig.4.

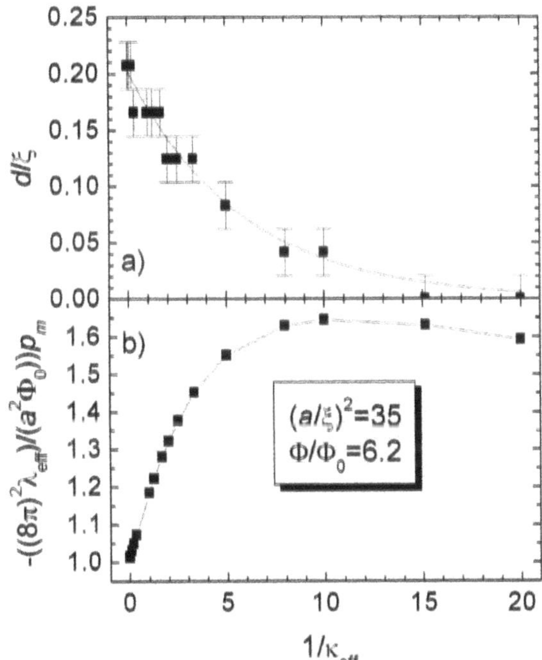

Figure 5. The vortex-antivortex distance a) and the magnetic moment of the sample b) as functions of the parameter $1/\kappa_{eff}$. In (a) error bars represent the grid spacing and the solid line the exponential fit discussed in the text.

Close to the H_{c2} line the repulsion of vortices from the boundaries and attraction of the 4 vortices to the antivortex stabilizes the phase with $m = 4 - 1$ and small vortex-vortex distances. At smaller value of ξ the repulsion from the boundaries decreases and one vortex annihilates with the antivortex. As a result, the repulsion between the remaining vortices increases, leading to an increase of the order parameter with a further decrease of ξ.

Increasing the external field up to $\Phi/\Phi_0 = 11.6$ leads to the stabilization of the phase with total vorticity $m = 7$. Near the H_{c2} line similar to the case with $m = 4 - 1$ the solution with $m = 8 - 1$ is realized (Fig.1d). When $(a/\xi)^2$

increases in a complete analogy to the case with $m = 4 - 1$ the phase transition to the phase with $m = 7$ and a similar order parameter takes place. Here also, both solutions with $m = 8 - 1$ and $m = 7$ coexists above $(a/\xi)_{crit}$ indicating that the transition is close to the first order. We believe that the situation is quite general for the case of arbitrary $m = 4l - 1$ for $l = 1, 2, 3, \ldots$.

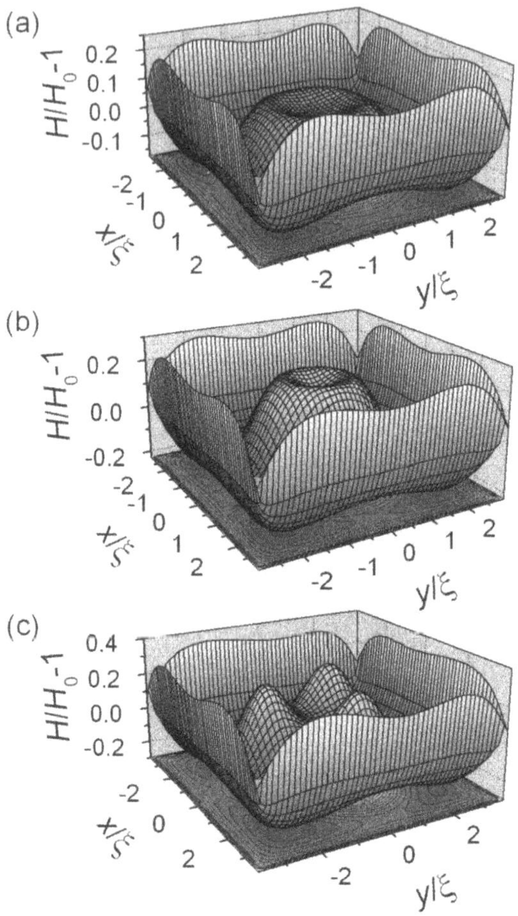

Figure 6. The magnetic field in the film in the case of (a) giant vortex with $m = 2$, (b) the antivortex solution with $m = 4 - 1$ and (c) three separate vortices with $m = 3$. Here H_0 is the external magnetic field.

At the end we would like to discuss the dependence of the stability of the antivortex phase at small κ_{eff}. According to the arguments of Ref.[8], at small

κ the vortex-vortex interaction changes the sign, making the antivortex phase more stable. As a result, the average distance between vortices in the middle of the square increases as well. In order to verify this conjecture for the thin film sample we plot in Fig.5 the vortex-antivortex distance r_0 as a function of $1/\kappa_{eff}$. The distance decreases with the decreasing κ_{eff}. For $\kappa_{eff} < 0.1$ the distance is smaller than the grid spacing a/N so we can not resolve separate vortices. We find that $r_0 \propto \exp(-\Lambda/\lambda_{eff})$ with $\Lambda \sim a$. The situation is just opposite to that reported in ref.[8]. We believe that in the case of the thin film of the square shape the reduction of κ_{eff} does not stabilize the phase with the antivortex.

It is interesting to note differences between samples of different shapes. For the cylindric shape the giant vortex phase with any vorticity is always stable close to the H_{c2} line[10, 11]. According to Ref.[8] for the mesoscopic triangle the giant vortex state with $m = 2$ is metastable and the solution with the antivortex ($m = 3 - 1$) is stable. For the case of the square shape the giant vortex solution with $m = 3$ is *never* stable for $\kappa_{eff} \geq 0.1$. For $\kappa_{eff} < 0.1$ the limited grid spatial resolution prevented us to distinguish the solution with the antivortex from the possibly (meta)stable giant-vortex solution.

Finally, let us discuss the possibility to detect the state with the antivortex experimentally. Calculation of the magnetic field in the sample shows that the magnetic field has a local minimum in the center of the sample also for the giant vortex solution with $m = 2$. The local minimum observed for the antivortex state with $m = 4 - 1$ is therefore not due to the antivortex formation (Fig.6) but due to a particular distribution of the current in the sample. Therefore, imaging of the magnetic field distribution cannot provide an evidence for the antivortex. The magnetic field for the multivortex solution with $m = 3$ has 3 well separated maxima that break the four fold rotational symmetry of the sample allowing a direct imaging of vortices. Since the antivortex state cannot be detected directly the observation of a hysteresis in the vicinity of the transition line from the $m = 4 - 1$ antivortex state to the $m = 3$ multivortex state could suggest that the symmetric phase is indeed the phase with the antivortex.

Acknowledgments

We wish to thank J. Bonca, A.S. Alexandrov and V.V. Moshchalkov for useful correspondence and discussions.

References

[1] Chibotaru L.F., Ceulemans A., Bruyndoncx V., Moshchalkov V.V., Nature **408**, 833 (2000).

[2] Abrikosov A.A., ZhETF, **32**, 1442 (1957).

[3] Chibotaru L.F., Ceulemans A., Bruyndoncx V., Moshchalkov V.V., Phys. Rev. Lett. **86**, 1323 (2001).

[4] Bonca J., V.V. Kabanov V.V., Phys. Rev. B**65**, 012509, (2002).

[5] Baelus B.J., Peeters F.M., Phys. Rev. B**65**, 104515, (2002).

[6] Melnikov A.S. et al, Phys. Rev. **B65**, 140503, (2002).

[7] Mertelj T., Kabanov V.V. Phys. Rev. B**67**, 134527, (2003).

[8] Misko V.R. et. al., Cond-mat/0203140.

[9] de Gennes P.G., 'Superconductivity of Metals and Alloys', Preus Books Publishing L.L.C. 1989.

[10] Schweigert V.A., Peeters F.M., Deo P.S., Phys. Rev. Lett. **81**, 2783 (1998).

[11] Schweigert V.A., Peeters F.M., Phys. Rev. Lett. **83**, 2409 (1999).

FANO EFFECT OF AN INTERACTING AHARONOV-BOHM SYSTEM CONNECTED WITH SUPERCONDUCTING LEADS

Anatoly Golub[1] and Yshai Avishai[1,2]
[1]*Department of Physics, Ben-Gurion University, Beer-Sheva, Israel*
[2]*Ilse Katz Center for Nanotechnology, Ben-Gurion University, Beer-Sheva, Israel*
agolub@bgumail.bgu.ac.il

Abstract The physics of a system consisting of an Aharonov Bohm (AB) interferometer with a single level interacting quantum dot (QD) on one of its arms, and attached to normal (N) and superconducting (S) leads is elucidated. Here the focus is directed mainly on N-AB-S junctions but the theory is capable of studying S-AB-S junctions as well. The interesting physics emerges under the conditions that both the Kondo effect in the QD and the the Fano effect are equally important. As the Fano effect becomes more dominant, the conductance of the junction is suppressed.

Keywords: Aharonov-Bohm interferometer, Fano effect, superconductivity, Quantum dot, strongly correlated electrons, Kondo effect.

1. Motivation

Transport through a mesoscopic Aharonov-Bohm (AB) ring (with an interacting quantum dot situated on one of its arms) weakly attached to normal (N) metallic leads is the subject of recent intensive experimental [1, 2, 3] and theoretical [4, 5, 6, 7, 8, 9] studies. To the above class of experiments one may also add STM measurements on a single magnetic atom adsorbed on a metallic surface [10]. The main result of the theoretical analysis [6, 7] indicates that in such N-AB-N junctions there is an interplay between two fundamental physical phenomena, namely, the Fano and the Kondo effects. The Fano effect is related to interference between the electron wave passing through the quantum dot (with a discrete level) and the wave travelling along the direct 'reference' channel characterized by its continuous spectrum. It results in an asymmetric shape of the conductance as a function of the applied bias or gate voltage. The Kondo effect is one of the simplest manifestations of many-body physics exhibiting strong correlations. It plays an important role in electron transport

A.S. Alexandrov et al. (eds.), Molecular Nanowires and Other Quantum Objects, 297–305.

through quantum dots, where the role of a magnetic impurity is played here by the presence of localized electrons [13, 14]. The interplay between these two effects causes the suppression of Kondo plateau with increasing transmission through the direct channel.

An analysis of the conductance of an AB interferometer when one of the leads is a normal metal and the other one is a superconductor (S) is the subject of the present research. The geometry of such N-AB-S junction is schematically displayed in Fig.1. We derive a general formula for the conductance of N-AB-S and S-AB-S junctions. As for the motivation, there is indeed a growing interest

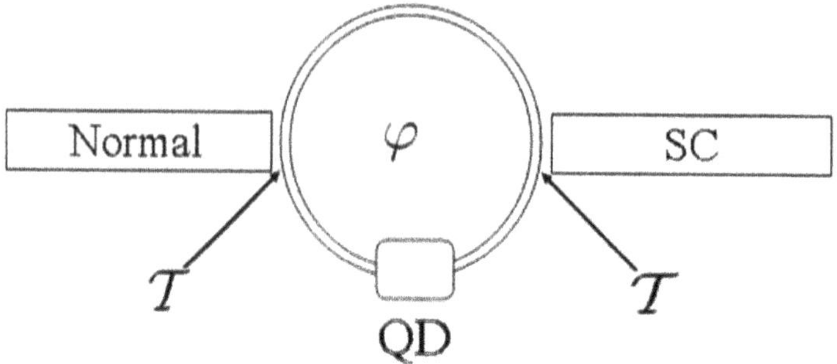

Figure 1. Schematic form of an N-AB-S junction.

in these types of superconducting junctions, especially in the light of recent experiments which involve carbon nanotubes as weak links[11, 18, 17]. They are characterized by relatively high Kondo temperature which helps the formation of Kondo plateau in conductance experiments. The Kondo finger-prints in such junctions is expected to survive even for cases where the superconducting gap approaches the Kondo temperature from below, that is, $\Delta \sim T_K$ (see Refs. [21, 20, 22]).

2. Model Hamiltonian and the current

In the system under consideration (see Fig. 1), transport of an electron between the two-dimensional electrodes (N on the left and S on the right) is possible via two paths, either through the QD, or else through the other 'direct' channel. The dynamics of the system is governed by the Hamiltonian

$$H = H_L + H_R + H_d + H_t + H_{LR}, \qquad (1)$$

in which H_j, ($j = L, R$) are the Hamiltonians of the electrons in the electrodes which depend on the electron field operators $\psi_{a\sigma}(\mathbf{r}, t)$ where $\mathbf{r} = (x, y)$ and

$\sigma = \pm$ is the spin index,

$$\mathrm{H}_j = \int dr[\psi^{\dagger}_{j\sigma}(\mathrm{r})\xi(\nabla)\psi_{j\sigma} - \alpha\psi^{\dagger}_{j\uparrow}(\mathrm{r})\psi^{\dagger}_{j\downarrow}\psi_{j\downarrow}(\mathrm{r})\psi_{j\uparrow}(\mathrm{r})]. \tag{2}$$

Here α is the BCS coupling constant and $\xi(\nabla) = -\nabla^2/2m - \mu$ with μ being the chemical potential at temperature T. The dot is represented by a single level Anderson impurity with energy $\epsilon_0 < 0$ and Hubbard repulsion U. Later on we will concentrate on the Kondo regime by setting $U \to \infty$ and assuming that $|\epsilon_0|$ exceeds any other energy scale except U. The Nambu representation for all fermion operators ($\psi_\sigma(\mathbf{r})$ for the lead electrons and c_σ for the QD electrons) is employed below (for example $c^\dagger \equiv (c_1^\dagger, c_2^\dagger) = (c_\uparrow^\dagger, c_\downarrow)$). The dot Hamiltonian then reads,

$$H_d = \epsilon c^\dagger \tau_z c + U n_\uparrow n_\downarrow, \tag{3}$$

while H_t, the tunnelling Hamiltonian from the leads to the dot takes the form

$$H_t = \sum_j \mathcal{T}_j c^\dagger \tau_z \psi_j(\mathbf{0}) + h.c., \tag{4}$$

where $\mathcal{T}_j$ is the tunnelling amplitude (symmetric junction with equal transfer matrix elements is assumed hereafter) and $n_\sigma = c_\sigma^\dagger c_\sigma$. The Pauli matrix τ_z acts in Nambu space. Finally, the occurrence of direct transport channel is represented by the term H_{LR} between left and right leads. The gauge is chosen such that the AB flux appears only in this direct term,

$$H_{LR} = \mathcal{W}\psi_R^\dagger(\mathbf{0})\tau_z F \psi_L(\mathbf{0}) + h.c., \tag{5}$$

where $F = \exp(i\varphi\tau_z)$ is the AB phase factor. Note that unlike the case of N-AB-N junctions, the phase factor is *non-Abelian*.

In the mean field slave boson approximation (MFA) which is employed below to handle the Kondo problem (in the limit $U \to \infty$), the tunnelling amplitude is modified by replacing the (slave) boson operators by their mean-field values, ($b \to < b >$) and the Hubbard repulsion term is dropped. This procedure imposes a constraint on the number of bosons and fermions and the Hamiltonian must now include a term which prevents double occupancy in the limit $U \to \infty$,

$$H_c = (\epsilon_r - \epsilon)(c^\dagger \tau_z c + b^\dagger b - 1), \tag{6}$$

where ϵ_r is the renormalized new level position which has the role of a Lagrange multiplier.

Formula for the current at one lead, say the right (superconducting), is derived from the basic relation $I = -ie < [n_R, H] >$. It is convenient to display the result of the commutation relation in terms of Keldysh Green's functions (GF) , namely,

$$I = \frac{e}{h}[\sqrt{\gamma}Tr(F^{\dagger}G^{K}_{LR} + H.c) + \pi\mathcal{T}NTr(G^{K}_{dR} + H.c)], \qquad (7)$$

where N is the density of states in the leads (assumed to be the same in both), $\gamma = \pi^2\mathcal{W}^2N^2$ and the trace operation acts in Nambu and energy spaces ($Tr[A(\omega)] \equiv \sum_i \int A_{ii}(\omega)d\omega$). Other notations are $G^{K}_{LR} = G^{<}_{LR} + G^{>}_{LR}$ (similarly for G^{K}_{dR}), where $G^{<}_{LR}(t) = i < \psi^{\dagger}_{R}\psi_{L}(t) >$ and $G^{<}_{dR}(t) = i < \psi^{\dagger}_{R}d(t) >$.

Employing equation of motion method for all Keldysh GF it is possible to express the current in terms of the dot GF. We single out the current through the direct channel by writing $G^{K}_{LR} = G^{K}_{LdR} + G^{K}_{dir}$, with,

$$\begin{aligned} G^{K}_{dir} &= 2\sqrt{\gamma}[g_{LL}F\tau_z g_R]^K && (8)\\ G^{K}_{LdR} &= \Gamma[g_{LL}(1 + 2\sqrt{\gamma}F\tau_z g_R)\hat{G}(1 + 2\sqrt{\gamma}g_L F\tau_z)g_{RR}]^K \\ G^{K}_{dR} &= \mathcal{T}[\tau_z\hat{G}(1 + 2\sqrt{\gamma}g_L F\tau_z)g_{RR}]^K. && (9) \end{aligned}$$

Here $\Gamma = 2\pi N\mathcal{T}^2$ stands for tunneling rate through the dot, $\hat{G} = \tau_z G\tau_z$, G is the dot GF, $g_{L,R}$ are generally nonequilibrium GF of the leads (left,right). The other two GFs include multiple scattering events and have the form,

$$\begin{aligned} g_{LL} &= (1 - 4\gamma g_L F\tau_z g_R\tau_z F^{\dagger})^{-1}g_L \\ g_{RR} &= (1 - 4\gamma g_R F^{\dagger}\tau_z g_L\tau_z F)^{-1}g_R. && (10) \end{aligned}$$

Each GF ($g_{L,R}$, g_{LL}, g_{RR},and G) has a standard 2×2 Keldysh matrix structure with elements $G_{11} = G^R$, $G_{22} = G^A$, $G_{21} = 0$, $G_{12} = G^K$. The superscript K stands for Keldysh component of the matrix product occurring in the square brackets of Eq.(7-9). The set of equations (7-10) together with the expression (7) for the current formally completes our task . They give an expression for the current through a closed AB interferometer in terms of the GF of the (strongly interacting) QD. The above formalism has been tested for the case of linear conductance with normal leads on both sides (N-AB-N junction) and the result agrees with pertinent calculations [6].

These rather general expressions are now employed for elucidating a particular case of interest, namely, the zero temperature limit of the linear conductance in N-AB-S junctions. Specifically, as in Fig. (1) the left lead is a normal metal biased with an external voltage V whose GF is, $g^{R,A}_{L} = -\pm i/2$, $g^{K}_{L} = -i(\tanh(\omega/2T) + \delta f\tau_z)$; $\delta f = eV/2T(\cosh(\omega/2T))^{-2}$. On the other hand, the right lead is an unbiased (s-wave) superconductor with gap Δ whose GF at equilibrium has the standard form [22].

3. Mean Field Approximation and Conductance

Within the MFA the algorithm starts with calculations pertaining to a geometrically identical system albeit with *noninteracting* QD ($U \to 0$). Then, at the end, the replacement $\Gamma \to T_k, \epsilon \to \epsilon_r$ is executed in the conductance formula. These two quantities (the Kondo temperature $T_K = \Gamma b^2$ and the effective renormalized position of the level) are evaluated by solving two mean field equations similar to those derived in Refs. [20, 22]. At zero bias one is free to adopt the Matsubara form of these equations. The first one, $2\pi(\epsilon_r - \epsilon) + \Gamma Tr(\mathcal{G}\Lambda) = 0$, follows from the extremum requirement of the effective action over ϵ_r. The second one, $2\pi T_K + \Gamma Tr(\mathcal{G}\tau_z) = 0$, reflects the single occupation condition. Here

$$\Lambda(\omega) = i\omega - \epsilon_r\tau_z + (\tau_z - i\sqrt{\gamma}\frac{\omega}{|\omega|}F)\mathcal{G}_{RR}(\tau_z - i\sqrt{\gamma}\frac{\omega}{|\omega|}F^\dagger) \quad (11)$$

$$\mathcal{G}^{-1}(\omega) = i\omega - \epsilon_r\tau_z - T_K\Lambda, \quad (12)$$

$$\mathcal{G}_{RR}(\omega) = -\frac{i}{2d(\omega)}(\omega + \gamma\frac{\omega}{|\omega|}\sqrt{\omega^2+\Delta^2} - i\tau_x\Delta), \quad (13)$$

where $d(\omega) = (1+\gamma^2))\omega^2\sqrt{\Delta^2+\omega^2} + 2\gamma\omega^2|\omega|$. Expressions (12,13) for the dot Matsubara GF were obtained from the effective action derived from the Hamiltonian of the N-AB-S system. The self-consistence equations were solved for different values of the direct transmission parameter γ, while the level position ϵ was scanned over a wide range. The Anderson model parameters are tuned such that the QD is found in the Kondo regime. The half width of the normal electrode density of states is served as an energy unit and the tunnelling rate through the dot is fixed at $\Gamma = 0.15$. The superconducting gap Δ is set at a value 0.0005. For the majority of level positions ϵ the inequality $\Delta < T_K$ is obeyed and the MFA is hence justified.

To obtain the zero bias differential conductance for the noninteracting dot the expressions for the GF G^R, G^A, G^K are found for this case and are then directly substituted into equations (7-9). It is important to note that any element in these GF which is related to the superconductor electrode must be evaluated at energies $|\omega| < \Delta$ (this is not the case for the self consistence equations). For a dot modelled by a simple resonance energy level, the formulae for the GF acquires the form (compare Eq.(11)),

$$G = \{\omega - \epsilon\tau_z - \Gamma[(\tau_z + 2\sqrt{\gamma}g_L F)g_{RR}(\tau_z + 2\sqrt{\gamma}g_L F^\dagger)]\}^{-1}$$

The retarded component of this equation is

$$G^R = [g + \tau_z g_z + \tau_x g_x + \tau_y g_y]^{-1}, \quad (14)$$

where $g = \omega + \Gamma\frac{i}{2}(1 - P(1-\gamma))$, the coefficients of the Pauli matrices are $g_z = -\epsilon - \sqrt{\gamma}\Gamma P\cos(\varphi)$, $g_x = -\frac{1}{2}\Gamma Q(1+\gamma\cos(2\varphi))$ and $g_y = \frac{1}{2}\Gamma Q\gamma\sin(2\varphi)$. On the Fermi surface the relations $P = -\gamma/(1+\gamma^2)$, $Q = 1/(1+\gamma^2)$ hold. Inserting these GF into equations (7-9) for the current one arrives at an expression for the zero bias conductance. After lengthy calculations the conductance is obtained and checked to be an even function of the flux, as is dictated by Onsager relations (although it is not immediately apparent from the complicated expressions).

$$\sigma_{NS} = \frac{4e^2}{h}(T_W + \frac{\Gamma}{2}\mathcal{N}(\varphi)\sum_{j=0}^{j=4} T_j\cos(j\varphi)) \tag{15}$$

Here $T_W = 4\gamma/(1+\gamma^2)^2$ is the background (direct) NS transmission. Introducing the quantity $T_{NN} = 4\gamma/(1+\gamma)^2$ which is the analogous quantity (direct transmission) for the N-AB-N junction[6] then $T_W = T_{NN}^2/(2-T_{NN})^2$. This is precisely the relation between transmission coefficients for N-BB-N and N-BB-S junctions (here BB is a "black-box" representing any non-interacting scatterer) suggested in Ref.[23], using Landauer scattering matrix approach. The coefficients T_W and T_j in equation (15) do not depend on flux. Note, however, that the normalization factor $\mathcal{N}$ is an even function of the flux,

$$\begin{aligned}
[\mathcal{N}(\varphi)]^{-1/2} &= 2(1+\gamma^2)^3[\Gamma^2(1+\gamma) + 2\epsilon^2(1+\gamma^2) \\
&\quad -4\Gamma\epsilon\gamma^{3/2}\cos(\varphi) + \Gamma^2\gamma\cos(2\varphi)] \\
T_0 &= -8\Gamma[\Gamma^2(-1+2\gamma^2+8\gamma^3+5\gamma^4) + \\
&\quad 16\gamma\epsilon^2(-1+\gamma-\gamma^2+\gamma^3+2\gamma^4)] \\
T_1 &= 64\sqrt{\gamma}\epsilon[\Gamma^2(-1-2\gamma^2+ \\
&\quad 4\gamma^3+5\gamma^4) + 4\gamma\epsilon^2(-1+\gamma^4)] \\
T_2 &= -64\Gamma\gamma[\Gamma^2\gamma^2(1+\gamma) + \epsilon^2(-1+5\gamma^4)]
\end{aligned}$$

The last two coefficients are $T_3 = 128\sqrt{\gamma}\gamma^4\Gamma^2\epsilon$ and $T_4 = -16\gamma^4\Gamma^3$. Here the result of calculations of conductance as function of gate voltage (15) (with $\Gamma = 0.04$ and zero AB phase) is shown on Fig.2. A typical Fano asymmetry form in the case $\gamma = 0.1$ is easily detected.

As was noted above, in the MFA, Γ and ϵ appearing in the above equations should respectively be replaced by T_K and ϵ_r, which, in turn, are obtained through the solutions of the self-consistence equations. The conductance of an N-AB-S junction as function of the level position ϵ is displayed in figure 3 and as a function of the flux in figure 4.

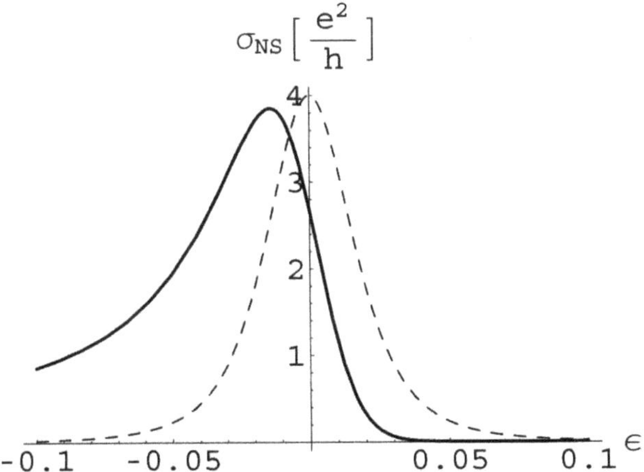

Figure 2. Linear conductance of an N-AB-S junction (see Fig. 1) at $T = 0$ and zero magnetic field as function of level position ϵ at $\gamma = 0.1$ (solid line) and $\gamma = 0$(dash line)

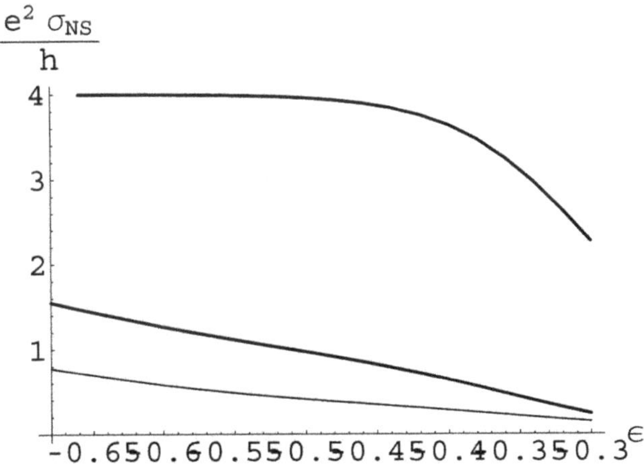

Figure 3. Linear conductance of an N-AB-S junction (see Fig. 1) at $T = 0$ and zero magnetic field as function of level position ϵ at $\gamma = 0.1$ and $\gamma = 0.2$. The case $\gamma = 0$ is displayed in the upper curve, and found to close to the unitary limit $\sigma = 4e^2/h$. for most of the energy range.

4. Discussion

If direct tunnelling is completely suppressed ($\gamma = 0$) the electron trajectory passes solely through the QD. This is a single channel Coulomb blockade situation for an N-QD-S junction. For level energies used in the numerical calculations one can expect Kondo behavior and Kondo plateau in the differential

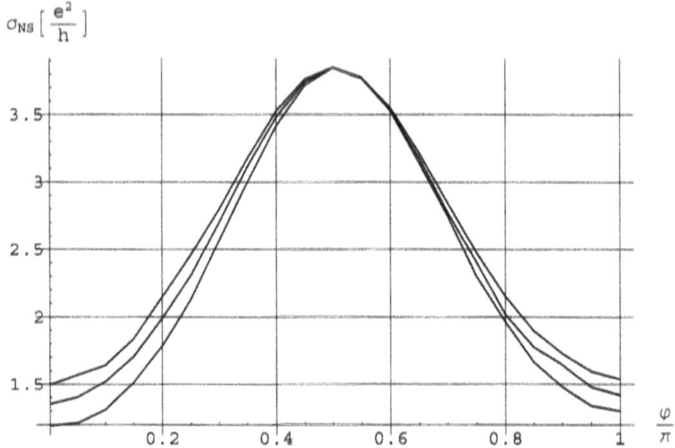

Figure 4. Linear conductance of an N-AB-S junction (see Fig. 1) at $T = 0$ versus AB phase for different level positions (ϵ=-0.67 top line, -0.63 middle line, -0.57 bottom line. The parameters are tuned to drive the quantum into the Kondo regime, and for all curves we set $\gamma = 0.1$

conductance at zero bias[20]. Figure 3 indeed reflects such a dependence of the linear conductance. In this figure the conductance is plotted as function of the level position at zero AB flux and different background transmissions ($T_{NN} = 0.33, 0.55$, or $\gamma = 0.1, 0.2$). The conductance for zero transmission ($T_{NN} = 0$ were also calculated and the plateau expected for a clean NS contact at $T_{NN} = 0$ ($G = 4e^2/h$) is obtained (it is displayed by the upper curve on Fig.3). However,when the direct channel is opened ($\gamma \neq 0$) the Fano effect comes into play and there is a clear suppression of the Kondo effect. The conductance at its plateau step is attenuated although, in this region of energies, the Kondo effect still survives. This also follows from solution of the MF equations (for $\gamma = 0.1$) yielding an effective Kondo temperature T_K which is smaller by 25 to 30 percents than in the case $\gamma = 0$. Except for the scale, the situation here is reminiscent of the Fano Kondo effect in normal junctions [6, 7]. Such behavior of the conductance may serve as an indication of Kondo correlations in the dot.

Having passed the stringent test of obtaining a conductance which is an even function of the flux, figure 4 manifests a remarkable behavior of the conductance as function of the AB flux. Several different level positions are considered, but for each energy there is a maximum very close to the Unitary limit $G = 4e^2/h$ at $\varphi = \pi/2$. This property characterizes solely the Kondo physics.

Acknowledgments

This research is partly supported by grants from the Israeli Science Foundation (ISF) and the American Israeli Binational Science foundation (BSF). We would like to thank Moshe Shechter for very helpful discussions.

References

[1] A.Yacoby, M. Heiblum, D. Mahalu, and H. Shtricman, Phys. Rev. Lett. **74**, 4047 (1995); Y. Ji,*et al.* Science **290**, 779 (2000).

[2] W. G. van der Wiel *et al.*, Science **289**, 2105 (2000).

[3] K. Kobayashi, H. Aikawa, S. Katsumoto and Y. Iye, Phys. Rev. Lett. **88** 256806 (2002).

[4] J. König and Y. Gefen , Phys. Rev. Lett. **86**, 3855 (2001).

[5] U. Gerland *et al.* , Phys. Rev. Lett. **84**, 3710 (2000).

[6] W. Hofstetter, J. König, and H. Schoeller , Phys. Rev. Lett. **87**, 156803 (2001).

[7] B. R. Bulka and P. Stefanski, Phys. Rev. Lett. **86**, 5128 (2001).

[8] M. A. Davidovich *et al.*, Phys. Rev. B **55**, R7335 (1997).

[9] O. Entin-Wohlman *et al.*, Phys. Rev. Lett. **88**, 166801 (2002).

[10] Y. Manassen *et al.*, Phys. Rev. Lett. **62**, 2531 (1989); H. Manoharan, Nature ,**416**,24 (2002).

[11] M.R. Buitelaar , N. Nussbaumer, and C. Schönenberger Phys. Rev. Lett. **89**, 256801 (2002).

[12] A. C. Hewson, *The Kondo Problem to Heavy Fermions* (Cambridge University Press, Cambridge, 1993).

[13] L. I. Glazman and M. E. Raikh, Pis'ma Zh. Eksp. Teor. Fiz. **47**, 378 (1988) [JETP Lett. **47** 452 (1988)]; T. K. Ng and P. A. Lee, Phys. Rev. Lett. **61**, 1768 (1988).

[14] Y. Meir, N. S. Wingreen, and P. A. Lee, Phys. Rev. Lett. **70**, 2601 (1993).

[15] D. Goldhaber-Gordon *et al.*, Nature **391**, 156 (1998); S. M. Cronenwett *et al.*, Science **281**, 540 (1998);

[16] M. A. Kastner, Comm. Cond. Matt. **17**, 349 (1996).

[17] J. Nygard, D. H. Cobben, and D. E. Lindelof, Nature **408**, 342 (2000).

[18] A. Yu. Kasumov *et al.*, Science **284**, 1508 (1999).

[19] R. Fazio and R. Raimondi, Phys. Rev. Lett. **80**, 2913 (1998); **82**, 4950 (E)(1999).

[20] P. Schwab and R. Raimondi, Phys. Rev. B **59**, 1637 (1999).

[21] A. A. Clerk, V. Ambegaokar, and S. Hershfield, Phys. Rev. B **61**, 3555 (2000).

[22] Y. Avishai, A. Golub, and A. D. Zaikin, cond-mat/0111442 Phys. Rev. B**67**(RC), 041301 (2003); Y. Avishai, A. Golub, and A.D. Zaikin, Europh.Letter **55** 397 (2001).

[23] C. W. J. Beenakker Phys. Rev. B **43**, 134515 (1991).

SPIN-DEPENDENT ELECTRONIC TRANSPORT THROUGH MOLECULAR DEVICES

Bogdan R. Bułka,[1] Tomasz Kostyrko,[2] Stanisław Lipiński[1] and Piotr Stefański[1]
[1]*Institute of Molecular Physics, Polish Academy of Sciences, 60-179 Poznań, ul. M. Smoluchowskiego 17, Poland*
[2] *Institute of Physics, A. Mickiewicz University, ul. Umultowska 85, 61-614 Poznań, Poland*
bulka@ifmpan.poznan.pl

Abstract A role of interference and electronic correlations in a transport through molecular devices is considered within an extended Anderson model. In magnetic systems correlations reduce the value of the magnetoresistance, which can even change its sign in some cases. Transport studies through a two-atomic molecule showed a series of voltage ranges with characteristic current dependences. For strong Coulomb interactions the studies predict bistable current solutions.

Keywords: electronic transport, molecular devices, Kondo resonance, interference, spin-dependent transport, magnetic nanostructures

1. Introduction

Most electronic devices use the charge of an electron to their operation. It seems, however, to be very promising to use also the spin degree of freedom of the electron in microelectronics [1]. The most spectacular example is the magnetic random access memory (MRAM), which has been developed recently by IBM and Infineon Technologies AG into the high-speed 128 Kbit chip [2]. MRAM is constructed from thin-magnetic multilayers and uses the giant magnetoresistive (GMR) effect discovered in the late 1980s [3]. A great progress in nanotechnology attracts much interest in magnetic semiconducting nanostructures and molecular systems in hope that their properties can be applied in spintronic devices.

In this paper we would like to present our recent studies of the coherent electronic transport in magnetic nanostructures, showing the interplay of the charge and the spin degree of freedom. Electronic correlations are relevant for the transport and they can lead to many body effects like the Kondo resonance, which is caused by resonant scattering of conducting electrons on localized

A.S. Alexandrov et al. (eds.), Molecular Nanowires and Other Quantum Objects, 307–318.

spins. These effects were observed in nonmagnetic semiconducting nanostructures [4], carbon nanotubes [5] as well as in single molecule systems [6,7]. The challenging problem is how these phenomena are influenced by injection of spin polarized electrons from magnetic electrodes. Some our results on the Kondo resonance in magnetic nanostructures will answer to that question.

In the coherent transport regime electrons exhibit their wave nature and interference processes play an important role. One can observe the Fano resonance (as in the transport through the quantum dots (QD) strongly coupled to the electrodes [8]) and the Fabry-Perrot etalon (in carbon nanotubes [9]). We would like to study these phenomena within an extended single impurity Anderson model. Although the model gives results in good qualitative agreement with transport measurements through systems with a single QD and simple molecules, its extension is needed for multi QD systems or for complex molecules. Here, we would like to present studies of electronic transport through short atomic chains in a strongly nonequlibrium regime, i.e. when a high voltage is applied.

The paper is organized as follows. In section 2 the single impurity Anderson model and a derivation of the current are presented. We extend the model and consider a formation of the Kondo resonance accompanied by the Fano resonance in strongly coupled quantum dots. The results of studies of a role of electronic correlations in magnetic nanostructures will be presented at the end of section 2. Section 3 is devoted to some aspects of the electronic transport through a two-atomic molecule. An analysis on the current bistability is addressed as well.

2. Transport within a single impurity Anderson model

This chapter is devoted to theoretical studies of the transport through nanostructures in presence of strong electronic correlations and formation of the Kondo resonance. These phenomena have been observed experimentally in various physical systems: quantum dots, carbon nanotubes and in single molecules [4-7]. It suggests a universal mechanism and description in the framework of the same model. The simplest model is given by the single impurity Anderson Hamiltonian

$$H = \sum_{k,\alpha,\sigma} \varepsilon_{k\alpha} c^{+}_{k\alpha,\sigma} c_{k\alpha,\sigma} + \sum_{\sigma} \varepsilon_0 c^{+}_{0\sigma} c_{0\sigma} + U n_{0\uparrow} n_{0\downarrow} + \sum_{k,\alpha,\sigma} t_\alpha \left(c^{+}_{k\alpha,\sigma} c_{0\sigma} + h.c. \right). \quad (1)$$

The first term describes electrons in the left ($\alpha = L$) and the right ($\alpha = R$) electrode, the second and the third one correspond to electrons in the nanostructure, the fourth term describes contact between the electrodes and the nanostructure. In most cases it is enough to take into account only one energy level

ε_0 and the onsite Coulomb interaction U of two electrons with the opposite spin orientation $\sigma = \uparrow$ and $\sigma = \downarrow$. The current is determined from the time evolution of the occupation number n_L for the electrons in the left electrode J = -e <dn_L/dt >. Here, e denotes the charge of the electron. Using the Hamiltonian (1) one gets

$$J = \frac{ie}{\hbar} \sum_{k,\sigma} \left[t_L \left\langle c^{+}_{kL,\sigma} c_{0\sigma} \right\rangle - c.c. \right] \ , \tag{2}$$

where the thermal averages are expressed by the nonequilibrium Green functions of the Kelsdysh type [10]. We used the equation of motion approach and the slave boson technique in order to find the Green functions. These methods allowed us to take into account electronic correlations and the Kondo resonance.

Fano resonance in strongly coupled quantum dots

Transport measurements through QD strongly coupled to the electrodes showed [8] strong temperature dependent asymmetric peaks of the conductance $\mathcal{G}$ measured as a function of the gate voltage. The data indicated the Fano resonance. As it is well known, the Fano resonance occurs in various systems and results from quantum interference between degenerate continuum states and an evanescent (discrete) state. In transport through nanostructures, traveling electronic waves represent the continuum state system needed for the Fano resonance. There is also an electron wave localized at the QD, which results from the geometry of the confining potential well of the QD and the electrodes. Interference between both waves leads to the Fano resonance. For some values of the gate voltage (when an odd number of electrons occupy the QD) one can also observe the Kondo resonance.

In order to describe the Fano resonance associated with the Kondo resonance one has to extend the model (1) by addition of the bridge term [11]

$$H_{bridge} = \sum_{k,k',\sigma} \left[t_{LR} \, c^{+}_{kL,\sigma} c_{k'R,\sigma} + h.c. \right], \tag{3}$$

which describes a direct transmission of electrons between the electrodes with the transmission rate t_{LR}. Independently the similar model was proposed by Heemeyer [12], who studied it by means of the mean-field slave boson method in the limit $U \rightarrow \infty$. In the present work we use the slave-boson approach proposed by Kotliar and Ruckenstein [13] for finite U.

Fig.1 presents the conductance $\mathcal{G}$ as a function of the position of the local energy level ε_0 with respect to the Fermi level E_F. In order to get the background of the conductance $\mathcal{G}$ =1.24 e^2/h (close to the experimental value [8]) the direct transmission parameter is taken t_{LR} = 310 meV. The solid curve represents the symmetric coupling of the QD to the electrodes. It exhibits a large peak (at eV = 2 meV) with a small dip at the shoulder and a large dip (at eV = 0).

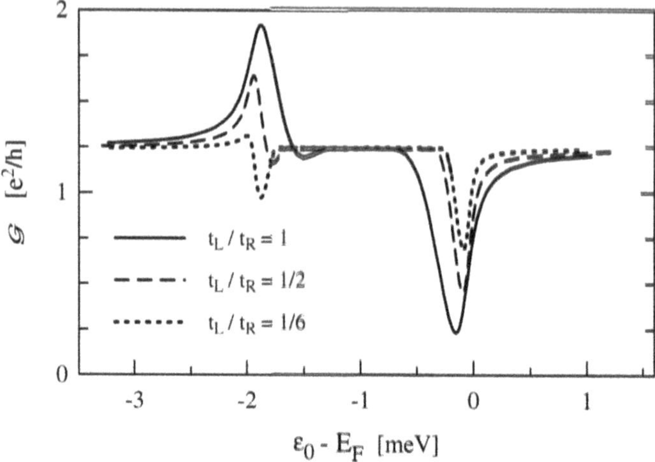

Figure 1. Conductance $\mathcal{G}$ vs. relative position of the local energy level $\varepsilon_0 - E_F$ for the QD with the symmetric and asymmetric couplings to the electrodes: $t_L/t_R = 1$ (solid curve), $t_L/t_R = 1/2$ (dashed curve) and $t_L/t_R = 1/6$ (dotted curve). The direct transmission rate is $t_{LR} = 310$ meV, $t_R = 2.5$ meV for all the cases, U = 2 meV and the temperature T = 0.5 K.

The separation of the peak and the large dip is $U = 2$ meV, which corresponds to the energy of adding of the second electron into the QD. The asymmetry of $\mathcal{G}$ decreases and the dips are transformed into two peaks for $t_{LR} \to 0$. Fig.1 shows also that the conductance characteristics depend on the coupling asymmetry. The dotted curve is plotted for $t_R = 2.5$ meV and $t_L = 0.41$ meV. This curve is very similar to the experimental one [8], with the same features: two asymmetric dips and a small bump of the left hand side. Moreover, our temperature studies showed a strong temperature dependence of both dips below the Kondo temperature in agreement with experimental data.

Kondo resonance in magnetic nanostructures

All up today known transport measurements through single molecules were performed in systems with paramagnetic electrodes. In this chapter we would like to convince the reader that it would be interesting to replace the electrodes by ferromagnetic metals. In such system one can expect new effects due to the interplay between two conducting channels for electrons with the opposite spin orientations $\sigma = \uparrow$ and $\sigma = \downarrow$. Different aspects of the problem have been recently theoretically studied in literature [14]. We want to focus ourselves on a role of electronic correlations on the spin-dependent transport and a formation of the Kondo resonance.

The magnetic system can be described by the single impurity Anderson Hamiltonian (1) with the spin-dependent electron density of states in the electrodes $\rho_{\alpha\sigma}$ (α = L,R). The transfer rate $\Gamma_{\alpha\sigma} = \pi t_\alpha^2 \rho_{\alpha\sigma}$ is thus spin-dependent. The nonequilibrium Green functions were determined within the slave boson approach developed by Coleman [15], which is applicable in the limit $U \to \infty$. For the temperature $T = 0$ one can get analytical results. In the empty state regime (i.e. when $\varepsilon_0 \gg E_F$) the conductance is expressed as

$$\mathcal{G} = \frac{e^2}{h} \sum_\sigma \frac{4\Gamma_{L\sigma}\Gamma_{R\sigma}}{\varepsilon_0^2}. \tag{4}$$

The magnetoresistance is defined as the relative difference of the conductance for the parallel (P) and the antiparallel (AP) configuration of the polarization in the electrodes $MR = (\mathcal{G}_P - \mathcal{G}_{AP})/\mathcal{G}_P$. In the empty state regime we get

$$MR = \frac{2P_L P_R}{1 + P_L P_R} . \tag{5}$$

Here, the relation connecting the magnetic polarization P_α of the electrode with the electronic density of states $\rho_{\alpha\sigma}$ has been used $P_\alpha \equiv (n_{\alpha\uparrow} - n_{\alpha\downarrow})/(n_{\alpha\uparrow} + n_{\alpha\downarrow}) = (\rho_{\alpha\uparrow} - \rho_{\alpha\downarrow})/(\rho_{\alpha\uparrow} + \rho_{\alpha\downarrow})$ (where $n_{\alpha\sigma}$ denotes the average number of electrons with the spin σ. The formula (5) is the same as derived by Julliere [16], what can be expected for uncorrelated transport.

In the Kondo regime (i.e. when the local energy level is far below the Fermi energy $\varepsilon_0 \ll E_F$) the conductance is given by

$$\mathcal{G} = \frac{e^2}{h} \sum_\sigma \frac{4\Gamma_{L\sigma}\Gamma_{R\sigma}}{\Delta_\sigma^2} , \tag{6}$$

where $\Delta_\sigma = \Gamma_{L\sigma} + \Gamma_{R\sigma}$ denotes the level broadening. In this regime the analytical formula for *MR* is more complex; therefore we present it for some limited cases. If the magnetic polarization of both electrodes has the same value ($P_L = P_R = P$) one gets

$$MR = \frac{P^2(1 - 3\alpha^2 + \alpha^2 P^2 + \alpha^4 P^2)}{(1 - \alpha^2 P^2)^2} , \tag{7}$$

where $\alpha = (t_L^2 - t_R^2)/(t_L^2 + t_R^2)$ is the asymmetry factor between the left and the right coupling. In the limit of a large asymmetry ($\alpha = 1$) one has also the simple analytical formula

$$MR = -\frac{2P_L P_R}{1 - P_L P_R}. \tag{8}$$

Fig. 2 shows *MR* for the whole range of ε_0, from the Kondo regime through the mixed valence range to the empty state regime (from the left to the right hand side). Two cases are presented: for the symmetric and the asymmetric couplings. These results suggest strong influence of electronic correlations on *MR*, which can even change its sign. The effect could be used in a magnetoresistive device, in which sensitivity of the device on a magnetic field would be tuned by the gate voltage changing the position of ε_0.

Using the slave-boson approach one can also determine the local magnetic moment $m_0 = n_{0\uparrow} - n_{0\downarrow}$, which is related with the total number of electrons n_0 by

$$\sin(\pi m_0) = \frac{\Delta_\uparrow - \Delta_\downarrow}{\Delta_\uparrow + \Delta_\downarrow} \sin(\pi n_0) . \tag{9}$$

The relation (9) yields $m_0 \to 0$ for $n_0 \to 1$, which corresponds to the formation of the singlet state in the Kondo regime ($\varepsilon_0 \ll E_F$). It suggests appearance of the Kondo resonance in magnetic nanostructures. This statement is justified for small polarization P_α in the electrodes and breaks down for $P_\alpha \to 1$, when the transport is strongly dominated by one channel.

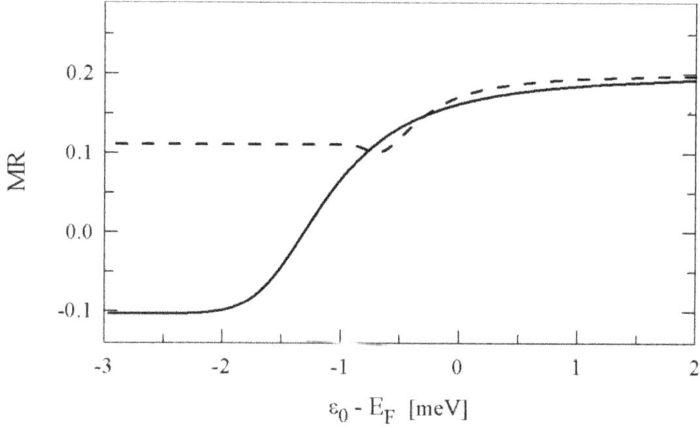

Figure 2. Magnetoresistance MR as a function of the relative position of the local energy level $\varepsilon_0 - E_F$ plotted for the symmetric couplings to the electrodes $t_L = t_R = 1.5$ meV (dashed curve) and for the asymmetric couplings $t_L = 1.5$ meV, $t_R = 3$ meV (solid curve).

3. Transport through linear atomic chains

Let us now consider two-atom linear molecule and next a multi-atom linear chain. The problem becomes more complex than discussed above, because an appropriate model is described now by the two (multi) impurity Anderson Hamiltonian, which has to include interactions between the impurities in a nonequilibrium situation. In this work we neglect electronic correlations and concentrate ourselves on discussion for a large source-drain potential. The

Hamiltonian describing the molecule is

$$H_{mol} = \sum_{i,j,\sigma} (\varepsilon_i \, \delta_{ij} + t) \, c^+_{i\sigma} c_{j\sigma} + U \sum_i n_{i\uparrow} n_{i\downarrow} \, , \tag{10}$$

where the summation is restricted to the nearest neighbour atoms and the site energy includes a uniform drop of the potential $\varepsilon_i = \varepsilon_i^0 + e\eta V$ [1-2i/(N+1)]/2. Here, ε_i^0 denotes the atomic level at equilibrium $V = 0$, N is the number of atoms in the chain and the coefficient $0 \leq \eta \leq 1$ describes the screening of the external electrostatic potential at the molecule. Since our studies are performed within the mean-field approximation, the interaction term in (10) is reduced to a shift of the local site energy $\tilde{\varepsilon}_i = \varepsilon_i + U\Sigma_\sigma n_{i\sigma}/2$. The thermal averages $n_{i\sigma}$ for the local number of electrons are self-consistently determined by means of the Keldysh Green functions.

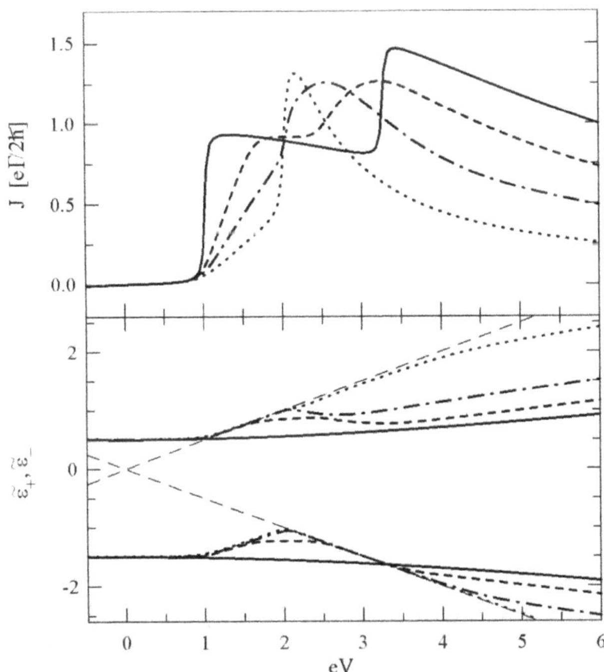

Figure 3. The current- voltage characteristics (upper figure) and the position of the self-consistent bonding $\tilde{\varepsilon}_-$ and the antibonding $\tilde{\varepsilon}_+$ levels (bottom figure) for various Coulomb interactions (U = 0 – solid curves, U = 1eV – dashed curves, U = 2 eV – dashed-dotted curves and for U = 4 eV – dotted curves). The position of the chemical potential in the left and the right electrodes is presented by the thin dashed lines in the bottom figure. The other parameters are $\tilde{\varepsilon}_1^0 = \tilde{\varepsilon}_2^0$ = - 0.5 eV, $\Gamma_L = \Gamma_R = \Gamma$ = 0.01eV, t =1 eV and η = 1.

Fig.3 shows typical current-voltage characteristics for various U and for the local energy $\tilde{\varepsilon}_1^0 = \tilde{\varepsilon}_2^0$ lying in the gap between the bonding $\tilde{\varepsilon}_-$ and the antibonding $\tilde{\varepsilon}_+$ levels at $V = 0$. The curves exhibit steps at the voltages, where the effective bonding $\tilde{\varepsilon}_-$ or the antibonding $\tilde{\varepsilon}_+$ level crosses the chemical potential in the left μ_L = eV/2 or in the right electrode $\mu_R \equiv$ -eV/2. The curves exhibit the negative differential (NDR) effect, for which the current drops with an increase of the voltage. The NDR effect is caused by an increase of the bonding-antibonding gap, which occurs for the unscreened potential at the molecule (for $\eta > 0$).

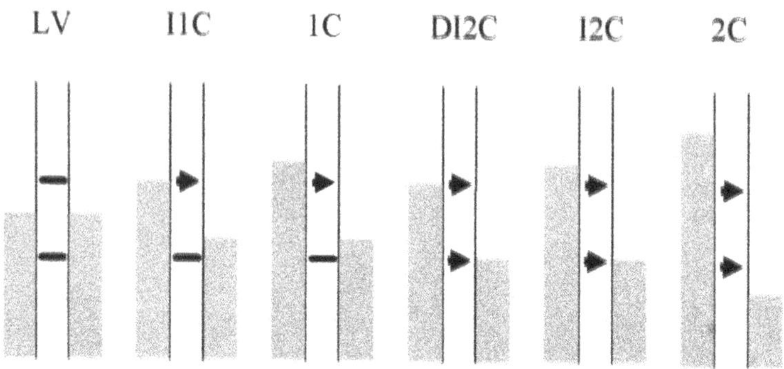

Figure 4. Schematic presentation of the evolution of the positions of the self-consistent energy levels with the voltage for the two-atom system. The gray bars show Fermi seas of the source and the drain electrodes. The horizontal strokes represent the positions of the antibonding and the bonding levels, while the arrows indicate levels active in transport for the given voltage range.

The bottom part of Fig.3 presents the plots for $\tilde{\varepsilon}_-$ and $\tilde{\varepsilon}_+$. Detailed studies showed [17] that one can distinguish six different voltage ranges schematically shown in Fig.4. The first region is the low-voltage (LV) regime, in which the chemical potentials μ_L, μ_R lie in the bonding-antibonding gap and there is no current flow. With an increase of V the chemical potential of the source electrode reaches the antibonding level ($\mu_L \equiv \tilde{\varepsilon}_+$), in which the systems enters the incomplete one-channel (I1C) transport. In this range the self-consistent state $\tilde{\varepsilon}_+$ is pinned to μ_L (see bottom Fig.3). For a higher V, in the one-channel region (1C), electron propagate still through the antibonding state, which now is below the source chemical potential ($\mu_L < \tilde{\varepsilon}_+$). In some voltage range both the levels $\tilde{\varepsilon}_+$, $\tilde{\varepsilon}_-$ can be pinned to μ_L, μ_R. This range is called two-channel incomplete region (DI2C). A further increase of V leads to the incomplete two-channel transport (I2C), in which the bonding state is pinned to the chemical potential of the drain electrode ($\tilde{\varepsilon}_- = \mu_R < \tilde{\varepsilon}_+ < \mu_L$). For the still higher voltage both the states get into the voltage window between source and the drain chemical potentials and they fully participate in transport. This range is called two-channel transport (2C).

In the case of the multi-atom systems a further increase of the voltage draws consecutive molecular levels into the voltage window. In general the system goes through the sequence of the ranges: ...InCT-nCT-I(n+1)CT..., where n denotes the number of molecular levels active in the charge transport. Each molecular level within the voltage window is occupied by one electron on the average (in the weak voltage limit). This is unlike the molecular levels outside the window, which can be only approximately empty or doubly occupied. Therefore with the change of the number of molecular levels inside the voltage window, the total electronic charge at the molecule alternates between even and odd numbers.

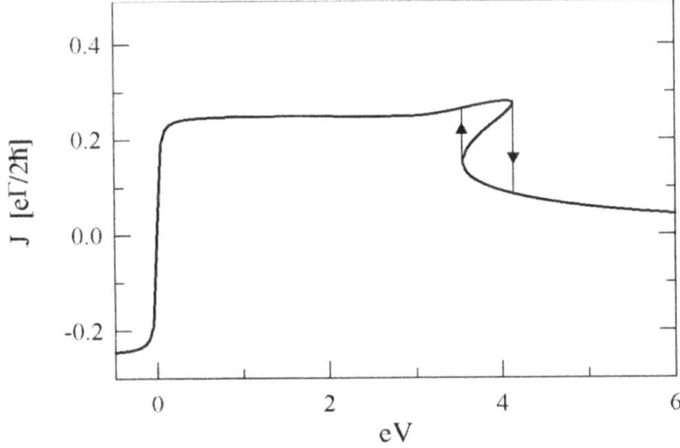

Figure 5. The current-voltage characteristic for large Coulomb interactions U = 12 eV and for the antibonding energy state below the Fermi level position at V = 0 (for $\tilde{\varepsilon}_1^0 = \tilde{\varepsilon}_2^0$= -3 eV).

Let us now consider the situation for a large U and for the antibonding level above the Fermi level in the electrodes at $V = 0$ (e.g. for $\tilde{\varepsilon}_1^0 = \tilde{\varepsilon}_2^0$= -3 eV). The current-voltage characteristic is presented in Fig.5. For small voltages the system is in the I1C transport regime, in which $\tilde{\varepsilon}_+$ is pinned to μ_L. Such a configuration of the self-consistent levels leads to a resonant transmission and a sharp increase of the current. For higher values of the voltage, $eV > 2$ eV, both the energy levels participate in the transport. Within the I2CT region one can find three solutions for a given voltage. The system shows bistability - the current can be for a given V either large or small. Sweeping the voltage up and down one could observe a current hysteresis (indicated in Fig.5). The origin of the bistability comes from a competition between polarization processes when two conducting channels participate in the transport. In this voltage range the electronic polarization has two solutions: with a small or with a large value.

The total energy of electrons at the molecule is different for both cases, and therefore, one can expect the hysteresis with sharp transitions (like a first-kind phase transition).

Recently, the electronic transport through the two-atom system was considered by Aguado and Langreth [18] and by Orellana et al [19] within the slave boson approach for $U \to \infty$. Their studies neglected however the shift of the self-consistent levels due to the external field (i.e. they assumed $\eta = 0$). The bistability effect was also found for an incoherent sequential tunneling transport through semiconductor superlattices with weakly coupled quantum wells [20]. In this case the origin of the bistability is the same as in our molecular system.

4. Summary

We have shown the main features of the electronic transport through molecular systems and semiconducting nanostructures. The Anderson model and the nonequilibrium Green function formalism are suitable for a description of these phenomena. The coherent transmission of electronic waves has been considered, for which interference and correlations play an important role. We have shown that both mechanisms can be studied within the bridge model [11], in which the Kondo resonance occurs together with the Fano resonance. Our calculations are in qualitative agreement with the experimental results [8] in strongly coupled quantum dots. The Kondo resonance was observed recently in transport through single molecules [6,7]. We hope that due to further technological progress the Fano resonance will be also seen in molecular devices.

The single impurity Anderson model can be, with a simple extension, applied to magnetic systems as well. We have shown that electronic correlations strongly modify the magnetoresistance, which can be even negative in systems with asymmetric couplings to electrodes. One can expect that this effect will be applied in magnetoresistive nanodevices, in which one can change sensitivity of the system on a magnetic field by change of the gate voltage potential. In our opinion the Kondo resonance should be observable in magnetic nanostructures as well.

Some results on transport through multi-atomic molecular systems have been also presented. We used the multi-impurity Anderson model solved within the mean-field approximation. One can distinguish a series of the voltage ranges characteristic for the transport. A relevant process is the pinning of the molecular energy levels to the chemical potential in the electrodes. The studies predict a strong increase of the electronic polarization for higher voltages, i.e. in the two-channel transport range. For stronger Coulomb interactions the bistability effect is expected in this voltage range. The effect is interesting for its potential applications in electronics, e.g. as switching or logical elements. Our molecular system is similar to semiconductor multiwell structures, where the multistability has been already found in vertical electronic transport [21]. We believe that the bistability will be soon found in molecular systems.

Acknowledgments

This work was supported in part by the State Committee for Scientific Research (Poland) within the project no. PBZ KBN 044 P03 2001 and by the Centre of Excellence for Magnetic and Molecular Materials for Future Electronics within the EC Contract no. G5MA-CT-2002-04049.

References

[1] Waser, R., (2003), Ed. *Nanoelectronics and information technology*, Wienheim, Wiley = VCH Verlag.

[2] IBM press release, June 9, 2003 (see IBM web page http://www.ibm.com/us/).

[3] Babich M.N., Broto J.M., Fert A., Nguyen Van Dau and Peroff F., (1988), *Phys. Rev. Lett. 61*, pp. 2472-2475; Binasch G., Gruenberg P., Saurenbach F. and Zinn W., (1989) *Phys. Rev. B39*, pp.4828-4830.

[4] see for review: Kouwenhoven L., and Glazman L. I., (2001), *Physics World, 14*, pp.33-38; and references therein.

[5] Nygard J., Cobden D.H., and Lindelof P.E., (2000), *Nature* 408, pp.342-346.

[6] Liang W.J., Shores M.P., Bockrath M., Long J.R., and Park H., (2002), *Nature 417*, pp.725-729.

[7] Park J., Pasupathy A.N., Goldsmith J.I., Chang C., Yaish Y., Petta J.R., Rinkoski M., Sethna J.P., Abruna H.D., McEuen P.L., and Ralph D.C.,(2002), *Nature 417*, pp. 722-725.

[8] Gores, J., Goldhaber-Gordon, D., Heemeyer, S., Kastner, M. A., Shtrikman, H., Mahalu, D. and Meirav, U., (2000), *Phys. Rev. B 62*, pp.2188-2194.

[9] Liang W.J., Bockrath M., Bozovic D., Hafner J.H., Tinkham M., and Park H., (2001), *Nature 411*, pp.665-669.

[10] Ferry D. K., (1997), *Transport in nanostructures*, Cambridge, Cambridge University Press; Haug, H. and Jauho, A.-P. (1998), *Quantum Kinetics in Transport and Optics of Semiconductors*, Springer Verlag.

[11] Bulka, B. R. and Stefanski, P., (2001), *Phys. Rev. Lett. 86*, pp.5128-5132.

[12] Heemeyer, S. (2000), Ph.D. thesis, Massachusetts Institute of Technology.

[13] Kotliar G. and Ruckenstein A. E. (1986), , *Phys. Rev. Lett. 57*, pp. 1362-1366.

[14] Sergueev N., Sun Q.F., Guo H., Wang B.G., Wang J., (2002), *Phys. Rev. B 65*, art. no. 165303; Lopez R. and Sanchez D., (2003), *Phys. Rev. Lett. 90*, art. No. 116602; Bulka B.R. and Lipinski S., (2003), *Phys. Rev. B 67*, art. No. 024404; Martinek J, Utsumi Y., H. Imamura H., Barnas J., Maekawa S., König J., and G. Schön, (2002), e-Print archive, http://arxiv.org/, cond-mat/0210006; Martinek J., M. Sindel M., Borda L., Barnas J., König J., Schön G. and von Delft J.,(2003), e-Print archive, http://arxiv.org/, cond-mat/0304385.

[15] Coleman P,(1987), *Phys. Rev. B 35*, pp. 5072-5116.

[16] Julliere, M.,(1975), *Phys. Lett. 54A*, pp. 225-227.

[17] Kostyrko T. and Bulka B. R. (2003), *Phys. Rev. B 67*, art. No. 205331.

[18] Aguado R. and Langreth D.C.,(2000) *Phys. Rev. Lett. 85*, pp.1946-1949.

[19] Orellana P.A., Lara G.A. and Anda E.V.,(2002) *Phys. Rev. B65*, art. No. 155317.

[20] Prengel F., Wacker A., and Scholl E., (1994) *Phys. Rev, B50*, pp.1705-1712; Aguado R., C. Platero C, Moscoso M. and Bonilla L.L.,(1997) *Phys. Rev. B55*, pp.16053-16056.

[21] Grahn H.T., Haug R.J., Muller E. and Ploog K., (1991) *Phys. Rev. Lett. 67*, pp.1618-1621.

QUANTUM INTERFERENCE AND SPIN-SPLITTING EFFECTS IN $Si_{1-X}Ge_X$ p-TYPE QUANTUM WELL

V.V. Andrievskii[1], I.B. Berkutov[1], T. Hackbarth[2], Yu.F. Komnik[1], O.A. Mironov[3,4], M. Myronov[3], V.I. Litvinov[5], and T.E. Whall[4]

[1] *B. I. Verkin Institute for Low Temperature Physics and Engineering of NAS of Ukraine, 47, Lenin Ave. Kharkov, 310164, Ukraine*

[2] *DaimlerChrysler Research Center, Wilhelm-Runge Str. 11, D-89081 Ulm, Germany*

[3] *Department of Physics, University of Warwick, Coventry CV4 7AL, United Kingdom*

[4] *International Laboratory of High Magnetic Fields and Low Temperatures, 53-421 Wroclaw, Poland*

[5] *WaveBand Corporation, 17152 Armstrong ave, Irvine CA, USA*

O.A.Mironov@warwick.ac.uk

Abstract The magnetoquantum and quantum-interference effects of two-dimensional hole gas in $Si/SiGe$-based heterostructures with $Si_{1-X}Ge_X$ quantum wells were studied in the temperature range $0.35 - 70K$ in a magnetic field up to $11T$. In high magnetic fields ($B > 1.5T$) the magnetic field dependencies of the samples resistances exhibit the Shubnikov-de Haas oscillations. The positive growth and maximum of magnetoresistances are observed in weak magnetic fields limit ($B < 0.1T$), which can be attributed to the influence of weak localization of the charge carriers when the inelastic scattering time τ_φ and spin orbit scattering time τ_{SO} have close values. It is shown that in the heterostructures studied splitting of the spin states occurs due to the influence of the perturbing potential (Rashba mechanism). The corresponding quantum times and spin splitting values are calculated.

Keywords: SiGe heterostructure; Weak localization; Spin-Splitting

1. Introduction

Recently much attention has been concentrated on the low-dimensional conductors exhibited quantum effects at low temperatures. One example of those conductors is two-dimensional hole gas (2DHG) in quantum wells (QWs) formed in modulation-doped semiconducting heterostructures. In this paper the kinetic properties of selectively boron-doped $Si_{1-X}Ge_X$ QWs with 2DHG

A.S. Alexandrov et al. (eds.), Molecular Nanowires and Other Quantum Objects, 319–328.

were investigated at low temperatures. Magnetoquantum Shubnikov - de Haas (SdH) oscillations, quantum - interference / weak localization (WL), and hole-hole interaction (HHI) effects were observed at $T = 0.35K - 35K$ temperature range. Besides, spin-orbit effects showed up in weak magnetic fields where the spin degeneracy was lifted.

The extent of the effects was compared for two metamorphic heterostructures with $Si_{0.2}Ge_{0.8}$ (sample I) and $Si_{0.05}Ge_{0.95}$ channels (sample II) as QWs with metallic-like 2DHG. The alloy of these Ge -compositions are on different sides of the cross-over point with value $X = 0.85$ at which the band structure changes from the Si - type to the Ge - type due to increasing X. The samples were grown by the solid-source molecular-beam epitaxy technique. A $Si_{1-y}Ge_y$ buffer layer of a linearly graded Ge composition $0 < y < 0.3$ (for sample I) and up to $y = 0.63$ (for sample II) was grown on an n-type $Si(001)$ substrate. On the top of buffer layer, subsequent layers were formed, one upon the other, in a succession: *(i)* a $10nm$ (sample I) and $11nm$ (Sample II) thick $Si_{1-X}Ge_X$ channels, *(ii)* a $7nm$ (Sample I) and $10nm$ (Sample II) undoped $Si_{1-y}Ge_y$ "spacers", *(iii)* $10nm$ $Si_{1-y}Ge_y$ layer with boron doping concentration 2×10^{18} cm^{-3} for both samples, and, finally, *(iv)* a $3nm$ Si-cap coating. Because of valance band offset between Si and $Si_{1-X}Ge_X$ and single-sided remote doping of QW, asymmetric (triangular-like) QW profile appear for 2DHG at the heterointerface, closed to boron-doped layer. Energy levels of QW are occupied by 2DHG owing to the boron impurity atoms presence in the remote $Si_{1-y}Ge_y$ layer. Since "spacer" from the QW separates the layer with boron impurities, the ionized impurity scattering of the 2DHG mobile charge carriers is essentially decreases.

The sample shape for magnetotransport measurements was as "double cross/ Hall bar", i.e. a narrow ($\sim 0.5mm$) strip with two pairs of narrow potential contacts about 1.5 mm apart. The measurements of diagonal and off-diagonal resistance components were carried out at the temperatures $0.35K - 70K$ and in the magnetic fields up to $11T$.

2. 2DHG characteristic parameters

The temperature dependencies of the zero magnetic field resistance (per square) for samples I and II are shown in Figure 1. Above $T \sim 35K$ and $T \sim 20K$ (for sample I and II respectively) the resistance decreases with temperature decreasing, i.e. the "metallic" behavior is observed. However, in these temperature regions the resistance has a minimum after which it starts growing despite the lowering temperature. These anomalous temperature dependencies pointed us to the effects of weak localization and quasiparticle (hole-hole) interaction, which are as explanation to the appearance of the resistance quantum corrections [1-5].

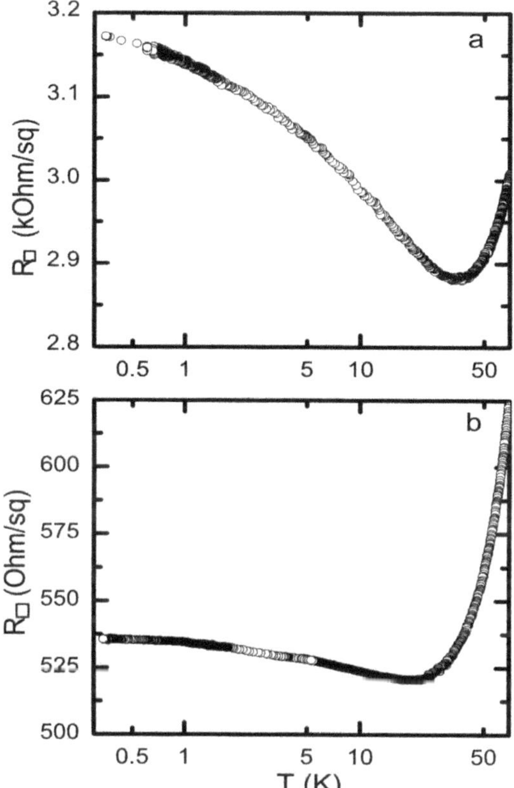

Figure 1. Temperature dependencies of sheet resistance $R_\square$ for sample I (a), and II (b).

The diagonal and Hall components of the samples resistances are shown in Figure 2 as a function of magnetic field B at temperature $T \sim 0.35K$. The magnetoresistance (MR) curves taken in strong field B exhibit the SdH oscillations whose amplitudes decrease appreciably as the temperature rises. Quantum Hall steps are clearly seen in the Hall resistance data.

The analysis of experimental data permits us to extract the main 2DHG characteristics. The 2DHG concentrations obtained from measured Hall coefficient are $p_H = 1.6 \times 10^{12}\ cm^{-2}$ (sample I) and $p_H = 1.76 \times 10^{12}\ cm^{-2}$ (sample II). The temperature and magnetic field changes in the SdH oscillations amplitudes provide the information about the effective mass and the concentration of the charge carriers (see Ref. [6] for calculation method description).

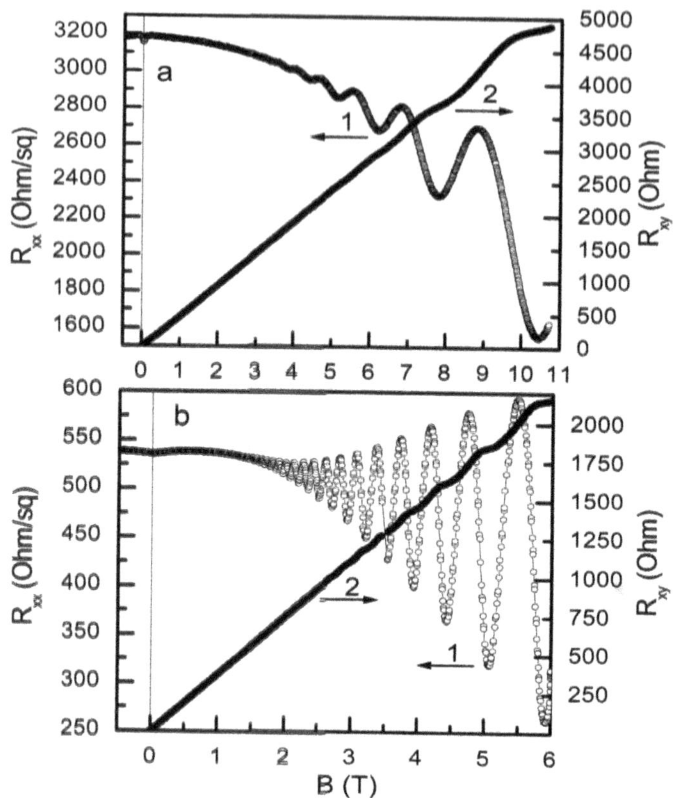

Figure 2. Magnetic field dependence of the diagonal R_{xx} (curve 1) and off-diagonal R_{xy} (curve 2) components of resistance $R_{\square}$ for samples I (a) and II (b) at temperatures 0.346 and 0.344 K, respectively.

The following values were obtained: $m^* = 0.16\, m_0$ (m_0 is the free electron mass), $p_{SdH} = 1.46 \times 10^{12}\ cm^{-2}$ for sample I and $m^* = 0.156\, m_0$, $p_{SdH} = 1.62 \times 10^{12}\ cm^{-2}$ for sample II.

The estimated m^* and p_H enable us to calculate (from electric conductivity of the channel) the elastic scattering times $\tau = 1.27 \times 10^{-13}\ s$ and $\tau = 6.05 \times 10^{-13}\ s$, and mean free paths $\ell = 29.08\ nm$ and $\ell = 14.94\ nm$ (using $R_0 = 2800\ Ohm/sq$ and $R_0 = 520\ Ohm/sq$ for samples I and II respectively), as well as the Fermi velocity V_F and energy ε_F, hole mobility μ, and diffusion coefficient D. The following relations for 2D system were used: $V_F = \frac{\hbar}{m}\sqrt{2\pi p}$, $\varepsilon_F = \frac{\pi \hbar^2 p}{m^*}$, $D = \frac{1}{2}\, V_F^2\, \tau$, and $\mu = \frac{1}{R_0 e p_H}$. The results obtained were: $V_F = 2.29 \times 10^7\ cm/s$ and $V_F = 2.47 \times 10^7\ cm/s$; $D =$

$33.4\ cm^2/s$ and $D = 184.4\ cm^2/s$; $\varepsilon_F = 23.97\ meV$ and $\varepsilon_F = 26.98\ meV$; $\mu = 1.396 \times 10^3\ cm^2/Vs$ and $\mu = 6.819 \times 10^3\ cm^2/Vs$ for sample I and II respectively. So, we can estimate the energy of the ground level $E_1 = \frac{\pi^2 \hbar^2}{2\, a^2\, m^*}$, where a is the QW width. $E_1 = 23.5\ meV$ ($a = 10\ nm$ for sample I) and $E_1 = 19.9\ meV$ ($a = 11\ nm$ for sample II) appear to be close to the Fermi energy value. This prompts the conclusion that only one quantum level is occupied by 2DHG in each QW samples.

3. Magnetoresistance Analysis

The peculiar features of the MR in weak fields B (see Figure 3) are the result of WL and HHI effects. The analysis of the quantum corrections to the conductivity permits us to derive information about the characteristic relaxation times τ_φ and τ_{SO}. The τ_φ is the phase relaxation (dephasing) time due to inelastic scattering and τ_{SO} is the time of spin orbit relaxation.

The change of the conductivity for the 2D system in the perpendicular magnetic field is induced by the WL effect and can be described as [4]:

$$\Delta\sigma_B^L = \frac{e^2}{2\pi^2\hbar}\left\{\frac{3}{2}f_2\left(\frac{4eBD\tau_\varphi^*}{\hbar}\right) - \frac{1}{2}f_2\left(\frac{4eBD\tau_\varphi}{\hbar}\right)\right\}, \qquad (1)$$

where $\tau_\varphi^{-1} = \tau_{\varphi 0}^{-1} + 2\tau_S^{-1}$, $(\tau_\varphi^*)^{-1} = \tau_{\varphi 0}^{-1} + \frac{4}{3}\tau_{SO}^{-1} + \frac{2}{3}\tau_S^{-1}$, $f_2 = ln(x) + \Psi\left(\frac{1}{2} + \frac{1}{x}\right)$, Ψ is the logarithmic derivative of the Γ-function,

$$f_2(x) = \begin{cases} \frac{x^2}{24} & , x \ll 1 \\ ln(x) + \Psi\left(\frac{1}{2} + \frac{1}{x}\right) & , x \gg 1 \end{cases}.$$

The characteristic field $B_0^L = \hbar/(4eD\tau_\varphi)$ corresponds to the change of the function $f_2(x)$ from quadratic to logarithmic. The change in the quantum correction in the magnetic field can be analyzed using the relation $-\Delta\sigma_B(B) = \frac{R(B)-R(0)}{R(B)R_\square(0)}$, where $-\Delta\sigma_B(B)$ reflects the MR variation. For sample I in the lowest magnetic fields MR is positive (see Fig 3a.) and has a specific initial region, where resistance grows up to maximum. Then it starts to decrease slowly, changing into negative MR. This shape of the MR curve is typical for the WL effect [3, 4] when τ_φ and τ_{SO} are comparable each other. The MR maximum value decreases rapidly as the temperature rises. This MR maximum appearance is due to the competing contributions of the opposite-in-sign terms in Eq.1 to the total weak localisation quantum correction. Those two terms appear when the spin states on the adjoint trajectories forming the interference contribution to the conductivity are taken into account. The first term corresponds to the triplet spin state (total spin $J = 1$), the other describes the singlet state ($J = 0$). The triplet state is characterized by the total momentum projection ($M = 0, \pm 1$), which changes randomly due to spin-orbit scattering. The spin-orbit scattering

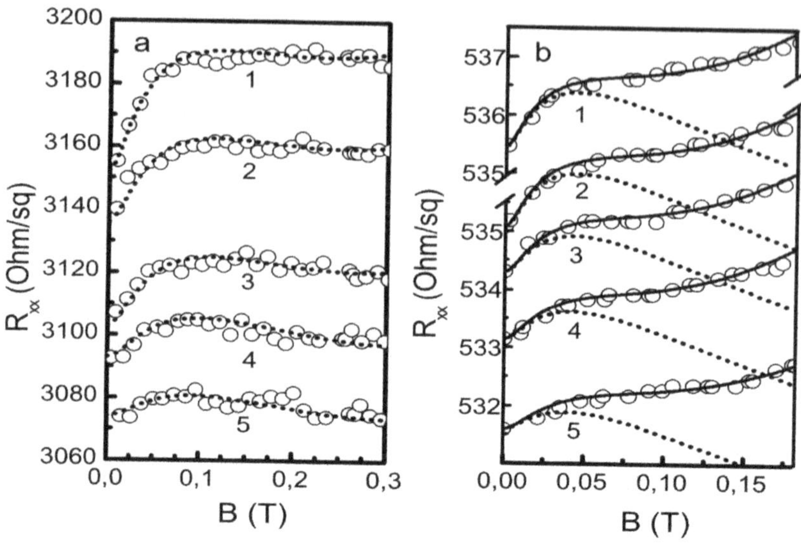

Figure 3. Magnetic field dependencies diagonal R_{xx} component of resistance $R_\square$ at temperatures: for sample I (a) $0.346K$ (1), $0.754K$ (2), $1.44K$ (3), $2.05K$ (4), $3.6K$ (5); for sample II (b) $0.355K$ (1), $0.7K$ (2), $1.1K$ (3), 1.56 K (4), $2.73K$ (5). The dash lines describe the WL contribution to MR that was calculated according to Eq.1.

suppresses coherence in the triplet spin state. The contribution of the triplet term to the conductivity forms negative MR since its interference enhances the resistance in zero magnetic field in comparison with classical magnitude. The singlet term in Eq.1 has negative sign and the wave interference in the singlet state stimulates the conductivity instead of suppressing it.

As the magnetic field grows at $\tau_{SO} \leq \tau_\varphi$, the interference of the adjoint waves in the singlet spin state starts to decrease. As the results, the resistance increases with the field because such interference leads to a suppression of the resistance in zero magnetic field. A further growth of the magnetic field suppresses the interference of adjoint waves in the triplet state and thus causes negative MR. Note, the contribution of the triplet state is three times over that of the singlet state.

Computer fitting of the theoretical dependence Eq.(1) and experimental MR data permits us to extract τ_φ, τ_φ^*, and finally τ_{SO}. Such fitting is most efficient, when a $\sim B^2$-type term is accounted. We believe that this term accounts for the hole-hole interaction in the Cooper channel and corresponds to quasiparticle repulsion.

The magnetoconductivity correction related to the particle interaction in the Cooper channel is described as [4, 5]:

$$\Delta\sigma_B^C = -\frac{e^2}{2\pi^2\hbar}\lambda_B^C\varphi_2(\alpha); \qquad \alpha = \frac{2eBD}{\pi k_B T} \tag{2}$$

where

$$\varphi_2(\alpha) = \int_0^\infty \frac{t\,dt}{sinh^2(t)}\left(1 - \frac{\alpha t}{sinh(\alpha t)}\right)$$

$$\varphi_2(x) = \begin{cases} 0.3\alpha^2 & , \alpha \ll 1 \\ ln(\alpha) & , \alpha \gg 1 \end{cases}$$

and λ_B^C is the interaction constant. The characteristic field $B_0^C = \frac{\pi k_B T}{2eD}$ corresponds to the change of the functional dependence $\varphi_2(\alpha)$ from parabolic to logarithmic. For low magnetic fields ($B < B_0^C$) we can therefore use a parabolic approximation for $\Delta\sigma_B^C$ if $B_0^L \ll B_0^C$. Theoretical Eq.1 with estimated τ_φ and τ_{SO} (dash curves) is compared with the experimental data for sample I in Fig. 3a.

The spin-orbit effects in low magnetic fields are less evident in the experimental MR data for Sample II (see Fig. 3b). Meanwhile, against the background of the ascending MR curve in the magnetic fields $B < 0.1\ T$ we can notice feature with the initial steep MR growth. A similar MR growth was observed for sample I. In the fields $B > 0.1\ T$ the MR curves become smooth with the subsequent slight magnetoresistance growth up to $B \sim 0.6\ T$. This feature is connected with the implicit maximum data MR curve, as in case of sample I. The contribution of this MR maximum decreases rapidly as the temperature rises.

For sample II the fitting of the theoretical dependences and the experimental results is more complicated than for sample I. However, we have got good fit (solid curves in Fig. 3b) to the experimental results, using the known procedure and the assumption of an implicit maximum. The HHI Eq.2 was used for sample II in general form because $B_0^L \sim B_0^C$ in this case. This indicates the role of the HHI in this sample has become significant due to a higher 2DHG concentration. The dash curves in Fig. 3b show the contribution of WL with the effects of the spin-orbit interaction become evident. These curves are similar to the curves for sample I in which WL contribution to MR is dominated.

The obtained values for τ_φ, and τ_{SO} are shown in Figure 4. The dependence $\tau_\varphi \sim T^{-1/2}$ (solid curves in Fig. 4) is most convenient to describe the τ_φ temperature variation. The inelastic relaxation mechanism of such type is unknown at the moment, and this required further investigation. We suggest, that such $\tau_\varphi(T)$ dependence is quite possible. The argument in this favor is the solid curve for sample I, which crosses the straight line $\tau_{SO} = const$ at $T \sim 6\ K$. At

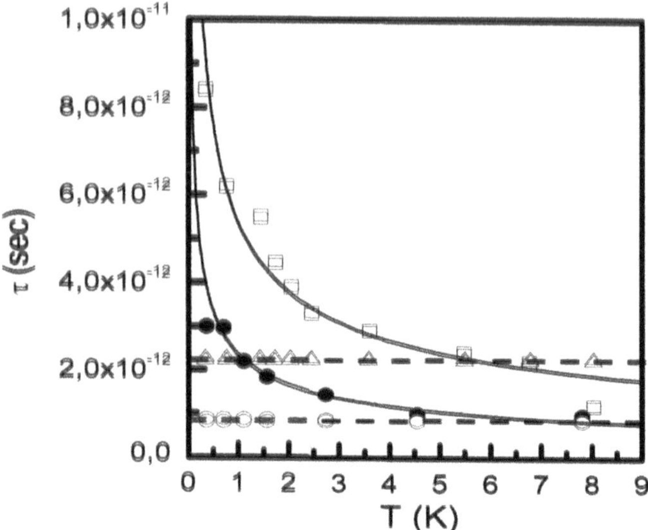

Figure 4. Temperature dependence of the inelastic scattering time τ_φ (□, •) and the spin-orbit scattering times τ_{SO} (△, ⊖) for samples I (triangles), and II (circles) respectively.

this temperature MR maximum (and the positive MR itself) disappear because the inequality $\tau_{SO} \ll \tau_\varphi$ reverses sign.

The calculated times of the spin-orbit relaxation are $\tau_{SO} = 2.24 \times 10^{-12}\ s$ for sample I and $\tau_{SO} = 8.5 \times 10^{-13}\ s$ for sample II.

4. Spin splitting and spin-orbit relaxation

The time of spin-orbit relaxation τ_{SO} obtained from the WL-related quantum corrections analysis can be used to calculate the value of spin splitting. On splitting, two hole spin subsystems with close parameters appear in bulk semiconductors and heterostructures. Splitting can occur for two reasons. It can be either due to the asymmetric crystal field existing in a bulk crystal without inversion center (Dresselhaus model [7]), or due to the nonuniform perturbing potential induced by the asymmetric potential well which is produced by the single-side QW doping used for heterostructures (Rashba model [8, 9]). In the first case the spin splitting is proportional to cubic wave vector $\sim k^3$. As it was shown in Ref. [10, 11], the quantum well in the crystal having no inversion center leads to a k-linear contribution to the effective hole Hamiltonian (resulting from averaging over the cubic terms along the quantization axis). Within the Rashba mechanism the spin splitting is linearly depend on the wave vector

$\sim k$. Note, the spin splitting follows the Rashba mechanism in our $Si_{1-X}Ge_X$ heterostructures since the Si and Ge are center-symmetrical crystals.

In systems with split spin states, spin-orbit relaxation realizes basically through the Dyakonov-Perel mechanism [12]. Spin splitting is equivalent to the magnetic field acting on the spin. As a result, the spin is forced to precess at the frequency Ω_0. On scattering, the direction of the charge carrier momentum changes and the precession axis turns. At the condition when

$$\Omega_0 \tau \gg 1 \tag{3}$$

the spin relaxation determined by equation

$$\tau_{SO}^{-1} \simeq \Omega_0 \tau, \tag{4}$$

where the precession frequency is $\Omega = \Delta/2\hbar$.

Assuming that the spin-orbit relaxation time τ_{SO} extracted from WL quantum corrections characterizes spin relaxation and using value calculated from Eq.4, we can obtain the spin splitting values Δ, which is $\Delta = 2.47\, meV$ for sample I and $\Delta = 1.8\, meV$ for sample II. Note that inequality $\Delta < \hbar/\tau$ following from Eq.3 is obeyed well for sample I. For sample II this inequality corresponds to the limiting case and Δ is a rough estimate.

The more detailed theory of weak localization in the case of p-type QWs with *strong* spin-orbit interaction has been recently developed in Ref. [13, 14]. Fitting of our experimental results to this elaborated theoretical treatment will be published elsewhere.

Acknowledgments

I.B.B. acknowledges for the support from National Academy of Science of Ukraine research fellowship for young scientists. This work was partially supported by INTAS-01-0184 project.

References

[1] P.A Lee and T.V. Ramakrishnan, Rev. Mod. Phys. **53**, 287 (1985).

[2] B.I. Altshuler and A.G. Aronov, in Electron - Electron Interaction in Disordered Systems. Modern Problems in Condensed Matter Science. **10**, A.L. Efros and M.P. Pollak (eds), Amsterdam, North-Holland, p.1 (1985)

[3] B.L. Altshuler, A.G. Aronov, M.E. Gershenson, and Yu.V. Sharvin, in Sov. Sci. Rev. **A9**, Schur, Switzerland, Harwood Academic Publisher Gmbh, p. 223 (1987)

[4] B.L. Altshuler, A.G. Aronov, A.I. Larkin, and D.E. Khmel'nitskii, Zh. Eksp. Teor. Fiz. **81**, 768 (1981) [Sov. Phys. JETP. **54**, 411 (1981)].

[5] B. L. Altshuler, A. G. Aronov, and P. A. Lee, Phys. Rev. Lett. **44**, 1288 (1980).

[6] Yu. F. Komnik, V.V. Andrievskii, I.B. Berkutov, S.S. Kryachko, M. Myronov, and T.E. Whall, Low Temp. Phys. **26**, 609 (2000).

[7] G. Dresselhaus, Phys. Rev. B **100**, 580, (1955).

[8] E.I. Rashba and V.I. Sheka, Solid State Physic (in Russian), Akad. Nauk SSSR, **2**, 162 (1959).

[9] Yu. A. Bychkov and E. I. Rashba, Pis'ma Zh. Eksp. Teor. Fiz. **46**, 66 (1984) [JETP Lett. 39, 78 (1984)].

[10] S.V. Iordanskii, Yu.B. Lyanda-Geller, and G.E. Pikus, Pis'ma Zh. Eksp. Teor. Fiz. **60**, 199 (1994) [JETP Lett. 60, 206 (1994)].

[11] F.G. Pikus and G.E. Pikus, Phys. Rev. B **51**, 16 928 (1995).

[12] M. I. Dyakonov and V. I. Perel. Zh. Eksp. Teor. Fiz. **60**, 1954 (1971) [Sov. Phys. JETP **33**, 1053 (1971)].

[13] N. S. Averkiev, L. E. Golub, and G. E. Pikus, JETP **86**, 780 (1998).

[14] S. Pedersen, C. B. Sorensen, A. Kristensen, P. E. Lindelof, L. E. Golub, and N.S. Averkiev, Phys. Rev. B **60**, 4880 (1999).

THE SIZE-INDUCED METAL-INSULATOR TRANSITION IN MESOSCOPIC CONDUCTORS

P. P. Edwards[a*], S. R. Johnson[b], M. O. Jones[a] and A. Porch[c*]
[a] *Inorganic Chemistry Laboratory, Oxford University, South Parks Road, Oxford, OX1 3QR*
[b] *School of Chemistry, The University of Birmingham, Edgbaston, Birmingham, B15 2TT*
[c] *School of Engineering, The University of Cardiff, Cardiff, CF24 0YF*
peter.edwards@chem.ox.ac.uk

Abstract We discuss the possibility - and experimental realisation - of a Size-Induced-Metal-Insulator-Transition (SIMIT) occurring entirely within individual particles of mesoscopic conductors. Thus, nanoscale particles of indium and gold exhibit electrical conductivities a factor of *ca.* 10^7 below that of the corresponding bulk metal. In this size regime, individual mesoscopic particles of these prototypical metallic elements of the Periodic Table behave as low conductivity insulators or non-metals. In addition to its fundamental importance in condensed matter science, an understanding of the SIMIT in isolated mesoscopic conductors may allow for an unprecedented degree of control and design of microelectronic device materials based on mesoscopic conductors. For instance, by changing the physical size of a mesoscopic conductor any value of electrical conductivity - and associated complex dielectric function - can be adjusted between a metal and an insulator.

Keywords: size-induced metal-insulator-transition, nanoscale particulate matter, microwave cavity perturbation, conductivity

1. Introduction

Metal nanoparticles in the size regime from a single atom up to conglomerates of the order of a few tens of nanometers challenge conventional notion in that their characteristic physical size represent *the* key variable governing their physical and chemical properties [1-8]. Particles within this size regime are thus intermediate between the microscopic size regime (nm) on the one hand, and the macroscopic, bulk state of matter on the other. For metallic elements of the Periodic Table, therefore, the inevitable consequence of ever-reducing the size of an individual piece of particulate matter must be the ultimate cessation of metallic conduction. This clearly leads to the intriguing possibility of

A.S. Alexandrov et al. (eds.), Molecular Nanowires and Other Quantum Objects, 329–342.

a Size-Induced Metal-Insulator Transition (SIMIT) occurring entirely within a particle of mesoscopic conductor [9,10]. The problem of the SIMIT is represented schematically in Figure 1, which attempts to highlight the key features of the *Macroscopic*, *Mesoscopic* and *Microscopic* regimes of metallic and non-metallic nanoparticles [6, 11, 12]. The notion of such an electronic phase transition occurring entirely within a single grain of a nominally metallic element, some of the accompanying basic science and its possible experimental realisation represents the basic theme of this paper.

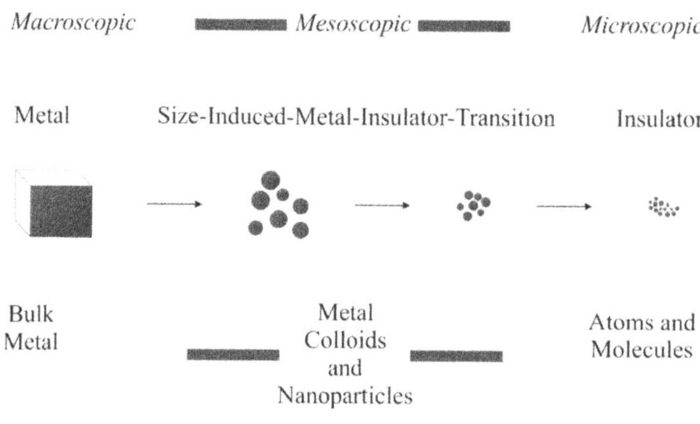

Figure 1. A representation of the successive division of a single grain of metal.

2. Size - Dependent Electronic Structure of Metal Nanoparticles

At a temperature of absolute zero (T = 0K) the highest occupied electronic energy state of a metal resides at the Fermi Energy, E_F. Importantly, above E_F, there exist an infinite number of infinitesimally-separated and empty electronic energy levels which can easily be populated upon the application of an external electric field, even at low temperatures [13].

In contrast, when a metal is sufficiently finely divided to mesoscopic or microscopic dimensions, the notion of an electronic energy continuum breaks down and the continuum of electronic energy levels now gives way to a manifold of discrete energy levels with an average separation, $\delta_K = E_F/N$, where N is the total number of atoms contained within the metal particle. When this level separation, the so-called Kubo energy gap [7], becomes comparable with

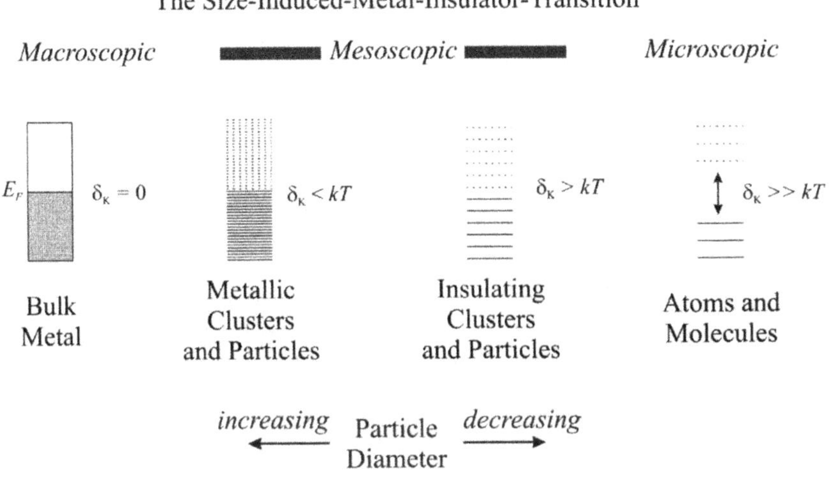

Figure 2. The effect on electronic structure of the division of a metal.

the (ambient) thermal energy, kT, the electronic energy levels now become discrete, rather than continuous. This process will ultimately determine the metallic or insulating status of an individual particle. A representation of the size-dependent electronic energy level structure of metal particles is given in Figure 2. A collection of high resolution electron micrographs of nanoparticles of three metals is given in Figure 3 [14].

In Figure 4, we show the size (particle diameter) dependence of the (average) energy level spacing of the prototypical metallic element, gold. Here the relevant values of the Kubo energy gap are given in degrees Kelvin. In addition, the finite size of the particle will also lead to a significant number of gold atoms being located on the surface of the cluster or particle. We show also in Figure 4 the computed percentage, P_S, of surface atoms for gold particles as a function of particle diameter [11].

For $E_F \approx$ 10eV and $N \approx 10^3$, $\delta_K \approx$ 120K, a quantity which is of the same order-of-magnitude as thermal energy kT at room temperature. For $N = 10^5$, δ_K is equivalent to a temperature of *ca.* 1K. One opinion therefore argues that genuine 'metallic' properties (*i.e.* high conductivity *etc.*) can only be sustained in particles at finite temperatures when $\delta_K < kT$, a situation enabling the facile creation of electronic charge carriers *via* thermal excitation. From this viewpoint, the following simple criteria; would presumably define the experimental parameters for a SIMIT in a finite, mesoscopic gold particle at temperatures above T = 0K.

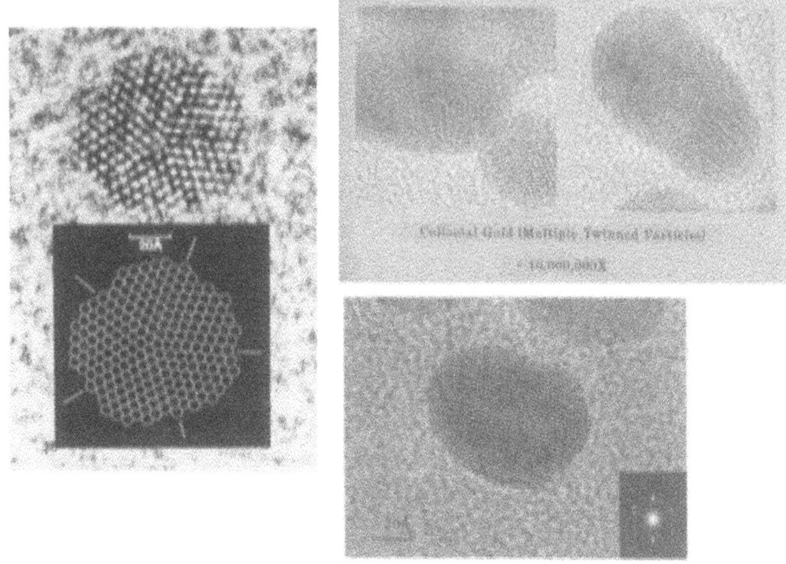

Figure 3. High resolution electron micrographs of silver, gold and palladium nanoparticles [14].

$$\delta_K < kT: (\textit{Metal}) \text{ and } \delta_K \gg kT: (\textit{Insulator})$$

Of course, as we move closer and closer to the microscopic regime, we should become increasingly concerned at the validity of any approaches derived from what one might term the continuum physics of macroscopic bulk metals. Conversely, any theoretical treatment originating from the quantum - chemistry viewpoint ('*Atoms and Molecules*' regime in Figure 1) becomes increasingly unrealistic as we enter the mesoscopic and macroscopic regimes. Mesoscopic conductors of controlled particle size, well-separated from each other in an insulating matrix, constitute ideal systems for investigating the possibility of a SIMIT. Measuring the d.c. conductivity of such isolated mesoscopic metal particles represents a challenging and key experiment and necessitates the use of innovative experimental techniques. The direct contact measurement of the I-V characteristics of individual particles represents one such approach. Here we highlight the novel method of contactless microwave absorption measurements which are applicable to a wide range of material systems.

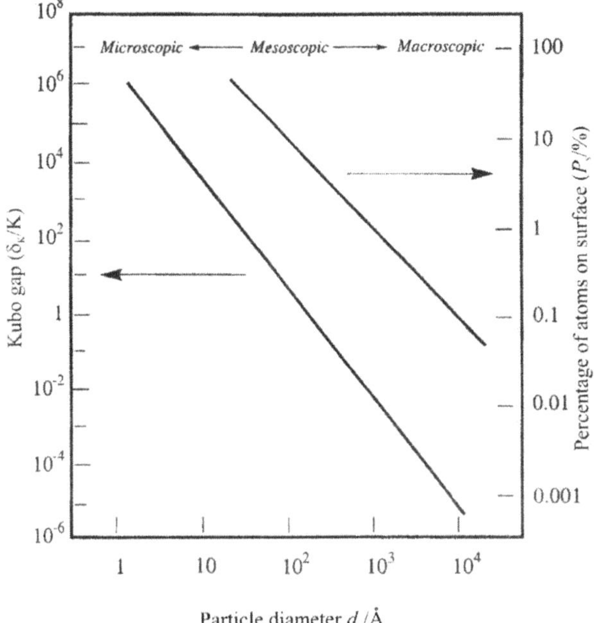

Figure 4. Particle diameter dependence of the electronic energy level structure of particulate gold.

3. Electrical Conductivity of Mesoscopic Gold Particles from Microwave Loss Studies

The stored and dissipated electrical energy within mesoscopic-scale metal particles are deduced from direct measurements of the change of resonant frequency and resonant bandwidth on inserting a sample into the microwave resonator (see section 4 below), enabling extraction of the electronic conductivity of the individual particles [15, 16]. This analysis depends crucially on the ratio of the electromagnetic skin depth, δ, to the particle diameter d. In terms of the angular frequency, ω, and the bulk d.c. conductivity, σ, the skin depth is defined as $\delta = \sqrt{2/\mu_o\omega\sigma}$ (for gold at 3 GHz this is around 1.3 μm). Spherical particles can be thus categorised into two limiting regimes according to the magnitude of the particle diameter.

(a) Large particle limit, $d \gg \delta$. Here the electromagnetic fields are confined to the surface of the particle and are described by classical skin effect theory. The microwave power dissipation per unit surface area of each particle is proportional to the particles' surface impedance $R_S = 1/\sigma\delta \equiv \sqrt{\mu_o\omega/2\sigma}$, which tends to zero in the limit of infinite conductivity.

(b) Small particle limit, $d \ll \delta$, of particular interest when considering conducting nanoparticles. The electromagnetic fields now penetrate the sample fully and for conducting particles placed in a microwave electric field, the microwave power dissipation, to the first approximation, is proportional to $\sigma E_{in}^2 V_s$ where E_{in} is the magnitude of the electric field within the particle and V_s is the total volume of the particles. Performing microwave measurements in this size regime therefore provides a direct, but contactless determination of the electrical conductivity of the individual particle.

A spherical particle of low conductivity within the limit $d \ll \delta$ can be treated as a lossy dielectric, producing a uniform internal electric field E_{in} within the particle in response to a uniform applied field E_0; such a quasistatic approximation gives rise to the following relationship between the internal and applied fields

$$E_{in} = E_0/(1 + \tfrac{1}{3}(\varepsilon - 1))$$

where the depolarising factor of a spherical particle is assumed to equal 1/3. The relative permittivity ε is now complex, and is written conventionally $\varepsilon = \varepsilon_1 - i\varepsilon_2$, where ε_1 quantifies the stored electrical energy and ε_2 the energy dissipation, related to conductivity via $\varepsilon_2 = \sigma/\omega\varepsilon_0$.

Microwave measurements are carried out using the cavity perturbation method, where the change in resonant bandwidth is measured on inserting the sample into a copper hairpin resonator. Provided that any resulting perturbation is small (valid if the volume of the sample is much less than the volume of the resonator), one finds that the fractional change in resonant bandwidth,$\Delta\omega_B/\omega_B$, is [17]

$$\frac{\Delta\omega_B}{\omega_B} \approx Q\alpha\frac{|E_{in}|^2}{E_0^2}\varepsilon_2 \approx \frac{Q\alpha\varepsilon_2}{(1+\frac{1}{3}(\varepsilon_1-1))^2+\frac{1}{9}\varepsilon_2^2} = \frac{9Q\alpha\varepsilon_2}{(2+\varepsilon_1)^2+\varepsilon_2^2} \quad (1)$$

where α is the fraction of the total electrical energy stored within the sample and Q is the unloaded quality factor of the resonator. Eqn. (1) thus allows the electrical conductivity to be determined. Two points should be noted here concerning the nature of measurements on gold nanoparticles like those described in Section 5. The first point concerns the inevitable particle size distribution. Provided that all of the particles remain in the small particle limit (*i.e.* $d \ll \delta$), the measured microwave losses only depend on the *total volume* of the sample and not on the details of the size distribution itself. The total volume of gold is determined straightforwardly in separate experiments. The second point concerns local field corrections associated with the presence of neighbouring particles and also the solvent, stabilisers *etc.*, surrounding the gold particles. Although these corrections can be applied in a rigorous manner, for dilute volume fractions (*i.e.* well-separated particles) in weakly polar solvents, they can be ignored to a good approximation. Combining the effects of local field corrections and the uncertainties associated with grain shapes leads to an overall uncertainty of a factor of around 2 when calculating the conductivity of mesoscopic conducting particles using this method.

Assuming that the quasi-static result of Eqn.(1) is valid at microwave frequencies for conductivity values as high as bulk gold (*i.e.* $\sigma = 4.3 \times 10^5 \Omega^{-1}\text{cm}^{-1}$ at room temperature), then even within the small particle limit (which will always be valid at 3 GHz for sub-micron gold particles) there are high conductivity ($\varepsilon_2 \gg \varepsilon_1$) and low conductivity ($\varepsilon_2 \ll \varepsilon_1$) limits. If an individual gold particle exhibits its full (*i.e.* bulk) conductivity (*i.e.* there is no size-induced suppression of conductivity) then $\varepsilon_2 = \sigma/\omega\varepsilon_0 \approx 2.6 \times 10^7 \gg 1$ at 3 GHz. Assuming that $\varepsilon_1 \approx 1$, Eqn.(1) reduces to

$$\frac{\Delta\omega_B}{\omega_B} \approx 9\alpha Q/\varepsilon_2 \qquad (\textit{high conductivity in the small particle limit}) \quad (2)$$

Typically, in our experiments $\alpha = 0.01$ and $Q \approx 2000$, so the fractional change in resonant bandwidth on sample insertion into the resonator is very small in this limit (of the order of 10 parts in a million), certainly too small to measure with any degree of confidence. Thus, highly conducting, small particles would therefore appear almost lossless according to this quasistatic analysis.

Conversely, if within the small particle limit the conductivity were to be suppressed enormously due to the Size-Induced Metal-Insulator Transition, then we could enter the low conductivity limit ($\varepsilon_2 \ll \varepsilon_1$, *i.e.* $\sigma \ll \omega\varepsilon_0$), and Eqn.(2) reduces to

$$\frac{\Delta\omega_B}{\omega_B} \approx \alpha Q\varepsilon_2 \qquad (\textit{low conductivity in the small particle limit}) \quad (3)$$

At 3 GHz, $\omega\varepsilon_0 = 1.7 \times 10^{-3}\Omega^{-1}\text{cm}^{-1}$; if the conductivity were, for example, a tenth of this then $\varepsilon_2 \approx 0.1$ and this would cause the resonant bandwidth to double (*i.e.* it is easily measurable). The fractional changes in resonant bandwidth as a function of conductivity in these low and high conductivity limits of a sample of gold nanoparticles is shown in Figure 5 based on Eqn.(2) for our 3 GHz hairpin resonator measurements.

In the present experimental configuration, the microwave system measures bandwidths accurately to at least 1 part in 10^3, allowing the conductivity of individual nanoparticles to be determined between the range of 10^{-7} up to around $10^3\Omega^{-1}\text{cm}^{-1}$, denoted by the dotted vertical lines on Figure 5. Unfortunately,

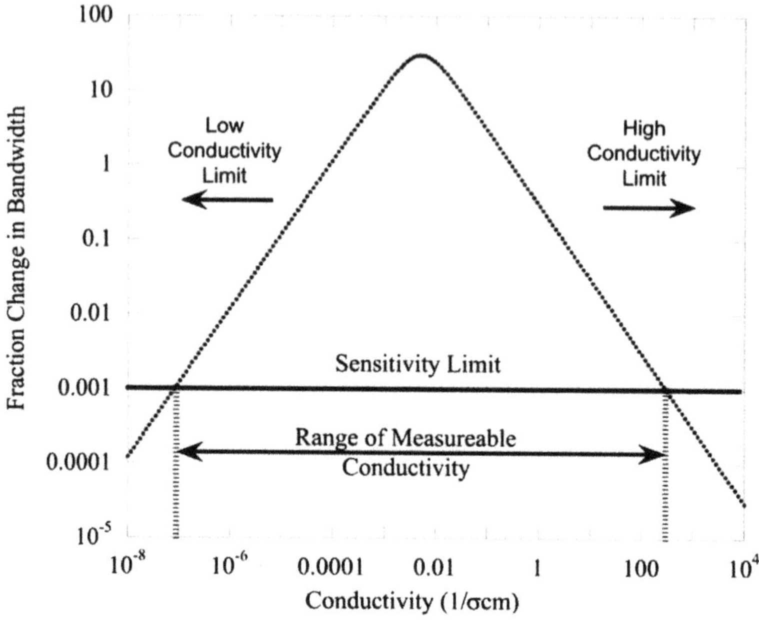

Figure 5. Fractional change in bandwidth $\Delta\omega_B/\omega_B$ vs. conductivity (1/σcm.)

we note that a single measurement of resonant bandwidth provides two possible values of conductivity (*e.g.* zero bandwidth would imply either zero or infinite conductivity!) depending on the conductivity limit (low or high). Since such small particles are completely penetrated by the microwave field, no additional information can be gleaned from the shift in resonant frequency on sample insertion, as is normally the case when measuring dielectric samples. However, since we would only expect to observe size-induced reductions in conductivity on decreasing the particle size, a unique value of conductivity can be deduced from whether the fractional bandwidth increases or decreases when performing this size reduction. Some experimental results on gold sols based on this analysis will be presented in Section 5.

4. The Experimental Method

Microwave measurements (around 3GHz) are carried out using a HP8720A network analyser. Microwaves are coupled in and out of the cavity using loop terminated co-axial transmission lines, and the transmitted microwave power is measured as a function of frequency. Samples are generally run in sealed

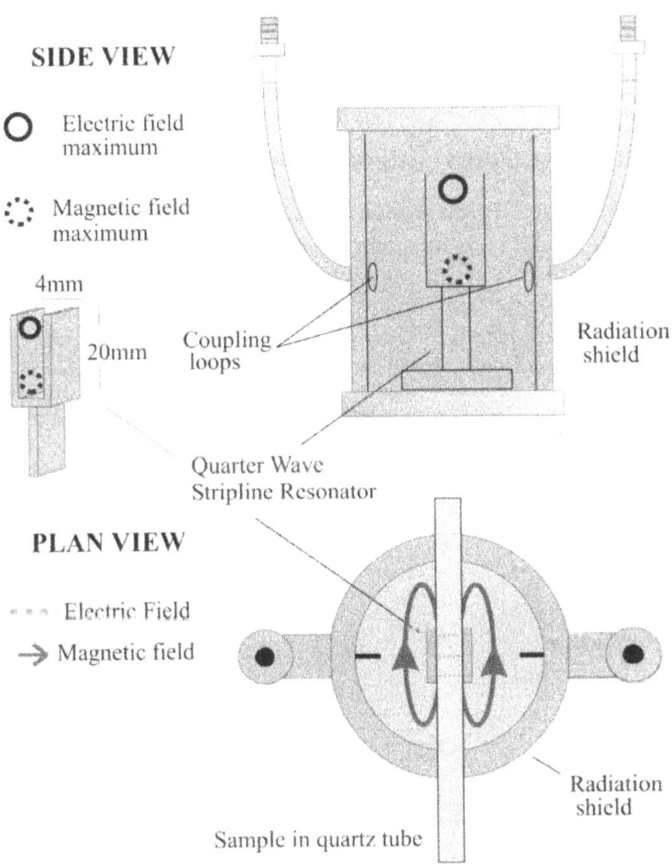

Figure 6. A schematic diagram of the copper stripline resonator.

Spectrosil tubes. The microwave resonator technique involves inserting the sample in a region of high microwave electric field within a $\lambda/4$ copper stripline ("hairpin") resonator. The tubes are passed through the hairpin via holes in the radiation shield. The sample data were measured relative to the empty cavity containing the empty quartz tube [15, 16].

The microwave interaction with the thin-walled quartz tube is small and thus it not cause significant alteration to the electric field in its vicinity, and neither does it contribute significantly to the microwave losses. The hairpin cavity has an unloaded quality factor of 2000 at room temperature. We note that this Q

factor is large for such a small cavity volume (~ 0.8 cm^3), resulting in a highly sensitive measurement of the nanoparticle response owing to the large sample filling factor in the cavity (*i.e.* α, as defined in Eqn.(1), is around 0.01). The samples can be placed at the antinode of the electric field E (at the open end of the hairpin), where the field magnitude is almost uniform.

5. Microwave Experiments on Mesoscopic Conductors

A superficially very simple system for the study of confinement effects and the potential SIMIT is a suspension or assembly of nanoscale colloidal metal particles [18].

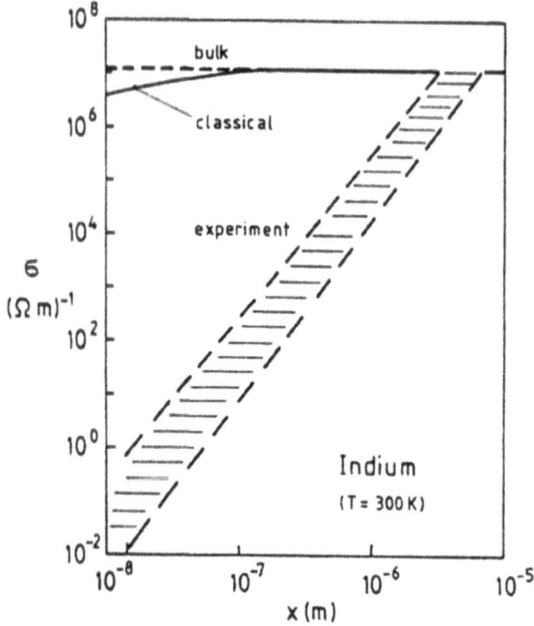

Figure 7. The size-dependent quasi-d.c. conductivity of mesoscopic indium particles versus particle size x (shaded area). For comparison, the bulk conductivity and the 'classical effect' (no confinement effects) are also shown. Taken from Nimtz *et al* [19].

Considerable effort across many fields is currently being expended in the preparation of colloidal gold particles [8], for this venerable area of chemistry potentially holds the key to the preparation of well-controlled, monodispersed mesoscopic particles [14]. From an experimental point of view, the small metal particles must satisfy stringent requirements in order to give meaningful results. Thus, their sizes must be small - patently within the colloidal (mesoscopic) size regime (Figure 1) - and well defined. Equally important, the size distribution must be as narrow as possible. In pioneering experiments, Marquardt, Nimtz and co-workers have reported the observation of a SIMIT in tiny crystals of indium via a novel microwave method [19]. In these particles, the conductivity decreases rapidly with particle diameter below a few μm in diameter (see Figure 7).

In relation to our own studies on nanoscale particles of gold, our initial experiments revealed that the total concentration of gold metal in a colloidal solution was too small (*ca.* 10mg) to allow for meaningful interpretation of microwave loss data. By moving to a situation in which gold particles are self-assembled in isolated particles in 2D islands upon solvent evaporation, higher concentrations of mesoscopic particles are produced [20].

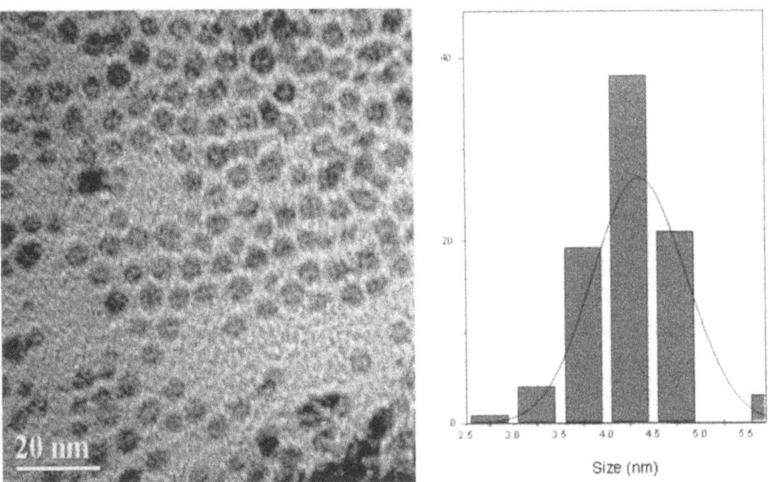

Figure 8. A electron micrograph of HS-C_6H_4-F stabilised nanoparticles and associated size distribution.

The synthesis of the gold nanoparticles was performed using the following materials: hydrogen tetrachloroaurate, 99.999%(chloroauric acid); sodium borohydride, 98%; methanol, 98% (HPLC grade); Diethyl Ether, 99%; HS-C_6H_4-F, 99%; HS-C_6H_4-Cl, 95%; HS-C_6H_4-Br, 95%; HS-C_6H_4-CH_3, 97%; HS-C_6H_4-OCH_3, 97%. All obtained from Aldrich. The solutions of all reagents were employed using standard volumetric techniques.

The nanoparticles stabilised with HS-C_6H_4-X, where X = F, Cl, Br and OCH_3, were synthesised using a route described previously [21], with the following ratios; $HAuCl_4.3H_2O$: Thiol (1:1), $HAuCl_4.3H_2O$: $NaBH_4$ (1:10).

Nanoparticles stabilised by HS-C_6H_4-CH_3 were synthesised using another, previously described, route [22]. The nanoparticles formed were soluble in polar solvents and can be routinely solubilised and precipitated from solution.

The mean diameters of the nanoparticles are as follows, 4.4 nm (sd = 0.6 nm), X = CH_3; 5.5 nm (sd = 0.8nm), X = OCH_3; 4.2 nm (sd = 0.5nm), X = Br; 4.5 nm (sd = 0.6 nm), X = Cl; 5.4 nm (sd = 0.4 nm), X = F. The 'polar' nature of the stabilising molecules is a possible explanation for this 'island' formation upon evaporation of the solvent, and not agglomeration of the gold cores (Figure 8). For these systems the implicit assumption is that the radii of the nanoparticles are much less than the skin depth of the fields within the nanoparticles, leading to the complete penetration of the field within the sample.

Gold Nanoparticle System	Mean diameter (nm)	Fractional increase in bandwidth ($\Delta\omega_B/\omega_B$)	Upper limit of conductivity ($\Omega^{-1}cm^{-1}$)
HS-C_6H_4-CH_3	4.4±0.6	26	0.009 ± 0.005
HS-C_6H_4-OCH_3	5.5±0.8	23	0.011 ± 0.006
HS-C_6H_4-Br	4.2±0.5	21	0.012 ± 0.006
HS-C_6H_4-Cl	4.5±0.6	19	0.014 ± 0.006
HS-C_6H_4-F	5.4±0.4	2.9	≈0.10 (≈0.0001 lower limit)

Table 1. Measured electrical conductivities of individual mesoscopic gold particles

Given the particle size data in Figure 8, this assumption is reasonable, remembering that the skin depth in bulk gold at 3 GHz is about 1.3 μm. The conductivity of the gold particles was calculated using Eqn.(1), yielding two values for the conductivity, for which we take the upper limit as a worse-case scenario; these are listed in Table 1. In fact, referring to Figure 5, for all but one of the samples the upper and lower values of conductivity are actually very similar owing to the fact that their conductivity values are close to the peak in the microwave dissipation. The contributions to the bandwidth from the stabilising molecules have been measured in separate experiments and found to be negligible. Hence, in all cases we have assumed that the bandwidth changes upon sample insertion are due entirely to the finite conductivity of the central gold nanoparticles. Generally, we conclude that the conductivity values for the stabilised gold nanoparticles are very low, around a factor of 10^7 smaller than the bulk value, and similar in magnitude to values reported for gold nanoparticles stabilised by other short aromatic thiols. Further experiments on metal colloids having different (but fixed) sites and filling factors are needed to add the general results for In and Au (Figure 7 and Table 1).

The continuous SIMIT demonstrated for In [19] - and not yet completely demonstrated for Au - is in agreement with the prediction of Gor'kov and Eliashberg [9] for the size range indicated by the solid line in Figure 7. These authors discussed the problem of a SIMIT in terms of the localization of the gas of conduction electrons in a metal through the (finite) size-induced confinement of the electron wave packet.

6. Summary

It appears that the SIMIT may be a universal quantum-size or confinement effect for all isolated mesoscopic conductors. Remarkably, by varying the size of the individual mesoscopic particles, any value of their conductivity - and associated complex dielectric functions (not discussed in detail here) - can be adjusted between the *full extremes of an insulator and a metal*. Besides its significance for fundamental studies of the mesoscopic state of matter (Figure 1), the SIMIT offers promising prospects and consequences for modern microelectronic applications. Effectively, the intriging possibility thus exists of manipulating any desired conductivity between a metal and an insulator in the mesoscopic range of conductors. Add to this the size-dependent dielectric response of mesoscopic metal particles and one has all the necessary ingredients for exotic and useful materials properties - *materials in which size matters* [8].

References

[1] G. Mie, *Ann. Phys.* (Leipz) **25** (1908) 377.

[2] H. Frohlich, *Physica* (Utr.) **4** (1937) 406.

[3] R. Kubo, *J. Phys. Soc. Jap.* **17** (1962) 975.

[4] L. P. Gor'kov and G. M. Eliashberg, *Zh. Eksp. Teor. Fiz.* **48** (1965) [English translation: *Sov. Phys.-JETP,* **21** (1965) 940].

[5] J. A. A. Perenboom, P. Wyder and F. Meier, *Physics Reports*, **78** (1981) 175.

[6] (a) M. R. Harrison and P.P. Edwards in *"The Metallic and Nonmetallic States of Matter"*, Eds. P. P. Edwards and C. N. R. Rao, Taylor and Francis, London, 1985. (b) Peter P. Edwards, *Proceedings of the INSA Golden Jubilee Symposium on Solid State Chemistry*, Indian Nation Science Academy, New Delhi, 1986 265-291.

[7] W. P. Halperin, *Revs. Mod. Phys.*, **58** (1986) 533.

[8] (a) C. N. R. Rao, G. U. Kulkarni, P. J. Thomas and P. P. Edwards, *Chem. Soc. Rev.*, **29** (2000) 27. (b) C. N. R. Rao, G. U. Kulkarni, P. J. Thomas and P. P. Edwards, *Chem. Eur. J.*, **8** (2002) 29.

[9] L. P. Gor'kov and G. M. Eliashberg, *Sov. Phys.-JETP*, **21** (1965) 940.

[10] G. Nimtz, P. Marquardt and H. Gleiter, *J. Cryst. Growth*, **86** (1988) 66.

[11] P. P. Edwards, R. L. Johnson and C. N. R. Rao in *"Metal Clusters in Chemistry"* Eds. P. R. Raithby *et al*, Wiley-VCH, Vol. 3 (2000) 1455.

[12] P. P. Edwards, R. L. Johnson, D. P. Tunstall, F. Hensel and C. N. R. Rao, *Sol. State Phys.*, **52** (1999) 229.

[13] N.W. Ashcroft and D. Mermin, "*Solid State Physics*" Thomson Learning, (1976).

[14] (a) D. G. Duff, A. C. Curtis, P. P. Edwards. D. A. Jefferson, B. F. G. Johnson, A. I. Kirkland and D. E. Logan, *Angew. Chem. Int. Ed.*, **26** (1987) 676. (b) A. I. Kirkland, P. P. Edwards, D. A. Jefferson and D. G. Duff, *Annual Reports C, The Royal Society of Chemistry* (1991) 247.

[15] P. A. Anderson, A. R. Armstrong, A. Porch, P. P. Edwards and L. J. Woodall, *J. Phys. Chem.*, **B101** (1997) 9892.

[16] M. Edmondson, A. Porch, P. A. Anderson and P. P. Edwards, *Z. Phys. Chem.*, **217** (2003) 939.

[17] B. I. Bleaney and B. Bleany, *"Electricity and Magnetism"*, Oxford University Press, 3rd edition. Vol. 1, (1989).

[18] P. P. Edwards, C. N. R. Rao, A. Porch, M. O. Jones, S. R. Johnson, G. U. Kulkarni and P. J. Thomas, *in preparation*.

[19] (a) P. Marquardt, L. Börngen, G. Nimtz, H. Gleiter, R. Sonnberger and J. Zhu, *Phys. Letts.*, **114A** (1986) 39. (b) G. Nimtz, P. Marquardt and H. Gleiter, J. *Crystal Growth*, **86** (1988) 66. (c) P. Marquardt and G. Nimtz, *Festkörperprobleme* **29** (1989) 317.

[20] (a) For a recent review see M. Brust, C. J. Kiely, *Colloids and Surfaces A: Physicochemical and Engineering Aspects* **202** (2002) 175. (b) C. J. Kiely, J. Fink, M. Brust, D. Bethell, D. J. Schiffrin, *Nature* **396** (1998) 444. (c)

[21] S. R. Johnson, S. D. Evans, R. Brydson, *Langmuir* **14** (1998) 6639-6647.

[22] S. R. Johnson, S. D. Evans, S. W. Mahon, A. Ulman, *Langmuir* **13** (1997) 51-57.

AN OPEN-BOUNDARY, TIME-DEPENDENT TECHNIQUE FOR CALCULATING CURRENTS IN NANOWIRES

David R. Bowler[1,2] and Andrew P. Horsfield[1]

[1] *Department of Physics and Astronomy, University College London, Gower Street, London WC1E 6BT, United Kingdom*

[2] *London Centre for Nanotechnology, University College London, Gower St, London WC1E 6BT*

david.bowler@ucl.ac.uk

a.horsfield@ucl.ac.uk

Abstract In this paper we briefly review the current state of models for computing electrical conduction in nanoscale devices, highlighting the progress made, but also some limitations still present. We then summarise our recent novel theory that allows the simultaneous evolution of the electronic and ionic degrees of freedom to be modelled within the Ehrenfest approximation in the presence of open boundaries. We describe our practical implementation using tight binding and use this theory to investigate steady-state conduction through an atomic scale device. We then use the model to investigate two systems not accessible with other contemporary techniques: the response of a nano-device to a rapidly varying external field, and non-adiabatic molecular dynamics in the presence of a current.

Keywords: nanowires, electrical conduction

1. Introduction

Advances in experimental techniques mean that it is now possible to measure the current flowing through individual molecules, atomic-scale wires and other systems where carriers have a quantum mechanically coherent history through the sample. When modelling such systems it is conventional to consider the system (or device) to be in contact with two macroscopic reservoirs, with some thermal equilibrium population of carriers deep within them. The connection between the device and the reservoirs is made by some form of tapering lead.

The theory and calculation of conductance for such nanoscale and mesoscopic systems was transformed by two key observations due to Landauer:

- The potential drop across a conductor can be viewed as arising from the self-consistent build-up of carriers, rather than the current arising from

A.S. Alexandrov et al. (eds.), Molecular Nanowires and Other Quantum Objects, 343–354.

the applied electric field [1] (this paper has been reprinted in a more accessible journal [2])

- The conductance of a device can be calculated from the electron transmission through it [3]

A brief overview of these ideas and their development is given in a review [4].

If we have a device with a number of transverse eigenstates, then the conductance (G) can be found using:

$$G = \frac{2e^2}{h} \mathrm{Tr}\left(tt^\dagger\right), \tag{1}$$

where t is the transmission matrix for the device. This formula can be derived by taking the zero-frequency limit of the Kubo formula [5], though more generally, it can be derived for systems with interacting (or non-interacting) electrons using the non-equilibrium Keldysh formalism [6]. One important effect on transmission and the calculation of conductance considered is the narrowing of the leads [7, 8] which has a significant effect on the scattering states.

This formalism has been developed and extended in many more directions than can be addressed here. Here we just note one extension that we consider particularly important: it has been generalised to multi-terminal samples [9].

The accurate calculation of the transmission matrix is perhaps the key problem. It can be found in terms of Green's functions and terms coupling the leads to the device [6]:

$$\mathrm{Tr}\left(tt^\dagger\right) = \mathrm{Tr}\left[\Gamma_L G^r \Gamma_R G^a\right], \tag{2}$$

where $\Gamma_{R(L)}$ is the coupling to the right (left) lead and $G^{r(a)}$ is the retarded (advanced) Green's function for the device.

Various electronic structure techniques have been applied to the calculation of the Green's functions, the transmission coefficients and the current in the system. They include tight binding methods [10, 11, 12, 13, 14, 15, 16, 17, 18], some of which have been extended to include the important effects of self-consistency [13, 17, 18, 19, 20]. There are now increasing numbers of calculations based on density functional theory (DFT) techniques [21, 22, 23, 24, 25, 26, 27, 28, 29] exploiting different basis functions.

Using these formalisms it is possible to calculate some of the effects of the current on the device, notably current-induced forces which can be found from tight binding [30] and DFT [31]. However, any molecular dynamics carried out on the atoms under the influence of these forces is necessarily adiabatic. There have been some initial attempts to model the effect of current-induced heating [32, 33], though these are perturbative in nature. The charging and deformation of molecules (including formation of defects such as solitons) has been considered to some extent [17, 18]. However, these are all adiabatic solutions, following the Born-Oppenheimer approximation. There are also recent theories which address the fundamental questions of how electronic structure is modified in the presence of a current [34, 35].

There are, however, problems with the present modelling techniques. There is good evidence that charged defects will have important roles in charge transport for conjugated polymers [18, 36] and solid state wires [37, 38], which involve highly non-linear effects due to electron-phonon coupling. These require a method that can handle the electrons and ions on an equal footing. A problem with wider implications is that first principles methods based on *static* DFT are challenged by the well-documented errors that it introduces for excited states. It is still an open question, but it is entirely possible the current carrying states may not be correctly modelled. A technique based on time-dependent density functional theory would have the advantage of standing on solid foundations.

To overcome the limitations of current methods, a different approach is required in which the ions and electrons evolve together in time in a consistent matter. Moving into the time domain allows us to exploit the benefits of time-dependent DFT for describing excited states and transient effects, and to introduce non-adiabatic terms which make possible modelling of heating of the ions as a result of the current flow.

There are a number of possible approaches to modelling non-adiabatic effects of widely varying complexity [39, 40, 41]. Here we consider the simplest in which the ions move along unique classical trajectories on which the atomic forces are determined by the Hellmann-Feynman [31] forces (the Ehrenfest, or mean field, approximation). That is, we neglect all quantum contributions to ionic motion.

The Ehrenfest method for closed systems has a long history, but we have extended it to open systems to allow an electric current to flow [42]. Below we describe a time-dependent formalism that is suitable for tight binding models implemented using density matrices. We have focused on tight binding because it is the simplest quantum mechanical model of electron motion that can deliver quantitative results [43]. We favour density matrices over wave functions and Green's functions because they provide a very compact description of the state of all the electrons [44, 45], and have proven very useful in the static description of materials in the context of linear scaling methods [46]. However, it is important to note that they have one particular limitation, namely that those parts of the density matrix treated explicitly must have a finite range if they are to be used in practical calculations. This is elaborated on below.

2. Physical Model

The paradigmatic system that we use to construct a model of current flow is shown in Fig. 1. It consists of a capacitor in series with a resistor, forming a complete circuit. For times $t \leqslant 0$, an external potential is applied to the left-hand side of the circuit so that there is an excess of electrons on this side and a deficit on the right-hand side. Most of the net charge will appear on the capacitor so as to minimise the total energy. Formally, this applied external potential arises from a chemical potential for the electrons that differs on the left and right of the system. This can be shown simply by minimising the total energy subject to the constraint that there are more electrons on the left than on the right.

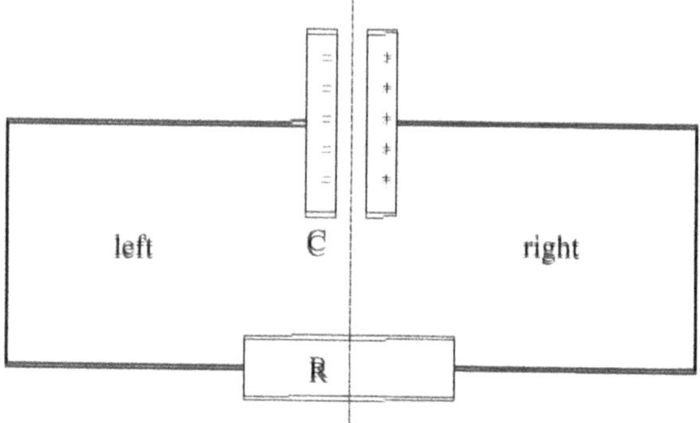

Figure 1. This circuit is the paradigm used to and build a model to describe the creation of an electric current. The capacitor represents the non-equilibrium source of charge, and the resistor the device through which we wish to drive the current.

At time $t = 0$, the external potential is removed and the charge is now free to move, and in the process will attempt to remove the imbalance in the charge. This leads to a current flow through the resistor, witch in turn produces a potential drop across it. Let us define Ψ_0 to be the many-body wave function for time $t \leqq 0$ and let the many-body Hamiltonian for $t \geqq 0$ be $\hat{\mathbf{H}}$. The wave function for $t \geqq 0$ ($\Psi(t)$) is then given by $\Psi(t) = \exp(\hat{\mathbf{H}}t/i\hbar)\Psi_0$, provided $\hat{\mathbf{H}}$ does not depend on time. Note that this Hamiltonian includes both ionic and electronic degrees of freedom, and so is able to describe the full response of the ions to the electronic current.

In the absence of dissipation we will obtain oscillatory solutions whose frequency spectrum is governed by the energy spectrum of the Hamiltonian. However, if $RC >> t >> \hbar/W$ (where W is the range of eigenvalues of $\hat{\mathbf{H}}$ contributing to Ψ_0), there will be a quasi-steady-state. It is this time range in which we are interested.

We now make the following important observation which allows us to perform practical calculations. In the quasi-steady-state regime, the potential in the wires does not vary strongly, and most of the potential drop therefore occurs across the resistor. This allows us to focus on the resistor alone, and to treat the capacitor and most of the wire as an external charge source or sink, which can be modelled by open boundary conditions. This is very similar in spirit to the Landauer approach in which only the transmission coefficient for the *device* needs to be evaluated.

The above formalism cannot be implemented directly because of the huge computational cost associated with such a many-body problem. The first simplifying step is to reduce the many-body electron problem to a single particle one. Provided we are willing to treat the ions in a mean field approach, which we are in this case, then we are free to describe the electrons using time-dependent density functional theory [47]. The key equations that we need are:

$$\begin{aligned} n(\vec{r},t) &= \sum_n f_n |\psi_n(\vec{r},t)|^2 \\ \hat{H}_{ks}\psi_n(\vec{r},t) &= i\hbar\frac{\partial}{\partial t}\psi_n(\vec{r},t) \\ \hat{H}_{ks} &= \hat{T} + \hat{V}_{eI} + \hat{V}_{Ha}[n] + \hat{V}_{xc}[n] \\ M_I\frac{\mathrm{d}^2\vec{R}_I}{\mathrm{d}t^2} &= -\vec{\nabla}_I V_{II} - \int \mathrm{d}\vec{r}\, n(\vec{r},t)\vec{\nabla}_I V_{eI} \end{aligned} \tag{3}$$

Here, $n(\vec{r},t)$ is the charge density, f_n is the orbital occupancy, $\psi_n(\vec{r},t)$ is an eigenfunction of the Kohn-Sham hamiltonian $\hat{H}_{ks}$, $\hat{T}$ is the kinetic energy operator, $\hat{V}_{eI}$ is the electron-ion interaction, $\hat{V}_{Ha}$ is the Hartree (electrostatic) interaction, $\hat{V}_{xc}$ is the exchange and correlation potential, $\hat{V}_{II}$ is the ion-ion repulsion, M_I is the mass of ion I and $\vec{R}_I$ is its position. Note that $\hat{V}_{xc}$ is nonlocal in time. It is certainly possible to work directly with these equations (once suitable approximations for $\hat{V}_{xc}$ have been made). However, we prefer to work with the single particle density matrix $\rho(\vec{r},\vec{r}')$, where

$$\rho(\vec{r},\vec{r}') = \sum_n \psi_n(\vec{r}) f_n \psi_n^*(\vec{r}'). \tag{4}$$

It is straightforward to write down the equation of motion for the density matrices using Eqs (3) and (4)and to recast the equation of motion for the ions:

$$i\hbar\frac{\partial\hat{\rho}}{\partial t} = [\hat{H}_{ks},\hat{\rho}] \tag{5}$$

$$M_I\frac{\mathrm{d}^2\vec{R}_I}{\mathrm{d}t^2} = -\vec{\nabla}_I V_{II} - Tr\{\hat{\rho}\vec{\nabla}_I\hat{H}_{ks}\} \tag{6}$$

where we have moved to operator notation.

Continuing in the spirit of reducing complexity to increase computational efficiency, we approximate DFT by tight binding. Further, if we use orthogonal tight binding we can replace operators in our previous equations with matrices. We will thus continue to use the operator notation with the new understanding that the operators will be represented by tight binding matrices.

As we indicated earlier on, we would like to concentrate our calculations only on the device and treat the environment in an implicit manner. To do this we separate the system into the device and the environment, which means we must divide Eq. (5) into components corresponding to device (designated by the subscript D), environment (designated by the subscript E) and the coupling between the two. We also introduce a damping term for the environment which makes it behave as a nearly equilibrium bath of electrons. This produces the following equations:

$$\begin{aligned} i\hbar\frac{\partial\hat{\rho}_D}{\partial t} &= [\hat{H}_D, \hat{\rho}_D] + (\hat{H}_{DE}\hat{\rho}_{ED} - \hat{\rho}_{DE}\hat{H}_{ED}) \\ i\hbar\frac{\partial\hat{\rho}_{DE}}{\partial t} &= \hat{H}_D\hat{\rho}_{DE} - \hat{\rho}_D\hat{H}_{DE} + \hat{H}_{DE}\hat{\rho}_E - \hat{\rho}_{DE}\hat{H}_E \\ i\hbar\frac{\partial\hat{\rho}_E}{\partial t} &= [\hat{H}_E, \hat{\rho}_E] + (\hat{H}_{ED}\hat{\rho}_{DE} - \hat{\rho}_{ED}\hat{H}_{DE}) \\ &- 2i\hbar\Gamma(\hat{\rho}_E - \hat{\rho}_{ref}). \end{aligned} \tag{7}$$

There is a closed form solution for the density matrix for the environment. If we assume that $\hat{H}_E$ is independent of time and define the driver terms $\hat{G}_E$ and $\hat{G}_E^{(0)}$ by $i\hbar\hat{G}_E = (\hat{H}_{ED}\hat{\rho}_{DE} - \hat{\rho}_{ED}\hat{H}_{DE})$ and $0 = [\hat{H}_E, \hat{\rho}_E(0)] + i\hbar\hat{G}_E^{(0)} - 2i\hbar\Gamma(\hat{\rho}_E(0) - \hat{\rho}_{ref})$, we find the following solution for Eq. (7) for the environment:

$$\hat{\rho}_E(t) = \hat{\rho}_E(0) + \int_0^t \mathrm{d}x\, \hat{O}(x)\left(\hat{G}_E(t-x) - \hat{G}_E^{(0)}\right)\hat{O}^\dagger(x) \tag{8}$$

where $\hat{O}(t) = \mathrm{e}^{-\Gamma t}\mathrm{e}^{\hat{H}_E t/i\hbar}$. For the parts of the density matrix belonging to the device and its coupling to the environment we treat the time evolution explicitly.

As a testbed for this formalism we now develop a very simple model system. It consists of two semi-infinite leads attached to a device. The lead on the left is at a different potential from that on the right. Each lead is represented by a linear chain of atoms with one orbital per atom. There are therefore two parameters that characterise the Hamiltonian for each lead: the on-site energy (a) and hopping integral between the nearest neighbour sites (b). We take b to be the same on the left and on the right. The bias is applied through the difference in onsite energies on the two sides: $a_L - a_R$.

The non-locality in time of Eq. (8) adds considerably to the cost of performing a calculation. However, from Fig. 2 we see that the evolution operator decays rapidly with time. We could thus approximate this by a function the goes strictly to zero outside some cut-off time. This is consistent with keeping the damping term in Eq. (7). If the evolution operator is truncated in time, it becomes truncated in space as well. This corresponds to the fact that a wave packet can travel only a limited distance in a finite time.

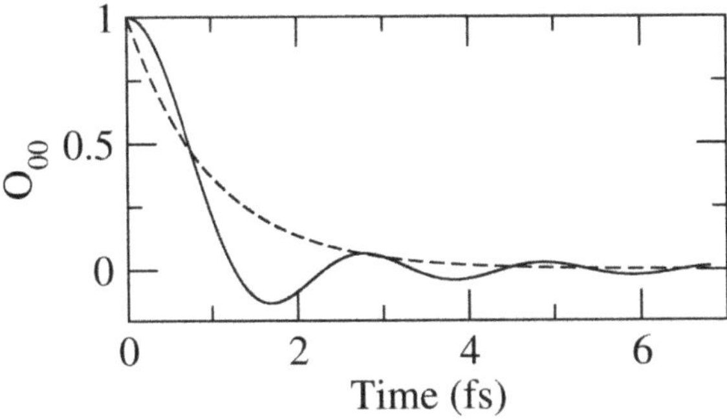

Figure 2. The solid line is the variation with time of the matrix element of the time evolution operator corresponding to the first atom in the environment. There is no damping ($\Gamma = 0$). The decay corresponds to the propagation of a wavepacket down the wire. The dashed line is the exponential damping factor with $\Gamma = 1.0fs^{-1}$.

The final quantities that we need to define in order to completely characterise the environment are the initial and reference density matrices ($\hat{\rho}_E(0)$ and $\hat{\rho}_{ref}$), both of which we take to be equal to the density matrix for the infinite wire (with the device present) in its ground state in the absence of a bias.

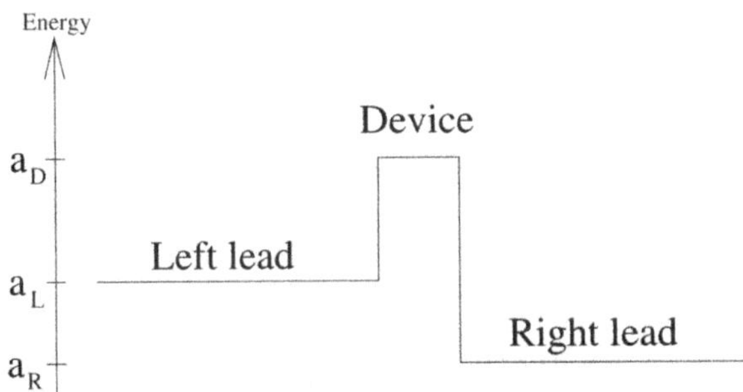

Figure 3. The energy profile for the model system. The energy axis on the left shows the positions of the onsite energies in the left lead (a_L), the right lead (a_R) and the device (a_D).

To complete our model system we need to introduce a device. The simplest device consists of only one atom. If we give this atom a high on-site energy it behaves as a barrier to current flow. The one-dimensional potential profile for this system is given in Fig. 3.

3. Results

As a check on the method described here, we can compute the conductivity of this system using the Landauer method. For this we need the transmission coefficient which can be found straightforwardly from Schrödinger's equation.

In our tight binding formalism, the wave function on the left-hand side has the form $\psi_n = e^{ikn} + Re^{-ikn}$, where ψ_n is the wave function evaluated at site n. On the right it has the form $\psi_n = Te^{iqn}$. Applying Schrödinger's equation ($\sum_j H_{ij}\psi_j = \varepsilon\psi_i$) to the left and right leads and the device gives $T = 2\sin(k)/(\sin(q) + \sin(k) + i(a_R + a_L - 2a_D)/2b)$ where a_L and a_R are the on-site matrix elements on the left and right respectively, and a_D is that for the device atom. For the special case of a half filled band and infinitessimal bias we get the following conductivity

$$g = \frac{2e^2}{h}\left(1 + \left[\frac{a_D - a_L}{2b}\right]^2\right)^{-1}. \tag{9}$$

If we have a bias of 0.1V, and the hopping and barrier height are both 1eV, we get a current of about 6.2μA.

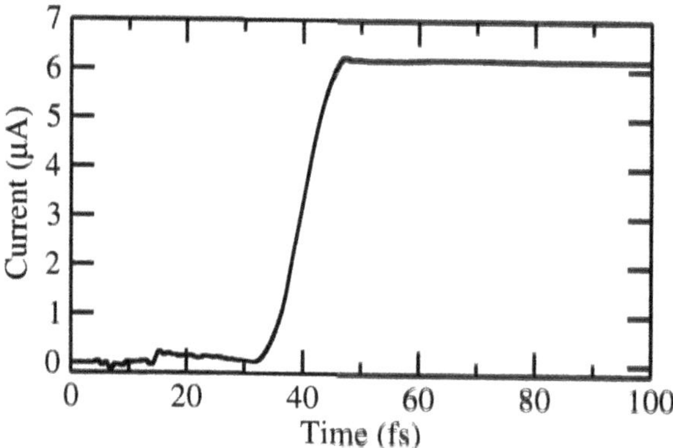

Figure 4. The variation of current into the device as a function of time. Note that it reaches a stable steady state. For the first 30 fs no bias was applied. The bias of -0.1V was then turned on over a period of 10 fs.

The time dependence of the current is shown in Fig. 4. We see that our time dependent scheme leads to stable steady currents.

One class of phenomena that our time-dependent formalism allows us to study is transient effects in the presence of rapidly changing external potentials. In Fig. 5 we show the effect of applying a voltage to the device atom, rather like the gate voltage in a field-effect transistor. The current responds smoothly and is reduced when the voltage is applied.

The transient effects seen in the current are proportional to the first derivative of the gate voltage. This is easy to understand by considering a hydrodynamic analogy. If you had some kind of large piston at the base of a canal full of water, then the rising gate voltage would correspond to the piston moving upwards from the base of the canal. This will displace water, with the amount being

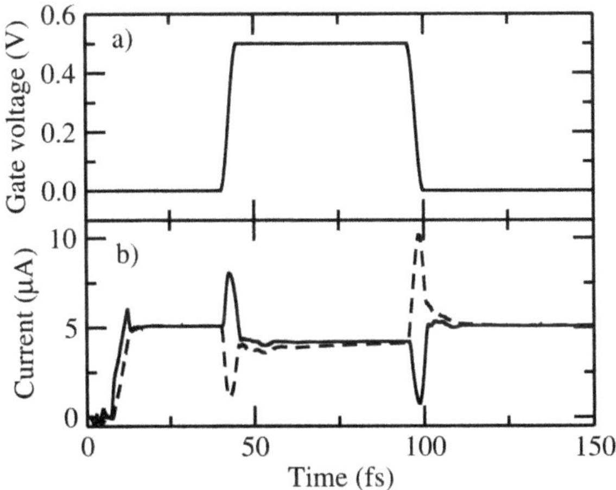

Figure 5. The current into (solid line) and out of (dashed line) the device while a gate-like voltage is applied. The voltage is turned on at 40 fs, over 5 fs, held for 50 fs, and then turned off, again over 5 fs. (a) Voltage applied. (b) Current.

displaced per second being proportional to the rate at which the piston rises. Locally the displaced water will look like a current.

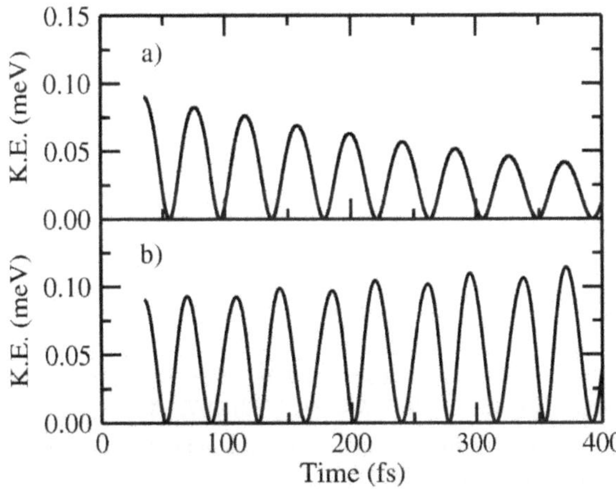

Figure 6. Plots showing kinetic energy of single atom device over time with bias (a) -0.1V (b) -1.0V. Molecular dynamics was started after a steady-state current was achieved, at 60 fs.

To investigate heating effects, once a steady-state current has been achieved we perform molecular dynamics on the device atom. The position of the atom is allowed to evolve according to Eq. (6). To monitor the heating we follow the evolution of the kinetic energy of the device atom with time. In Fig. 6 we show the effects of two different applied biases. Fig. 6(a) shows that with a bias of

-0.1V, we have cooling, while Fig. 6(b) shows that a bias of -1.0V gives gradual heating of the system. These heating results will be discussed in more detail in future work.

In conclusion, we have presented a time-dependent technique which allows evolution of electronic and ionic degrees of freedom for an open system. It offers a number of possible improvements over static methods. These include the ability to study transient behaviour and non-adiabatic processes, as well as possibly providing a framework for an improved density functional description of current carrying electrons. We have implemented the method using tight binding, and demonstrated transient behaviour, and both heating and cooling for a one dimensional metallic wire.

Acknowledgments

The financial support of the IRC on nanotechnology for APH and of the Royal Society for DRB, and very useful conversations with Tchavdar Todorov, Andrew Fisher and Marshall Stoneham are gratefully acknowledged.

References

[1] Landauer, R. (1957). Spatial variation of currents and fields due to localized scatterers in metallic conduction. *IBM J. Res. Dev.*, 1:223–231.

[2] Landauer, R. (1996). Spatial variation of currents and fields due to localized scatterers in metallic conduction. *J. Math. Phys*, 37(10):5259–5268.

[3] Landauer, R. (1970). Electrical resistance of disordered one-dimensional lattices. *Phil. Mag.*, 21:863–867.

[4] Imry, Y. and Landauer, R. (1999). Conductance viewed as transmission. *Rev. Mod. Phys.*, 71(2):S306–S312.

[5] Fisher, D. S. and Lee, P. A. (1981). Relation between conductivity and transmission matrix. *Phys. Rev. B*, 23(12):6851–6854.

[6] Meir, Y. and Wingreen, N. S. (1992). Landauer formula for the current through an interacting electron region. *Phys. Rev. Lett.*, 68(16):2512–2515.

[7] Landauer, R. (1989). Conductance determined by transmission: probes and quantised constriction resistance. *J. Phys.: Condens. Matter*, 1(12):8099–8110.

[8] Payne, M. C. (1989). Electrostatic and electrochemical potentials in quantum transport. *J. Phys.: Condens. Matter*, 1(8):4931–4938.

[9] Büttiker, M. (1986). Four-terminal phase-coherent condcuctance. *Phys. Rev. Lett.*, 57(14):1761–1764.

[10] Todorov, T. N. (2002). Tight-binding simulation of current-carrying nanostructures. *J. Phys.: Condens. Matter*, 14:3049–3084.

[11] Samanta, M. P., Tian, W., Datta, S., Henderson, J. I., and Kubiak, C. P. (1996). Electronic conduction through organic molecules. *Phys. Rev. B*, 53(12):R7626–7629.

[12] Sanvito, S., Lambert, C. J., Jefferson, J. H., and Bratkovsky, A. M. (1999). General Green's-function formalism for transport calculations with spd hamiltonians and giant magnetoresistance in Co- and Ni-based magnetic multilayers. *Phys. Rev. B*, 59(18):11936–11948.

[13] Yaliraki, S. N. and Ratner, M. A. (1999). Molecule-interface coupling effects on electronic transport in molecular wires. *J. Chem. Phys.*, 109(12):5036–5043.

[14] Mujica, V., Kemp, M., and Ratner, M. (1994). Electron conduction in molecular wires. I. A scattering formalism. *J. Chem. Phys.*, 101(8):6849–6855.

[15] Sirvent, C., Rodrigo, J. G., Vieira, S., Jurczyszyn, L., Mingo, N., and Flores, F. (1996). Conductance step for a single-atom contact in the scanning tunneling microscope: Noble and transition metals. *Phys. Rev. B*, 53(23):16086–16090.

[16] Nardelli, M. B. (1999). Electronic transport in extended systems: Application to carbon nanotubes. *Phys. Rev. B*, 60(11):7828–7833.

[17] Emberly, E. G. and Kirczenow, G. (2000). Multiterminal molecular wire systems: A self-consistent theory and computer simulations of charging and transport. *Phys. Rev. B*, 62(15):10451–10458.

[18] Emberly, E. G. and Kirczenow, G. (2001). Current-driven conformational changes, charging, and negative differential resistance in molecular wires. *Phys. Rev. B*, 64:125318.

[19] Mujica, V., Roitberg, A. E., and Ratner, M. (2000). Molecular wire conductance: Electrostatic potential spatial profile. *J. Chem. Phys.*, 112(15):6834–6839.

[20] Tian, W., Datta, S., Hong, S., Reifenberger, R., Henderson, J. I., and Kubiak, C. P. (1998). Conductance spectra of molecular wires. *J. Chem. Phys.*, 109(7):2874–2882.

[21] Lang, N. (1995). Resistance of atomic wires. *Phys. Rev. B*, 52(7):5335–5342.

[22] Ness, H. and Fisher, A. J. (1997). Nonperturbative evaluation of STM tunneling probabilities from ab initio calculations. *Phys. Rev. B*, 56(19):12469–12481.

[23] Kobayashi, N., Aono, M., and Tsukada, M. (2001). Conduction channels of Al wires at finite bias. *Phys. Rev. B*, 64:121402(R).

[24] Brandbyge, M., Mozos, J.-L., Ordejón, P., Taylor, J., and Stokbro, K. (2002). Density-functional method for nonequilibrium electron transport. *Phys. Rev. B*, 65(16):165401.

[25] Taylor, J., Guo, H., and Wang, J. (2001). Ab initio modeling of quantum transport properties of molecular electronic devices. *Phys. Rev. B*, 63(24):245407.

[26] Damle, P. S., Ghosh, A. W., and Datta, S. (2001). Unified description of molecular conduction: From molecules to metallic wires. *Phys. Rev. B*, 64:201403.

[27] Palacios, J. J., Pérez-Jiménez, A. J., Louis, E., and Vergés, J. A. (2001). Fullerene-based molecular nanobridges: A first-principles study. *Phys. Rev. B*, 64:115411.

[28] Di Ventra, M. and Lang, N. D. (2001). Transport in nanoscale conductors from first principles. *Phys. Rev. B*, 65:045402.

[29] Nardelli, M. B., Fattebert, J.-L., and Bernholc, J. (2001). O(N) real-space method for ab initio quantum transport calculations: application to carbon nanotube-metal contacts. *Phys. Rev. B*, 64:245423.

[30] Todorov, T. N., Hoekstra, J., and Sutton, A. P. (2000). Current-induced forces in atomic-scale conductors. *Phil. Mag. B*, 80(3):421–455.

[31] Di Ventra, M. and Pantelides, S. T. (2000). Hellmann-Feynman theorem and the definition of forces in quantum time-dependent and transport problems. *Phys. Rev. B*, 61(23):16207–16212.

[32] Todorov, T. N. (1998). Local heating in ballistic atomic-scale contacts. *Phil. Mag. B*, 77(4):965–973.

[33] Montgomery, M. J., Todorov, T. N., and Sutton, A. P. (2002). Power dissipation in nanoscale conductors. *J. Phys.: Condens. Matter*, 14(13):5377–5389.

[34] Baer, R. and Neuhauser, D. (2003). Many-body scattering formalism of quantum molecular conductance. *Chem. Phys. Lett.*, 374(5-6):459–463.

[35] Kosov, D. S. (2003). Kohn-Sham equations for nanowires with direct current. *J. Chem. Phys.*, 119(1):1–5.

[36] Ness, H. and Fisher, A. J. (1999). Quantum inelastic conductance through molecular wires. *Phys. Rev. Lett.*, 83(2):452–455.

[37] Bowler, D. R. and Fisher, A. J. (2001). Small polaron formation in dangling-bond wires on the Si(001) surface. *Phys. Rev. B*, 63:035310.

[38] Todorovic, M., Fisher, A. J., and Bowler, D. R. (2002). Diffusion of a polaron in dangling bond wires on Si(001). *J. Phys.:Condens. Matter*, 14(7):L749–L755.

[39] Kohen, D., Stillinger, F. H., and Tully, J. C. (1998). Model studies of nonadiabatic dynamics. *J. Chem. Phys.*, 109(12):4713–4725.

[40] Billing, G. D. (1999). Time-dependent quantum dynamics in a Gauss-Hermite basis. *J. Chem. Phys.*, 110(12):5526–5537.

[41] Kapral, R. and Ciccotti, G. (1999). Mixed quantum-classical dynamics. *J. Chem. Phys.*, 110(18):8919–8929.

[42] Horsfield, A. P., Bowler, D. R., and Fisher, A. J. (2003). An open-boundary Ehrenfest model of current induced heating in nanowires. *Phys. Rev. Lett.*, submitted.

[43] Goringe, C. M., Bowler, D. R., and Hernández, E. (1997). Tight-binding modelling of materials. *Rep. Prog. Phys.*, 60(12):1447–1512.

[44] McWeeny, R. (1960). Some recent advances in density matrix theory. *Rev. Mod. Phys.*, 32:335–369.

[45] He, L. and Vanderbilt, D. (2001). Exponential decay properties of Wannier functions and related quantities. *Phys. Rev. Lett.*, 86(23):5341–5344.

[46] Goedecker, S. (1999). Linear scaling electronic structure methods. *Rev. Mod. Phys.*, 71(4):1085–1123.

[47] Calvayrac, F., Reinhard, P.-G., Suraud, E., and Ullrich, C. A. (2000). Nonlinear electron dynamics in metal clusters. *Physics Reports*, 337(6):493–579.

ELECTRONIC STATES OF NANOSCOPIC CHAINS AND RINGS FROM FIRST PRINCIPLES: EDABI METHOD

E.M. Görlich[1], J.Kurzyk[2], A. Rycerz[1], R. Zahorbeński[1], R.Podsiadły[1], W. Wójcik[2], and J. Spałek[1]

[1]*Marian Smoluchowski Institute of Physics, Jagiellonian University, ulica Reymonta 4, 30-059 Kraków, Poland*

[2]*Institute of Physics, Technical University, ulica Podchorążych 1, 30-084 Kraków, Poland*

ufspalek@if.uj.edu.pl

Abstract We summarize briefly the main results obtained within the proposed *EDABI* method combining **E**xact **D**iagonalization of (parametrized) many-particle Hamiltonian with **Ab I**nitio self-adjustment of the single-particle wave function in the correlated state of interacting electrons. The properties of nanoscopic chains and rings are discussed as a function of their interatomic distance R and compared with those obtained by Bethe ansatz for infinite Hubbard chain. The concepts of renormalized orbitals, distribution function in momentum space, and of Hubbard splitting as applied to nanoscopic systems are emphasized.

Keywords: Nanoscopic Systems, Electronic Properties, Correlated Systems, EDABI Method

1. Introduction

Recent development in computing techniques, as well as of analytical methods, has lead to a successful determination of electronic properties of semiconductors and metals based on LDA [1], LDA+U [2] and related [3] approaches. Even strongly correlated systems, such as V_2O_3 (undergoing the Mott transition) and high-temperature superconductors, have been treated in that manner [4]. However, the discussion of the metal-insulator transition of the Mott-Hubbard type is not as yet possible in a systematic manner, particularly for low-dimensional systems. These difficulties are caused by the circumstance that the electron-electron interaction is comparable, if not stronger than the single-particle energy. In effect, the procedure starting from the single-particle picture (band structure) and including subsequently the interaction via a *local* potential might not be appropriate then. In this situation, one resorts to

A.S. Alexandrov et al. (eds.), Molecular Nanowires and Other Quantum Objects, 355–375.

parametrized models of correlated electrons, where the single-particle and the interaction-induced aspects of the electronic states are treated on equal footing (in the Fock space though)[5]. The single particle wave functions are contained in the formal expressions for model parameters. We propose to combine the two efforts in an exact manner, at least for model systems.

Our method of approach to the electronic states grew out of the following question: Can one *complete* the procedure starting from a parametrized model by determining the single-particle wave functions in the resultant correlated state *a posteriori*? In other words, we determine *first* the energy of interacting particles in terms of the microscopic parameters rigorously and only then optimize this energy with respect to the wave functions contained in those parameters by deriving the *self-adjusted wave equation* for this state. Physically, the last step amounts to allowing the single-particle wave functions to relax in the correlated state. This method has been overviewed in number of papers [6, 7, 8, 9, 10], so we present here examples of its application to low-dimensional and nanoscopic systems. We start with the analysis of the infinite Hubbard chain and then compare the results with those for finite chains. We also discuss briefly small hydrogenic ring of $N = 6$ atoms. Also, throughout the paper we are using adjustable Wannier composed of atomic or Gaussian functions, which are determined explicitly from the minimization of the system ground state energy as a function of interatomic distance. The paper describes various ground-state characteristics of simple monoatomic chains and rings.

2. Exact diagonalization combined with wave-function determination: formal aspects

As our method contains both many-particle and single-particle (wave-function) aspects, both treated in a rigorous manner, it may be useful to summarize the *basic principle* behind it. First, we start with the standard expression of the many-particle Hamiltonian in the Fock space

$$\hat{H} = \int d^3\mathbf{r}\hat{\Psi}^{\dagger}(\mathbf{r})H_1(\mathbf{r})\hat{\Psi}(\mathbf{r})+$$

$$\frac{1}{2}\int d^3rd^3r'\hat{\Psi}^{\dagger}(\mathbf{r})\hat{\Psi}^{\dagger}(\mathbf{r}')H_2(\mathbf{r}-\mathbf{r}')\hat{\Psi}(\mathbf{r}')\hat{\Psi}(\mathbf{r}), \tag{1}$$

where H_1 and H_2 are the Hamiltonian for one and one pair of particles, and

$$\hat{\Psi}(\mathbf{r}) = \sum_i w_i(\mathbf{r})\begin{pmatrix} a_{i\uparrow} \\ a_{i\downarrow} \end{pmatrix} \equiv \sum_i w_i(\mathbf{r})a_i, \tag{2}$$

is the field operator, $\{w_i(\mathbf{r})\}$ is the single-particle basis of wave-functions (complete, but otherwise *arbitrary*), ad $a_{i\sigma}$ is the annihilation operator of the particle in the single-particle state $|i\sigma >$ represented by $w_i(\mathbf{r})$ and the spin quantum number $\sigma = \pm 1$. The only approximation we make in our whole analysis is that instead of taking the summation over a complete set $\{i\}$ of single-particle states (and transition between them), we limit ourselves to a finite subset of M states. This means that we are solving a *model many-body system* rather than the complete problem (*Hubbard* or *extended Hubbard models* are classic examples representing one-orbital-per-atom).

The essential step in our analysis follows from taking the finite single-particle basis, which amounts to limiting the occupation-number representation space to a space of finite dimension. To minimize the error in estimating the ground-state energy of many-particle system we calculate first *all* configurations in the limited Fock subspace and then optimize the orbitals in the interacting (correlated) ground state. In this manner, the second quantization takes care of counting various many single-particle microconfigurations enforced by the interaction between them, whereas the wave-function optimization adjusts each of them to the milieu of all others. In brief, second-quantization aspect addresses the question *how* they are distributed among the single-particles state and the first-quantization optimization tells us how their states look like once they are there.

One should also address the problem of many-body vs. single-particle wave function. In the wave mechanics of the interacting system only the N-particle wave function $\Psi(\mathbf{r}_1, \ldots \mathbf{r}_N)$ has a sense. However, in the second quantization the single-particle wave function appears explicitly in the expression for the field operator, the remaining part is the evaluation of various microconfigurations, with proper weights characterized by their energy (no entropy appears as they form a single coherent state). In other words, the microconfiguration counting in the occupation-number representation replaces the determination of N-particle Hilbert space. But then, we are faced with the single-particle states determination, on which the counting is performed. Explicitly, the N-particle state $|\Phi_0>$ in the Fock space can be defined as

$$|\Phi_0\rangle = \frac{1}{\sqrt{N!}} \int d^3\mathbf{r}_1 \ldots \mathbf{r}_N \Psi_0(\mathbf{r}_1, \ldots, \mathbf{r}_N) \hat{\Psi}^\dagger(\mathbf{r}_1) \ldots \hat{\Psi}^\dagger(\mathbf{r}_N)|0\rangle, \tag{3}$$

where $|0\rangle$ is the vacuum state. The N-particle wave function is then determined from

$$\Psi_0(\mathbf{r}_1, \ldots, \mathbf{r}_N) = \frac{1}{\sqrt{N!}} \langle 0|\hat{\Psi}(\mathbf{r}_1) \ldots \hat{\Psi}(\mathbf{r}_N)|\Phi_0\rangle. \tag{4}$$

Expanding $|\Phi_0>$ in the basis involving M single-particle states, i.e.

$$|\Phi_0\rangle = \frac{1}{\sqrt{N!}} \sum_{j_1,\ldots,j_N=1}^{M} C_{j_1 \ldots j_N} a^\dagger_{j_1} \ldots a^\dagger_{j_N} |0\rangle, \tag{5}$$

we obtain the N-particle wave function in the form

$$\Psi_0(\mathbf{r_1},\ldots\mathbf{r_n}) = \frac{1}{N!}\sum_{i_1,\ldots,i_N=1}^{M}\sum_{j_1,\ldots,j_N=1}^{M}\langle 0|a_{i_N}\ldots a_{i_1}a_{j_1}^{\dagger}\ldots a_{j_N}^{\dagger}|0\rangle$$

$$C_{j_1\ldots j_N} w_{i_1}(\mathbf{r_1})\ldots w_{i_N}(\mathbf{r_N}). \tag{6}$$

The many-body coefficients $C_{j_1\ldots j_N}$ will be calculated from either the direct diagonalization or the Lanczos algorithm, whereas the wave functions $\{w_i(\mathbf{r})\}$ will be determined from the *renormalized (self-adjusted) wave equation*, which is set up in the following manner. First, we substitute (2) into (1) and obtain the formal expression for the ground state energy

$$E_G \equiv \langle H\rangle = \sum_{ij\sigma} t_{ij}\langle a_{i\sigma}^{\dagger}a_{j\sigma}\rangle + \frac{1}{2}\sum_{ijkl\sigma_1\sigma_2} V_{ijkl}\langle a_{i\sigma_1}^{\dagger}a_{j\sigma_2}^{\dagger}a_{l\sigma_2}a_{k\sigma_1}\rangle, \tag{7}$$

where the microscopic parameters

$$t_{ij} = \int d^3\mathbf{r} w_i^{\star}(\mathbf{r})H_1(\mathbf{r})w_j(\mathbf{r}),$$

and

$$V_{ijkl} = \int d^3\mathbf{r_1} d^2 d^3\mathbf{r_2} w_i^{\star}(\mathbf{r_1})w_j^{\star}(\mathbf{r_2})V(\mathbf{r_1}-\mathbf{r_2})w_k(\mathbf{r_1})w_l(\mathbf{r_2}),$$

contain the single-particle wave functions and the averages are $\langle a_{i\sigma}^{\dagger}a_{j\sigma}\rangle \equiv \langle\Phi_0|a_{i\sigma}^{\dagger}a_{j\sigma}|\Phi_0\rangle$, etc. Second, in the situation when we work with definite number of particles (or else, if the chemical potential can be regarded as constant), then we can determine $\{w_i(\mathbf{r})\}$ by setting Euler equations for each of them, i.e. by minimizing the functional $F = E_G\{w_i(\mathbf{r}), \nabla w_i(\mathbf{r}) - \sum_i \lambda_i \int d^3\mathbf{r} w_i^{\star}(\mathbf{r} w_i(\mathbf{r}.$ Such a procedure leads to the equation

$$\frac{\delta E_G}{\delta w_i^{\star}(\mathbf{r})} - \nabla\frac{\delta E_G}{\delta(\nabla w_i^{\star}(\mathbf{r}))} = \lambda_i w_i(\mathbf{r}) \tag{8}$$

A direct solution of this equation is very difficult to achieve. Therefore, we define the starting wave functions $w_i^{(0)}(\mathbf{r}) = \sum_{j=1}^{M}\beta_{ij}\Phi_j(\mathbf{r};\alpha)$, where β_{ij} are the mixing coefficients and $\Phi_j(\mathbf{r};\alpha)$ are the atomic wave functions of the size α^{-1}. In result, the renormalized wave equation reduces to the minimization of (7) with respect to α (there is only one size α^{-1} when we take orbitals of the same type for each atomic site in the system). The whole EDABI procedure is schematically summarized in Fig. 1.

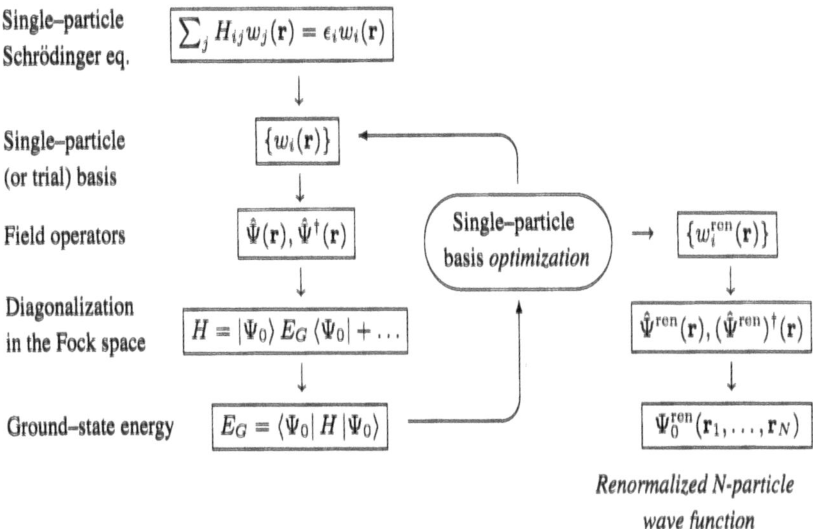

Figure 1. Flowchart of the EDABI method described in this paper. The top line is missing when the adjustable Gaussian basis is used.

3. Electron states for the Hubbard chain and a comparison with nanochains

We implement first the EDABI method to the case of the linear chain composed of N atoms, which obeys periodic boundary conditions. In the simplest situation we have one valence electron per atom, as is in the case of monoatomic chains composed of Na, K or Cs. However, for the sake of simplicity, we consider here the chain composed of hydrogen atoms; this means only that we compose the Wannier function of the conduction band from 1s-like atomic states of an adjustable size (there is no principal problem in considering ns-like states, with $n > 1$).

As mentioned earlier, we start from the many-body model of interacting electrons. For this purpose, we consider an extended Hubbard model, as represented by the Hamiltonian

$$H = \epsilon_a \sum_{i\sigma} n_{i\sigma} + t \sum_{i\sigma} \left(a^\dagger_{i\sigma} a_{i+1\sigma} + h.c. \right)$$

$$+ U \sum_i n_{i\uparrow} n_{i\downarrow} + \sum_{i<j} K_{ij} n_i n_j + \sum_{i<j} V_{ion}(R_i - R_j) \tag{9}$$

The first term represents the atomic energy ($\epsilon_a \equiv \langle w_i | H_1 | w_i \rangle$, where H_1 is the Hamiltonian for a single particle in the system and $w_i \equiv w_i(\mathbf{r})$ is the Wannier state centered on site i in that system). The second term is the so-called hopping term with the hopping integral $t \equiv \langle w_i|H_1|w_{i\pm1}\rangle$(we use the tight-binding approximation and disregard more distant hoppings). The next two terms represent respectively the intra- and inter- atomic parts of the Coulomb interaction between electrons, with $K_{ij} \equiv \langle w_i w_j|V_{12}|w_i w_j\rangle$, $U = K_{ii}$, and where $V_{12}(\mathbf{r}-\mathbf{r}')$ is the classical Coulomb interaction for pair of electrons. The last term expresses the classical repulsion between the ions located at sites i and j. Also, the single particle operator H_1 contains both kinetic energy and the Coulomb attractive interaction of ions (it is sufficient to take $6 \div 10$ ionic Coulomb wells surrounding the electron located on site i to reproduce to a good accuracy the effective values od ϵ_a and t).

For a detailed analysis it is convenient to rewrite (9) in the equivalent form, which for the case of one electron per atom reads

$$H = \epsilon_a^{eff} N_e + t\sum_{i\sigma}\left(a_{i\sigma}^{\dagger}a_{i+1\sigma} + h.c.\right) + U\sum_i n_{i\uparrow}n_{i\downarrow} + \sum_{i<j} K_{ij}\delta n_i \delta n_j, \quad (10)$$

where $\delta n_i \equiv 1 - n_i$, $N_e = \sum_{i\sigma} n_{i\sigma}$ is the number of electrons (here equal to the number of atoms), and

$$\epsilon_a^{eff} \equiv \epsilon_a + \frac{1}{N}\sum_{i<j}\left(K_{ij} + \frac{e^2}{|R_i - R_j|}\right) \quad (11)$$

is the effective atomic energy, which includes the compensating repulsive interactions to provide correctly the atomic limit. We disregard the last term in (10), since we would like to relate the nanochain results to those for the Hubbard chain for $N \to \infty$ [11]. Under these circumstances, Hamiltonian (10), we consider here explicitly, represents only the Hubbard model with the energy part ϵ_a^{eff} providing a proper neutral-atom limit for $R \to \infty$. Eq. (10) then can be diagonalized exactly with the help of Bethe ansatz [12]. Explicitly, the expression for the ground state energy in terms of microscopic parameters ϵ_a^{eff}, t, and U takes the form

$$\frac{E_G}{N} = \epsilon_a^{eff} + 4t\int_0^{\infty} d\omega \frac{J_0(\omega)J_1(\omega)}{\omega[1+\exp(-\omega U/2t)]}, \quad (12)$$

where $J_n(x)$ is the Bessel function. One should note that the energy expression is not an additive function of terms $\sim t$ and $\sim U$, as the solution is of nonperturbative nature.

As we have stressed earlier, the Lieb-Wu solution (12) does not represent the final step of the analysis as it still contains parameters, which in turn are expressed through the one-particle (Wannier) functions $\{w_i(\mathbf{r})\}$. Therefore, we minimize the energy *functional* $E \equiv E\{w_i(\mathbf{r}), \nabla w_i(\mathbf{r})\}$ with respect to $\{w_i(\mathbf{r})\}$. Such a procedure leads to the Euler variational principle for *renormalized* or *self-adjusted* wave functions in the correlated state [5]. Only after solving that equation and calculating explicitly the parameter values for given

R we obtain the energy of the correlated ground state as a function of the lattice parameter. This last step completes the theoretical analysis of the Hubbard model.

Technically, we compose the Wannier functions of atomic 1s-like functions of variable size, with respect to which we minimize E. Explicitly, in the spirit of tight-binding approximation we can write that

$$w_i(\mathbf{r}) = \beta\psi_i(\mathbf{r}) - \gamma[\psi_{i+1}(\mathbf{r}) + \psi_{i-1}(\mathbf{r})], \tag{13}$$

where β and γ are the mixing coefficients determined from the orthonormality condition $\langle w_i | w_j \rangle = \delta_{ij}$ and

$$\psi_i(\mathbf{r}) = (\alpha^3/\pi)^{1/2} \exp(-\alpha|\mathbf{r} - \mathbf{R}_i|), \tag{14}$$

with α being the adjustable parameter. Substituting (13) to the expressions for ϵ_a, t, and U, and then those expressions to (12) we obtain the physical ground state energy as the minimal energy with respect to α for given lattice constant $R \equiv |R_i - R_{i\pm1}|$.

In Fig. 2 we plot the optimized ground state energy with respect to α as a function of interatomic distance. We have also added there the corresponding values obtained using the Gutzwiller ansatz (GA)[13] and the Gutzwiller wave-function (GWF)[14] approximations. Obviously, the approximate solutions [13, 14] represent the upper estimates for the system energy. In the inset we have plotted the interaction-to-bandwidth ratio U/W; one sees that the electrons are strongly correlated (i.e. $U/W > 1$) for realistic values of R ($a_0 \simeq 0.53$ Å is the Bohr radius). Obviously, the chain composed of hydrogen atoms is not stable, as the energy E_G (per atom) is above -1Ry. However, this is not an issue here, since we are considering only a model situation (such a chain could be stabilized on a substrate, but then we should have to include the trapping potential, in addition to the periodic potential composed of the protonic Coulomb wells). It would be interesting to repeat this analysis for 2s and 3s orbitals representing the conduction band of a quantum wire composed of Li and Na respectively; this does not present a major obstacle.

In Fig. 3 we draw the Wannier function centered on site "0" for the hydrogen chain: the solid line represents the self-adjusted Wannier function in the tight-binding approximation, whereas the dashed curve is the usual Wannier function calculated in TBA. For comparison, the dotted line is the usual 1s-type atomic function. All the curves were drawn along the chain direction. Although the differences in the first two cases do not seem crucial, the values of microscopic parameters: ϵ_a^{eff}, t, U and K differ remarkably. The determined values of those parameters versus R/a_0 are provided in Table 1 (the values of E_G and α are also listed there). One notices numerically, that $U/W > 1$, where $W = 4|t|$, for $R \geq 2a_0$.

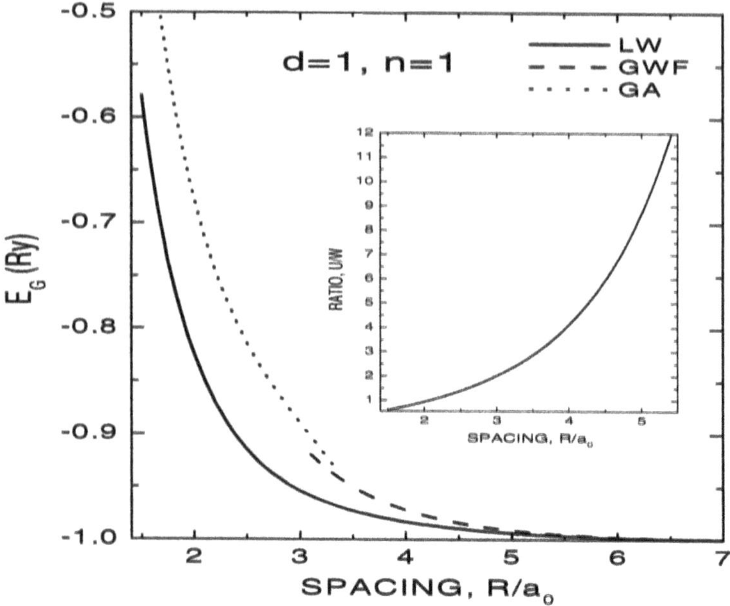

Figure 2. Optimized ground state energy as a function of interatomic distance - solid line, Gutzwiller ansatz(GA) solution - dotted line, Gutzwiller wave-function (GWF) - dashed line.

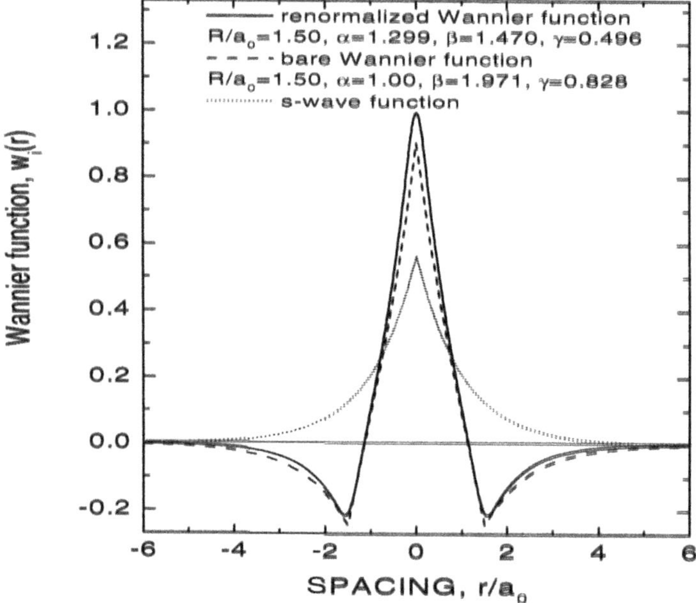

Figure 3. The shape of the optimized Wannier function - solid line, non-optimized Wannier function - dashed line, 1s atomic wave function - dotted line (for comparison).

R/a_0	$\alpha_{\min}a_0$	ϵ_a^{eff}	t	U	K	E_G/N
1.5	1.806	0.9103	-1.0405	2.399	1.695	0.0665
2.0	1.491	-0.1901	-0.5339	1.985	1.172	-0.5179
2.5	1.303	-0.6242	-0.3076	1.722	0.889	-0.7627
3.0	1.189	-0.8180	-0.1904	1.553	0.713	-0.8800
3.5	1.116	-0.9104	-0.1230	1.440	0.596	-0.9391
4.0	1.069	-0.9559	-0.0815	1.365	0.513	-0.9693
4.5	1.039	-0.9784	-0.0546	1.317	0.451	-0.9848
5.0	1.022	-0.9896	-0.0370	1.288	0.403	-0.9926
6.0	1.013	-0.9977	-0.0165	1.269	0.334	-0.9982
7.0	1.001	-0.9995	-0.0072	1.252	0.286	-0.9996
8.0	1.001	-0.9999	-0.0031	1.251	0.250	-0.9999
10.0	1.000	-1.0000	0.0003	1.250	0.200	-1.0000

Table 1. Optimized inverse orbital size, microscopic parameters and the ground–state energy for $N = 10$ atoms calculated in Slater–type basis, as a function of interatomic distance. Intersite Coulomb repulsion K_1 is included on the mean–field level in ϵ_a^{eff}, Hubbard U term is treated exactly. Single–particle potential contains *six* Coulomb wells.

The detailed electronic properties of the chain have been discussed separately [15]. Probably, the most interesting result of relevance to this workshop is the conclusion that the values of the energy per atom (and of microscopic parameters as well) are almost the same when obtained either from the solution for $N \to \infty$ (discussed above) and from the numerical solution for, say, $N = 10$ atoms [16]. This statement is illustrated in Table 2, where the optimal inverse size α_{min} of the atomic wave functions composing $w_i(\mathbf{r})$ and that of ground-state energy have been listed as a function the interatomic distance. The three columns in each category represent respectively the following results: (i) when single-site and two-site parameters have been calculated for 1s-like functions and 3- and 4-site terms (in the atomic basis) have been calculated in the contracted STO-3G basis; (ii) and (iii) represent *all* calculations in the Gaussian basis. The results are pretty close, independently of the trial atomic basis selected to represent $w_i(\mathbf{r})$. However, there is one restriction, namely the boundary conditions selected in the $N \to \infty$ and $N = 10$ cases must coincide (here they are selected as periodic b.c.). The importance of the results displayed in Table 2 relies on the fact that in this manner we can apply analytic results obtained for the infinite chain to the finite chains (nanowires) if only realistic single particle wave functions are taken into account. Obviously, in Table 2 the comparison was made for the case of one electron per atom only (the Lieb-Wu solution provides then the insulating state). We plan calculating the properties of Li and Na nanowires starting from this prescription. Parenthetically, since in the infinite chains we have *charge-spin separation* and other *non-Fermi liquid effects* [17]; they should appear also in some form in non-half filled nanoscopic

R/a_0	α_{min} $N=\infty$ 1s/3G	$N=\infty$ 3G	$N=10$ 3G	E_G $N=\infty$ 1s/3G	$N=\infty$ 3G	$N=10$ 3G
1.5	1.2985	1.3062	1.3094	-0.5788	-0.5527	-0.5684
2.0	1.1753	1.2038	1.2047	-0.8246	-0.8104	-0.8154
2.5	1.0924	1.1203	1.1203	-0.9152	-0.9123	-0.9139
3.0	1.0485	1.0683	1.0672	-0.9540	-0.9560	-0.9567
4.0	1.0212	1.0212	1.0203	-0.9832	-0.9840	-0.9841
5.0	1.0109	1.0062	1.0054	-0.9939	-0.9900	-0.9901
6.0	1.0055	1.0020	1.0027	-0.9981	-0.9914	-0.9914
7.0	1.0021	1.0005	1.0000	-0.9995	-0.9917	-0.9917
8.0	1.0005	1.0001	1.0000	-0.9999	-0.9917	-0.9917
10.0	1.0002	1.0001	1.0000	-1.0000	-0.9917	-0.9917

Table 2. Renormalized values of the inverse size (α) of the atomic wave-function (columns $2 \div 4$) and the corresponding values of the ground state energy (columns $5 \div 7$), both versus the lattice parameter R.

chains, which should also be regarded as strongly correlated systems from the start. This last question will be taken up again in Sec.5.

4. Nanoscopic H_N rings

As a second example we consider consider a ring composed of $N = 6$ hydrogen atoms arranged planarly (the stable H_4 clusters arranged spatially have been considered elsewhere [16, 5]). In this situation, the periodic boundary conditions (PBC) are the *physical* condition for the system geometry. In Fig. 4 we plot the profile of the wave function located around the exemplary atom in the hexagon H_6. This density contains renormalized orbitals and the calculations of the ground states involve (in principle) $\binom{12}{6} = 924$ 6-particle states in the occupation number representation spanned on $M = 12$ states and containing 6 Wannier functions of adjustable size. The space profiles are useful for the determination of the density function profile $n(\mathbf{r}) \equiv \langle \hat{\Psi}^\dagger(\mathbf{r})\hat{\Psi}(\mathbf{r})\rangle$, where $\hat{\Psi}(\mathbf{r})$ is the field operator spanned on those 6 Wannier states. In effect, we have that

$$n(\mathbf{r}) = \sum_{i\sigma} |w_i(\mathbf{r})|^2 \langle n_{i\sigma}\rangle + \sum_{ij\sigma}{}' w_i^\star(\mathbf{r}) w_j(\mathbf{r}) \langle a_{i\sigma}^\dagger a_{j\sigma}\rangle, \qquad (15)$$

where the primed summation means that $i \neq j$. The second part represents the part which does not appear in the Hartree-Fock (single determinant) approximation for the many particle wave function. In Fig. 5 we display the electron density profiles (normalized to unity) for the $N = 6$ atoms arranged in the hexagon, as is also in the translationally invariant along the ring density profile $n(\mathbf{r})$.

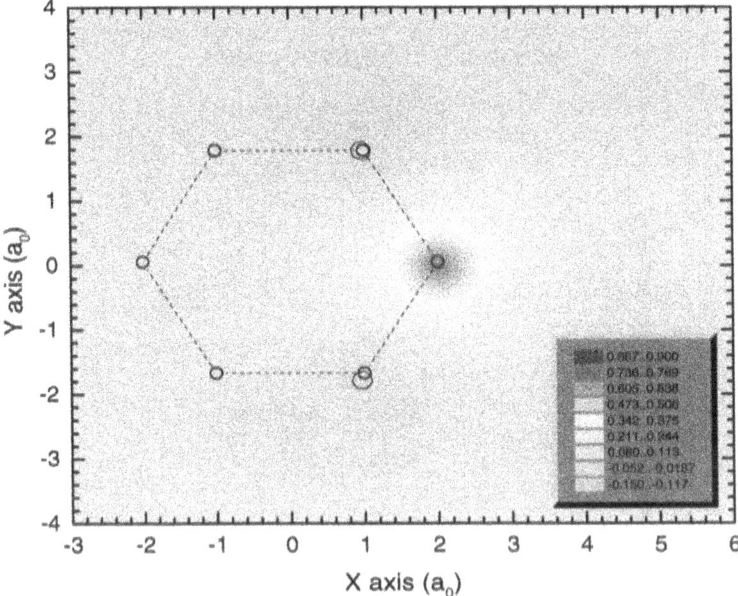

Figure 4. Spatial wave-function profiles for a selected site of H_6 cluster of hexagonal shape. Note the negative value on the neighboring site to the central atom.

An interesting feature of the spectrum of electronic states arises when the distance between the atoms increases. Namely, the spectrum decomposes into well defined Hubbard subbands, as shown in Fig. 6 (more appropriately, they represent *manifolds* corresponding to the *subbands* when $N \to \infty$). The lowest manifold (I) corresponds to the configuration with approximately singly occupied orbitals (highest occupied Wannier orbitals) whereas the manifolds II-IV correspond respectively to the states with one to three double occupancies. This division into the well separated manifolds for larger R is even better seen for the clusters of $N = 4$ and 5 atoms [8]. One should mention that the states considered here represent the excited states calculated with the help of Lanczos procedure [5], repeated many times until the configuration with the minimal energy and the optimal single-particle wave function size are reached simultaneously.

5. Further features of the results: collective properties of nanochains

5.1 One electron per atom case: localization threshold

In the previous Sections we illustrated the applications of the EDABI method, in which *the interaction among particles is dealt with first.* This is because, in

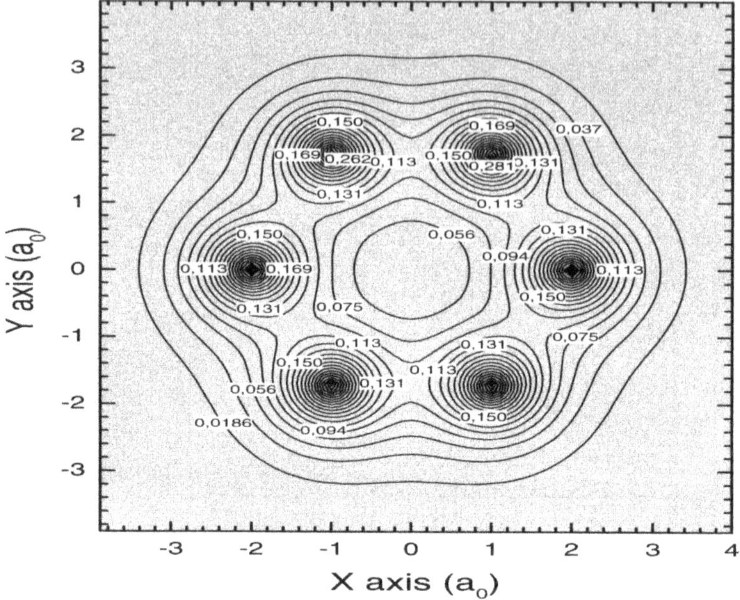

Figure 5. Exact density profiles for electrons in a hexagonal ring of atoms.

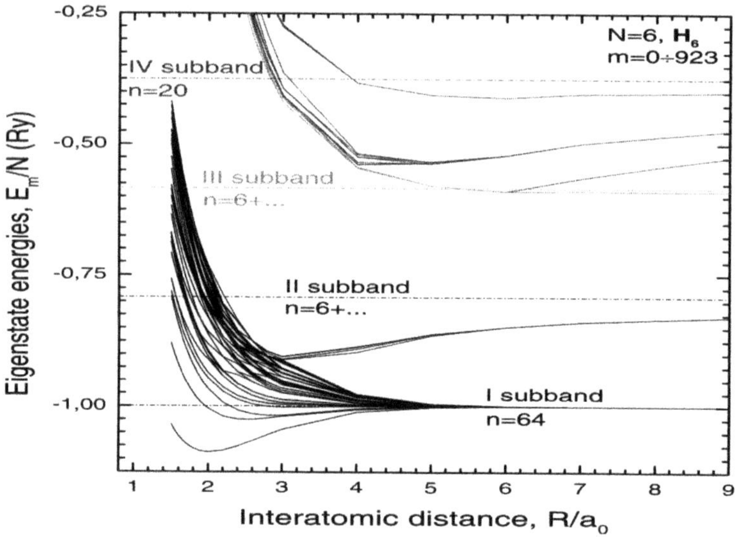

Figure 6. Decomposition of the system energies into Hubbard subbands for H_6 cluster, plotted as a function of interatomic distance. The horizontal lines represent the atomic limit values for the levels $\frac{Ul}{N}$, with $l = 0, 1, 2,$ and 3.

most cases, the interaction parameters (coupling constants) represent the largest energy scale in the system. The first of the examples (*the Hubbard chain*) represents the situation, for which an analytic expression for the ground state energy exists [12], whereas the case of H_N *rings* must be treated numerically all the way through [18]. In applications of this method to the extended three-dimensional systems one will have to resort to the approximate treatments of the model Hamiltonian in the Fock space. This last problem poses a real challenge for the future. In the remaining part of this brief review we concentrate on the collective properties of the nanochain.

We have concentrated first on the basic quantum-mechanical features of the system such as the ground-state energy or the renormalized single-particle wave function in the milieu of other particles. In Fig. 7 we plot the exact ground state energy of a chain of $N = 6 \div 10$ atoms and compare it with that obtained in the Hartree-Fock approximation for the Slater antiferromagnetic state. The starting Hamiltonian is of the form (10). As one can see the Hartree-Fock energy represents an upper estimate, as it should be. Additionally, the curve M represents the "metallic" approximation, for which the correlation function $< \delta n_i \delta n_j >$ has been taken for the 1D electron gas on the lattice. On the contrary, INS represents the energy of the Heisenberg-Mott state in the mean-field approximation. The state of the system crosses over from the Slater metallic-type state to the localized-spin-type of state. This is seen explicitly when we calculate the evolution of the spin-spin correlation function with the increasing interatomic distance, as displayed in Fig. 8. Well defined oscillations of $< \mathbf{S}_i \cdot \mathbf{S}_j >$ are seen for even ($N = 12$) number of atoms, which become more pronounced with the increasing N (the frustration effects appear for odd N). What is much more important, the autocorrelation part $< \mathbf{S}_i \cdot \mathbf{S}_i >=< \mathbf{S}_i^2 >= (3/4)(1-2 < n_{i\uparrow} n_{i\downarrow} >)$ evolves from the value close to the free-electron value $< \mathbf{S}_i^2 >= (3/4)(1 - 2 < n_{i\uparrow} >< n_{i\downarrow} >) = 3/8$ to the atomic-limit value $< \mathbf{S}_i^2 >= (1/2)(1/2 + 1) = 3/4$. This evolution provides a direct evidence of the crossover from delocalized to the localized regime.

Other properties such as the electrical conductivity [16] and the statistical distribution in momentum space ($n_{k\sigma}$) have also been addressed [9]. Here the question emerges whether the *quantum nano-liquid* of electrons in a nanochain resembles at all the Landau-Fermi liquid or if it is rather represented by the Tomonaga-Luttinger scaling laws [17]. The answer is not yet settled, as within our method we can deal only with up to N=16 hydrogen atoms assembled into a linear chain. However, one can make some definite statements for the particular cases. Namely, for the half-filled case (one electron per atom) the modified Fermi distribution is a good representation of the $n_{k\sigma}$ for smaller R values; with the increasing atom spacing it is smeared out above critical spacing $R = R_c \approx 3.4a_0$ [10], which characterizes a *crossover* from the case with extended states to the state of localized electrons on atoms, as detailed below. In

a/a_0	D^*_{14}	D^*_{12}	D^*_{10}	D^*_8	D^*_6	D^*_∞	$\sigma(D^*_\infty)/D^*_\infty$
1.5	0.9225	0.9420	0.9563	0.9727	0.9822	0.8008	0.019
1.6	0.8879	0.9162	0.9378	0.9612	0.9754	0.7175	0.029
1.7	0.8419	0.8817	0.9130	0.9459	0.9667	0.6148	0.043
1.8	0.7826	0.8365	0.8805	0.9256	0.9552	0.4967	0.064
1.9	0.7095	0.7794	0.8389	0.8992	0.9406	0.3728	0.092
2.0	0.6245	0.7105	0.7875	0.8660	0.9222	0.2567	0.129
2.1	0.5315	0.6310	0.7265	0.8254	0.8996	0.1606	0.172
2.2	0.4338	0.5431	0.6549	0.7755	0.8714	0.0899	0.228
2.3	0.3403	0.4523	0.5766	0.7179	0.8379	0.0455	0.287
2.4	0.2554	0.3631	0.4937	0.6524	0.7982	0.0207	0.352
2.5	0.1839	0.2812	0.4109	0.5813	0.7526	0.0087	0.420
2.6	0.1269	0.2096	0.3315	0.5065	0.7009	0.0033	0.489
2.7	0.0840	0.1508	0.2595	0.4312	0.6441	0.0012	0.566
2.8	0.0536	0.1049	0.1972	0.3586	0.5836	0.0004	0.641
2.9	0.0315	0.0706	0.1456	0.2914	0.5208	0.0001	0.920
3.0	0.0196	0.0461	0.1047	0.2314	0.4575	0.0000	–

Table 3. Normalized Drude weight D^*_N, the extrapolated value D^*_∞, and its relative error for 1D half–filled system with long–range Coulomb interactions.

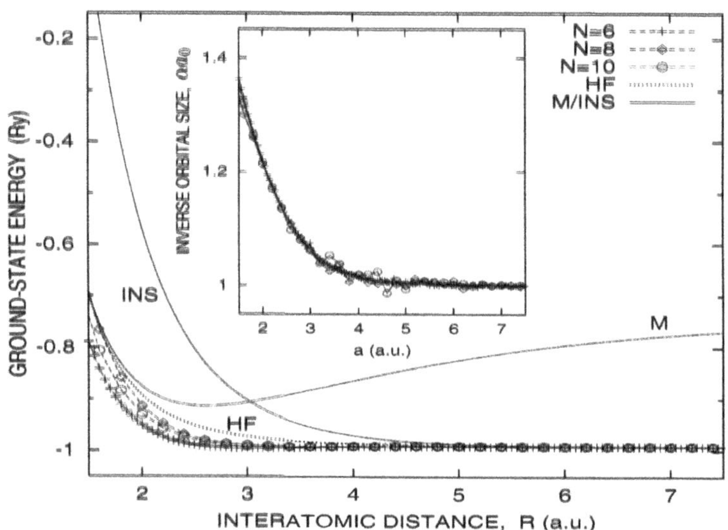

Figure 7. Ground state energy per atom vs. R for the linear chain with $N = 6 \div 10$ atoms with periodic boundary conditions. The STO-3G Gaussian basis for representation of atomic orbitals forming the Wannier function has been used. The inset provides a universal behavior of the inverse size α of the orbitals. For details see main text.

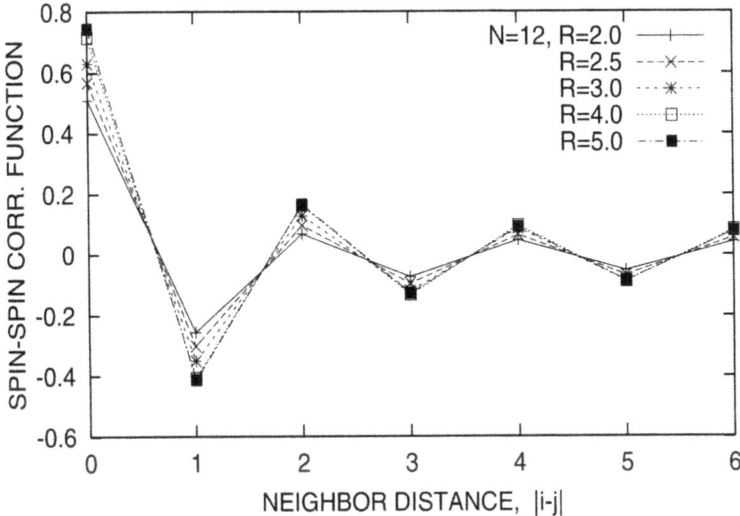

Figure 8. Spin-spin correlation function $\langle \mathbf{S}_i \cdot \mathbf{S}_j \rangle$ vs. the distance $|i-j|$ between the atomic sites, for different lattice constant R and for $N = 12$ atoms. Due to periodic boundary conditions only the distance up to $|i-j| = 6$ is relevant. The continuous line is guide to the eye. A quasi-antiferromagnetic arrangement is clearly seen, particularly for larger R.

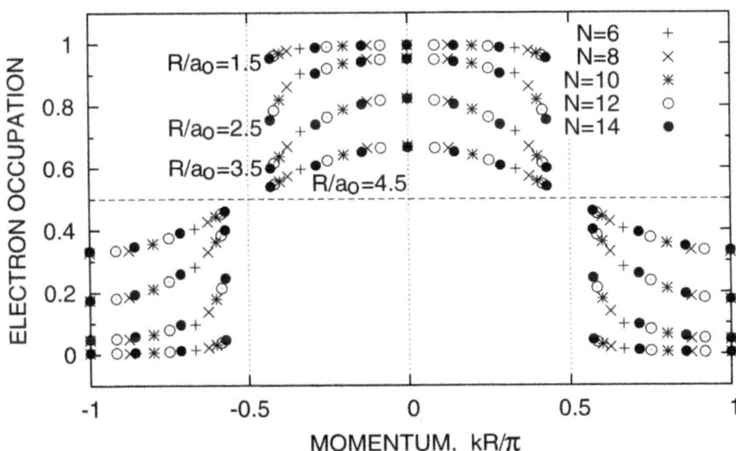

Figure 9. Evolution of the statistical momentum distribution $n_{k\sigma}$ with the increasing inter-atomic distance R. For the nanochain of $N = 6, 10$, and 14 the periodic boundary conditions provide the energy minimum, whereas for $N = 8$ and 12 antiperiodic boundary conditions are appropriate. For details see main text.

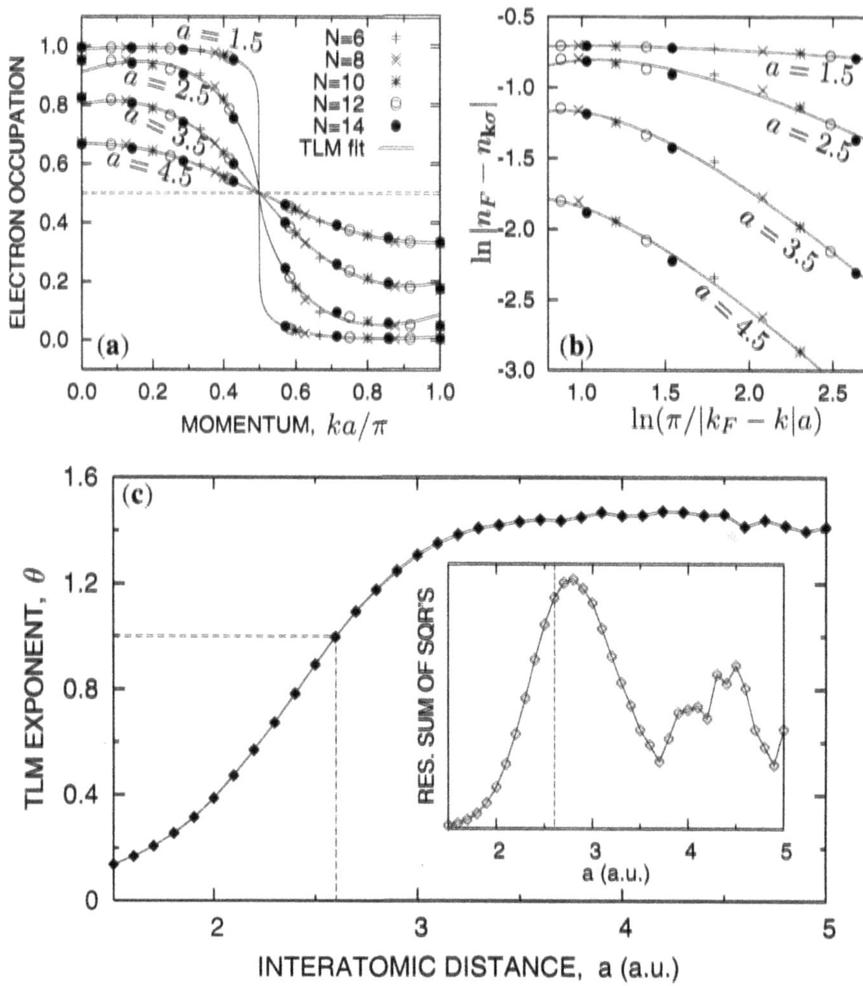

Figure 10. Luttinger–liquid scaling for a *half-filled* 1D chain of $N = 6 \div 14$ atoms with *long-range* Coulomb interactions: (a) momentum distribution for electrons in the linear and (b) log-log scale, continuous lines represent the fitted singular expansion in powers of $\ln(\pi/|k_F - k|a)$ (see main text for details); (c) Tomonaga–Luttinger model exponent θ vs. lattice parameter a (specified in a_0) and (in the *inset*) the corresponding residual sum of squares. The solid lines in Figures (a) and (b) represent the TLM fitting of Eq. (16).

Fig. 9 we exhibit this evolution on the example of the distribution function (i.e. electron occupation in the momentum space); a clear universality is observed for $N = 6 \div 14$ atoms. The adjustable Gaussian (STO-3G) single-particle basis has been used in the analysis. The existence of the Fermi ridge quasi-discontinuity for a small R is very suggestive in this case and is positioned near the Fermi wave-vector $k_F^\infty = \pm\pi/(2R)$, corresponding to that in Landau-Fermi liquid, with with $N \to \infty$. For $N = 6, 10, 14$ periodic boundary conditions (PBC) provide the minimal ground state energy, whereas for $N = 8$ and 12 the anti-periodic boundary conditions (ABC) lead to the lower energy. However, one clearly sees the absence of the points at the Fermi momentum k_F^∞. This is because, for example, the Fermi points for $N = 14$ are located at $k_F R/\pi = \pm 3/7$, whereas they are located at $k_F R/\pi = \pm 5/12$ for $N = 12$, close to the values $\pm 1/2$ in both cases. One would have to apply a renormalization group approach [19] for the states close to k_F (i.e. perform the analysis for larger number $N \sim 10^2$ atoms) to determine the precise evolution of the distribution function, this time with the system size N. Nevertheless, the results for $N \leq 14$ atoms represent those for a true nanoscopic system.

Before addressing the question of localization directly, we would like to address the question whether the computed distribution displayed in Fig. 9 can be fitted into the Tomonaga-Luttinger mode, with the logarithmic scaling corrections included [17]. The statistical distribution near the Fermi point can be represented by

$$\ln |n_{k\sigma} - n_F| = -\theta \ln z + b \ln \ln z + O(1/\ln z), \tag{16}$$

with $z \equiv \pi/|k - k_F|$. Here θ is a non-universal (interaction-dependent) exponent (it yields the nonexistence of fermionic quasiparticles, since its residue vanishes as $Z_k \sim |k - k_F|^\theta$ with $k \to k_F$. The corresponding electron-momentum distribution is depicted in Fig. 10a in the linear, and in Fig. 10b in the log-log scale. The R dependence of the exponent θ is shown in Fig. 10c and crosses the value $\theta = 1$ for $R = R_c \approx 2.6a_0$ corresponding to the localization threshold [17]. This threshold is about 30% smaller than the corresponding value ($R_c \approx 3.4a_0$) when the almost-localized Fermi-liquid view was taken [10]. One should also note that the Luttinger scaling does not reproduce well the occupancies farther away either way from the Fermi point. Thus the results concerning $n_{k\sigma}$ do not provide a definite answer as to the exact nature of the *nanoliquid* composed of $N \sim 10$ electrons, although it is absolutely amazing that they have such a nice and simple scaling properties.

To address the question of the electron localization directly, we have calculated the optical conductivity $\sigma(\omega)$, which can be written in the form $\sigma(\omega) =$

$D\delta(\omega) + \sigma_{reg}(\omega)$, where the regular part is

$$\sigma_{reg}(\omega)\frac{\pi}{N}\sum_{n\neq 0}\frac{|<\Psi_n|j_p|\Psi_0>|^2}{E_n - E_0}\delta(\omega - (E_n - E_0)), \tag{17}$$

whereas the *Drude weight (the charge stiffness)* D is given by

$$D = \frac{\pi}{N} < \Psi_0|T|\Psi_0 > -\frac{2\pi}{N}\sum_{n\neq 0}\frac{|<\Psi_n|j_p|\Psi_0>|^2}{E_n - E_0}, \tag{18}$$

with T being the hopping term as in (9) and j_p the current operator defined as $j_p = it\sum_{ij\sigma}(a^{\dagger}_{j\sigma}a_{i\sigma} - h.c.)$. Here $|\Psi_n>$ is the system eigenstate corresponding to the eigenvalue E_n. For a finite system of N atoms D is always nonzero due to nonzero tunnelling rate through a potential barrier of finite width. Because of that, the finite-size scaling with $1/N \to 0$ must be performed on D. We use the following parabolic extrapolation

$$\ln D^{\star}_N = a + b(1/N) + c(1/N)^2,] \tag{19}$$

where $D^{\star}_N = -(N/\pi)D/ < \Psi_0|T|\Psi_0 >$ denotes the normalized Drude weight for the system of N sites. We observe that $0 \leq D^{\star} \leq 1$ and hence can be regarded as an order parameter for the transition to the localized (atomic) states. In Table 3 we plot the weights $D^{\star}_N$, the extrapolated values $D^{\star}_{\infty}$, and its relative error for 1D half-filled system with long-range Coulomb interactions included. What is very important, the value of $D^{\star}_N$ drops by two orders of magnitude when R changes in the range by factor of two (between $1.5a_0$ and $3a_0$). Note also that $D^{\star}_{\infty}$ is within its error for $R \simeq R_c \approx 2.3a_0$, close to the value obtained from the Tomonaga-Luttinger scaling for $n_{k\sigma}$, as one would expect. The result for $D^{\star}_N$ is probably telling us how far can we go quantitatively when discussing the localization in nanowires. These results will be detailed in a separate publication.

5.2 Quarter-filled case

For the quarter-filled case (i.e. when every second atom contributes a valence electron) the distribution $n_{\mathbf{k}\sigma}$ is more smeared out, as shown in Fig. 11. The ground state for $R \geq 3a_0$ is then well represented by the charge-density-wave state, since then the density-density correlation function $\langle(n_i - \overline{n})(n_j - \overline{n})\rangle$ exhibits the oscillatory behavior, as shown in Fig. 12. The onset of the charge-density wave state in that case is invariably due to the long-range Coulomb interaction $\sim K_{ij}$, which is reduced in that state. The charge-density wave order parameter defined as

$$\Theta_{CDW} \equiv \frac{1}{N}\sum_m(-1)^m < (n_i - \overline{n})(n_j - \overline{n}) > \tag{20}$$

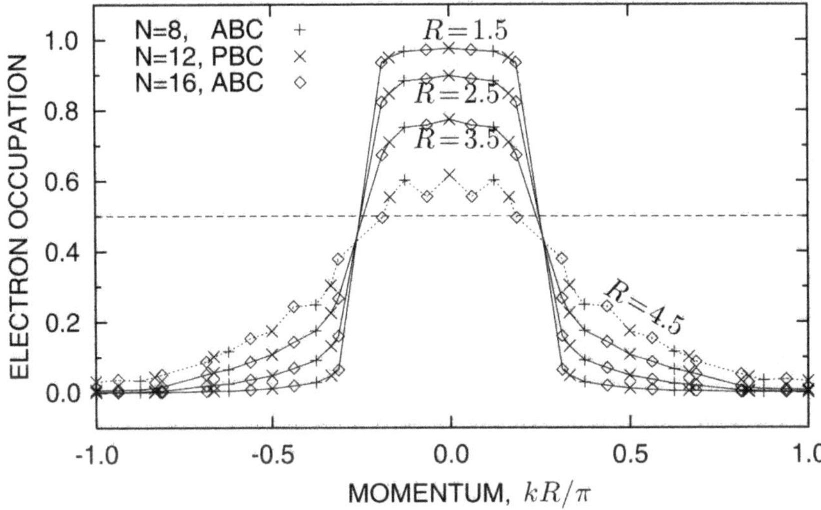

Figure 11. Momentum distribution $n_{k\sigma}$ for electrons on a chain of $N = 8 \div 16$ atoms in the *quarter-filled* band case ($N_e = N/2$). Lines are drawn as a guide to the eye only. Values of the lattice parameter R are specified in units of a_0. PBC and ABC denote periodic and antiperiodic boundary conditions, respectively. The dashed line marks occupation $n_{k\sigma} = 1/2$.

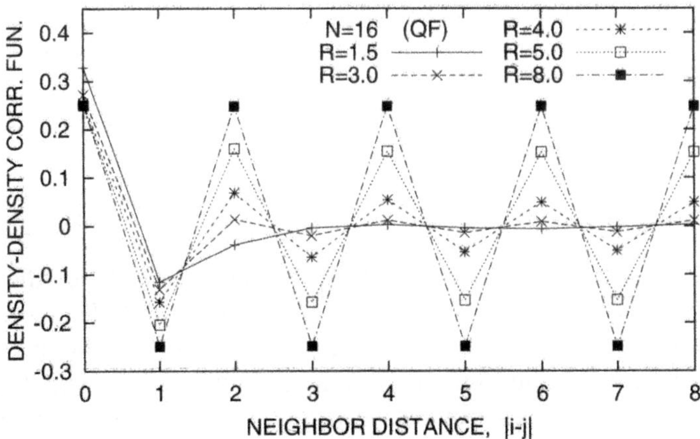

Figure 12. Charge-density distribution for the *quarter-filled* ($N_e = N/2$) nanochain as viewed by density-density fluctuation correlation function $\langle \Delta n_i \Delta n_j \rangle$ vs. relative distance $|i - j|$.

reaches its maximal value $1/4$ for $R \geq 8a_0$. Let us stress again, the form of the statistical distribution (and its feasibility) for $N \sim 10$ atoms is a very interesting question by itself, since this is *the regime of nanoscience*. Our results show that even in that regime one should be able to see the signatures of the phase transition to the spin- or charge- density wave states.

6. Conclusions

The EDABI method (**E**xact **D**iagonalization combined with**AB I***nitio* orbital readjustment)provides the *exact* ground state energy of the model systems considered (*Hubbard chain, nanoscopic chains and rings*) as a function of interatomic distance. It also provides reliably other ground-state dynamical characteristics for nanoscopic systems: spin and charge correlation functions, the spectral density and the density of states, as well as the system conductivity. Not all the characteristics have been presented in this overview [5, 10, 16, 17]. Furthermore, the method can also be extended to nonzero temperatures. Finally, it should be underlined again that our method of approach is particularly suited for strongly correlated systems, where the interaction and the single-particle parts should be treated on the same footing. The exact numerical results in a model situation can also serve as a test for approximate analytic treatments. The analysis with inclusion of long-range Coulomb interactions of the distribution function $n_{k\sigma}$ in either Fermi-like or Tomonaga-Luttinger categories suggests the existence of the *crossover* transition to the localized state with the increasing interatomic distance R. This state is either the spin- or the charge- ordered for the number of electrons $N_e = N$ and $N/2$, respectively. In the small R limit the state can be rendered as quasi-metallic in the sense that the quasimomentum $\hbar k$ can be regarded as a good quantum number, even though the level structure is discrete. The difference between the short-chain and infinite-chain situation is due to the circumstance that to form extended states here the electrons tunnel through a barrier of finite width (of the length $L = NR$).

Acknowledgments

The work was supported by the State Committee for Scientific Research KBN, Grant No. 2P03B 050 23. The two authors (A.R & J.S.) acknowledge respectively the junior and the senior fellowships of the Foundation for Science (FNP). We are also grateful to the Institute of Physics of the Jagiellonian University for the support for computing facilities used in part of the numerical analysis.

References

[1] P.C. Hohenberg, W. Kohn, and L.I. Sham, in *Adv. Quantum Chemistry*, edited by S.B. Trickey (Academic, San Diego, 1990) vol. 21, pp. 7-26; W. Temmerman et al., in *Elec-*

tronic Density Functional Theory: Recent Progress and New Directions, edited by J.F. Dobson et al. (Plenum, New York, 1998) pp. 327-347.

[2] V.I Anisimov, J. Zaanen, and O.K. Andersen, Phys. Rev. B **44**, 943 (1991); P. Wei and Z.Q.Qi,*ibid.* **49**, 10864(1994).

[3] A. Svane and O. Gunnarson, Europhys. Lett. **7**, 171 (1988); Phys. Rev. Lett. **65**, 1148(1990)

[4] S. Ezhov et al., Phys. Rev. Lett. **83**, 4136(1999); K. Held et al., *ibid.* **86**, 5345 (2001).

[5] J. Spałek et al., Phys. Rev. B **61**, 15676 (2001); A. Rycerz and J. Spałek, *ibid.* B **63**, 073101(2001); B **65**, 035110(2002); J Spałek et al., submitted to Phys. Rev. B.

[6] J. Spałek et al., Acta Phys. Polonica B **31**, 2879(2000); *ibid.*, B **32**, 3189 (2001).

[7] A. Rycerz et al., in *Lectures on the Physics of Highly Correlated Electron Systems VI*, edited by F. Mancini, (AIP Conf. Proc. No. 629, New York, 2002) pp. 213-223; *ibid.* (AIP Conf. Proc. No. 678, New York, 2003) pp. 313-322.

[8] J. Spałek et al., in *Concepts in Electron Correlation*, Proc. of the NATO Adv. Res. Workshop, edited by A.C. Hewson and V. Zlatić (Kluwer, Dordrecht, 2003) pp. 257-268.

[9] J. Spałek et al., in *Highlights of Condensed Matter Physics*, (AIP Conf. Proc. No. 695, New York, 2003)pp. 291-303.

[10] J. Spałek and A. Rycerz, Phys. Rev. B **64**, 161105 (2001)(R).

[11] If we want to include it, then it is irrelevant for the case of one electron per atom, since then pure spin-density-wave correlations set in. However, if the system is non-neutral ($N_e \neq N$), then the charge-density-wave state may become stable (c.f. Fig. 12 in Sec. 5).

[12] E.H. Lieb and F.Y. Wu, Phys. Rev. Lett. **20**, 1443(1968). For overview see: M. Takahashi, *Thermodynamics of one-dimensional solvable models* (Cambridge Univ. Press, 1999) ch. 6.

[13] M.C. Gutzwiller, Phys. Rev. **137**, A1726(1965) and references therein.

[14] W. Metzner and D. Vollhardt, Phys. Rev. B **37**, 7382 (1988).

[15] J. Spałek, J. Kurzyk, W. Wójcik, E.M. Görlich, and A. Rycerz, submitted for publication.

[16] A. Rycerz, Ph. D. Thesis, Jagiellonian University, Kraków - 2003 (unpublished).

[17] J. Solyom, Adv. Phys. **28**, 201(1979); J. Voit, Rep. Prog. Phys. **57**, 977(1995).

[18] R. Zahorbeński and J. Spałek, unpublished.

[19] S.R. White, Phys. Rep. **301**, 187 (1998); R. Shankar, Rev. Mod. Phys. **66**, 129 (1994)

ULTRAFAST REAL-TIME SPECTROSCOPY OF LOW DIMENSIONAL CHARGE DENSITY WAVE COMPOUNDS

J. Demsar[1], D. Mihailovic[1], V.V. Kabanov[1], and K. Biljakovic[2]
[1] *J. Stefan Institute, Jamova 39, 1000 Ljubljana, Slovenia*
[2] *Institute for Physics, Bijenicka 46, HR-10000 Zagreb, Croatia*
jure.demsar@ijs.si

Abstract We present a femtosecond time-resolved optical spectroscopy (TRS) as an experimental tool to probe the changes in the low energy electronic density of states as a result of short and long range charge density wave order. In these experiments, a femtosecond laser pump pulse excites electron-hole pairs via an interband transition in the material. These hot carriers rapidly release their energy via electron-electron and electron-phonon collisions reaching states near the Fermi energy within 10-100 fs. The presence of an energy gap in the quasiparticle excitation spectrum inhibits the final relaxation step and photoexcited carriers accumulate above the gap. The relaxation and recombination processes of photoexcited quasiparticles are monitored by measuring the time evolution of the resulting photoinduced absorption. This way, the studies of carrier relaxation dynamics give direct information of the temperature-dependent changes in the low energy density of states. Here we present the application of the femtosecond time-resolved optical spectroscopy for studying changes in the low energy electronic density of states in low dimensional charge density wave systems associated with various charge density wave (CDW) transitions and review some recent experiments on quasi 1D and 2D CDW compounds.

Keywords: femtosecond time-resolved spectroscopy, low dimensional charge density waves

1. Introduction

Femtosecond time-resolved optical spectroscopy has been shown in the last couple of years to present an excellent alternative to the more conventional time-averaging frequency-domain spectroscopies for probing the changes in the low energy electronic structure in strongly correlated systems [1, 2]. In these experiments (see Figure 1), a femtosecond laser pump pulse excites electron-hole pairs via an interband transition in the material. In a process which is similar in most materials including metals, and superconductors, these hot carriers

A.S. Alexandrov et al. (eds.), Molecular Nanowires and Other Quantum Objects, 377–392.

rapidly release their energy via electron-electron and electron-phonon collisions reaching states near the Fermi energy within 10-100 fs. Further relaxation and recombination dynamics, determined by measuring photoinduced changes in optical properties (reflectivity, transmissivity or, in case of time-resolved terahertz spectroscopy, far infrared conductivity) as a function of time after photoexcitation, depends strongly on the nature of the low-lying electronic spectrum. In particular, the experimental technique was found to be sensitive to opening of the superconducting gap, appearance of a short-range and long range charge-density wave order [3, 4], and changes in the electronic specific heat and electron-phonon coupling associated with the heavy fermion behavior [5], just to mention a few. What is particularly important is the fact, that even though the probe photon wavelength in these experiments ranges from THz [6, 7, 8] (enabling direct measurement of photoinduced conductivity dynamics), mid-IR to several eV [1,9-16], the dynamics is in many instances the same [6, 7, 8], supporting the idea [1] that the photoinduced reflectivity (transmissivity) dynamics is determined by relaxation and recombination processes of quasiparticles in the vicinity of Fermi energy.

Since the optical penetration depth in these materials is on the order of 100 nm, the technique is essentially a bulk probe. Moreover, since the effective shutter speed is on the order of a picosecond, the technique is particularly useful to probe the systems with (dynamic) spatial inhomogeneities. In this case different local environments (that appear frozen on the timescale of picoseconds) give rise to different components in measured photoinduced reflectivity (transmissivity) traces. As the different components can have different time scales [17, 18], temperature[1, 3], photoexcitation intensity, and probe polarization [17] or wavelength dependences, they can be easily extracted.

Furthermore, due to the fast effective shutter speed of the technique (≈ 1 ps) one can expect to observe short lived fluctuations that would appear frozen on this timescale. Indeed, the experiments on quasi-1D charge density wave compound $K_{0.3}MoO_3$ [3] suggest that above the transition temperature to the 3D ordered CDW state the technique is sensitive to the presence of short range 3D fluctuations of CDW order.

2. Experimental details

In the experiments discussed below, a Ti:sapphire mode-locked laser operating at a 78 MHz repetition rate and pulse length of 50-70 fs was used as a source of both pump and probe pulse trains. The wavelength of the pulses was centered at approximately $\lambda \approx 800$nm (1.58eV) and the intensity ratio of pump and probe pulses was about 100:1. The pump and probe beams were crossed on the sample's surface, where the angle of incidence of both beams was less than 10°. The diameters of the beams on the surface were $\sim 100\mu$m for the pump beam and $\sim 50\mu$m for the probe beam. The typical energy density of pump pulses was $\sim 0.1 - 1\ \mu J/cm^2$, which produced a weak perturbation of the electronic system with the density of thermalized photoexcited carriers on the order of $10^{-4} - 10^{-3}$ per unit cell (the approximation is based on the assumption that each photon with energy $\hbar\omega$ creates $\hbar\omega/\Delta$ thermalized photoexcited carriers, where $\Delta \approx 50$ meV is of the order of the CDW gap [3]). The train of the pump pulses was modulated at 200kHz with an acousto-optic modulator and the small photoinduced changes in reflectivity or transmission were resolved out of the noise with the aid of phase-sensitive detection. The pump and probe beams were cross-polarized to reduce scattering of pump beam into the detector (avalanche photodiode). This way, photoinduced changes in reflectance of the order of 10^{-6} can be resolved. A detailed description of the experimental technique can be found in Ref. [2].

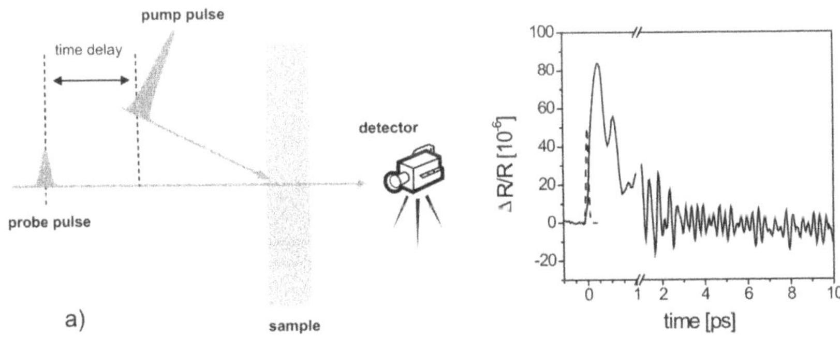

Figure 1. a) Schematic of the experimental technique: pump pulse excites the system under investigation, while the resulting changes in optical properties (reflectivity, transmission) as a function of time are investigated by a suitably delayed probe pulse. Panel b) presents the photoinduced reflectivity change in quasi-2D charge density wave compound 1T-TaS_2 obtained at 10 K. The dashed line represents the excitation pulse.

3. Photoexcited quasiparticle dynamics in narrow-gap materials

In this section we review the basic ideas of the theoretical model [1] adopted to associate the measured amplitude and the relaxation dynamics of the photoinduced transients with the corresponding changes in the low-energy electronic structure in these narrow-gap materials.

The basis of the model is shown in Figure 2. A pump pulse with photon energy $\hbar\omega$ (e.g. 1.5 eV) excites carriers from occupied states below E_F to unoccupied states in bands $\hbar\omega$ above - schematically shown by **a)**. The initial phase of the photoexcited carrier relaxation after absorption of the pump laser photon proceeds rapidly. The photoexcitation is followed by carrier thermalization through electron-electron scattering with a characteristic time $\tau_{e-e} \sim \frac{\hbar E_F}{2\pi E^2}$ for intraband relaxation, where E is the carrier energy measured from the Fermi energy E_F. In the initial thermalization process, a quasiparticle avalanche multiplication due to electron-electron collisions takes place as long as τ_{e-e} is less than the electron-phonon (e-ph) relaxation time τ_{e-ph} which is $\approx$ 100 fs in these materials. Therefore, in the absence of gap in the density of states the photoexcitation is followed by rapid carrier relaxation resulting in slightly elevated electron-phonon temperature within $\sim$ 100 fs.

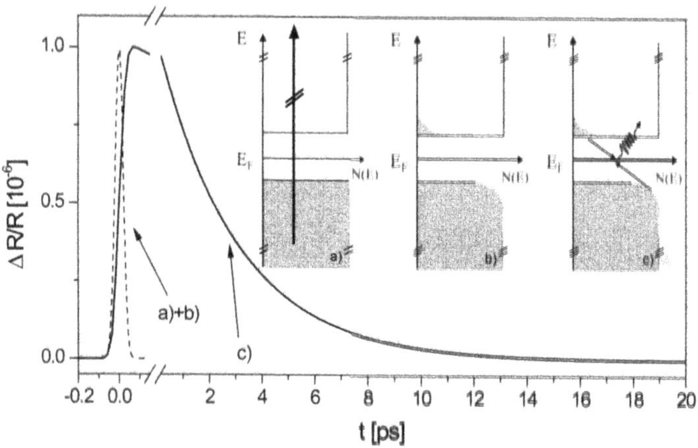

Figure 2. Schematic of the photoexcitation, carrier relaxation and photoinduced absorption processes in materials with a small energy gap in DOS. Interband photoexcitation (step a) is followed by an ultrafast initial e-e thermalization (step b). The gap in the density of states creates a relaxation bottleneck. The probe pulse measures through the photoinduced absorption (step c) the time evolution of the photoexcited carrier density.

When the gap with magnitude 2Δ is present in the low energy density of states, phonons with energies less than 2Δ cannot contribute to the relaxation of carriers just above the gap, therefore the situation is strongly modified and a bottleneck in the relaxation occurs after $t \sim 100$ fs. As a result quasiparticles accumulate near the gap, forming a non-equilibrium distribution shown by **b)**. Since typical values of the energy gap in CDW's (and cuprates) are of the order of $2\Delta \sim 30$ - 100 meV, each photon thus creates 20 $\sim$ 40 quasiparticles given by $\hbar\omega/2\Delta$. The final relaxation step across the gap is strongly suppressed [19], since the high energy phonons emitted by bi-particle recombination (shown by **c)**) have enough energy to further excite electron-hole pairs. Therefore the quasiparticles together with high frequency phonons (with $\hbar\omega_q > 2\Delta$) form a near-steady state distribution. The recovery dynamics of this system is governed by the decay $\hbar\omega_q > 2\Delta$ phonon population, governed either by the diffusion out of the probed volume or by anharmonic decay to $\hbar\omega_q < 2\Delta$ phonons. Since in CDW compounds the energy gap is on the order of 30 - 100 meV, the phonons in question are optical phonons, whose anharmonic decay times are on the order of picoseconds the later mechanism was found to be dominant (diffusion takes place on the timescale of 100's of picoseconds).

Considering the probe process, we are measuring photoinduced changes of transmission or reflectivity on the picosecond timescale. Since we are dealing with weak perturbations, we can assume that the photoinduced transmission $\Delta T/T$ (or reflectivity $\Delta R/R$) is in the linear approximation proportional to photoinduced absorption $\Delta A/A$. Through the Fermi golden rule the photoinduced absorption is due to changes in the *initial or final state* carrier density. As the lifetime of quasiparticles high above E_F is of the order of 10 fs, we can assume that the main changes in absorption involve photoexcited carriers just above the gap as initial or final states for absorption. Since the probe laser photon energy $\hbar\omega$ is typically well above the plasma frequency, we make an approximation that the photoinduced absorption is given by the Fermi golden rule, with photoinduced quasiparticles above the gap as initial states and unoccupied states at $\hbar\omega$ above the Fermi energy as final states - see **c)** in insert to Figure 2. The amplitude of the photoinduced absorption is thus proportional to the photoexcited quasiparticle density n_{pe} and by measuring the photoinduced transmission $\Delta T/T$ (or reflectivity $\Delta R/R$) the temporal evolution of the photoexcited carrier density n_{pe} is probed. The photoinduced transmission amplitude is weighted by the dipole matrix element and the joint density of states, so $\Delta T/T \propto -n_{pe}\rho_f |M_{ij}|^2$, where n_{pe} is the photoexcited carrier density, ρ_f is the density of (final) unoccupied states, and M_{ij} is the dipole matrix element.

Based on the above arguments, that after initial e-e and e-ph thermalization processes quasiparticles and $\hbar\omega_q > 2\Delta$ phonons are is quasi-thermal equilibrium, and that the amplitude of the transient is proportional to the photoinduced quasiparticle density, the one-to-one relation between the temperature dependence of the photoinduced transient amplitude and the amplitude of the low-energy gap in the DOS can be found. Assuming that the energy gap is isotropic, one can approximate non-equilibrium phonon (n_{ω_q}) and quasiparticle (f_ε) distribution functions as follows [1, 21]:

$$n_{\omega_q} = \begin{cases} \frac{1}{\exp(\frac{\hbar\omega_q}{k_B T})-1} \quad , & \hbar\omega_q < 2\Delta \\ \frac{1}{\exp(\frac{\hbar\omega_q}{k_B T'})-1} \quad , & \hbar\omega_q > 2\Delta \end{cases} \tag{1}$$

$$f_\varepsilon = \frac{1}{\exp(\frac{\varepsilon}{k_B T'}) + 1}, \tag{2}$$

where T is the lattice temperature and T' is the temperature of quasiparticles and high frequency phonons with $\hbar\omega_q > 2\Delta$. The number of photoexcited quasiparticles n_{pe} can be calculated as the difference between the numbers of thermally excited quasiparticles (per unit cell) after and before photoexcitation characterized by temperatures T' and T. The number of photoexcited carriers $n_{pe}(= n_{T'} - n_T)$ can be obtained directly considering energy conservation [1].

As an illustration of the calculation, let us assume that $\Delta = \Delta^p$ is temperature *independent* and large in comparison to $k_B T$. Since the magnitude of the gap Δ^p is of the order of several 10 meV, which corresponds to temperatures of a few hundred kelvins, we can assume that quasiparticles are non-degenerate and f_ε can be approximated as $f_\varepsilon \sim \exp(-\varepsilon/k_B T)$. Similarly we can approximate $n_{\omega_q} \sim \exp(-\hbar\omega_q/k_B T)$. When considering the temperature-independent (pseudo-)gap, we take the quasiparticle density of states given by

$$N(E) = \begin{cases} 0\ , & E < \Delta^p \\ N(0)\ , & E > \Delta^p \end{cases}. \tag{3}$$

Strictly speaking the model density of states corresponds to the real gap, however Δ^p could be understood also as an energy where the relaxation of photoexcited quasiparticles in inhibited. E.g., the density of states below Δ^p could be finite but the relaxation through these states is suppressed (e.g. relaxation through localized states) therefore they are not available for the relaxation.

Further, we assume that the phonon spectral density is constant at large frequencies ($\hbar\omega_q > 2\Delta^p$). In this case the quasiparticle energy and the energy of high frequency phonons at temperature T are given by

$$E_T = 2N(0)\,\Delta^p T \exp(-\Delta^p/k_B T)\ ;\ \ E_T^{ph} = \frac{2\nu\Delta^p T}{\Omega_c}\exp(-2\Delta^p/k_B T)$$

respectively. Here ν is the number of high frequency phonon modes (per unit cell) and Ω_c is the phonon cut-off frequency. Since we assume that after photoexcitation the high energy phonons and quasiparticles are described by the same temperature (T'), we write the conservation of energy as

$$\mathcal{E}_I = \left(E_{T'} + E_{T'}^{ph}\right) - \left(E_T + E_T^{ph}\right) \tag{4}$$

where $\mathcal{E}_I$ is the energy density per unit cell deposited by the incident pump laser pulse. Since quasiparticle density is given by $n_T = 2N(0)\,T\exp(-\Delta^p/k_B T)$ [21], by making the approximation that $k_B T \ll \Delta$, Eq.(4) can be rewritten in terms of quasiparticle densities at temperatures T' and T

$$(n_{T'} - n_T)\,\Delta^p + \left(n_{T'}^2 - n_T^2\right)\frac{\nu\Delta^p}{2\hbar\Omega_c N(0)^2 k_B T'} = \mathcal{E}_I\ . \tag{5}$$

There are two limiting cases to be considered with respect to the ratio of photoexcited vs. thermally excited quasiparticle densities, n_{pe}/n_T.

In the low temperature limit $n_{pe} \gg n_T$, since n_T is exponentially small. In this case $n_{T'} \gg n_T$ and by equaling n_T to 0 in Eq.(5) one gets the quadratic equation for $n_{T'}$ ($\simeq n_{pe}$). Since n_{pe} is small, one can neglect the quadratic term in Eq.(5) obtaining

$$n_{pe} \left(= n_{T'} - n_T\right) = \mathcal{E}_I/\Delta^p \,. \tag{6}$$

It follows that in the low temperature limit the photoinduced signal amplitude ($\approx n_{pe}$) is independent of temperature and its magnitude is proportional to photoexcitation intensity $\mathcal{E}_I$.

The second limiting situation is the case when $n_{pe} \ll n_T$ (high temperature limit). Then, taking into account that $n_{pe} = (n_{T'} - n_T) \ll n_T$ and $n_T = 2N(0)k_BT\exp(-\Delta^p/k_BT)$, the number of photogenerated quasiparticles at temperature T is given by

$$n_{pe} = \frac{\mathcal{E}_I/\Delta^p}{1 + \frac{2\nu}{N(0)\hbar\Omega_c}\exp(-\Delta^p/k_BT)} \,. \tag{7}$$

It is important to stress that Eq.(7) includes also the solution of Eq.(5) in the low temperature limit given by Eq.(6).

Similar derivation can be applied to determine the T-dependence of the number of photoexcited carriers in the case of a temperature dependent mean–field-like gap $\Delta_c(T)$ such that $\Delta_c(T) \to 0$ as $T \to T_c$. This results in a slightly modified expression for n_{pe} that again contains both the low and the high temperature limits

$$n_{pe} = \frac{\mathcal{E}_I/(\Delta_c(T) + k_BT/2)}{1 + \frac{2\nu}{N(0)\hbar\Omega_c}\sqrt{\frac{2k_BT}{\pi\Delta_c(T)}}\exp(-\Delta_c(T)/k_BT)} . \tag{8}$$

Note that in Eqs.(7) and (8) the explicit form of $n_{pe}\,(T)$ depends only on the ratio $k_BT/\Delta\,(0)$, showing that the intensity of the photoresponse is a universal function of k_BT/Δ as long as the particular functional form of temperature dependence of Δ is the same. The only parameter in Eqs.(7) and (8) is the dimensionless constant $\frac{2\nu}{N(0)\hbar\Omega_c}$, which can be estimated for each compound studied to at least an order of magnitude. Therefore by fitting the T-dependence of the PI amplitude of the transient the value of the gap Δ can be determined quite accurately.

The above derivation has been extended also for the case of an anisotropic gap with nodes [1], which could be used to describe the quasiparticle relaxation dynamics in cuprates, where the vast amount of data suggests d-wave order parameter with nodes[20]. However, the experimental results on cuprates were found to be at odds with the simple d-wave picture. In particular, linear intensity dependence of the photoinduced transient amplitude and its peculiar temperature dependence - see Figure 3 are inconsistent with simple d-wave case scenario. The fact that these experiments do suggest large (more or less isotropic) gap in the density of states, can be however due to particularities of dynamics in cuprates. In other words, since time-resolved techniques measure the fastest channel for relaxation, it is possible that carrier relaxation from antinodes (in direction of maximum of the gap) to nodes is much slower process than recombination to the condensate.

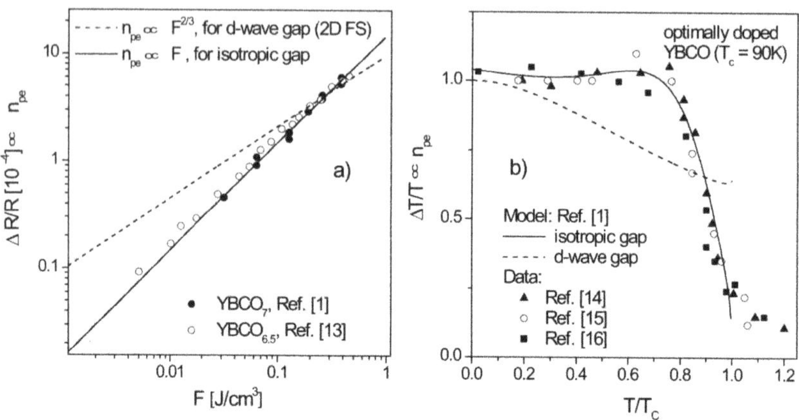

Figure 3. a) Photoexcitation intensity (F) dependence of the photoinduced transient amplitude from Refs. [1, 13]. The data, presented on log-log plot show linear dependence over two orders of magnitude in F (linear fit is given by solid line). On the other hand, the expected intensity dependence for anisotropic gap in 2D gives $F^{2/3}$ dependence [1] (best fit shown by dashed line). b) the experimental T-dependence of photoinduced transient amplitude from [14, 15, 16] compared to the prediction for the case of large isotropic gap (solid line), and anisotropic gap with nodes (dashed), given by Eqs. 6 and 12 from Ref. [1] respectively.

To complete the description of the theoretical model for carrier relaxation dynamics in narrow-gap systems, we should briefly discuss the relaxation dynamics. We have suggested that the lifetime of photoexcited quasiparticle density is governed by anharmonic decay of high frequency phonons. Using the kinetic equations for phonons taking into account phonon-phonon scattering [22]we have obtained the following expression for the rate of recovery of the photoexcited state [1].

$$\tau^{-1} = \frac{12\Gamma_{\omega} k_B T' \Delta(T)}{\hbar\omega^2} . \tag{9}$$

The relaxation time for the temperature dependent gap $\Delta_c(T)$ is expected to show a divergence $\tau \propto 1/\Delta_c(T) \to \infty$ due to the gap closing as T_c is approached from below. This can be easily understood considering phase space arguments. Namely, in the case of a mean-field-like gap, upon increasing temperature closer to T_c, the specific heat of high frequency phonons increases while the specific heat of $\hbar\omega_q < 2\Delta$ phonons is decreasing, giving rise to $\propto 1/\Delta_c(T)$ increase in relaxation time [1].

4. Carrier relaxation dynamics in quasi-1D CDW compounds.

In these section we briefly review recent experimental results on femtosecond time-resolved spectroscopy on CDW compounds [3, 4]. In particular, we focus on the results on quasi-1D CDW compound $K_{0.3}MoO_3$, while for the details on the studies of quasi-2D CDW are given elsewhere [4].

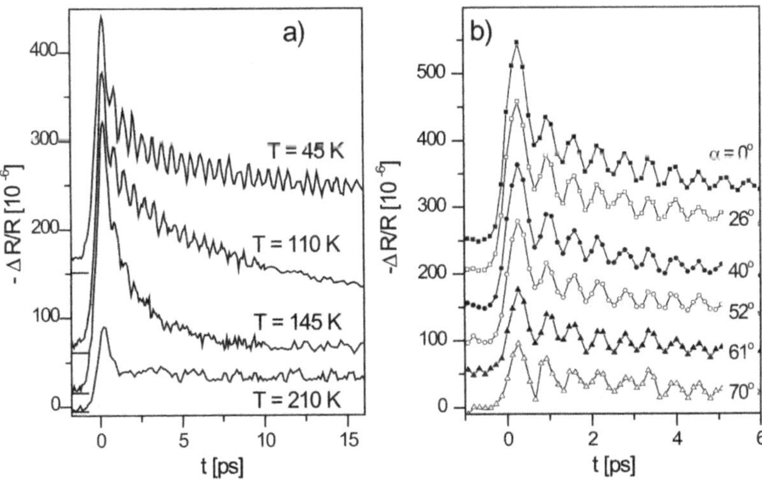

Figure 4. a) Photoinduced reflectivity traces taken on $K_{0.3}MoO_3$ at temperatures below and above $T_c^{3D} = 183$ K. b) The dependence of the photoinduced signal on the probe polarization, with α being the angle between the polarization of probe beam and the crystal [102] axis. The data was taken at 100 K.

Molybdenum oxides $A_{0.3}MoO_3$, where A is a monovalent metal like K, Rb, or Tl - also called blue bronzes due to their shiny blue appearance - are well known for their interesting electronic properties arising from their one-dimensional (1D) chain structures [23]. $K_{0.3}MoO_3$ crystallizes in a monoclinic unit cell [25]. The structure contains rigid units comprised of clusters of ten distorted MoO_6 octahedra, sharing corners along monoclinic b-axis. This corner sharing provides an easy path for the conduction electrons along the chain direction. The chains also share corners along the [102] direction and form infinite slabs separated by the potassium cations. The [102] direction together with [010] direction form a cleavage plane. At room temperature $K_{0.3}MoO_3$ is a highly anisotropic one-dimensional metal with conductivity ratios $\sigma_b : \sigma_{2a-c} : \sigma_{2a+c} = 30 : 1 : 0.05$ (it is a quasi 1D metal).[23] Upon cooling, blue bronzes become susceptible to a Peierls instability on the 1D chains causing fluctuating local CDW ordering. Upon further cooling, inter-chain interactions cause the CDWs on individual chains to become correlated, eventually undergoing a second-order phase transition to a three-dimensionally (3D) ordered state below $T_c^{3D} = 183$ K. The formation of a 3D CDW ordered state is concurrent with the appearance of a gap Δ_{CDW} in the quasiparticle excitation spectrum, while the collective excitations of the 3D CDW state are described by an amplitude mode (AM) and a phase mode (phason) [24].

Figure 4a) presents photoinduced reflectivity $\Delta R/R$ as a function of time at different temperatures. Below T_c^{3D}, an oscillatory component is observed on top of a negative induced reflection, the latter exhibiting a fast initial decay followed by a slower decay - see Figure 5 a). As T_c^{3D} is approached from below, the oscillatory signal disappears, while the fast transient signal remains observed well above T_c^{3D}, as shown by the trace at 210K. For a quantitative analysis, we separate the different components of the signal according to their temperature dependence and probe polarization anisotropy.

Figure 5 presents the decomposed reflectivity transient taken at T = 45 K. Panel a) presents the signal with the oscillatory component subtracted. The logarithmic plot enables us to clearly identify two components with substantially different lifetimes, one with $\tau_s \simeq 0.5$ ps, and the other with $\tau_p \gtrsim 10$ ps at low T. Their amplitudes and relaxation times are analyzed by fitting the recovery dynamics with two exponential decay. Since the two components have very different temperature dependencies they are attributed to the quasiparticle recombination (sub-picosecond component) and an overdamped *phason* relaxation (10 ps component), respectively[3]. Importantly, the ps transient has a pronounced probe polarization anisotropy with respect to the crystal axes - plotted in panel c). Panel b) presents the oscillatory component, whose Fourier spectrum shows a peak at $\nu_A = 1.7$ THz. In contrast to the transient signal, the amplitude of the oscillatory signal is independent of polarization - see panel d).

Collective modes response

The oscillatory component, that has been observed also in the experiments on the quasi-2D CDW compounds [4], has a pronounced T-dependence. The frequency ν_A shows clear softening as T_c is approached from below, while damping $\Gamma_A = 1/(\pi\tau_A)$ increases - see Fig. 6. The measured ν_A and Γ_A

closely follow the expected behavior for the *amplitude mode* and are in good agreement with previous spectroscopic neutron [26] and Raman data [27].

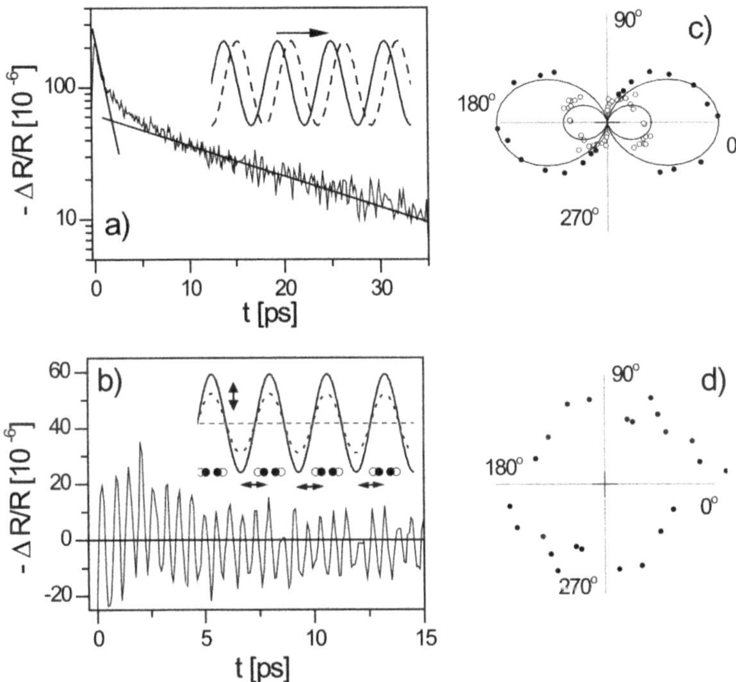

Figure 5. Below T_c^{3D} the response consists of three distinct components: sub-picosecond and 10 ps timescale dynamics shown on a semi-log plot in a) and b) the oscillatory transient with frequency 1.7 THz. While the sub-picosecond response is attributed to the PI absorption on quasiparticles, the oscillatory component and the 10 ps transient are due to photoexcitation of collective modes (amplitudon and phason). The collective modes (q = 0) are schematically shown in insets to panels a) and b)), where solid lines and dots represent the unperturbed carrier density and the ionic positions while dashed lines (open symbols) represent the effect of photoexcitation. c) The amplitude of the fast transients as a function of probe polarization with respect to the crystal [102] direction below (solid) and above (open circles) $T_c^{3D} = 183$ K. d) c) The amplitude of the oscillatory signal as a function of probe polarization with respect to the crystal [102] direction.

The amplitudon is of A_1 symmetry and involves displacements of ions about their equilibrium positions Q_0, which depend on the instantaneous surrounding electronic density $n(t)$. Since the excitation pulse is shorter than h/ν_A, the excitation (and subsequent ultrafast e-e thermalization) may be thought of as a δ-function-like perturbation of the charge density and the injection pulse acts as a time-dependent displacive excitation of the ionic equilibrium position $Q_0(t)$. The response of the amplitudon to this perturbation is a modulation of the reflectivity $\Delta R_A/R$ of the form $A(T)e^{-t/\tau_A}\cos(\omega_A t + \phi_0)$ by the displacive excitation of coherent phonons (DECP) mechanism, known from femtosecond experiments on semiconductors [28].

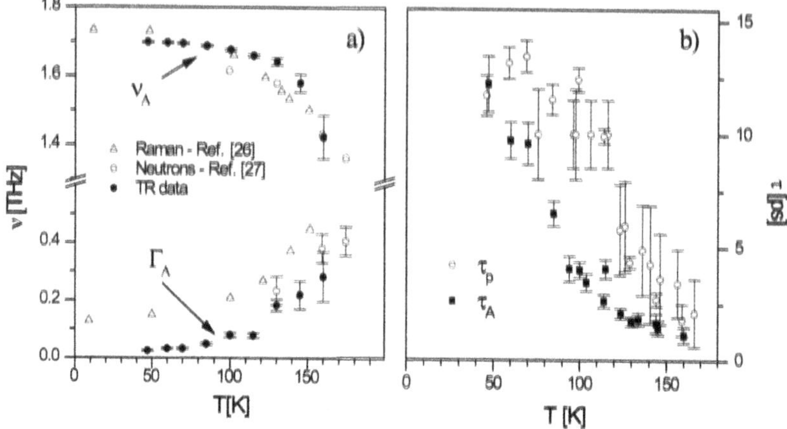

Figure 6. a) The oscillation frequency ν_A (full circles) and damping constant $\Gamma_A = 1/(\pi\tau_A)$ as functions of temperature (full circles). The data from Refs. [26] (open circles) and [27] (open triangles) are also included for comparison. b) The phason damping constant τ_p as a function of temperature (open circles), compared to the amplitudon decay time τ_A is also plotted for comparison (solid squares).

While the association of the oscillatory component to the photoexcited amplitude mode is straight-forward, the association of the 10 ps transient with the overdamped phase mode needs further clarification. In equilibrium, the phason mode is expected to be pinned [24] and at a finite frequency $\omega_p > 0$ [29], but in non-equilibrium situation such as here, where the excess carrier kinetic energy may easily exceed the de-pinning energy, the mode may be de-pinned. In this case we may expect an *overdamped* reflectivity transient that can be written as $\mathcal{P}(T)e^{-t/\tau_p}\cos(\omega_p t + \phi)$, with $\omega_p \to 0$, but with the damping constant which is expected to be similar to that of the amplitude mode $\tau_p \simeq \tau_A$, i.e. ~10 ps [30]. In Fig. 6 b) we plot T-dependence of τ_p. At T =50 K $\tau_p = 12 \pm 2$ ps in agreement with the $\Gamma = 0.05 \sim 0.1$ THz linewidths of the pinned phason mode in microwave and IR experiments [23, 26, 31]. Indeed, comparing τ_p and τ_A [also plotted in Fig. 6 b)], at 50K $\tau_p \simeq \tau_A$ but the fall-off at higher temperatures appears to be faster for τ_A than for τ_p.With increasing temperature τ_p is approximately constant up to 100 K and then falls rapidly as $T \to T_c^{3D}$. The decrease of τ_p near T_c^{3D} is consistent with increasing damping due to the thermal phase fluctuations arising from coupling with the lattice and quasiparticle excitations.

Quasiparticle response

Let us now turn to the transient (sub-picosecond) reflectivity signal, attributed to the photoinduced absorption from the photoexcited quasiparticles. The T-dependence of the amplitude of the reflectivity change and the rel. time, determined by the single exponential decay fit to the data, is presented in Fig.7.

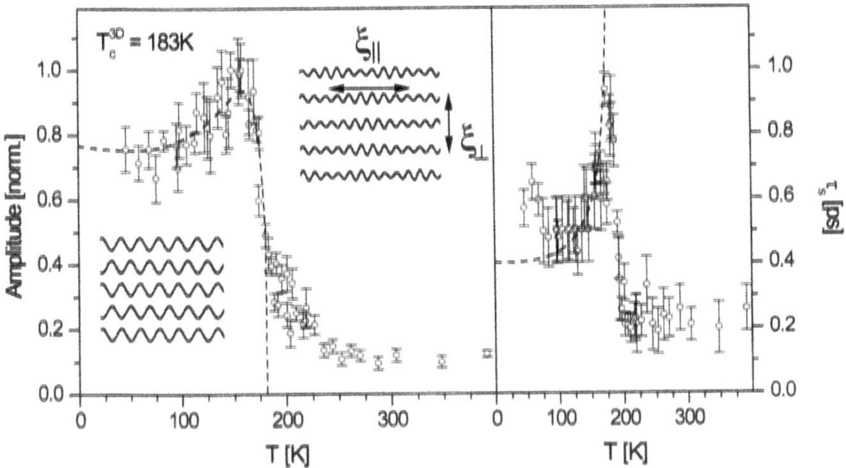

Figure 7. The temperature dependence of a) the amplitude and b) the relaxation time of the sub-picosecond relaxation component. The amplitude exhibits a sharp drop at the second order phase transition to the 3D ordered CDW state at T_c^{3D} = 183 K. Above 183 K the amplitude gradually decreases to ~250 K, and remains constant above that temperature. The behavior above T_c^{3D} is a manifestation of a fluctuating presence of short range segments with 3D order. This is schematically presented in insets shown by the charge density modulation along the chains at $T < T_c^{3D}$ (long range 3D order) and at $T > T_c^{3D}$ (short range 3D fluctuations when $\xi_\perp$ exceeds the interchain separation). b) The temperature dependence of the CDW state recovery dynamics. Dashed lines present the fits using Eqs.(8,9) respectively.

The amplitude of the PI transient has a pronounced temperature dependence shown in Figure 7a). Upon increasing the temperature, it first slightly increases, followed by a sharp drop at the second order phase transition to the 3D ordered CDW state at T_c^{3D}. Above 183 K the amplitude gradually decreases to ~ 250 K, and remains constant above that temperature all the way up to the highest temperatures measured (~ 400 K). On the other hand, relaxation time τ_s is roughly constant far below T_c^{3D}, shows a quasi-divergence when T_c^{3D} is approached from below, and then drops to constant value of ~ 0.2 ps.

The T-dependence of the photoinduced signal amplitude below T_c has been fitted with Eq.(8). We have used a BCS functional form for the T-dependence of the gap consistent with the T-dependence of lattice distortions [23]. Plotting Eq.(8) as a function of temperature in Fig. 7a), we find that the amplitude obtained from the fits to the data agrees remarkably well with the model for $T < T_c$: amplitude is nearly constant up to nearly 100 K, then increases slightly and then drops very rapidly near T_c^{3D}. Using the value of the dimensionless

[3] J. Demsar, K. Biljakovic, D. Mihailovic, *Phys. Rev. Lett.* **83**, 800 (1999); see also recent work in the high excitation regime by A.A. Tsvetkov et al., *Acta Physica Polonica B* **34**, 387 (2003).

[4] J. Demsar, H. Berger, L. Forro, D. Mihailovic, *Phys. Rev. B* **66**, 041101 (2002).

[5] J. Demsar *et al.*, *Phys. Rev. Lett.* **91,** 027401 (2003).

[6] R.D. Averitt *et al.*, *Phys. Rev. B* **63**, 140502 (2001).

[7] R.D. Averitt *et al.*, *Phys. Rev. Lett.* **87**, 017401 (2001).

[8] J. Demsar *et al.*, *Phys. Rev. Lett.* **91**, 267002 (2003); J. Demsar *et al.*, *Int. J. Mod. Phys.* **17**, 3675 (2003).

[9] P. Gay *et al.*, *J. Low Temp. Phys.* **117**, 1025 (1999).

[10] D.C. Smith *et al.*, *Physica C* **341-348**, 2219 (2000).

[11] J. Demsar *et al.*, *Phys. Rev. B* **63**, 54519 (2001).

[12] M.L. Schneider *et al.*, *Europhys. Lett.* **60** 460 (2002).

[13] G.P. Segre *et al.*, *Phys. Rev. Lett.* **88**, 137001 (2002).

[14] C.J. Stevens *et al.*, *Phys.Rev.Lett.* **78**, 2212 (1997).

[15] S.G. Han, Z.V. Vardeny, O.G. Symko, G. Koren, *Phys. Rev. Lett.* **65**, 2708 (1990).

[16] J. Demsar *et al.*, *Europhys. Lett.* **45**, 381 (1999).

[17] D. Dvorsek *et al.*, *Phys. Rev. B* **66**, 020510 (2002).

[18] J. Demsar *et al.*, *Phys. Rev. Lett.* **82**, 4918 (1999).

[19] A. Rothwarf, B.N. Taylor, *Phys. Rev. Lett.* **19**, 27 (1967).

[20] C.C. Tsuei and J.R. Kirtley, *Rev. Mod Phys.* **72**, 969 (2000) and the references therein.

[21] G.M. Eliashberg, *Zh. Exsp. Theor. Fiz.* **61**, 1274 (1971), A.G. Aronov and B.Z. Spivak, *J. Low Temp. Phys.* **29**, 149 (1977).

[22] J.M. Ziman, *Electrons and Phonons* (Oxford University Press, London 1960), E.M. Lifshitz, L.P. Pitaevskii, *Physical kinetics*, (Butterworth-Heinemann ,Oxford, 1995).

[23] G. Grüner, *Rev.Mod.Phys.* **60**, 1129 (1988).

[24] G. Grüner, *Density Waves in Solids*, (Addison-Wesley, 1994).

[25] J. Graham and A.D. Wadsley, *Acta Cryst.* **20**, 93, (1966).

[26] J.P. Pouget et al., *Phys. Rev.* **B 43,** 8421 (1991).

[27] G. Travaglini, I. Mörke, P. Wachter, *Sol.State.Comm.* **45**, 289 (1983).

[28] H.J. Zeiger *et al.*, *Phys. Rev. B* **45**, 768 (1992).

[29] T.W. Kim *et al.*, *Phys. Rev.* **B 40,** 5372 (1989).

[30] E. Tutiš, S. Barišić, Phys. Rev.B **43**, 8431 (1991).

[31] L. Degiorgi *et al.*, *Phys.Rev.B* **44**, 7808 (1991).

[32] H. Requardt *et al.*, *J.Phys.: Cond. Matt.* **9**, 8639 (1997).

[33] P.B. Allen, *Phys. Rev. Lett.* **59**, 1460 (1987).

[34] S. Girault, A.H. Moudden, J.P. Pouget, *Phys. Rev.* **B 39,** 4430 (1989).

NORMAL METAL COLD-ELECTRON BOLOMETER: RESPONSE, NOISE, AND ELECTRON COOLING

Michael Tarasov and Leonid Kuzmin
Institute of Radio Engineering and Electronics RAS, and M. Lomonosov Moscow State University
tarasov@hitech.cplire.ru

Abstract In the last decade superconducting bolometers became the most sensitive radiation detectors of Sub-mm, Infrared, and Optical radiation with an estimated ultimate sensitivity down to 10^{-20} W/Hz$^{1/2}$. At present the most developed superconducting bolometer is a transition-edge sensor (TES). However, the TES performance is limited by excess noise, overheating by dc feedback power, and background power load. To avoid this overheating, a new concept of a cold-electron bolometer with direct electron cooling has been proposed. The unique thermal transport properties of a SIN tunnel junction can be used to remove hot electrons from the absorber in order to cool electrons in absorber well below the lattice. The optimal realization of this sensor proved to be a two junction cold-electron bolometer with capacitive coupling to the antenna by tunnel junctions. Recently the electron cooling from 300 mK to 100 mK have been demonstrated in samples with improved trapping of hot quasiparticles from superconductor. Experimental dc noise equivalent power NEP=10^{-17} W/Hz$^{1/2}$ is consistent with optical NEP measured using black-body radiation source inside the cryostat. Influence of SIN junction leakage resistance on bolometer performance has been studied both experimentally and numerically. With the use of such bolometer we measured radiation of Josephson junction at submillimeter waves.

Keywords: hot electron bolometers, direct electron cooling, submm-wave radiation

1. Comparison of bolometers

At present there are four main concepts of superconducting bolometers, namely Transition Edge Sensor (TES), Normal metal Hot Electron Bolometer with Andreev mirrors (ANHEB), Capacitively Coupled Normal metal Hot Electron Bolometer (CCNHEB), and CCNHEB with direct electron cooling – Cold Electron Bolometer (CEB). The most developed superconducting bolometer is TES, however its performance is limited by excess noise, overheating by dc feedback power, and background power load. The most severe problem for TES is additional overheating by dc power for electro-thermal feedback, which prevents electron temperature from deep cooling and limits the performance

A.S. Alexandrov et al. (eds.), Molecular Nanowires and Other Quantum Objects, 393–404.

of bolometer. The normal metal hot-electron bolometer NHEB [1] consists of a normal-metal absorber strip of small volume connected as a matched load to a planar antenna, and a pair of Superconductor – Insulator - Normal metal (SIN) tunnel junctions for temperature sensing. The electron gas in the absorber strip receives thermal energy from the high-frequency currents induced in the antenna. At low temperatures the electron-phonon interaction time is much longer than the electron-electron interaction time, so that the electrons achieve thermal equilibrium at electron temperature that is higher than the lattice temperature (hot-electron effect). In order to avoid energy loss through diffusion of the electrons into the antenna the absorber strip is contacted via superconducting electrodes, since the Andreev effect prohibits energy transport from the normal metal to the superconductor at an NS-interface. Such bolometer with Andreev reflection is abbreviated as ANHEB. The electron temperature in the absorber strip of the ANHEB is measured by SIN junction, which is formed as a strip across normal-metal absorber.

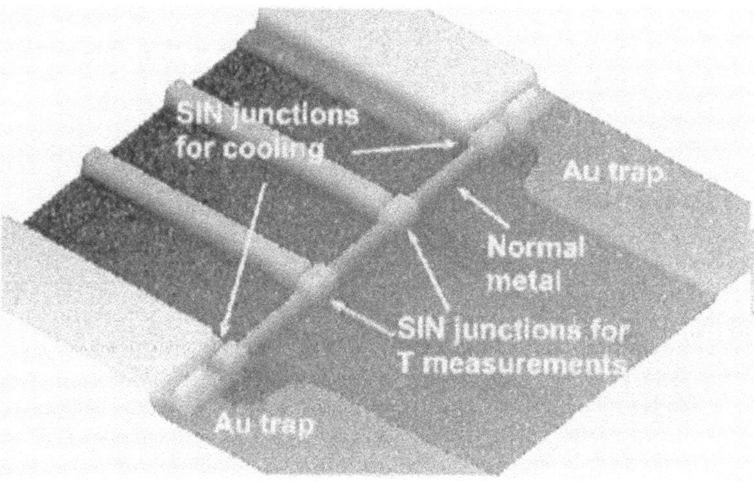

Figure 1. View of the central part of CCNHEB. Length of the normal metal strip is 10 μm.

Normal metal hot electron bolometer with capacitive coupling (CCNHEB) was proposed in [2] and experimentally demonstrated in [3]. The difference from ANHEB is that direct contact between a superconducting electrode and a normal metal strip is substituted by a tunnel junction (see Fig. 1). CCNHEB is a further development of the concept of ANHEB. It was proposed to avoid frequency and energy limitation of ANHEB in which Andreev mirrors are effective for relatively long absorbers and energies below the superconductor energy gap. Another advantage of CCNHEB is much easier layout in which the same tunnel junctions provide both thermal insulation and temperature sensing. To improve CCNHEB performance we suggest using electron cooling by a superconductor-insulator-normal metal (SIN) tunnel junction [4]. The direct electron cooling was demonstrated in [5] and further developed in [6]. Contrary to additional dc electro-thermal feedback overheating of TES, in our Cold Electron Bolometer (CEB) heating is replaced by effective electron cooling. It means principle breakthrough in realization of supersensitive detectors. The electron temperature could be returned to phonon temperature or lower in real operation mode with the background power load. All incoming power is removed from supersensitive absorber to the next stage of readout system. The CEB can be used for operation with realistic background power load for radio-astronomical applications.

2. Temperature sensing

For estimation of the actual electron temperature the natural way is to fit IV curve of a real junction by ideal SIN tunnel junction IV curve

$$I(V,T) = \frac{k}{eR}\sqrt{11TT_c} \cdot \exp\left(-\frac{1.76T_c}{T}\right) \cdot \sinh\left(\frac{eV}{kT}\right) \tag{1}$$

in which T is temperature, T_c is critical temperature of superconductor, e – electron charge, k – Boltsman constant, V – voltage. A subgap leakage current can affect both shape of IV curve and electron temperature of the bolometer. Such subgap residual conductivity can be due to defects of the tunnel barrier, normal inclusions in superconductor, etc. Another mechanism of subgap conductivity is two-electron tunneling when two normal electrons can be converted into a Cooper pair. Such current will be much smaller compared to ordinary single-electron current [7]. It was mentioned in [8] that reducing insulator thickness and resistivity of junction leads to increase of Andreev reflections and subgap leakage.

Another parasitic effect is back-tunneling that can dominate at temperatures below 200 mK [9]. Without trap the electron temperature saturates and cooling power is reduced due to hot electrons that return back to the normal metal. Some experiments [10] demonstrated temperature saturation below 300 mK when hot electrons are confined in the normal metal strip. For practical junctions very important can be presence of a pinhole-type defect in tunnel barrier. Such defect is easy determined by residual conductivity at zero bias. For typical junction with normal resistance 1 kΩ at 250 mK the dynamic resistance at zero bias should be over 1 MΩ and we can take as example the leakage resistance of about the same 1MΩ.

The electron temperature under absorbed power is estimated as

$$T_e = \left(T_{ph}^5 + \frac{P}{\Sigma V}\right)^{1/5} \tag{2}$$

in which P is absorbed power, Σ=3$\cdot 10^9$ Wm^{-3}K^{-5} is metal parameter, V=0.18 μm^3 is volume of absorber. For zero phonon temperature at standard dc bias of 400 μV the temperature is increased by about 200 mK. It means that even small probe current increase electron temperature over equilibrium phonon temperature and calibration of electron temperature becomes rather tricky. Let us consider the finite-bias resistance ratio as electron temperature probe. Adding to equation (1) the parallel conductance of leakage current one can obtain the temperature dependence of zero-bias resistance of real junction

$$rds = \left[\left(\sqrt{\frac{2k_bT}{\pi\Delta}} \frac{\exp\left(\frac{\Delta}{k_bT}\right)}{\cosh\left(\frac{eV}{k_bT}\right)} \right)^{-1} + \frac{1}{R_s} \right]^{-1}, \tag{3}$$

in which R_n is normal resistance and R_s is shunting leakage resistance. The illustration to this can be calculation of the dynamic resistance for the same junction in dependence on temperature for bias voltages 0, 200 and 300 μV presented in Fig. 2. One can see clear saturation of zero-bias thermometer at temperatures below 200 mK. But for lower temperatures we can use still not saturated region of bias voltages where the dynamic resistance is small compared to leakage shunting resistance. So when temperature is below 200 mK the resistance ratio should be taken for bias voltage 250 μV, and below 100 mK bias should be 300 μV.

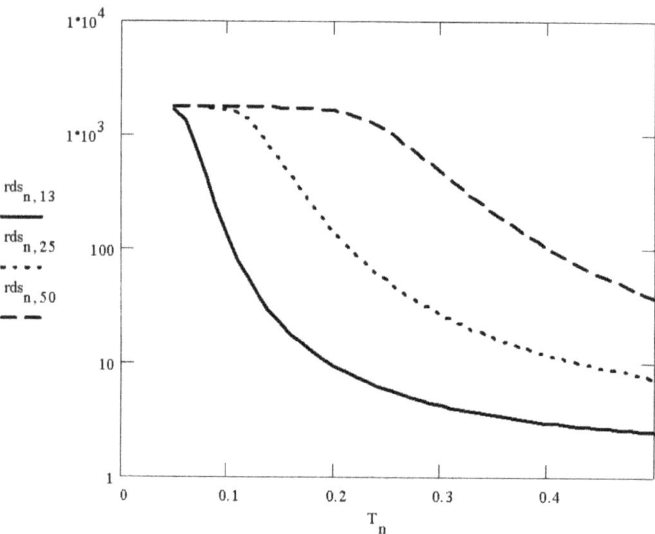

Figure 2. Dependencies of dynamic resistance on bath temperature for bias voltages 0 μV (dashed), 200 μV (dotted) and 300 μV (solid).

3. Samples design and fabrication

Samples were fabricated by means of direct e-beam writing with absorber made of bilayer of Cr and Cu to match the impedance of antenna to the absorber resistance. Electrodes were made of Al and tunnel junctions are formed over the Al oxide layer. Bolometers are integrated with log-periodic and double-dipole planar antennas. Typical chip MF46 was fabricated on oxidized Si substrate by shadow angle evaporation. Aluminum electrodes were 60nm thick, oxided 2 minutes in oxygen at 10^{-1} mBar. Bolometer absorber was a bilayer of 50 nm Cr and 10 nm Cu. Wiring, antenna, contact pads were of 10 nm Cr, 40 nm Au, 10 nm Pd. Bolometer connected to contact pads 6-16 was integrated in log-periodic antenna and the rest to a double-dipole antennas.

4. Estimations of the CCNHEB Sensitivity by a black body radiation source

Noise Equivalent Power (NEP) of the bolometer can be determined as:

$$NEP = V_n/S \tag{4}$$

in which V_n is the total voltage noise referred to the output of the bolometer and $S$is the bolometer's responsivity (output voltage change per unit of input power change). The output noise is a sum of many contributions: thermal fluctuation noise, shot noise in the tunnel junctions, noise from thermal flow fluctuations in the junctions, amplifier noise, and the photon noise. In our experiments the output noise has been dominated by the noise of a semiconductor amplifier that is V_n =3 nV/Hz$^{1/2}$ in the white noise region. The responsivity is given by

$$S = \frac{dV}{dT} \cdot \frac{dT}{dP} = \frac{dV}{dT} \cdot \frac{1}{G} \quad , \tag{5}$$

where $T$is the electron temperature in the absorber strip, dV/dT is the temperature responsivity of the tunnel junctions, and $G$is the thermal conductance between electron and phonon systems in the absorber strip. The electron-phonon cooling rate is given by (2) and

$$G = 5\Sigma\Delta T^4. \tag{6}$$

In experiment we observed dV/dT=1.7 mV/K that corresponds to responsivity S=1.3$\cdot 10^8$ V/W. For amplifier noise V_n=3nV/Hz$^{1/2}$ the technical noise equivalent power is TNEP=2.3$\cdot 10^{-17}$and intrinsic bolometer self noise V_n=0.8 nV/Hz$^{1/2}$ it brings NEP=V_n/S=10^{-17}W/Hz$^{1/2}$.

For microwave evaluation of bolometer we use a black body radiation source comprising of thin NiCr film deposited on thin sapphire substrate and suspended on nylon support fibers [11]. Source was placed in front of CCNHEB attached to extended hemisphere sapphire lens. Received radiation power is $\Delta P=\eta*k*\Delta T*\Delta f=4\cdot 10^{-13}$ W and with responsitivity $1.1\cdot 10^8$ V/W one can get voltage response 40 μV. Measured voltage response value 20 μV is in reasonable correspondence to this value 40 μV. Experimental temperature and optical response dependencies are presented in Fig. 3.

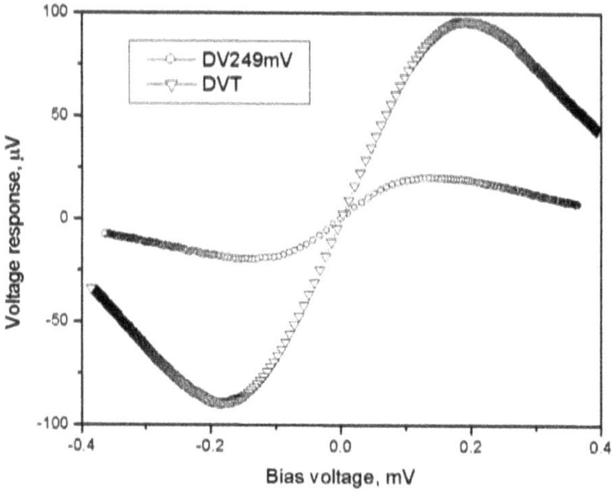

Figure 3. Voltage response of CCNHEB for temperature variation (triangles) and for microwave signal from cryogenic radiation source (circles)

5. Mmeasurements with Josephson junctions

For experiment we use the direct connection of substrate with Josephson oscillator to the substrate with receiver. When planar antenna is placed on dielectric substrate with high refraction index about ten, the main lobe of the beampattern is directed into the substrate. In this case most of radiation from Josephson oscillator is directed to the antenna with bolometer. Log-periodic antennas used in both oscillator and receiver structures are identical and designed for frequencies 100-1000 GHz.

Dependencies of bolometer voltage response are presented in Fig. 4. Applying magnetic field one can suppress the critical current of Josephson junction that leads to decrease of output power and frequency range according to Josephson equations. When critical current is suppressed below 2 μA the response voltage is clear proportional to the square of the JJ bias current. This brings clear evidence that we can separate Josephson radiation at frequencies below 1 THz and infrared radiation of overheated junction for bias voltages over 1 mV.

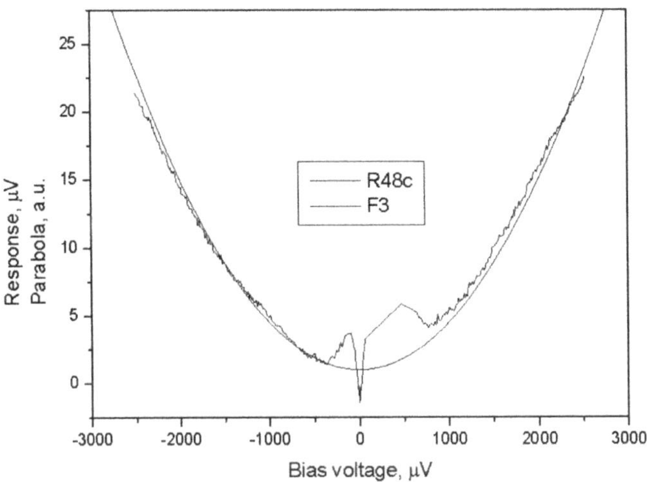

Figure 4. Bolometer response measured with a Josephson junction radiation source. Solid parabola is fitting for Joule heating.

For Josephson oscillations we can estimate the maximum available power as $P_{osc}=0.1 \cdot I_c V_c = 2 \cdot 10^{-9}$ W. Misalignment of antennas, mismatch of beampatterns (spillover losses), mismatch of impedances can bring at the intermediate bias voltage and critical current the total attenuation of the maximum power up to 30 dB that corresponds to the available power at bolometer of about 10^{-12} W. The estimated above bolometer responsivity is $S=1.1 \cdot 10^8$ V/W that brings the maximum voltage response to this power about $1.1 \cdot 10^{-4}$ V. In our experiments we measured the voltage response up to 10 μV. The order of magnitude difference in response can be explained by non-ideal characteristics of Josephson junction (excess current) and overheating that reduces output power.

Response to non-Josephson radiation at parabolic part of response curve can be attributed to measurements of submm and IR radiation.

If we take as approximation a model of overheating in JJ by [4] for variable thickness microbridge

$$T_m = \sqrt{T_b^2 + 3\left(\frac{eV}{2\pi k}\right)^2}, \tag{7}$$

it brings the equivalent electron temperature at 1 mV bias of about 3 K. Taking into account that IR radiation is spread in 4π solid angle and the bolometer at a distance over 1 mm in dielectric can absorb a small part of this radiation, the measured increase in temperature of 5 mK looks reasonable. Now we should take into account that this power is radiated and then received. It means that Plank's radiation law should be applied

$$P_r = \frac{hf}{e^{\frac{hf}{kT}} - 1} 0.3f \tag{8}$$

for which maximum of radiation is obtained at hf≈kT. If we apply the Plank's formula to equation (1) neglecting the phonon temperature

$$P_{rad} = \frac{0.6}{4\pi^2} \cdot \frac{e^2 V^2}{h} \tag{9}$$

it brings the square-law voltage dependence, as in experiment.

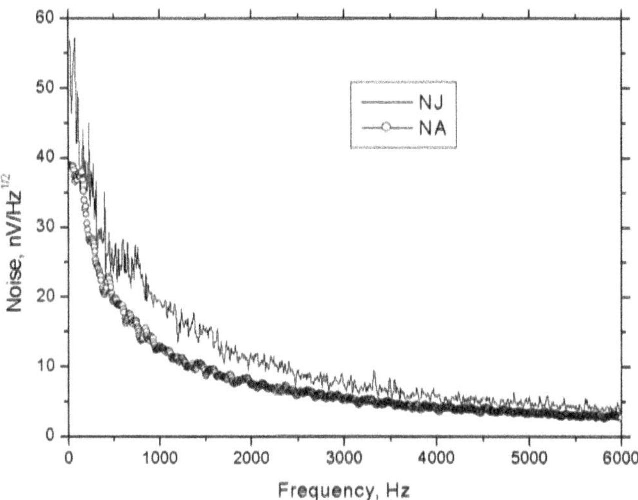

Figure 5. Voltage noise measured for amplifier (circles) and with bolometer at maximum response.

6. Noise measurements

To clarify noise mechanisms in mesoscopic structures at millikelvin temperatures we performed measurements of a low-frequency noise in several specific cases. First is our bolometer with tunnel junctions at temperature above T_c of aluminum. It shows clear shot noise dependence and a Johnson noise background. The next was bolometer below 1 K, noise measurements bring us NEP for our bolometer in the white noise region in dependence on bias voltage. We also measured frequency dependence of output noise at bias point with largest response (see Fig. 5).

7. Electron cooling

For electron cooling in our experiments we use external tunnel junctions (see Fig. 1) with normal metal traps and for temperature sensing – central junctions. The illustration for our estimations could be example of dynamic resistance of sensor junctions at different cooling voltages measured in dilution refrigerator at 20 mK and 250 mK. One can see in Fig. 6 that the largest resistance of 45 MΩ is observed only at cooler zero bias. With any other voltage via refrigerator the resistance is below 37 m. Large increase of sensor resistance at 250 μV with cooling bias 400 μV can be explained as cooling down from overheated level of over 100 mK. The same dependence measured at 250 mK shows increase of a zero-bias sensor resistance from 12 MΩ to 36 MΩ with increase of cooling bias. It means that achieved by cooling the electron temperature difference is approximately the same in both cases. The resistance of junction in this case is not so much shunted by leakage and temperature can be estimated with high accuracy from the resistance ratio dependence calculated for shunted SIN junction according to equation (3).

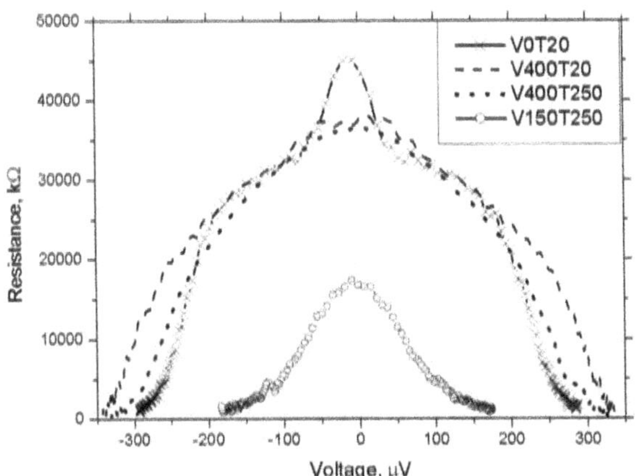

Figure 6. Dynamic resistance measured at 20 mK (crosses, dashed) and at 250 mK (dots, circles).

At very low bias close to zero the IV curve differs from theoretical one and resistance maximum could be attributed to Coulomb blocade in a small absorber connected via tunnel junctions. The performance of bolometer is strongly affected by external overheating from the background power load and also from normal conducting channels in the tunnel barrier itself. The actual electron temperature without electron cooling can have excess level above the phonon temperature of the order of 100 mK. We can numerically model both processes of overheating via shunting resistance plus background power load and electron cooling. Accurate relation for electron cooling is rather complicated integral equation, but for bias voltages close to the energy gap it can be presented by a simplified analytic expression

$$P_{cool}\left(\tau, V\right) = \frac{\sqrt{2\pi\Delta k_b \tau}}{2eR_N}\left(\frac{\Delta}{e} - V\right)\exp\left(-\frac{\Delta - eV}{k_b\tau}\right) \qquad (10)$$

and the effective electron temperature τ is determined from equation

$$\left(T_{ph}^5 - \tau^5\right)\Sigma\Lambda = P_{cool}\left(\tau\right) - P_{bgn} - \frac{V^2}{R_s} \qquad (11)$$

in which T_{ph} is phonon temperature, V is dc bias voltage, R_s is shunting resistance, P_{bgn}=0.5hfΔf=6$\cdot 10^{-14}$ is background power, Σ=3$\cdot 10^{9}$ is material parameter, Λ=1.8$\cdot 10^{-19}$ is absorber volume. Calculation for leakage resistance of 3 MΩ brings cooling by ΔT=160 mK at 250 mK. Estimations for electron temperature deduced from zero bias and 300 μV sensor biases are presented in Fig. 7. It shows that finite bias of sensor allows to obtain the same temperature as in numeric estimation.

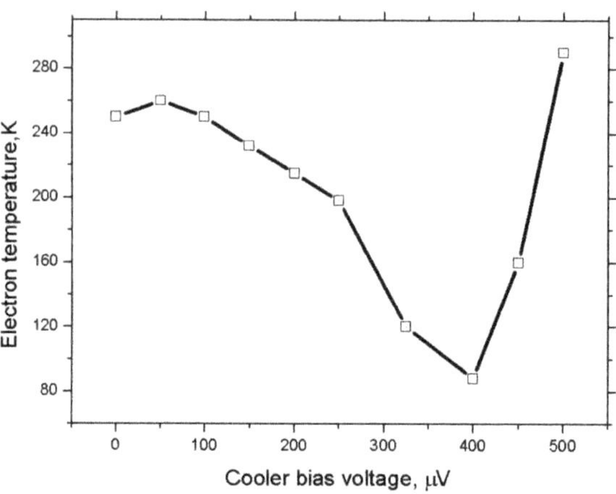

Figure 7. Electron temperature measured for 300 μV sensor bias.

8. Conclusion

A novel concept of the Cold-Electron Bolometer (CEB) based on strong direct electron cooling of the absorber has been proposed. This concept is purposed to overcome the main contradiction of supersensitive detectors – overheating by background power load due to high sensitivity of the detector. Besides that, additional dc overheating for electro-thermal feedback in TES (transition-edge sensor) will be replaced by effective electron cooling in CEB. The electron temperature could be reduced to phonon temperature and below in real operation with power load. All incoming power is removed from supersensitive absorber to the next stage of readout system. We have designed, fabricated and experimentally studied a capacitive coupled normal metal hot electron microbolometer integrated with planar antenna. Sensitivity dV/dT=1.7 mV/K, responsivity $S=1.1 \cdot 10^8$ V/W and NEP=10^{-17} W/Hz$^{1/2}$ and time constant below 1 μs makes CEB a strong competitor to transition edge sensor (TES) bolometer. Electron cooling allows increasing the dynamic resistance and response of the bolometer. We observed cooling by about 160 mK at bath temperature 250 mK. For estimations of effective electron temperature the ratio of finite-bias resistance to the normal resistance can be used.

Acknowledgments

Authors acknowledge support from INTAS-01-686, STINT

References

[1] M.Nahum, J.M.Martinis, Ultrasensitive hot electron microbolometer, Appl. Phys. Lett., v. 63, N 22 (1993), pp. 3075-3077.

[2] L.Kuzmin, On the concept of a hot-electron microbolometer with capacitive coupling to the antenna, Physica B, 284-288 (2000), 2129-2130.

[3] M.Tarasov, M.Fominsky, A.Kalabukhov, L.Kuzmin, Experimental study of a normal-metal hot electron bolometer with capacitive coupling, JETP Letters, v. 76, N 8, 2002, pp. 507-510.

[4] L.Kuzmin, I.Devyatov, D.Golubev, Cold-electron bolometer with electronic microrefrigeration, Proc. Of SPIE, v. 3465, pp. 193-199 (1998).

[5] M.Nahum, T.M.Eiles, J.M.Martinis, Electronic microrefrigerator based on a normal-insulator-superconductor tunnel junction, Appl. Phys. Lett., v. 65, N 24, 3123-3125 (1994).

[6] M.Leivo, J.Pecola, D.Averin. Efficient Peltier refrigeration by a pair of normal metal/insulator/superconductor junctions, Appl. Phys. Lett., v. 68 (14), 1996-1998 (1996).

[7] F.W.J.Hekking, Yu.Nazarov, Subgap conductivity of a superconducting-normal tunnel interface, arXiv:cond-mat/9302034 v1, 23 Feb. 1993, pp. 1-12.

[8] A.Bardas, D.Averin, Peltier effect in normal-metal-superconductor microcontacts, Phys. Rev. B, v. 52, N 17 (1995),12873-12877.

[9] J.Jochum, C.Mears, S.Golwala, et al., Modelling the power flow in normal conductor-insulator-superconductor junctions, J.Appl. Phys., v. 83, N 6, (1998), 3217-3224.

[10] D.Quirion, F.Lefloch, M.Sanquer, Transport and heating effect in proximity superconduct ing structures, Physica E, 12 (2002), pp. 934-937.

[11] Haller/Beeman Assoc. Inc., http://www.haller-beeman.com

MAGNETIC SWITCHING IN THE PEROVSKITE NANO-DEVICES

Janus Baszyński
Institute of Molecular Physics, Polish Academy of Sciences, 60-179 Poznań, ul. M. Smoluchowskiego 17, Poland
jbasz@ifmpan.poznan.pl

Abstract The magnetic switching processes in mechanical "break" junctions (*MBJs*), made from $La_{0.7}Sr_{0.3}MnO_3$ perovskite ceramic, are studied as a function of the DC current (up to 10^9 A/cm^2) passing through the nanoconstruction. The current-voltage ($I - V$) curves of the *MBJs* are non-ohmic with a parabolic form of the differential conductance (dI/dV) versus the voltage, typical for an electron tunnelling process. By fitting $I - V$ curves to the Simmons theory the barrier width ($0.5 \div 1.2$) nm and height ($1 \div 3$) eV of the junctions and their effective tunnel area ($1.7 \div 2.1$) x10^{-11} cm^2 were estimated.

In the generally accepted "double exchange" model, the magnetic field aligns the local Mn-t_{2g} spins ferromagnetically and facilitates hopping of the conduction electrons Mn-e_g between adjacent Mn-ions sites. Due to the close relation between transport properties and the magnetization in manganite compounds, the resistivity measurements provide an excellent indirect method to characterize the magnetic switching observed in atomic scale constrictions. Based on this, we interpret the jumps in conductance of *MBJs*, in integer multiples of e^2/h, as due to the configuration reorientation of the magnetization of the Mn-ions clusters at the surface in the devices.

This switching can be understood in terms of the exchange of the angular momentum between the spin-polarized current and magnetic moments of the cluster situated between the electrodes or on their surfaces.

Keywords: perovskite tunnelling devices, current induced switching, magnetic nanostructures, electronic transport

1. Introduction

In 1996, Slonczewski [1] and also Berger [2] predicted that the magnetization of a magnetic layer can be reversed by the injection of a spin polarized current and the spin transfer to the layer. A spin-polarized current traversing a magnetic multilayer can, through exchange interactions, alter the orientations of ferromagnetic moments, producing domain reversal or other magnetic dynamics.

A.S. Alexandrov et al. (eds.), Molecular Nanowires and Other Quantum Objects, 405–414.

An experimental confirmation of the reversed of the magnetic moment without the application of an external field, *e.g.* the observed current-induced instabilities and resistivity jumps, are of considerable interest for magnetic switching of micro-devices.

Experimental verification of the spin torque has been carried out in magnetic nanowires [3,4], spin valve pillar structures [5-7], point-contact geometry [8], and perovskite magnetic tunnel junctions [9]. The convincing observation of these experiments is that there exists a critical current density above which the magnetization can be switched forth and back.

The main features of Freitas and Berger [10] and Koo *et al.* [11] results are: the direction of the domain wall (DW) motion is reversed when the direction of the current pulses is reversed, and that the order of magnitude of the current pulses needed to move the DW is about 10^7 A/cm^2.

J.Grollier *et al.* [12] observed in spin valves Co/Cu/Py the displacement of the DW at zero magnetic fields, in opposite directions for opposite DC current directions and with considerably lower current densities of the order of 10^6 A/cm^2.

Only the spin transfer mechanism, first proposed by Berger, is consistent with these experimental results and, particularly, can explain the reversal of the DW motion where opposite currents are included.

Recently, current-induced switching of magnetic moments and resistivity jumps have been also observed in artificial manganite trilayers [9,13]. In such systems current-induced resistivity jumps result from spin-polarized current-induced switching of the magnetic moments between parallel and antiparallel orientations, the latter having a higher resistance than the former one [1,9,14,15].

The magnetically ordered regions may act as ferromagnetic electrodes in a spin-polarized tunnelling arrangement in which the magnetic tunnel barrier plays the role of a spacer. The electrons moving between the magnetic metallic domains separated by a nonmagnetic spacer become spin polarized by the local moment of the first domain. These electrons tend to align the magnetic moment in the second domain to the direction parallel to their own spin polarization. The switching of the magnetic moment occurs at sufficiently high density of spin-polarized current ($> 10^7$ A/cm^2) exceeding some threshold current.

The convincing observation of these experiments at metallic structures is that there exists a critical current density above which the magnetization can be switched back and forth.

Manganite perovskites, half-metallic ferromagnets where the spin polarization of electrons at the Fermi level is 100% [16,17], from a class of materials of great interest for their potential application in spin electronics. However, in relation to devices such as spin-polarized tunnel junctions, perovkites are of great interest as injectors of fully spin-polarized electrons into the junction [18-20].

The magnetic and transport properties of these materials have been explained on basis of the double exchange mechanism (DE) introduced by Zener [21]. This model was successive developed by Anderson, Hasegawa [22] and de Gennes [23], to describe the essential interactions among Mn ions giving rise to the ferromagnetic properties of the perovskite. The DE is dependent on the electron hopping probability t between Mn^{3+} and Mn^{4+} ions. The t is affected by the relative alignment of the Mn^{3+}/Mn^{4+} core spins (t_{2g}) owing to the strong Hund coupling between the t_{2g} and e_g electron spins. Thus an external magnetic field which aligns the t_{2g} spins, promotes the e_g charge hopping, resulting in increasing conductivity.

Another scenario of competing DE and superexchange is found at surfaces of ferromagnetic manganites [24]. The reduced lattice symmetry at the surface in the first two or three layers of perovskite cells (of $\sim$ 0.39 nm thickness) suppresses the DE. This probably causes the semiconducting nature of the surfaces, which is the origin of the tunnelling properties of the MCBJ in our experiments.

Both transport and magnetic properties are very sensitive to the doping level (i.e., Mn^{4+} content) and the degree of overlap between Mn^{3+}-Mn^{4+} $3d$ orbitals and O 2p orbitals [25].

In this paper we demonstrate the results of experiments in which the perovskite nanoconstruction can be switched, by jumps, between the high- and low-conductance states, by passing current through the devices. Resulting jumps of conductance G by multiples of e^2/h ($\Delta G \approx ne^2/h$)indicate the reversal of the magnetic moment of the Mn-ions cluster at the apexes of the junction electrodes. Morigaki *et al.* [26] found theoretically that for a ferromagnetic material the quantum of conductance should be equal to $G_0/2 = e^2/h$ due to the lifting of the spin degeneracy.

For the above reason this article presents our studies of the quantization switching conductance in a manganite perovskite nanoconstruction as a current-induced process.

2. Experimental details

The MCBJ (mechanical controllable break junction) was fabricated from the perovskite ceramic $La_{2/3}Sr_{1/3}MnO_3$ by breaking technique, which is a well-established way to study the conduction of nanostructures [27,28].

The current-voltage characteristics ($I - V$) of the perovskite MCBJ has been measured using a two-point contact method; this implies that the conductance measurement probes the narrowest region of the contact.

The voltage and the current were measured, at zero magnetic field at RT (temperature below T_C), applied by a specially built *homemade* electrical circuit. A good resolution of the electrical circuit (about 10^{-6} A and 10^{-6} V, respectively) enabled a reliable numerical calculation of the differential conductance R_d = dI/dV.

By fitting $I-V$ curves to the Simmons theory [29], the barrier width $d \sim (0.5 \div 1.2)$ nm, the average barrier height $\Phi \sim (1 \div 3)$ eV and the effective tunnel area $s \sim 0.002\ \mu\text{m}^2$ of the MCBJs were estimated. Differential conductance curves $(\mathrm{d}I/\mathrm{d}V)$ *vs.* V at RT confirm that a tunnel process exists for these devices. A better fit of the curves $(\mathrm{d}I/\mathrm{d}V)$ *vs.* V can be obtained by the Glazman-Matveev theory [30] for multistep tunnelling via localized states in an insulating barrier. On can conclude that tunnelling between ferromagnetic apexes occurs mainly via one or more localized states.

In the $I-V$ characteristic one can distinguish two different regions. At low voltages the current goes as V^2, what is consistent with suggested quantitative models for the current transport across the grain boundary. [17,31].

In our experiments however, the structure of the MCBJs and crystallographic orientation of the "apexes" cannot be controlled; so, each conductance measurement corresponds to a new magnetic arrangement. It is due to fact that the junction resistance as well as in the magnitude of the current switching is changed by mechanical way.

3. Results and discussion

The increasing or decreasing DC current (I) is applied to the MCBJ, and the variation of the resistance with this current is recorded. The results reported here were obtained at zero applied external magnetic field at RT (below $T_C \approx$ 350 K).

In Fig.1 and 2 we present the representative results of the switching of the MCBJs induced by applying the currents to our devices, and determining the resulting jumps of conductance G.

In Fig. 1 we display the variations of the conductance G of the MCBJ as a function of the DC current I (current sweeping steps equal 0.1 mA). Starting from a parallel configuration of the magnetization M_P of the electrodes (for I= 25 mA) and next decreasing the current towards the negative values, only a progressive and reversible small decrease or increase of the conductance, respectively, can be observed when current densities $|J|$ exceed the value of 10^8 A/cm^2.

In contrast, when $|J|$ decreases below a critical value ($\sim 10^8$ A/cm^2), the reversible jump of the conductance ($\Delta G \approx$ multiples e^2/h) is clearly seen, which corresponds to a transition from M_P to the M_{AP} (from parallel to antiparallel configuration of magnetization). This indicates the reversal of the magnetic moment of the Mn-ions cluster at surface of the electrodes. This supports a use of Zener model [21], which claims close correlation between magnetic and transport properties in manganite compounds. The observed values of the jumps of the conductivity are in accordance with theoretically results of Morigaki *et al.* [26], who found that for a ferromagnetic material the quantum of conductance should be equal to $G_0/2 = e^2/h$ due to the spin degeneracy lifting.

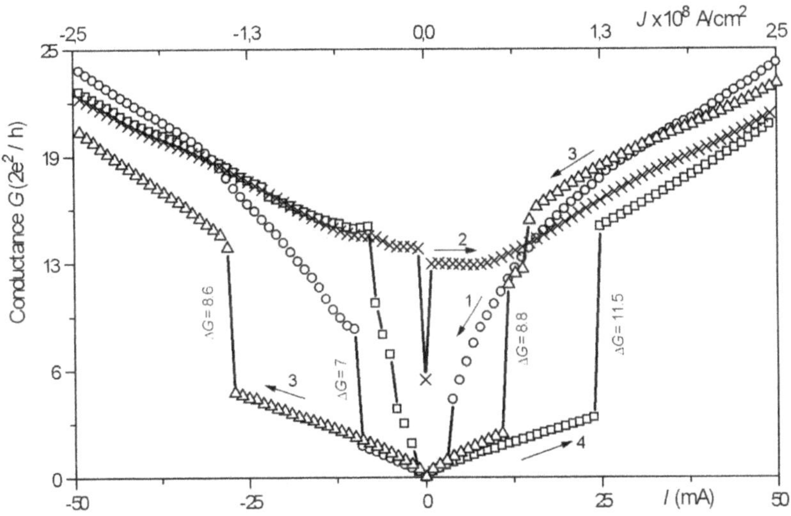

Figure 1. Conductance G dependence *vs.* current I at MCBJ at RT and zero external magnetic field. Arrows denoted the sweeping of the current. The jumps of conductance ΔG= n (e^2/h) for definite critical current I_C clearly visible.

In Fig. 2 we displayed the variation the conductance G for high current density ($J > 7$ x10^8 A/cm^2) of the MCBJ as a function of the DC current I for $H_{appl} = 0$. Starting from the antiparallel configuration of the magnetization $M_{AP}$of the electrodes (for I = 0 mA) and increasing the current towards the negative values, the progressive increase of the conductance, can be observed when current densities J drop below the value of 7x10^8 A/cm^2.

Current-induced switching observed when $|J|$ increases above any critical value ($<$ 9x10^8 A/cm^2). The reversible jumps of the conductance ($\Delta G \approx$ multiples e^2/h) are clearly seen, which correspond to a transition from M_{AP} to M_P and next to M_{AP} configurations. This indicates the reverse of the magnetic moment of the Mn-ions cluster at surface of the electrodes when the direction of the current passing through the MCBJ is reversed. This is in agreement with the Zener model [21], in which close correlation between magnetic and transport properties in manganite compounds is assumed.

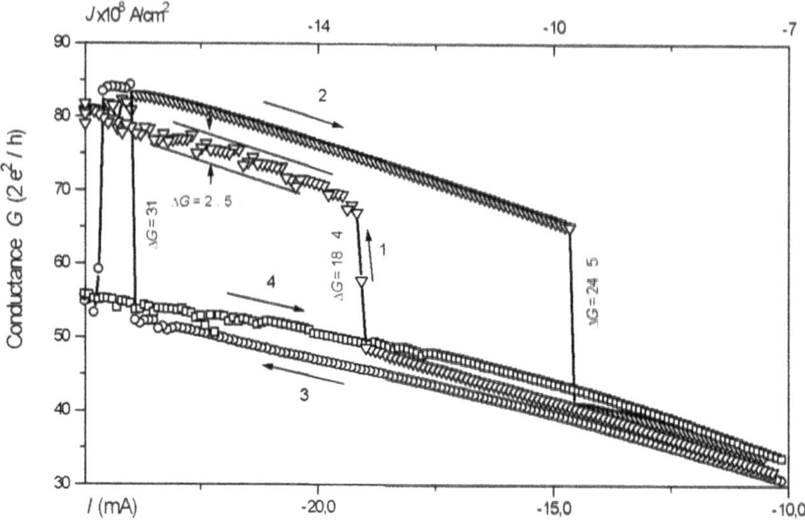

Figure 2. Conductance G dependence *vs.* current I at MCBJ at RT and zero external magnetic field. Arrows denoted the sweeping directions of the current. The jumps of conductance ΔG= n (e^2/h) for definite critical current I_C clearly visible.

The fluctuations of the conductivity ΔG= 2.5 x $(2e^2/h)$, between two values displayed at curve 1, are clearly visible for the current range (-19 → -25) mA. It is easy to imagine that MCBJ contains not just one but several active tunnel junctions, and that several parallel filaments constitute the global path.

The current-switching threshold depends of the measurements history, as it is seen in Fig.1 and 2. The hysteretic dependence on the current sweep is observed by comparing the jumps positions between the current up-sweep and current down-sweep graphs. The hysteresis region of the current-induced switching coincides with magnetic structures of the junction according to DW model [20].

Our data show reproducible, novel features that lead us to suggest a model of MCBJ switching that differs from the widely used to the metallic multilayered structures [10-13]. Our data suggest a quantum origin of current-induced switching of the perovskite MCBJ. According to the theoretically results of Morigaki *et al.* [26] for ferromagnetic materials, the jumps equal to multiples quantum of the conductivity e^2/h (Fig.1 and 2).

For the current-induced increase of conductance, an alignment of Mn spins within the barriers, mediated by a spin-polarized current of high density seems likely. Also, magnetic coupling of ferromagnetic clusters via the spin current [32,33] might be involved. Recently current-dependent switching of manganite cluster surrounded by ferromagnetic manganite layers has been described in that way for trilayers structures [9]. They are characterized by a current of high density between ferromagnetic regions of manganite (grains, or particles in tunnel junction).

Current-induced instabilities and resistivity jumps observed in our experiments (Fig. 1 and 2) closely resemble effects associated with current-induced resistivity switching in artificial magnetic multilayers [14,15].

The convincing observation of our experiments is that there exists the critical current density above which the magnetization at MVBJs can be switched back and forth.

Our results have confirmed some of the main features predicted by the theory: the current densities needed to switch such magnetic configuration are of the order of magnitude predicted by theory i.e. $> 10^7$ A/cm^2.

The observed switching at metallic constructions can result from two distinct physical mechanisms:

- the Oersted field (named vortex field) created by a large vertical current passing through the multilayers pillar that induces a vortex configuration [15,34]. This field contribution can be minimized by decreasing the pillar diameter below 500 nm [35].

- the spin-transfer torque mechanism, inducing the rotation of magnetization in a F/NM/F structure [1,2,36]

The circular Oersted field H_{cir}, which is strongest at the edge of the device, is associated with the current passing through MCBJ. This field can create vortex states in the magnetization of the surface electrodes, the helicity of which depends on the direction of the current flow. Due to the small distance between electrodes (few nm), the field is not uniform, and the fields produced by currents flowing in the electrodes can be also significant. According to Amperes Law [34], we estimate, that a 25 mA current is generated by H_{cir} field at the edge of a 50 nm diameter apexes, located between the electrodes, up to 2000 Oe (Fig. 2). Such a field is strong enough to rotate the magnetization of the MCBJ to form a circularly polarized orientation of the magnetization.

Applying Berger [37] relation to the field H_t corresponding to the spin transfer by a torque we estimate that a 5x10^7 A/cm^2 (Fig. 1) generates H_t of about 70 Oe. Considering of the values of coercivity field of the perovskite we see that this H_t value is sufficient to align partially the magnetization of the electrodes and change the conductance of the MCBJ.

In the data analysis of the switching in exchange biased metallic magnetic tunnel junctions, the interplay between spin-transfer effects and the contribution of current-generated circular magnetic field H_{cir} has been discussed. It is believed that the pillar diameter d must be less than 200-500 nm for spin-transfer effects to overcome the magnetic fields created by the current flowing through the GMR pillars since current-generated fields scale as $1/d$ and spin transfer effects scale as $1/d^2$ [38].

The estimated critical current for pure spin-transfer switching from the high resistance state to the low resistance state is $J_C \sim 10^6$-10^7 A/cm^2 [38]. When the vortex field is included, this switching current is lowered, the critical current $J_C \sim 10^6$ A/cm^2 is sufficient to induce a transition from the high resistance state to a vortex state.

The fact that spin transfer dominates the switching process at small scales, $d << 300$ nm (d is the wire diameter) at metallic nanostructures [39] is assumed for at our perovskite MCBJs, with junction diameter near 50 nm. Conductance jumps are induced by spin-transfer switching.

Our results enable us to deduce that this phenomenon occurs spontaneously for MCBJ like in point-contact. The polarized current can originate either from the fact that the contained apexes have identical magnetic orientation, or, magnetic domain walls may be expelled from the nanometric construction. Further studies to elucidate this point involving temperature or magnetic field dependence are in progress.

The microscopic origin of the observed current-induced quantum switching of perovskite nanoconstruction needs further clarification.

4. Summary

The following main conclusions can be derived from our experimental results and their analysis.

1 The experimental results confirm current-induced magnetization reversal of the Mn-ions clusters in semi-metal manganite perovskite nanoconstruction by spin transfer in the absence of the external magnetic field.

2 Our data show reproducible, novel features that lead us to suggest a model of switching that differs from the widely used to the metallic multilayers structures. The data suggest a quantum origin of current-induced switching of the perovskite MCBJ that according to the theoretically results Morigaki *et al.* [26] for ferromagnetic materials equals to multiples of conductivity quantum e^2/h (Fig.1 and 2).

These results open a wealth of new opportunities to get a deeper understanding of spin dynamics or spin control in nanostructures, and will have important implications for the development of future spintronics devices.

Acknowledgments

We thank W. Kowalski for technical assistance. The work was supported by the Centre of Excellence for Magnetic and Molecular Materials for Future Electronics within the European Commission Contract No. G5MA-CT-2002-04049.

References

[1] J. Slonczewski, *J. Magn. Magn. Mater. 159,* L1 (1996)

[2] L. Berger, *Phys. Rev. B 54*, 9353 (1996)

[3] J. E. Wegrowe, D. Kelly, Ph. Guitienne, Y. Jaccard, and J,-Ph. Ansermet, *Europhys. Lett. 45*, 626 (1999)

[4] J. E. Wegrowe, X. Hoffer, Ph. Guitienne, A. Fabian, L. Gravier, T. Wade, and J,-Ph. Ansermet, *J. Appl. Phys. 91*, 6806 (2002)

[5] J. A. Katine, F. J. Albert, R. A. Buhrman, E. B. Myers, and D. C. Ralph, *Phys. Rev. Lett. 84*, 3149 (2000)

[6] J. Grollier, V. Cros, A. Hamzic, J. M. George, H. Jaffers, A. Fert, G.Faini, J. Ben Youssef, and H. Legall, *Appl. Phys. Lett. 78*, 3663 (2001)

[7] J. Z. Sun, D. J. Monsma, M. J. Rooks, and R. H. Koch, *Appl. Phys. Lett. 81*, 2202 (2002)

[8] M. Tsoi, A. G. M. Jansen, J. Bass, W. C. Chiang, M. Seck, V. Tsoi, and P. Wyder, *Phys. Rev. Lett. 80*, 4281 (1998)
M. Tsoi, A. G. M. Jansen, J. Bass, W. -C. Chiang, V. Tsoi, and P. Wyder, *Nature (London) 406*, 46 (2000)

[9] J. Z. Sun, *J. Magn. Magn. Mater. 202*, 157 (1999)

[10] P. P. Freitas and L. Berger, *J. Appl. Lett. 57*, 1266 (1985)

[11] H. Koo, C. Krafft, and R. D. Gomez, *Appl. Phys. Lett. 81*, 862 (2002)

[12] J. Grollier, P. Boulenc, V. Cros, A.Hamziæ, A. Vaurès, A. Fert, and G. Faini, *Appl. Phys. Lett. 83*, pp 509 (2003)

[13] M. H. Jo, N. D. Mathur, J. E. Evetts, and M. G. Blamire, *Appl. Phys. Lett. 77*, 3803 (2000)

[14] E. B. Myers, D. C. Ralph, J. A. Katine, F. J. Albert, and R. A. Buhrman, *J. Appl. Phys. 87*, 5502 (2000)

[15] J. A. Katine, F. J. Albert, and R. A. Buhrman, *Appl. Phys. Lett. 76*, 354 (2000)

[16] W. E. Picket and D. J. Singh, *Phys. Rev. B 53*, 1146 (1996)

[17] C. Höfener, J.B. Philipp, J. Klein, L. Alff, A. Marx, B. Büchner, R, Gross, *Europhys. Lett. 50*, 681 (2000)

[18] M. Viret, M. Drouet, J. Nassar, J. P. Contour, C. Fermon, and A. Fert, *Europhys. Lett. 39*, 545 (1997)

[19] C. T. Tanaka, J. Nowak, and J. S. Moodera, *J. Appl. Phys. 81*, 5515 (1997)

[20] X. W. Li, Yu Lu, G. Q. Gong, Gang Xiao, A. Gupta, P. Lecoeur, J. Z. Sun, Y. Y. Wang, and V. P. Dravid, *J. Appl. Phys. 81*, 5509 (1997)

[21] C. Zener, *Phys.Rev. 82*, 403 (1951)

[22] P.Anderson and H. Hasegawa, *Phys. Rev. 100*, 675 (1955)

[23] P. de Gennes, *Phys. Rev. 118*, 141 (1960)

[24] M. J. Calderón, L. Brey, and F. Guinea, *Phys. Rev. B60*, 6698 (1999)

[25] J. Fontcuberta, B.Martinez, A. Seffar, S. Piñol, J. L. Garcia-Muñoz, and X. Obradors, *Phys. Rev. Lett. 76*, 1122 (1996)

[26] Y. Morigaki, H. Nakanishi, H. Kasai, A. Okiji, *J. Appl. Phys. 88*, 2682 (2000)

[27] J.Moreland and J.W.Ekin, *J.Appl.Phys. 58*, 3888 (1985)

[28] C. J. Muller, J. M. van Ruitenbeek, and L. J. de Jongh, *Physica C 191*, 485 (1992)

[29] J.C.Simmons, J.Appl.Phys. **34,** 1793 (1963); *J. Appl. Phys. 34*, 2581 (1963)

[30] L.I. Glazman and K.A. Matveev, *Zh. Eksp. Teor. Fiz. 94*, 332 (1988) *Sov. Phys. JETP 67*, 1276 (1988)

[31] N. K. Todd, N. D. Mathur, S. P. Isaac, J. E. Evetts, and M. G. Blamire, *J. Appl. Phys. 85*, 7263 (1999)

[32] C. Heide, R. J. Elliot, and N. S. Wingreen, *Phys. Rev. B 59*, 4287 (1999)

[33] J. C. Slonczewski, *Phys. Rev. B 39*, 6995 (1989)

[34] K. Bussmann, G. A. Prinz, S.-F. Cheng, and D. Wang, *Appl. Phys. Lett. 75*, 2476 (1999)

[35] F. J. Albert, J. A. Katine, R. A. Buhrman, and D. C. Ralph, *Appl. Phys. Lett. 77*, 3809 (2000)

[36] L. Berger, *J. Appl. Phys. 91*, 6795 (2002)

[37] L. Berger, *J. Appl. Phys 49*, 2156 (1978)

[38] Y. Liu, Z. Zhang, J. Wang, P. P. Freitas, and J. L. Martins, *J. Appl. Phys. 93*, 8385 (2003)

[39] Ya. B. Bazaliy, B. A. Jones, and S. C. Zhang, *J. Appl. Phys. 89*, 6793 (2001)

SPIN POLARIZED EFFECTS AT THE INTERFACE BETWEEN MANGANITES AND ORGANIC SEMICONDUCTORS

I. Bergenti, F. Biscarini, M. Cavallini, V. Dediu, M. Murgia, P. Nozar, G. Ruani, C. Taliani

ISMN-CNR, Via P. Gobetti, 101, 40129, Bologna, Italy

adediu@jolly.bo.cnr.it

Abstract Spintronics is a new branch of electronics based on purely quantum effects. Instead of carrier's charge transfer as in usual electronics, it evokes carrier's spin transfer. The search of new materials suitable for injecting and transferring carriers with a preferential spin orientation is of paramount importance for the development of spintronics. Here we report the results of the investigation of the spin polarization of the $La_{0.7}Sr_{0.3}MnO_3$ manganite films and the spin polarized injection from the manganite into π-conjugated organic semiconductors. The surface of the $La_{0.7}Sr_{0.3}MnO_3$ films at room temperature is composed of homogeneous ferromagnetic (FM) phase in which paramagnetic (PM) defects are embedded. The ferromagnetic phase is highly spin polarized showing a room temperature half-metallic behavior. A strong magnetoresistance (up to 30%) was measured on nanostructured planar hybrid junctions LSMO/sexithiophene/LSMO indicating both the spin polarized injection and the spin polarized transport up to distances of about 100 nm at room temperature.

Keywords: manganite, half-metal, spin polarization, spin polarized injection, organic semiconductors

1. Introduction

In spintronics the information is stored, transmitted and read via electrical carrier spin orientation [1]. Thus, its operation principles are directly related to the quantum nature of the particles. Spintronics devices are mainly based on the generation of carrier spin polarization in some active medium and the spin-polarized (SP) detection at the output. Recent research is mainly concentrated on development of new methods and materials for generation of SP carriers and, on the other hand, on investigation of new materials able to transport such a polarization coherently to distances up to 10^2-10^3 nm.

A.S. Alexandrov et al. (eds.), Molecular Nanowires and Other Quantum Objects, 415–424.

SP carriers can be generated in different ways either indirectly, via photoexcitation by circularly polarized light [2] or directly, via real space injection from a SP material [3]. The discovery of colossal magnetoresistance (CMR) in perovskite manganites opens the way to achieve 100% SP injectors [4].

SP coherent transport has been previously observed in different inorganic materials [2,3,5-7] mainly at low temperatures, although the room temperature experiment has been also reported recently [8]. The realization of room temperature devices is fundamental for spintronics development.

The work is mostly divided in two parts: the spin polarized tunneling spectroscopy of the manganite surface and direct spin polarized injection into organic thin films from LSMO electrodes. All reported experiments are performed at room temperature. The temperature dependence of the observed effects is still under investigation and will be reported elsewhere.

2. Spin polarization in ferromagnetic manganites

The half metallic properties (100% SP) of the $La_{0.7}Sr_{0.3}MnO_3$ have been demonstrated for the first time at 40 K by spin resolved photoemission spectroscopy [4]. The room temperature ($\sim$0.8T_c) spin polarization of this manganite is not clear so far, the literature data being spread in a wide interval [9-11]. It is authors opinion that this disagreement is also strongly induced by sample quality.

Epitaxial $La_{0.7}Sr_{0.3}MnO_3$ thin films have been prepared by Channel-Spark ablation on $NdGaO_3$ substrates [12]. In order to remove the over-oxygenation effects the best films were annealed at 400÷450°C in high vacuum after deposition. Such films exhibit high T_c($\sim$350÷370 K) and the resistivity lower than 10 mΩcm at 300 K. As-grown films had a lower T_c($\sim$300÷320 K) and higher resistivity. The micro-Raman analysis of films showed that the annealed films are essentially rhombohedral (structural phase showing the ferromagnetic metallic state [12], with orthorhombic inclusions (structural phase characterized by the strong Jahn-Teller distortion [13], i.e. an insulating paramagnetic phase), while the as-grown films consist of mixture (roughly 50/50%) of rhombohedral and orthorhombic phases. The magnetic homogeneity and spin polarized properties of the epitaxial $La_{0.7}Sr_{0.3}MnO_3$ film surfaces at room temperature were studied by spin polarized Scanning Tunneling Microscopy.

In spin polarized STM the maximum transmissivity of carriers between two spin polarized electrodes (tip and sample) separated by a tunnel junction occurs for their parallel SP and drops with increasing misalignment (spin-valve effect) [14]. In our experiments STM was operated in the parallel tip-sample plane magnetic configuration. For each bias voltage V_{bias}, we acquired both topography z(x,y) and *dI(x,y)/dV* (differential tunneling conductance) maps by applying a low-frequency modulation on the V_{bias} while holding the feedback.

We used electrochemically etched Ni wire tips, which have been shown to act effectively as an injector of carriers with strong spin polarization [15].

A typical morphology of a 50-nm thick film exhibits a smooth background with low density of outgrowths. The *dI(x,y)/dV* maps (Fig. 1) which are sensitive to local conductance show the coexistence of well defined high conductance (HC) and low conductance (LC) regions. The *dI(x,y)/dV* contrast between LC and HC regions amounts to more than one order of magnitude.

Figure 1. The *dI(x,y)/dV* maps for a 50-nm thick film: the bright part corresponds to the metallic ferromagnetic phase (high conductivity), the dark part corresponds to the insulating paramagnetic phase.

The LC defects (typical dimensions of the order of few microns) cover a few percent of the film surface in samples post-annealed in vacuum, while the as-grown films consist roughly of 50/50% mixture of HC and LC regions (similarly to the La-Ca-Mn-O samples studied earlier [16]. Both the HC and LC regions conductance is quite homogeneous and can be characterized by two different values. The difference of the conductance is due to the metallic nature of ferromagnetic HC regions and insulating nature of paramagnetic LC defects [16]. The results are in good agreement with micro-Raman studies, where a similar fraction of the surface was found to correspond to the orthorhombic paramagnetic phase.

The magnetic nature of the spectroscopic images becomes evident from the evolution of the local conductance versus V_{bias}. The normalized conductivity curves ($V/I*dI/dV = d(lnI)/d(lnV)$) in Fig. 2 clearly show different responses from the HC and LC regions. HC characteristics lie above the LC ones at any bias voltage. The LC curve is smooth and extrapolates to zero at $V_{bias} = 0$ indicating an insulating behavior. The HC curves on the other hand, exhibit a metallic and non-linear behavior with a few distinct features: a small narrow peak at 1 V, a sharp growth at 1.3-1.5 V, and two local maxima at 1.9 and 2.6 V.

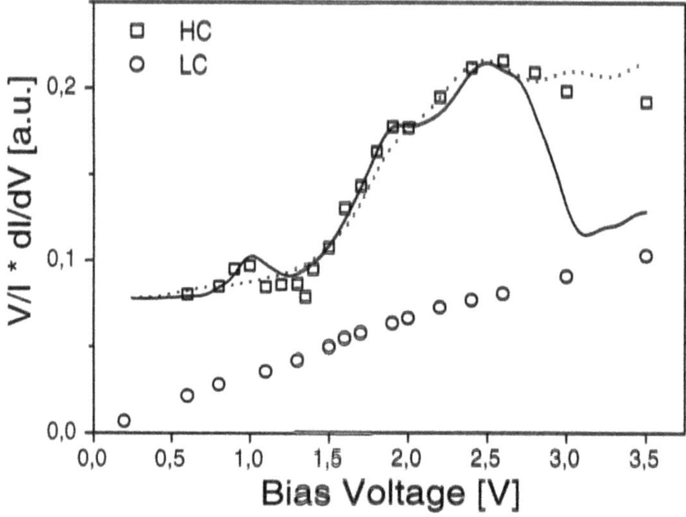

Figure 2. Logarithmic derivatives $d(lnI)/d(lnV) = V/I{*}dI(x,y)/dV$ acquired at $I = 1$ nA and different bias voltages: data are taken separately for high conductance (HC) and low conductance (LC) regions. The curves represent the results of deconvolution calculations.

Deconvolution of the manganite spin dependent density of states (SP DOS) was performed by taking into account the tip-sample current separated in spin-up and spin-down channels [17]. The theoretically calculated SP DOS for bulk [18] and surface [19]states for the Ni tip were used in our deconvolution.

The resulting manganite band structure is represented in Fig. 3. It provides the solid (bulk Ni DOS) and dashed (surface Ni DOS) fitting curves for $d(lnI)/d(lnV)$ in Fig. 2. A rather good agreement with experimental data is found for both cases.

The only sub-band with non-zero DOS at E_F is the spin-up polarized band (Fig. 3), which extends roughly up to 1.5 eV (the energies are calculated from E_F). This band is associated with the tail of the spin up $e_g t_{2g}$ band of Mn [4,20]. The spin down band is an overlap of two lorentzian sub-bands, and has non-zero DOS from 0.4 eV up to more than 4 eV. The band width is roughly 2 eV, and it is characterized by a maximum at 2.5 eV and a shoulder at 1.8 eV.

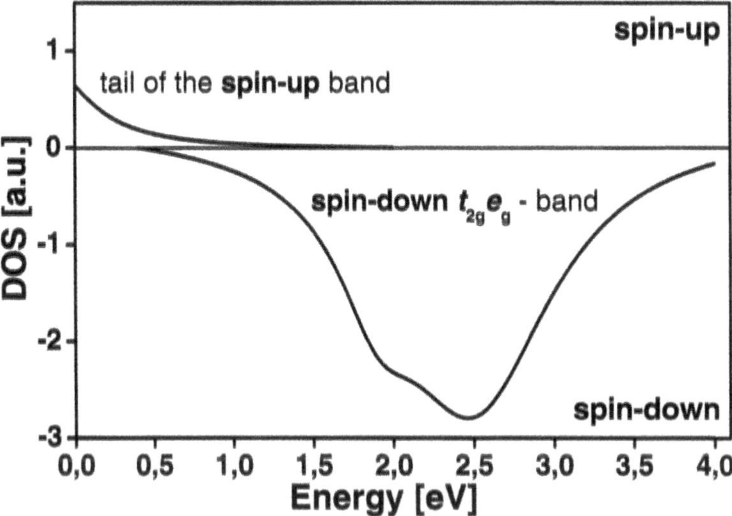

Figure 3. $La_{0.7}Sr_{0.3}MnO_3$ spin polarized DOS at 300K from deconvolution of the tunneling spectroscopy data.

Logarithmic derivatives $d(lnI)/d(lnV) = V/I*dI(x,y)/dV$ acquired at $I = 1$ nA and different bias voltages: data are taken separately for high conductance (HC) and low conductance (LC) regions. The curves represent the results of deconvolution calculations. The most striking feature of the manganite band structure is 100% polarization of charge carriers at E_F due to the fully developed gap in spin-down polarized DOS. We would like to draw the attention to the fact, that due to the high maximum in spin-down polarized DOS of Ni at E_F, our measurements are much more sensitive to the presence of the non-zero spin-down DOS than the spin-up DOS at E_F. To the best of our knowledge this is the first demonstration of the half metallic properties of manganites at room temperature. The shape and characteristic energies of derived bands are in good agreement with theoretical ab-initio calculations and optical data [20,21]. A remarkable agreement between proposed spin down band and $t_{2g}e_g$ spin up band calculated from spin polarized photoemission should be also noticed [4].

An important result of the SP-STM study is the high magnetic homogeneity of the annealed films. Within the accuracy of our experimental setup, that amounts to 0.1 nA/V for $dI(x,y)/dV$ maps, no phase separation in the ferromagnetic regions was detected. This is in accordance with the dynamical mean field calculations for the FM double exchange systems of this doping level (x=0.3) [21]. On the other hand, the charge ordered manganites exhibit a distinct electronic phase separation on the nanoscopic scale.

Spin polarized STM has a spatial resolution that allows us to investigate separately the ferromagnetic phase and the "non-magnetic" inclusions. This is the reason for the apparent disagreement between our results and some tunneling and photoemission experiments [9,10] that indicate the $La_{0.7}Sr_{0.3}MnO_3$ surface as only partially spin polarized at room temperature. In the later case all the magnetic defects are integrated in the final result, while the defect density, on the other hand, depends strongly on preparation procedure (see above).

Thus, completely spin polarized regions in $La_{0.7}Sr_{0.3}MnO_3$ films surface were found at room temperature. Apart from the low conductance defects no additional phase separation is detected in the dominating high conductance regions, indicating high magnetic homogeneity of $La_{0.7}Sr_{0.3}MnO_3$ down to the 50 nm lengthscale. We believe these properties are common for high quality homogeneous films of different ferromagnetic manganites.

3. Spin polarized injection from manganite into organic semiconductors

The spin polarized injection in the π-conjugated organic semiconductor sexithiophene (T_6) was investigated. T_6 is a rigid-rod organic semiconductor which is currently one of the materials of choice for the development of organic based electronic [22]. The thin film mobility ranges from 10^{-2} to 10^{-4} $cm^2V^{-1}s^{-1}$ depending on morphology [23,24]. Resistance measurements in magnetic fields up to 1 T show no intrinsic magnetoresistance (MR) on T_6 films.

A spin-valve experiment was performed on hybrid junctions LSMO/T_6/LSMO. LSMO epitaxial thin films were deposited on matching substrates ($NdGaO_3$, $SrTiO_3$).

Two electrode LSMO planar structures separated by a channel of a length w were fabricated by electron-beam lithography (Fig. 4). Each LSMO film contained six electrically separated couples of electrodes with different separations ranging from 70 to 500 nm. This permits us to study the effect of channel length on SP transport for the same organic film. T_6 thin films (100-150 nm thick) were deposited by molecular beam deposition [25] in order to cover the channel separating the electrodes and create electrical connection between them.

In the absence of the external magnetic field the electrodes have a random spin orientation with respect to each other. In the external magnetic field the spins in both electrodes orient parallel. For a direct contact between FM electrodes this results in a strong negative MR. A negative MR across the organic semiconductor would indicate both a SP injection into organic and a SP coherent transport between the electrodes.

Figure 5 shows the magnetic field dependence of the I-V characteristics for 140 nm, 200 nm and 400 nm channel lengths LSMO/T_6/LSMO junctions measured in the ambient atmosphere at room temperature. The LSMO film

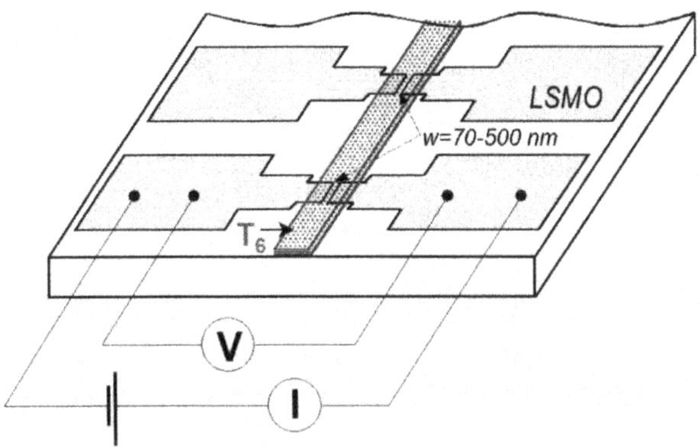

Figure 4. A schematic view of the hybrid junction (drawing not to scale) and the dc four-probe electrical scheme.

thickness is 100 nm, the T_6 film thickness is 150 nm. All I-V curves were nearly ohmic, indicating a negligible interface barrier for carrier injection.

Changing the mutual orientation of SP electrodes from random to parallel by applying a perpendicular magnetic field (3.4 kOe), induced a strong decrease of the device resistance for 100÷200 nm channel lengths, while no magnetoresistance was detected for 300÷500 nm channels. Fig. 5 reports the highest observed resistance decrease of about 30% measured on the 140 nm channel (lower resistance corresponds to higher slope of I-V curves), and for comparison the 400 nm channel junction. The saturation magnetization in perpendicular fields is reached approximately at 10 kOe for 5x10 mm^2 area films. The electrode ends are reduced to 2.5x10 μm^2 leading to lower perpendicular saturation fields, so that the 3.4 kOe field should lead to the magnetization value close to saturation. Any negative MR is unexpected in the bare T_6 material (see above) indicating that spin polarization is maintained inside the organic semiconductor. The inset in Fig. 5 shows the MR as a function of w measured in the same T_6 film. Disappearance of MR for longer channels indicates unequivocally that it is not caused by electrode/organic interface effects. At this point it is quite difficult to estimate exactly the characteristic spin diffusion length L_S, as the carrier injection may take place both horizontally (along the shortest path) and vertically in some random points where the organic film covers the injecting electrode (see Fig. 4). Nevertheless we estimate the room temperature L_S to be about 200 nm.

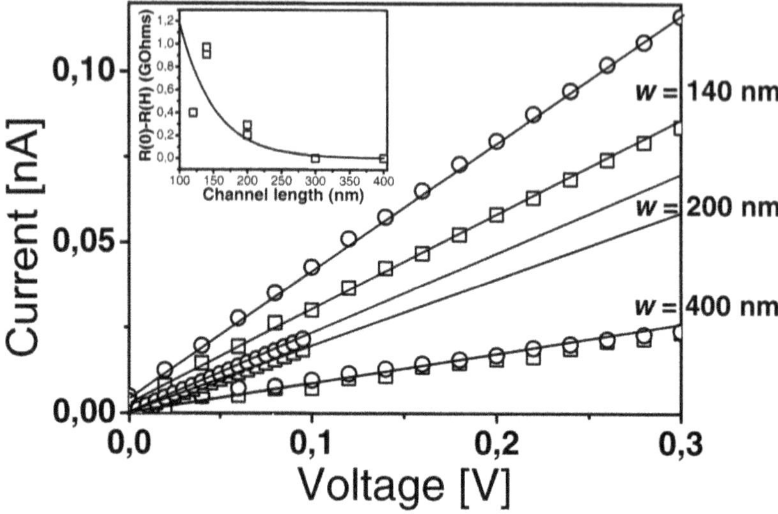

Figure 5. I-V characteristics of a $La_{0.7}Sr_{0.3}MnO_3/T_6/$ $La_{0.7}Sr_{0.3}MnO_3$ as a function of the magnetic field for 140, 200 and 400 nm channel lengths. Squares indicate the zero magnetic field measurements, while circles indicate the curves taken at 3.4 kOe. The inset indicates MR as function of w, where MR=R(0)-R(3.4 kOe).

By removing and re-depositing the organic film with suitable solvents the MR was repeatedly found. It ranges from 15% to 30 % for junctions with the channel length close to 100 nm (70-140 nm), decreases to 7-10% for the 200 nm channels, and disappears for longer channels.

The average value of 10^6 Ohm cm of the electrical resistivity measured on all channels indicates that T_6 films are indeed weakly doped by ambient oxygen[26]. The spin relaxation time estimated for the 10^{-4} $cm^2V^{-1}s^{-1}$ mobility is about 10^{-6} s. This value is in good agreement with electron paramagnetic resonance studies on π-conjugated polymers, where at room temperature the spin-lattice relaxation time T_1 was found to vary for different materials in the range 10^{-5}-10^{-7} s [27].

It is important to note that for a given junction the MR was independent on whether the field orientation was parallel or perpendicular.

Discussing results we shall say that the observed magnetoresistance is a direct evidence of a spin polarized transport across an organic material, although it does not permit to quantify the spin polarization amplitude. The total junction resistance consists of three different resistances: the LSMO resistance, the T_6 resistance and an unknown in this geometry LSMO/T_6 interface resistance. The LSMO resistance for our geometry is about 5 kOhms, while the total resistance is a few GOhms typical for the T_6film. The manganite films showed roughly

10-25% magnetoresistance on the scale of 10^3 Ohms. The interface resistance, as mentioned above, is roughly irrelevant for both total resistance and MR . Indeed, it is difficult to expect any MR at the LSMO/T_6 interface in the absence of the second electrode, as T_6 accepts equally both spin orientations. Thus, the resulting magnetoresistance is attributed to the spin polarized transfer across the organic material, where for the channel length $w<L_S$ the carrier injection probability is proportional to the acceptance probability by the second electrode. This decreases the resistance for the parallel electrode spin alignment and causes the MR signal. For a longer channel, where the spin polarization is lost, the total resistance does not depend on the electrodes mutual orientation. The small electrode MR of the order of 1kOhm, that should be still present in this case, is not detectable on the scale of 10^9 Ohms.

In the absence of magnetic impurities, the most important spin-flip mechanisms in the organic film are spin-orbit interaction and hyperfine interaction. In the π-conjugated organic semiconductor only p-orbitals are delocalized. The wavefunctions for these orbitals have zero amplitude on the nucleus site, minimizing the effect of the hyperfine interaction. As for the spin-orbit coupling, it is very weak in this kind of organic materials.

Strong evidences of the spin polarized injection and polarized spin transport through organic semiconductors (spin diffusion length of about 200 nm) was demonstrated for the first time. The observed MR response is comparable with the commercial magnetoelectronic devices. A higher MR may be achieved by improving the spin alignment contrast between the electrodes. We expect the spin diffusion length to increase significantly for higher mobility values. The ambient atmosphere and room temperature operation of this process opens the possibility for practical applications. The benefit of the spin polarized injection may also be applied in organic optoelectronics, as in the case of organic light emitting diodes (OLED), since the spin statistics of the radiative carrier recombination may be changed by inducing preferential formation of singlet excitons.

References

[1] G. A. Prinz, Science **282**, 1660 (1998).

[2] D. Hägele, M. Oestreich, W. W. Rühle, et al., Appl. Phys. Lett. **73**, 1580 (1998).

[3] M. Johnson and R. H. Silsbee, Phys. Rev. B. **37**, 5326 (1988).

[4] J.-H. Park, E. Vescovo, H.-J. Kim, et al., Nature **392**, 794 (1998).

[5] V. P. LaBella, D. W. Bullock, Z. Ding, et al., Science **292**, 1518 (2001).

[6] I. Malajovich, J. M. Kikkawa, D. Awschalom, et al., Phys. Rev. Lett. **84**, 1015 (2000).

[7] R. Fiederling, M. Keim, G. Reuscher, et al., Nature **402**, 787 (1999).

[8] H. J. Zhu, M. Ramsteiner, H. Kostial, et al., Phys. Rev. Lett. **87**, 016601 (2001).

[9] J.-H. Park, E. Vescovo, H.-J. Kim, et al., Phys. Rev. Lett. **81**, 1953 (1998).

[10] P. Lyu, D. Y. Xing, and J. Dong, Phys. Rev. B **60**, 4235 (1999).

[11] Y. Ji, C. L. Chien, Y. Tomioka, et al., Phys. Rev. B **66**, 012410 (2002).

[12] V. Dediu, J. Lòpez, F. C. Matacotta, et al., Phys. Stat. Sol. (b) **215**, 625 (1999).

[13] M. V. Abrashev, A. P. Litvinchuk, M. N. Iliev, et al., Phys. Rev. B **59**, 4146 (1999).

[14] R. Wiesendanger, *Scanning probe microscopy and spectroscopy* (Cambridge University Press, Cambridge, England, 1994).

[15] S. F. Alvarado and P. Renaud, Phys. Rev. Lett. **68**, 1387 (1992).

[16] M. Faeth, S. Freisem, A. A. Menovsky, et al., Science **285**, 1540 (1999).

[17] J. C. Slonczewski, Phys. Rev. B **39**, 6995 (1989).

[18] J. W. D. Connolly, Phys. Rev. **159**, 415 (1967).

[19] N. Papanikolaou, cond-mat/0210551 (2002).

[20] S. Satpathy, Z. S. Popovic, and F. R. Vukajlovi, Phys. Rev. Lett. **76**, 960 (1996).

[21] A. Chattopadadhyay, A. J. Millis, and S. Das Sarma, Phys. Rev. B **61**, 10738 (2000).

[22] A. Dodabalapur, L. Torsi, and H. E. Katz, Science **268**, 270 (1995).

[23] L. Torsi, A. Dodabalapur, L. J. Rothberg, et al., Phys. Rev. B **57**, 2271 (1998).

[24] P. Ostoja, S. Guerri, M. Impronta, et al., Adv. Mater. Opt. Electr. **1**, 127 (1992).

[25] M. Murgia, R. H. Michel, G. Ruani, et al., Synth. Met. **102**, 1095 (1999).

[26] D. Fichou and C. Ziegler, in *Handbook of Oligo- and Polythiophenes*, edited by D. Fichou (Wiley-VCH, Weinheim, 1998).

[27] V. I. Krinichnyi, Synthetic Metals **108**, 173 (2000).

CONTRIBUTING AUTHORS

A. S. Alexandrov

Department of Physics
Loughborough University
Loughborough LE11 3TU
U.K.
a.s.alexandrov@lboro.ac.uk

P. B. Allen

Dept. of Physics and Astronomy
State University of New York
Stony Brook, NY 11794-3800
USA
philip.allen@stonybrook.edu

A. F. Andreev

P. L. Kapitza Institute for Physical Problems,
ul. Kosygina 2, 117334, Moscow
Russia
andreev@kapitza.ras.ru

Y. Avishai

Department of Physics
Ben-Gurion University
Beer-Sheva 84105
Israel
yshai@bgumail.bgu.ac.il

J. Baszyński

Institute of Molecular Physics,
Polish Academy of Sciences
ul. M. Smoluchowskiego 17
60-179 Poznań
Poland
jbasz@ifmpan.poznan.pl

W. Belzig

Dept. of Physics and Astronomy
University of Basel
Klingelbergstr. 82, CH-4056 Basel
Switzerland
Wolfgang.Belzig@unibas.ch

I. B. Berkutov

B. I. Verkin Institute for Low Temperature Physics and Engineering of NAS of Ukraine,
47 Lenin Ave., Kharkov 310164,
Ukraine
Berkutov@ilt.kharkov.ua

J-P. Bourgoin

CEA Saclay, DSM/DRECAM/SCM,
91191 Gif-sur-Yvette
France
jbourgoin@cea.fr

D. R. Bowler

Dept. of Physics and Astronomy,
University College London
Gower Street, London WC1E 6BT,
U.K.
david.bowler@ucl.ac.uk

A. M. Bratkovsky

Hewlett-Packard Laboratories,
1501 Page Mill Road,
Palo Alto, California 94304
USA
alexmb@exch.hpl.hp.com

B. R. Bułka

Institute of Molecular Physics, Polish
Academy of Sciences,
ul. M. Smoluchowskiego 17,
60-179 Poznań
Poland
bulka@ifmpan.poznan.pl

V. Dediu

ISMN-CNR,
Via P. Gobetti 101
40129, Bologna
Italy
adediu@jolly.bo.cnr.it

J. Demsar

Dept. for Complex Matter
J. Stefan Institute,
Jamova 39, 1000 Ljubljana,
Slovenia
jure.demsar@ijs.si

J. T. Devreese

Theoretische Fysica van de Vaste
Stoffen (TFVS)
Universiteit Antwerpen
Universiteitsplein 1, B-2610
Antwerpen
Belgium
devreese@uia.ua.ac.be

P. P. Edwards

Inorganic Chemistry Laboratory
Oxford University
South Parks Road, Oxford, OX1 3QR
U.K.
peter.edwards@chem.ox.ac.uk

A. Erbe

Lucent Technologies, Bell Labs
600 Mountain Avenue
Murray Hill, NJ, 07974
USA
aerbe@lucent.com

A. Golub

Department of Physics,
Ben-Gurion University,
Beer-Sheva 84105
Israel
agolub@bgumail.bgu.ac.il

I. M. Grace

Department of Physics,
Lancaster University,
Lancaster, LA1 4YB
U.K.
i.grace@lancaster.ac.uk

S. Gredeskul

Department of Physics,
Ben Gurion University of the Negev,
Beer-Sheva 84105
Israel
sergeyg@bgumail.bgu.ac.il

M. L. H. Green

Inorganic Chemistry Laboratory
University of Oxford
South Parks Road, Oxford, OX1 3QR
U.K.
malcolm.green@chem.ox.ac.uk

V. V. Kabanov

Dept. for Complex Matter
J. Stefan Institute,
Jamova 39,1000 Ljubljana,
Slovenia
viktor.kabanov@ijs.si

K. Kikoin

Ben-Gurion University of the Negev,
Beer-Sheva 84105
Israel
kikoin@bgumail.bgu.ac.il

P. Kornilovitch

Hewlett-Packard Company,
MS 321A
Corvallis, OR 97330
USA
pavel.kornilovich@hp.com

C. J. Lambert

Department of Physics,
Lancaster University,
Lancaster, LA1 4YB
U.K.
c.lambert@lancaster.ac.uk

M. Lange

Nanoscale Superconductivity and Magnetism Group,
Laboratory for Solid State Physics and Magnetism,
K. U. Leuven,
Celestijnenlaan 200D, 3001 Leuven,
Belgium
martin.lange@fys.kuleuven.ac.be

Y. Oreg

Dept. of Condensed Matter Physics,
Weizmann Institute of Science,
Rehovot, 76100
Israel
yuval.oreg@weizmann.ac.il

F. M. Peeters

Departement Natuurkunde
Universiteit Antwerpen
Universiteitsplein 1 B-2610 Antwerpen
Belgium
peeters@uia.ua.ac.be

A. Ramšak

Jožef Stefan Institute and FMF
University of Ljubljana
Jamova 39, 1000 Ljubljana
Slovenia
anton.ramsak@fmf.uni-lj.si

T. Schimmel

Institute for Applied Physics,
University of Karlsruhe,
D-76128 Karlsruhe
Germany
Thomas.Schimmel@physik.uni-karlsruhe.de

A. Sidorenko

Institute of Applied Physics
LISES
MD2028 Kishinev
Moldova
sidorenko@lises.asm.md

J. Spałek

M. Smoluchowski Institute of Physics,
Jagiellonian University
ulica Reymonta 4, 30-059 Kraków
Poland
ufspalek@if.uj.edu.pl

M. Tarasov

Institute of Radio Engineering and
Electronics RAS, and M. Lomonosov
Moscow State University
Mokhovaya 11-7, Moscow, 101999
Russia
tarasov@hitech.cplire.ru

A. Troyanovsky

Kapitza Institute for Physical
Problems, Russian Academy of
Sciences
Kosygina 2, Moscow, 119334
Russia
troyan@kapitza.ras.ru

S. A. Trugman

Theoretical Division
Los Alamos National Laboratory
Los Alamos, New Mexico 87545
USA
sat@lanl.gov

I. K. Yanson

B. Verkin Institute for Low
Temperature Physics and
Engineering,
National Academy of Sciences,
310164, Kharkov
Ukraine
yanson@ilt.kharkov.ua

G-M. Zhao

Dept. of Physics and Astronomy
California State University at Los
Angeles
Los Angeles, CA 90032
USA
gzhao2@calstatela.edu

GPSR Compliance
The European Union's (EU) General Product Safety Regulation (GPSR) is a set of rules that requires consumer products to be safe and our obligations to ensure this.

If you have any concerns about our products, you can contact us on

ProductSafety@springernature.com

In case Publisher is established outside the EU, the EU authorized representative is:

Springer Nature Customer Service Center GmbH
Europaplatz 3
69115 Heidelberg, Germany

www.ingramcontent.com/pod-product-compliance
Ingram Content Group UK Ltd.
Pitfield, Milton Keynes, MK11 3LW, UK
UKHW021859190726
13853UKWH00003B/1338
* 9 7 8 9 4 0 1 5 6 9 8 1 1 *